Improving poultry meat safety and sustainability

It is widely recognised that agriculture is a significant contributor to global warming and climate change. Agriculture needs to reduce its environmental impact and adapt to current climate change whilst still feeding a growing population, i.e. become more 'climate-smart'. Burleigh Dodds Science Publishing is playing its part in achieving this by bringing together key research on making the production of the world's most important crops and livestock products more sustainable.

Based on extensive research, our publications specifically target the challenge of climate-smart agriculture. In this way we are using 'smart publishing' to help achieve climate-smart agriculture.

Burleigh Dodds Science Publishing is an independent and innovative publisher delivering high quality customer-focused agricultural science content in both print and online formats for the academic and research communities. Our aim is to build a foundation of knowledge on which researchers can build to meet the challenge of climate-smart agriculture.

For more information about Burleigh Dodds Science Publishing simply call us on +44 (0) 1223 839365, email info@bdspublishing.com or alternatively please visit our website at www.bdspublishing.com.

Related titles:

Achieving sustainable production of eggs Volume 1
Print (ISBN 978-1-78676-076-0); Online (ISBN 978-1-78676-078-4, 978-1-78676-079-1)

Achieving sustainable production of eggs Volume 2
Print (ISBN 978-1-78676-080-7); Online (ISBN 978-1-78676-082-1, 978-1-78676-083-8)

Achieving sustainable production of poultry meat Volume 1
Print (ISBN 978-1-78676-064-7); Online (ISBN 978-1-78676-066-1, 978-1-78676-067-8)

Achieving sustainable production of poultry meat Volume 3
Print (ISBN 978-1-78676-072-2); Online (ISBN 978-1-78676-074-6, 978-1-78676-075-3)

Optimising poultry flock health
Print (ISBN 978-1-78676-887-2); Online (ISBN 978-1-78676-889-6, 978-1-78676-890-2)

Reducing greenhouse gas emissions from livestock production
Print (ISBN 978-1-78676-439-3); Online (ISBN 978-1-78676-441-6, 978-1-78676-442-3)

Chapters are available individually from our online bookshop:
https://shop.bdspublishing.com

climate•SMART•publishing
IN AGRICULTURAL SCIENCE

BURLEIGH DODDS SERIES IN AGRICULTURAL SCIENCE
NUMBER 161

Improving poultry meat safety and sustainability

Edited by Professor Steven C. Ricke, University of Wisconsin-Madison, USA

burleigh dodds
SCIENCE PUBLISHING

Published by Burleigh Dodds Science Publishing Limited
82 High Street, Sawston, Cambridge CB22 3HJ, UK
www.bdspublishing.com

Burleigh Dodds Science Publishing, 1518 Walnut Street, Suite 900, Philadelphia, PA 19102-3406, USA

First published 2025 by Burleigh Dodds Science Publishing Limited

Library of Congress Control Number: 2025930490

British Library Cataloguing in Publication Data
A catalogue record for this book is available from the British Library

ISBN 978-1-80146-789-6 (print)
ISBN 978-1-80146-790-2 (ePub)
ISBN 978-1-80146-791-9 (PDF)
ISSN 2059-6936 (print)
ISSN 2059-6944 (online)

DOI: 10.19103/AS.2024.0146

Typeset by Deanta Global Publishing Services, Dublin, Ireland

Contents

Part 3 Postharvest control along the poultry supply chain

Series list

Title	Series number
Achieving sustainable cultivation of maize - Vol 1 From improved varieties to local applications Edited by: Dr Dave Watson, CGIAR Maize Research Program Manager, CIMMYT, Mexico	001
Achieving sustainable cultivation of maize - Vol 2 Cultivation techniques, pest and disease control Edited by: Dr Dave Watson, CGIAR Maize Research Program Manager, CIMMYT, Mexico	002
Achieving sustainable cultivation of rice - Vol 1 Breeding for higher yield and quality Edited by: Prof. Takuji Sasaki, Tokyo University of Agriculture, Japan	003
Achieving sustainable cultivation of rice - Vol 2 Cultivation, pest and disease management Edited by: Prof. Takuji Sasaki, Tokyo University of Agriculture, Japan	004
Achieving sustainable cultivation of wheat - Vol 1 Breeding, quality traits, pests and diseases Edited by: Prof. Peter Langridge, The University of Adelaide, Australia	005
Achieving sustainable cultivation of wheat - Vol 2 Cultivation techniques Edited by: Prof. Peter Langridge, The University of Adelaide, Australia	006
Achieving sustainable cultivation of tomatoes Edited by: Dr Autar Mattoo, USDA-ARS, USA and Prof. Avtar Handa, Purdue University, USA	007
Achieving sustainable production of milk - Vol 1 Milk composition, genetics and breeding Edited by: Dr Nico van Belzen, International Dairy Federation (IDF), Belgium	008
Achieving sustainable production of milk - Vol 2 Safety, quality and sustainability Edited by: Dr Nico van Belzen, International Dairy Federation (IDF), Belgium	009
Achieving sustainable production of milk - Vol 3 Dairy herd management and welfare Edited by: Prof. John Webster, University of Bristol, UK	010
Ensuring safety and quality in the production of beef - Vol 1 Safety Edited by: Prof. Gary Acuff, Texas A&M University, USA and Prof. James Dickson, Iowa State University, USA	011
Ensuring safety and quality in the production of beef - Vol 2 Quality Edited by: Prof. Michael Dikeman, Kansas State University, USA	012
Achieving sustainable production of poultry meat - Vol 1 Safety, quality and sustainability Edited by: Prof. Steven C. Ricke, University of Arkansas, USA	013
Achieving sustainable production of poultry meat - Vol 2 Breeding and nutrition Edited by: Prof. Todd Applegate, University of Georgia, USA	014

Acknowledgements

We wish to acknowledge the following for their help in reviewing particular chapters

- Chapter 1: Dr Young Min Kwon, University of Arkansas, USA
- Chapter 2: Professor John Maurer, Virginia Tech, USA
- Chapter 8: Professor Alessandra Piccirillo, University of Padua, Italy and Dr Erika Ganda, Penn State University, USA
- Chapter 9: Dr Jeferson Lourenco, University of Georgia, USA and Dr Steven Foley, FDA, USA
- Chapter 10: Dr Richard K. Gast, USDA-ARS, USA
- Chapter 16: Dr Greg Thoma, AgNext, Colorado State University, USA
- Chapter 20: Dr Todd Callaway, University of Georgia, USA
- We'd also like to Professor Marta Hernandez-Jover, Charles Sturt University, Australia for their help during the publishing process.

Introduction

The poultry industry continues to face challenges in ensuring product safety for the consumer. It must also do all it can to minimize its environmental impact in the face of climate change. The collection reviews key current research on ways of improving safety and sustainability along the poultry value chain.

Part 1 Zoonotic hazards

The first group of chapters review the last research on key pathogenic risks affecting poultry meat. As Chapter 1 points out, despite a marked decrease in *Campylobacter*-positive chicken carcasses, the incidence of Campylobacteriosis remains high, a problem potentially exacerbated by the move away from routine antibiotic use in poultry production. The chapter looks at vectors of transmission for *Campylobacter* in the poultry supply chain, and explores the range of strategies for enhancing poultry meat safety, including improved biosecurity measures, bacteriophages, probiotics, and natural antimicrobials, as well as chemical and physical treatments. The chapter discusses how these interventions, when applied optimally at various stages of the poultry production cycle, can significantly reduce contamination, leading to safer poultry products.

Building on issues raised in Chapter 1, Chapter highlights that antimicrobial-free strategies hold the key to sustainably controlling major pathogens like *Salmonella*. However, the widely acknowledged need to develop more integrative disease management options is hampered by large gaps in our knowledge of the basic biological processes that promote the establishment and maintenance of *Salmonella* infection. This chapter addresses these gaps in knowledge by focusing not just on the pathogen itself but on the microbial community in which *Salmonella* is embedded. This understanding makes it easier to select the right mix of techniques, including use of biotics, for effective pathogen control.

Chapter 3 discusses what we know about the various pathotypes of *Escherichia coli* and the risks they present not just to human health but as a cause of disease and morbidity in poultry. Based on a better understanding of the sources and dynamics of *E.coli* infection, it will be easier to target the right blend of biotics and organic acids to alter the gastrointestinal microbial population and thus reduce these pathogenic groups. Strategies that can reduce these important pathogens from poultry are critical to ensuring both efficient production and safety for consumers.

Part 2 Pre-harvest safety management on the farm

The next group of chapters addresses how poultry diseases can be better detected and how the implementation of biosecurity measures can prevent the spread of zoonoses. Chapters also analyse recent modifications in antimicrobial use (AMU), their impact on trends in antimicrobial resistance (AMR), as well as strategies to reduce antimicrobial resistance in poultry production whilst controlling the incidence of disease.

Building on Part 1, Chapter 4 begins by summarising the main pathogenic bacteria in poultry meat and eggs and how these pathogens are able to adapt to survive in poultry processing environments. The chapter then explores developments in faster and more accurate detection such as polymerase chain reaction (PCR) techniques and immunoassays. In particular, the chapter summarises development in high-throughput, low-cost methods with short response times and low detection limits for on-site use. These include optical biosensors (e.g. use of fluorescence, colorimetry, lateral flow, Surface Plasmon Resonance (SPR) and Surface Enhanced Raman Spectroscopy (SERS)) and electrochemical methods (e.g. immunosensors, aptasensors, genosensors and imprinted polymers).

Chapter 5 discusses how a better understanding of transmission routes for and trends in occurrence of zoonotic infection can be used to optimize biosecurity on poultry farms. The chapter explores the range of both structural and operational measures involved in an effective biosecurity regime, as well as monitoring and other techniques to assess effectiveness and compliance. As it shows, the right combination of biosecurity measures can create a safe and disease-free environment for birds, those involved in the industry and consumers.

Chapter 6 explores patterns of antimicrobial use (AMU) in poultry production, reviews recent modifications in AMU use and assesses their impact on trends in antimicrobial resistance (AMR). The chapter also examines the common antimicrobial-resistant bacteria and it evaluates strategies for managing and reducing AMR in poultry production, such as flock management practices and gastrointestinal colonization control

Chapter 7 reviews current research on alternatives to antibiotics to control disease in both preharvest and postharvest settings. It analyses the categories of biotics currently used, starting with what we know about the use and effectiveness of prebiotics such as non-digestible carbohydrates, fructooligosaccharides (FOS), galactooligosaccharides (GOS), isomaltooligosaccharide (IMO), yeast-derived components and mannan derivatives. The chapter also reviews trends in the use of postbiotics and synbiotics as well as potential candidates for future applications in poultry production.

Chapter 8 provides an overview of the particular safety issues affecting free-range poultry production systems, with a particular focus on *Campylobacter* and *Salmonella*. It also considers other infections such as Erysipelas and avian influenza which present a risk to both birds and farm staff. Finally, the chapter considers the particular biosecurity requirements needed in free-range systems.

Part 3 Post-harvest control along the poultry supply chain

The next group of chapters addresses how safety management and monitoring can be controlled at the post-harvest stages in the poultry supply chain by using a systematic approach. Chapter 9 begins by reviewing key challenges and solutions in ensuring effective safety management and monitoring in poultry slaughter operations. The chapter shows how safety management and monitoring can be effectively managed using a systematic approach with both facility-tailored and defined program plans. These include Good Manufacturing Practices (GMPs), Sanitation Standard Operating Procedures (SSOPs), and Hazard Analysis and Critical Control Points (HACCP) plans.

Building on Chapter 9, Chapter 10 highlights the key importance of official monitoring by government agencies of safety management in poultry processing operations. This can be challenging given the size of the sector as well as constraints in resources, personnel and time. The chapter reviews how best to perform targeted and high-quality meat inspection and food safety inspections in poultry processing. Key aspects include preparatory food chain information and ante-mortem inspection, ensuring the right knowledge and skills to perform food safety inspections, ensuring the right frequency and focus of food safety inspections, inspection techniques, non-compliance and official control actions, as well as documentation, analysis and disclosure of inspection results

Chapter 11 explores the use and effectiveness of various antimicrobials, including organic and inorganic acids, used as short-duration dips and sprays to reduce common pathogens (*Salmonella*, *Campylobacter jejuni*, and *Escherichia coli*) on raw chicken whole carcasses and meat portions. It analyses individual antimicrobials, summarises factors affecting antimicrobial efficacy and approaches to optimizing their use.

Chapter 12 discusses how a broader understanding of the microbial ecology of ready-to-eat (RTE) poultry meat products is needed to fully assess measures such as use of antimicrobials to extend shelf life. It explains how developments in next generation sequencing have emerged as a viable means to assess overall microbial ecology responses in food products. This chapter discusses how these approaches can be used to better understand and manage spoilage processes.

Part 4 Sustainability

The last group of chapters focuses on the environmental impact of poultry production. It addresses the potential use of more sustainable alternative feed ingredients. It then describes current poultry processing practices with their current outcomes and future opportunities to enhance the value of these byproducts.

As Chapter 13 highlights, addressing the environmental sustainability of poultry production systems is a multidimensional challenge that requires a comprehensive understanding of interconnected factors. Systems modeling-based life cycle assessment (LCA) offers a valuable approach to systematically navigate these complex interactions. The chapter reviews current applications of LCA to analyse the environmental impact of poultry production and to suggest potential mitigation strategies.

Chapter 14 discusses the potential use of more sustainable alternative feed ingredients in poultry feed. It addresses the role of exogenous feed enzyme additives (phytase, carbohydrase and protease) in counteracting anti-nutritional factors in alternative feeds and optimizing release of nutrients. Finally, the chapter assesses the role of prebiotics, probiotics, synbiotics and precision biotics in optimizing gut function for improved feed utilization.

Chapter 15 explores the microbiological quality and safety aspects of poultry litter, along with various treatment methods to mitigate risks associated with its use. It focuses on different aspects, from composting of poultry litter to anaerobic digestion. As the chapter shows, the application of well-managed poultry litter can contribute to significantly higher yields with a range of crops.

The last chapter addresses the challenge of how best to deal with the byproducts from poultry processing. The chapter describes current poultry processing practices along with the current streams for valorising byproduct materials. In particular, the chapter focuses on recalcitrant materials that otherwise go to waste. As the chapter shows, these have the potential to be repurposed as high-value compounds using biologics-based research. The emerging field of 'biologics' involves discovery of unique properties and functions associated with a particular molecule such as an antimicrobial peptide or tissues which can improve or benefit animals and/or humans, depending on the properties of the biologic candidate.

Part 1

Zoonotic hazards

Chapter 1

Campylobacter in poultry meat

Igori Balta and Iulia Adelina Bundurus, University of Life Sciences King Mihai I from Timisoara, Romania; Ozan Gundogdu, London School of Hygiene and Tropical Medicine, UK; Todd Callaway, University of Georgia, USA; Ioan Peț and Ducu Stef, University of Life Sciences King Mihai I from Timisoara, Romania; David McCleery, Agri-Food and Biosciences Institute, UK; Lavinia Stef, University of Life Sciences King Mihai I from Timisoara, Romania; Alastair Douglas, Agri-Food and Biosciences Institute, UK; and Nicolae Corcionivoschi, University of Life Sciences King Mihai I from Timisoara, Romania/Agri-Food and Biosciences Institute, UK

1 Introduction

Over the past half-century, global poultry meat production has grown dramatically, growing over 12-fold from 1961 to 2014 [67, 68]. Driven by the perception of a healthier, lower-fat nutritional profile, and less association with cardiovascular disease and cancer compared to red meat, poultry has become a preferred choice for more health-conscious consumers [49]. In the US, e.g., chicken consumption has significantly outpaced that of other meats, including beef and pork, achieving growth of 160% per capita between 1970 and 2022, while beef and pork have seen a 31% and 6% reduction, respectively [50, 51].

http://dx.doi.org/10.19103/AS.2024.0146.01

Major producers include the US (annual output of over 20 million tons), the European Union (EU) (around 19 million tons) China (18 million tons) and Brazil (13 million tons) [67]. Growth projections are from 82 million tons in 2005 to a potential 181 million tons by 2050, an average annual expansion of 1.82%. [68]. This growth has led to more intensive forms of poultry production and processing, whether in large, indoor barns containing thousands of birds or in large-scale, highly automated processing facilities [48]. However, these increases in the level and scale of production have also increased the potential risk of foodborne disease.

Foodborne diseases from pathogens such as *Campylobacter* can be traced back to a wide range of food and non-food sources [57]. However, data analysing foodborne illness in 2019 in the US implicate chicken in nearly two-thirds (64.7%) of non-dairy *Campylobacter* illnesses, and that 56.5% of human *Campylobacter* infections are attributable to contaminated poultry products, highlighting chicken as a significant source of campylobacteriosis [50, 54, 58, 59]. The name 'Campylobacter' is derived from ancient Greek, where 'kampylos' indicates 'curved' and 'baktron' indicates 'rod'. The *Campylobacter* genus is comprised of approximately 57 species, including a number of clinical and veterinary importance in causing disease (e.g. *Campylobacter jejuni* and *Campylobacter coli*). The genus is characterized by Gram-negative bacilli with distinctive spiral shapes and polar flagella enabling a corkscrew-like motion (Fig. 1) [1, 2].

The potential role of *Campylobacter* spp. as a human health risk can be traced back to the US in 1938 when the bacteria were first isolated during

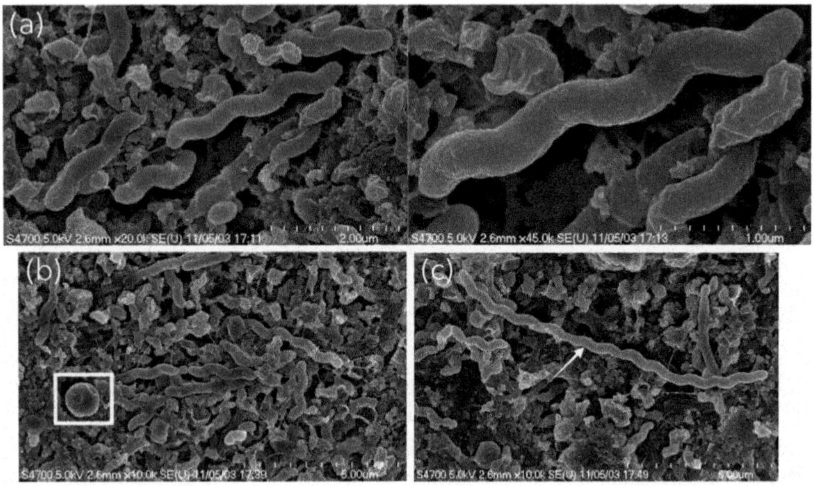

Figure 1 Morphology of *C. jejuni*. (a) helical-shaped *Campylobacter*; (b) viable but non-culturable form (coccoid) isolate is in the white frame box; (c) distinctive morphological deviation of a *Campylobacter jejuni* isolates (arrow).

Published by Burleigh Dodds Science Publishing Limited, 2025.

a disease outbreak related to milk [3]. These bacteria were identified in the 1950s in the blood samples of children suffering from diarrhoea [3, 4]. A critical turning point in understanding a causal relationship between *Campylobacter* spp. and disease in humans came in Belgium in 1972 when they were isolated from the blood and faeces of a young woman who was previously healthy but had suddenly developed acute febrile hemorrhagic enteritis [5]. In 1977, Skirrow isolated the same organism from an infant with diarrhoea, prompting further investigation into the pathogenic potential of these bacteria which has continued ever since [6, 7]. Key developments in identifying and understanding the significance of *Campylobacter* as a pathogen are shown in Fig. 2.

The World Health Organization (WHO), together with Foodborne Disease Burden Epidemiology Reference Group (FERG), reported in 2020 that, out of an estimated 550 million foodborne illnesses per year, *Campylobacter* is identified as one of the top four global contributors to diarrhoeal diseases [8]. Their data indicates that *Campylobacter* was responsible for 166 million annual cases of diarrhoea and 31 700 cases of Guillain-Barré Syndrome (a neurological disorder), a problem exacerbated by the widespread distribution of multidrug-resistant strains of the pathogen [9–11, 26]. The latter issue highlights the complex interplay between environmental, animal, and human health which requires an integrated 'One Health' approach to disease surveillance, control, prevention, and mitigation [20, 33].

Campylobacteriosis due to contamination from poultry and other sources thus represents a significant public health challenge, frequently topping the list of reported foodborne illnesses. Health authorities in Europe and the US report around 230 000 and over 1.5 million annual cases, respectively, with an economic burden estimated in the US at around $6.9 billion [52, 53]. Other

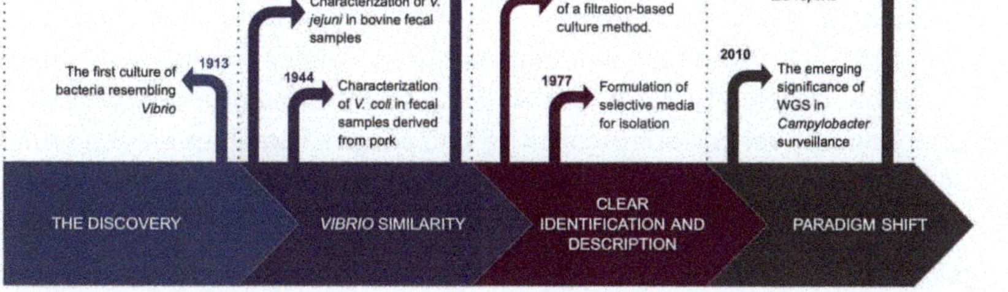

Figure 2 The taxonomy of *Campylobacter* has been shaped by a series of significant historical events.

studies estimate the economic impact of *Campylobacter* as ranging from $1.3 to $20 billion annually in the US [48]. Healthcare expenditures associated with treating Campylobacteriosis have been estimated at approximately $80 million in Canada, £51 million in the UK, and €82 million in the Netherlands [54]. In the US, *C. jejuni* and *C. coli* were responsible for 98% of reported human cases from 2015 to 2017, whilst the WHO estimates the global annual mortality rate attributed to *Campylobacter* to be approximately 37 600 [56].

The pathogen typically triggers diarroheal conditions in humans, though post-infection complications can be far-reaching, including septicaemia, hepatitis, cholecystitis, hemolytic uremic syndrome (HUS), meningitis, and even leading to foetal mortality [11, 12]. Approximately 99% of cases in humans manifest as isolated, sporadic incidents rather than forming part of recognized outbreaks, with a significant number remaining undiagnosed or unreported [3]. However, given less developed immune defences, young children from the age 0 to 5 are particularly susceptible to *Campylobacter* infections [13]. About 18% of asymptomatic school-aged children were recently discovered in one study to be carriers of *Campylobacter* species, namely *C. coli*, *C. jejuni*, and *Campylobacter lari*; of these, 84% of isolates displayed resistance to multiple drug classes [14]. Older and immunocompromised patients are also vulnerable to infection from *C. jejuni* and *C. coli* [15, 16]. The highest percentage of hospitalizations and fatalities predominantly occur amongst individuals aged 65 years and above [8]. The increasing incidence of disease in the 65 and over age group in countries such as Ireland, Australia, the UK, and Germany has been related to age-related immunosenescence and immunosuppression [27]. Evidence from acute gastroenteritis epidemiological surveillance in Australia, Canada, and the US indicates that the likelihood of experiencing severe symptoms and requiring hospitalization is greater among older people after *Campylobacter* infections [55].

A low dose of approximately 500 bacteria is capable of triggering clinical symptoms [17]. *Campylobacter* virulence factors include adhesion, chemotaxis, motility, response to oxidative stress, cytolethal distending toxin production, intricate regulation of iron uptake and invasion, lipopolysaccharide and capsular polysaccharide synthesis, and resistance to multiple drugs and bile acids [18–20]. Among the primary virulence determinants of *C. jejuni*, the cytolethal distending toxin (cdt A, B, C) complex stands out, facilitating the induction of host cell apoptosis. The bacterium's flagellum plays a crucial role in governing both motility and the secretion of invasive antigens [21]. The polysaccharide capsule (CPS) of *Campylobacter* significantly increases its virulence [22].

2 Transmission vectors for *Campylobacter* spp. in poultry supply chains

To curb the risk of human infection, it is crucial to decrease the *Campylobacter* burden both at the farm level and during other stages in the supply chain such as the slaughtering process [29]. As an example, it has been estimated that reducing *Campylobacter* spp. levels by 3 $\log_{10}$ units in chicken caeca (part of the digestive system) at slaughter could potentially reduce the risk of human infection by up to 90% [28]. A reduction in *Campylobacter* populations present on the neck and breast skin of poultry of 10^3 CFU per gram could reduce infection risk by 50% [23–25].

Pathogenic strains of *Campylobacter* such as *C. jejuni* and *C. coli* are widely found in the intestinal ecosystems of avian species, including poultry and many wild bird species [19, 21, 25]. Domestic pets can also be a reservoir for *C. jejuni* and *Campylobacter upsaliensis* [1, 31]. Surface water contaminated with faeces from wild animals and agricultural waste provides a reservoir for diverse *Campylobacter* strains [1, 32]. Wild birds acquire *C. jejuni* from sources such as contaminated water and waste, including refuse dumps, which they can then further disseminate [1, 18]. Migratory birds significantly contribute to the distribution of *Campylobacter* within the poultry sector and the broader environment. Examination of wild birds, particularly those in close proximity to poultry farms and live bird markets, has revealed the presence of diverse, virulent, antibiotic-resistant *C. jejuni* and *C. coli* strains, [18, 21]. Waterfowl, especially geese, act as significant vectors for transmitting *Campylobacter* spp., either directly through contact with poultry or indirectly via wider environmental contamination e.g. of reservoirs supplying drinking water [18, 34, 35].

Climatic and weather conditions can also play a role in promoting the spread of *Campylobacter* spp. As an example, the emergence of campylobacteriosis as the most common bacterial gastrointestinal infection in Northern European countries such as Sweden has been associated in part with a combination of increased rainfall and higher mid-to-late summer temperatures which both encourage *Campylobacter* growth and persistence in the environment and facilitate transmission [41, 42, 44]. Other EU countries such as Ireland, Hungary, and Slovakia, as well as the UK, have been found to experience a moderate increase in cases during the early summer associated with higher temperatures and, to a lesser extent, increased precipitation [43, 45]. Increased temperatures and higher rainfall have e.g. been associated with a greater risk of contamination of water sources, resulting in higher risks for both humans and livestock [46, 47, 67].

Figure 3 illustrates a number of ways *Campylobacter* can be transmitted from the wider environment to poultry flocks. Other sources include people visiting the farm, highlighting the importance of effective biosecurity measures [25].

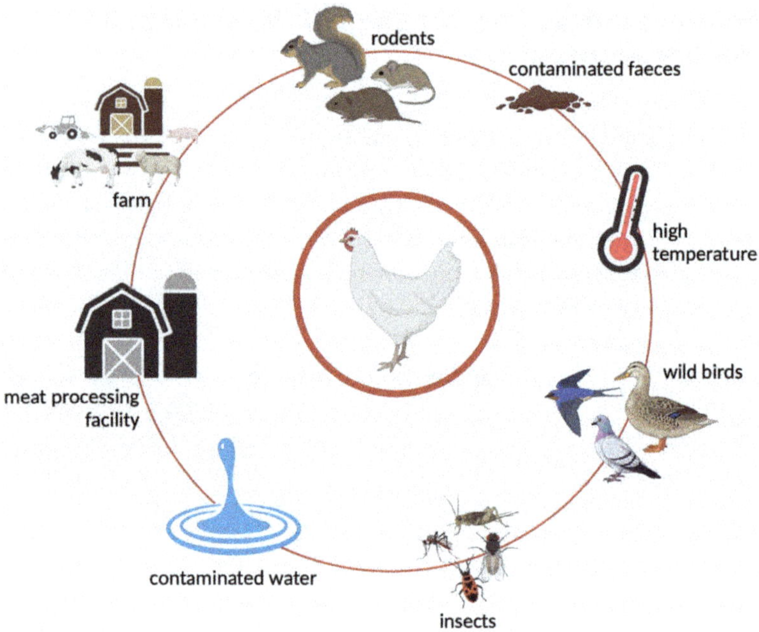

Figure 3 Potential pathways and risk factors through which *Campylobacter* contamination can occur in poultry production.

Organisms such as beetles, flies, amoebae, yeasts, and moulds have also been found to be significant vectors for the transmission of *Campylobacter* into poultry houses [2, 36]. The coexistence of *Campylobacter* cells with amoebae, yeasts, and moulds enhances pathogen survival. Microbial eukaryotes (e.g. the amoeba host *Acanthamoeba polyphaga*) can serve as environmental reservoirs of *Campylobacter* [2, 37]. Given the typical prevalence of eukaryotes in both drinking water systems and as microbial biofilms on farms, infected eukaryotes may play a pivotal role in the transmission and growth of *Campylobacter* spp. in poultry farms.

Insects can act as vectors for *Campylobacter* transmission, introducing potential contaminants into the broiler houses e.g. through poorly maintained buildings or via ventilation systems [3]. The lesser mealworm beetle and its larvae (*Alphitobius diaperunus*) e.g. have been identified as carriers of *C. jejuni* in poultry settings, with the ability to transmit *C. jejuni* not only within individual flocks but also enabling cross-contamination across successive rearing cycles. Houseflies and darkling beetles have also been identified as potential vectors for introducing *Campylobacter* to poultry flocks [36]. A study from Denmark reported a daily entry of approximately 900 flies into broiler houses, and UK research found an average of around 31 flies gaining entry every 2 h. As an example, the multi-drug resistant *C. jejuni* strain ST-1839 has been found in

Published by Burleigh Dodds Science Publishing Limited, 2025.

numerous fly-derived isolates from turkey farms, underlining the role of insects in the propagation of resistant strains [36, 39]. Implementing fly screens could substantially reduce this contamination route [36, 38].

Campylobacter colonization in poultry farms occurs predominantly via horizontal transmission between individual birds, often mediated by feed or drinking water contaminated by *Campylobacter* present in the environment [25, 67]. Factors affecting transmission include the age of birds, flock size, type of production system, water quality, and the presence of other animals on the farm [67]. *Campylobacter* has been found to be particularly prevalent in areas such as drinking water, feed, broilers' faeces and worker's footwear rather than from sources such as litter, airborne particulates, and farm buildings [25]. *Campylobacter* exhibits remarkable endurance in farm settings, surviving in slurry, manure, and dirty water for as long as three months [25, 74]. *Campylobacter jejuni* is able e.g. to adapt to conditions of nutrient deprivation and oxygen stress [74, 75]. Moreover, survival characteristics can be boosted when in the presence of other pathogens such as *Pseudomonadaceae*, since *Campylobacter* can latch onto biofilms produced by these other pathogens [74]. Vertical transmission of *Campylobacter* from hen to egg is not a significant route [25]. Once introduced to a chicken flock, *Campylobacter* can propagate rapidly, effectively colonizing most of the avian intestinal tract – including the caeca, small intestine, and crop (part of the oesophagus) – within a week [21]. An entire flock can be colonized within 2–13 days [30].

Maternal antibodies specific to *Campylobacter* can protect the gastrointestinal tract of fledgling broilers against bacterial colonization [25]. However, this protective effect can be reduced by a restricted diet in more intensive poultry-rearing systems which can result in a less varied and competitive microbiota in the gastrointestinal tract, creating a more conducive environment for the proliferation of *Campylobacter* [2]. Contemporary developments in the field of *Campylobacter* colonization control within avian intestinal tracts have unveiled an array of novel strategies. These include the use probiotics, the incorporation of organic or inorganic acids into feed, the application of bacteriocins, the deployment of bacteriophages, and the development of targeted vaccines to boost gut and immune function and prevent *Campylobacter* becoming established [19, 32, 76–80]. A synthetic peptide, oligo-acyl-lysyl C12K-2β12, has been investigated for its ability to inhibit *Campylobacter*, particularly given its ability to withstand both heat and acid conditions characteristic of subsequent processing environments [82].

The high population density typical of commercial broiler farming also accelerates pathogen transmission from bird to bird [3]. High densities of birds in indoor settings promote higher temperatures and humidity levels which promote *Campylobacter* growth, particularly in summer months [30]. A survey of 27 poultry farms across California, Oregon, Washington, and Idaho in the US found that farms with higher average wind speeds in the week before sampling (increasing the risk of pathogen transmission from the outside environment), combined with younger flocks (with lower immunity), higher flock densities and frequent rotations (increasing horizontal transmission) were most at risk of *Campylobacter* spp. infections [40].

3 Pre- and post-slaughter measures to prevent *Campylobacter* contamination in poultry processing

Recent model simulations highlight the benefit of combining different intervention measures against *Campylobacter* contamination. These suggest insect control as the most important measure, followed by enhanced farm sanitation and biosecurity measures to control access to the farm [29]. It was found e.g. that insect control measures such as fly screens/mesh could reduce the percentage of infected chickens from 51% to 26% and highly infected (>1000 CFU/g) chicken neck samples from 13% to 8% [3, 29]. The highest level of biosecurity can be found in modern broiler houses with minimized access points for rodents, birds, or insects, free of potential entry points such as floor, wall, or ceiling defects and with well-designed ventilation systems that prevent external contaminants from entering the poultry house [3]. Good flock management e.g. an 'all in, all out' approach where groups of birds are kept together throughout their time on the farm and segregated from other groups (rather than mixed e.g. by thinning and moving birds between flocks) is also important as are good hygiene practices such as changing clothes and footwear before entering the barn (the latter being more effective than boot dips alone) [3]. An integrated approach combining all these measures has been found to reduce the percentage of infected chickens to 5% and highly infected chicken neck samples to 2%.

Reported rates of contamination of poultry carcasses with *Campylobacter* show considerable variability worldwide with some countries (e.g. Iceland and Brazil) having historically much lower rates than countries such as the US and UK [24, 61–65]. The Food Safety and Inspection Service (FSIS) of the United States Department of Agriculture (USDA) has tracked a significant decrease in the prevalence of *Campylobacter*-positive chicken carcasses [60]. Initial figures from the 1994–1995 survey indicated a high prevalence of 88.2%. However, subsequent surveys from 2007 to 2008 and 2018 to 2019 demonstrated a substantial reduction to 46.6% and 18.3%, respectively contributing to a decline in human *Campylobacter*-associated infections [58, 60].

Results from Ireland also indicate a downward trend in *Campylobacter* prevalence in broilers, with a remarkable 45% reduction (from 98.3% to 53%) in contaminated carcasses and a 17% decrease (from 83.1% to 66%) in colonized batches of poultry. Despite this progress, the incidence of *Campylobacter* in humans has not seen a corresponding decline, with confirmed cases increasing from 1752 in 2008 to over 3000 in 2018. Consequently, poultry meat remains a significant source of *Campylobacter* contamination in Ireland [61]. There has been a similar trend in New Zealand where improved control measures have resulted in a 74% decrease in poultry-attributed campylobacteriosis cases [58] though without eradicating disease outbreaks related to the pathogen [66]. In the Canadian poultry sector, *Campylobacter* has been found to be present in about 24% of broiler chicken samples at the moment of slaughter, 33% of poultry samples during the processing stage, and up 42% of chicken samples available for retail sale, indicating the challenges and importance of controlling the pathogen at each stage of the supply chain, including proper handling, storage, and cooking of poultry meat [8, 67, 71].

The pre-slaughter transport phase entails moving broilers from the farm to the abattoir. Birds are placed into cages for transport, denied feed and water, and subjected to stress during travel, promoting cross-contamination from a single colonized broiler to others via contact with faeces or feathers [25]. The stress broilers experience during transportation increases the excretion of faecal matter which promotes contamination [2]. Measures such as cleaning and sanitizing contaminated crates before and after broiler transportation may have only a modest impact in preventing *Campylobacter* infection [70, 73].

Defeathering, trimming, and deboning can inadvertently foster an environment conducive to cross-contamination of *Campylobacter* species, rendering previously uncontaminated carcasses vulnerable [72]. De-feathering, evisceration, and portioning result in intestinal contents being expelled or leaked, releasing contaminated material particularly from the crop and the intestines [12]. Deboning can lead to bacterial proliferation on both equipment and within the slaughterhouse environment [72]. *Campylobacter* spp. has been found particularly on de-feathering and evisceration equipment, workers' gloves, shackles for holding carcasses, and conveyor belt equipment [25].

As many other pathogens, *C. jejuni* has shown a remarkable ability to attach to an array of materials such as glass, wood, stainless steel, or plastic, forming a biofilm that is resistant to cleaning [12]. *C. jejuni* e.g. adheres strongly to stainless steel interfaces, even in nutrient-deprived conditions and at temperatures of 20°C [73]. The use of antimicrobials and quorum sensing (QS) signal inhibitors can eradicate biofilms, though this is a risk of bacterial resistance developing through mechanisms such as horizontal gene transfer and spontaneous mutations [54]. Disinfectants such as sodium hypochlorite (NaOCl) are not always effective in eradicating pathogens tenaciously adhering to carcasses [12].

Ways of preventing or removing bacterial contamination of carcasses can be grouped into three categories [81]:

- physical interventions (hot water, cold plasma, steam irradiation, ultrasound, ultraviolet light, air chilling, freezing, and others);
- chemical interventions (natural antimicrobials, EOs, organic acids (e.g. lactic acid), chlorine, hypochlorite, electrolysed oxidizing water, and ozonated water); and
- biological interventions (bacteriophages and probiotics).

The application of 2% lactic acid on poultry carcasses can potentially lower the infection risk of *Campylobacter* by between 37% and 56% [24]. Employing acidified sodium chlorite (1200 mg/L) and trisodium phosphate (10–12%, at pH 12) in carcass treatment can diminish this risk by 75–96% and 67–84% [24]. Techniques such as the use of peracetic acid and water immersion chilling, a widely used procedure in the US, also contribute to pathogen reduction strategies [60].

4 The use of bacteriophages to control *Campylobacter* in the poultry gut

Bacteriophages, viruses targeting bacteria, and bacteriocins, antimicrobial peptides produced by probiotic lactic acid bacteria, have emerged as potential feed additives to suppress *C. jejuni* populations in the caecum of broilers [3, 25]. *Campylobacter* isolated and cultivated phages have been explored to control *Campylobacter* colonization within the gastrointestinal tract of live birds [80]. This strategy has also been extended to poultry carcasses to lower *Campylobacter* levels during processing. Phage therapy maintains the health of the broilers' intestinal microbiota while effectively tackling the pathogen [80]. Most phage research focuses on the *Myoviridae* family of *Campylobacter* phages, while studies involving the *Siphoviridae* family are comparatively uncommon [85]. In analysing the connections between phages and antibiotic-resistance genes, it was found that the development of phage resistance often coincided with heightened antibiotic sensitivity in bacteria [86]. This complex interaction suggests the potential for phage resistance to mitigate antibiotic resistance. Bacteriophage treatment has been shown to decrease caecal levels of *C. jejuni* in broiler flocks by approximately 5 logs CFU/g [87]. However, disadvantages include their low dilution rate in the gut [2].

Studies on broilers have shown that the deployment of bacteriophages (including but not limited to NCTC 12671, 12669, 12672, 12673, 12674, 12678, 12684, CAM-P21, Cj6, CP8, CP34, CP81, phiCcoIIBB35,phiCcoIIBB12, and phiCcoIIBB37) can appreciably reduce *Campylobacter* levels when administered 24–48 h before slaughter [24, 25, 80]. The introduction of individual bacteriophages or their combinations (like phage NCTC 12673, 12674, and 12678) has resulted in notable reductions of bacterial logarithmic counts in caecal content, ranging between 2.5 and 3.2 CFU/g. It was also shown that phage cocktail treatments (CP20 and CP30A), unlike broad-spectrum antibiotics, do not adversely affect the gut microbiota of broilers, thus preventing dysbiosis [25].

H-fiber, sourced from a prophage akin to *C. jejuni* Integrated Elements (CJIEs)-1, is a distinct receptor-binding protein [80]. Zampara et al. developed innolysins to specifically target *C. jejuni* by fusing this H-fiber with T5 endolysin [88]. The application of these innolysins, which merged the enzymatic function of endolysins with the binding ability of phage receptor proteins (in order to amplify their effect on Gram-negative bacteria), led to a 1.18 to 1.63 log reduction in *C. jejuni* CAMSA2147 on chicken skin refrigerated to 5°C [24, 88].

Richards et al. showed how a CP30A and CP20 phage cocktail significantly reduced intestinal *C. jejuni* populations in colonized birds over 5 days without impacting the diversity of caecal and ileal microbiota [89]. An evident effect was observed in the caeca, with bacterial counts reduced by 2.4 $\log_{10}$ CFU/g. Significant bacterial reductions were found in the ileum (1.36 $\log_{10}$ CFU/g) on day 2 and the colon (1.74 $\log_{10}$ CFU/g) on day 3. Both phages were detected in all three compartments of the chicken's gastrointestinal tract for 5 days, implying successful *in vivo* replication and coexistence without competition [89]. Whilst completely eradicating *Campylobacter* in the bird's gastrointestinal tract with phage therapy may be unachievable, these reductions are significant [80].

Others have identified CAM-P21, a broad-spectrum phage, exhibiting superior performance under varied stress conditions and a higher titer than *Myoviridae* family phages [85]. CAM-P21 significantly decreased viable *C. coli* counts by more than 2 logs after incubation at 42°C and 37°C *in vitro* over 12–24 h. In another study, oral administration of two field phages to experimentally infected broiler chickens aged 37 days significantly diminished Campylobacter levels by 2 $\log_{10}$ CFU/g at 40 days [24].

Companies such as PhageBiotech (Israel), Micreos (Netherlands), and Initralytix (USA) are developing commercially available phage-based products aiming to control pathogens including *Listeria*, *E. coli*, *Campylobacter*, *Salmonella*, and other species. Phage therapy can be improved in terms of antimicrobial precision and efficiency via CRISPR-Cas systems as demonstrated in an *in vivo* study by reducing *Clostridioides difficile* infection [90].

5 The use of probiotics in poultry feed

Modification of gut microbiota through competitive exclusion using probiotics and faecal microbial transplants (FMTs) has demonstrated remarkable potential in mitigating *Campylobacter* colonization [24]. Introducing a non-pathogenic strain into a niche occupied by a pathogenic *Campylobacter* strain can outcompete the latter, leading to its removal from the niche [28]. This strategy is often adopted with innovative probiotic species or strains to regulate pathogens such as *C. jejuni*.

Probiotic mixtures containing *L. paracasei* J.R, *L. rhamnosus* 15b, *L. lactis* Y, and *L. lactis*, have been found to result in as much as 5 $\log_{10}$ CFU/mL reduction in caecal *Campylobacter* counts [91]. Equally impressive outcomes were reported where the caecal load of *C. jejuni* in broiler chickens was decreased by up to 6 $\log_{10}$ [92]. This substantial reduction was achieved by administering a probiotic mixture comprising *Bifidobacterium animalis*, *Enterococcus faecium*, *L. salivarius*, *L. reuteri*, and *Pediococcus acidilactici*.

Partial success has been recently observed with the use of lactic acid bacteria [54]. Incorporating acidic feed additives and acids into drinking water has shown promising outcomes. Including the probiotic (Lavipan) in broiler chicken feed significantly mitigated invasion of *Campylobacter* spp. within the birds' gastrointestinal tract [93]. Regular oral intake of *L. salivarius* SMXD51 every 2–3 days from day 1 to day 35 of age can diminish caecal *Campylobacter* counts by 0.8 $\log_{10}$ at 14 days, with a more significant reduction of 2.8 $\log_{10}$ noted at 35 days [94]. The probiotic *E. coli Nissle* 1917 (EcN), either unencapsulated or encapsulated within chitosan-alginate microspheres, has been administered orally to chickens three times weekly for 2 weeks [95]. The authors observed a reduction in caecal *C. jejuni* colonization between 2 and 2.5 log CFU/g. The treatment also enhanced the chickens' growth performance, intestinal structure, and immunity without disrupting the gut microbiota [95]. EcN is a compelling case of a highly effective probiotic, distinguished by its unique Lipopolysaccharide (LPS) that invokes immunogenic responses without inducing immunotoxic effects [77]. The biological safety profile of EcN has been rigorously evaluated in animal studies and human trials [77].

Šimunović and collaborators have shown that *B. subtilis* PS-216 exhibited an inhibitory effect on *C. jejuni* growth, with a substantial reduction in sterile chicken litter (by 4.07 log) and in sterile intestinal content (by 2.26 log) [28]. *B. subtilis* PS-216 can proliferate even in nonsterile intestinal experimental conditions, suggesting its potential utility as a chicken probiotic that can be assimilated into the avian intestinal microbiota. *B. subtilis* PS-216 manifested the most potent antagonistic effect against *C. jejuni* NCTC 11168 under conditions mimicking those of the chicken's environment, specifically, 42°C, microaerobic atmosphere, and chicken litter medium [28]. *Lactobacillus gallinarum* PL53 (at a concentration of 10^8 CFU) has been found to diminish *C. jejuni* presence within the cloaca and caeca, which reinforces the potential of employing such microorganisms as suppressors of *Campylobacter* colonization during primary production [77, 96].

Published by Burleigh Dodds Science Publishing Limited, 2025.

6 The use of chemical agents for decontamination

The poultry industry routinely employs chemical decontamination methods to remove contaminants from carcasses. These methods involve using substances interchangeably referred to as sanitizers, disinfectants, sterilants, antiseptics, biocides, antimicrobials, or decontaminating agents [97]. In the US, the deployment of such processing aids is regulated and monitored by the Food and Drug Administration (FDA) as well as the USDA [97]. In contrast to the multiple chemical decontamination strategies permitted in the US, the EU has shown a more limited approach, currently only authorizing up to 5% lactic acid as a decontaminant [81]. Various compounds – including acetic, citric, peracetic, peroxyacetic, and lactic acids, along with acidified sodium chlorite, cetylpyridinium chloride, chlorine, vinegar, sodium hypochlorite, and trisodium phosphate – can contribute significantly to the reduction of *Campylobacter* spp. on poultry carcasses during the slaughter process [83, 84, 97, 98].

According to a recent meta-analysis, cetylpyridinium chloride, acidified sodium chlorite, and trisodium phosphate have demonstrated the most significant log reduction estimates ≈1.38 and 1.26 $\log_{10}$ CFU, respectively [97]. In contrast, treatments involving cetylpyridinium chloride and a blend of potassium hydroxide and lauric acid revealed lower reductions. A combination of ethanol, acidic calcium sulfate, lactic acid, propylene glycol, and sodium dodecyl sulphate resulted in an impressive log reduction estimate of ≈ 4.91 $\log_{10}$ CFU [97]. There is considerable interest in the commercial antimicrobial product Amplon, a blend of sulfuric acid and sodium sulfate [48]. The USDA-FSIS has allowed this product for spray, wash, and dip treatments. A recent investigation demonstrated a significant *Campylobacter* reduction, averaging a decrease of ≈ 1.53 log CFU per chicken, following a 15-s post-chill immersion in an Amplon solution [48].

Peracetic acid (PAA) is routinely employed in the US poultry processing industry, with concentrations typically between 200 and 2000 ppm during chilling, dependent on the duration of chilling [99]. This approach has demonstrated considerable efficacy in mitigating *Campylobacter* numbers in chicken carcasses, showing reductions ranging from ≈ 1.5 to ≈ 2.0 $\log_{10}$ CFU/g or ≈1.8 to 2.6 $\log_{10}$ CFU/mL in microbial load and a 50–100% decrease in prevalence [100]. Notably, no discernible negative impact on chicken quality was reported, with sensory evaluations indicating that 1200 ppm of PAA not only resulted in a larger *Campylobacter* reduction (≈2 $\log_{10}$ CFU/g) compared to 50 ppm chlorine but also maintained the sensory quality of the chicken. PAA has proven highly effective in post-chill dip treatments applied to chicken breasts [48]. One investigation reported reductions of 2.23 and 4.08 log CFU/chicken in *Campylobacter* and *Salmonella* following a 15-s immersion in a solution containing 750 ppm PAA [48].

Proteomic analysis indicates that *C. jejuni* strain 2704 rapidly reacts to PAA stress, with initial response occurring within 15 min and full adaptation between 30 and 45 min [100]. This response manifests as perturbations in the cytoplasmic redox state and central metabolism. Regulatory and homeostatic responses, coupled with adjustments to energy metabolism and flagellar/chemotaxis protein response, underpin *C. jejuni* survival under PAA exposure. Optimizing PAA application necessitates understanding how to bypass these adaptive responses for optimal reduction efficacy. Certain heat stress proteins (Lon, HtrA, HslR) suggest that heat stress response may offer cross-protection when *Campylobacter* is later exposed to oxidative PAA treatment [100].

7 The use of natural antimicrobials

Mixtures of natural antimicrobials, including components like organic acids and plant extracts, reduce the colonization or infection of epithelial [101–108]. Research has investigated the efficacy of a blend of natural antimicrobials (A3001) in limiting the adherence of T6SS$^{+/-}$ *C. coli* isolates (NC1 $^{hcp-}$, NC2 $^{hcp-}$, and NC3 $^{hcp+}$) to chicken skin and carcasses [19]. The antimicrobial mixture significantly diminished the colonization capacity of *C. coli* isolates on chicken skin and carcasses, resulting in an approximate 3-log reduction [19]. Following an *in vitro* infection model, a 0.5% mixture concentration led to significant reductions of adherence in the HCT-8 cell line and downregulated the *hcp* gene in the NC3 isolate, which indicates the presence of T6SS [19].

Olive leaf extract (E1), notably rich in oleuropein – a precursor to hydroxytyrosol, exhibits a significant capacity to inhibit the proliferation of *Campylobacter* [109]. E1 has potent antibacterial activity at a MIC range of 0.1–2 mg/mL, curtailing the growth of *C. jejuni* strains from 4.12 to 8.14 log CFU/mL while inducing a reduction between 6.99 and 7.99 log CFU/mL against *C. coli* strains [109].

Multi-omics analysis and chicken trials have shown that *C. jejuni* colonization notably increases levels of CREG-1, creatinine, and 3-[2-(3-Hydroxyphenyl) ethyl]-5-methoxyphenyl, while concurrently decreasing sphingosine, SP d18:1, high mobility group protein B3, and several phosphatidylcholines (PC P-20:0_16:0, PC 11:0_26:1, and PC 13:0_26:2) [110]. These essential proteins and lipids are implicated in inflammation and immunity in *C. jejuni*-colonized birds. The authors observed average colonization of *C. jejuni* for control groups showing positive results at approximately 6.5 log CFU/g of caecal content on day 14. Treating with an in-water supplement of 0.125% eugenol nanoemulsion (EGNE) reduced counts by around 2.5 log CFU/g compared to the positive controls. Additionally, they detected a rise in the blood serum concentrations of various metabolites, such as 5-hydroxyindole-3-acetic acid, pyridoxal, taurine, indole-3-lactic acid, monoolein stachydrine, taurousodeoxycholic acid, and methionine sulfoxide, after EGNE treatment [110]. Collectively, these metabolites appear to promote favourable responses in broilers through their antioxidant, anti-inflammatory, and antimicrobial properties [110].

According to *in vitro* studies, oregano species and essential oils (Eos) derived from them are rich in antimicrobial bioactive compounds such as carvacrol, p-cymene, thymol, and g-terpinene and are known to exhibit anti-campylobacter activity [111]. When EO derived from *Origanum compactum* was combined with tetracycline or ampicillin, more efficient inhibitory synergy effects were noted against multidrug-resistant and strong biofilm *Campylobacter* strains [111, 112]. Research simulating industrial environments has shown that a significant decrease in *C. jejuni* counts on chicken skin can be achieved – approximately 2.4–4 log10 CFU/sample [112].

Bio-responsive composite liposomes imbued with the EO from *Litsea cubeba* have exhibited impressive antibacterial activity [113]. With an inhibition zone of 23.5 mm, just 20% (v/v) of these composite liposomes can neutralize 99% of *C. jejuni*. After spraying, these liposomes demonstrated a prolonged capacity to impede bacterial proliferation for up to 5 days on refrigerated chicken meat inoculated with *C. jejuni*. The authors reported that these liposomes did not affect the food's colour and texture and could pave the way for the use of liposomes as potent antimicrobial agents [113].

Other research has indicated that edible film, derived from chitosan and pectin and incorporating 2% eugenol, can decrease *C. jejuni* loads by 2 and 3 log CFU/sample in chicken wingettes stored at 4°C for 7 days [49]. A combination of carvacrol or eugenol with chitosan considerably suppresses the expression of specific genes tied to motility (*motA*), stress response (*katA*), and quorum sensing (*luxS*) in *C. jejuni*, which may inhibit the bacteria's survival in poultry [49, 114].

Edible coatings derived from natural plant-based sources and enriched with natural antimicrobials are emerging as innovative solutions for enhancing product longevity and quality [97]. One study comprising 60 trials analysed the influence of eugenol-strengthened pectin and chitosan coatings [97]. The collective alteration in concentration was projected at 2.36 $\log_{10}$ CFU declines within a 95% CI, and significant heterogeneity was observed (I2 = 88%, $P <$ o.001) [97]. Carvacrol-infused chitosan and gum arabic films were evaluated across 69 trials from two studies, generating a cumulative effect estimated at 2.00 $\log_{10}$ CFU reductions and similar significant heterogeneity [97].

8 Physical decontamination treatments

Physical decontamination strategies involve measures to lower contamination in broiler carcasses or meat products by manipulating temperature or applying electrochemical, electromagnetic, or mechanical processes that disrupt the integrity of pathogenic cells [97]. These interventions include crust freezing, immersion chilling, forced air chilling, hot water treatment, vent plug application, rapid surface cooling, and steam-ultrasound treatment [98]. A total eradication of campylobacteriosis risk can be achieved via industrial-scale radiation or the cooking of poultry meat post-slaughter [54]. Compared to chemical methods, physical decontamination has a greater impact on prevalence, while chemical treatments more significantly affect bacterial concentration [98]. Decontamination interventions applied through immersion exhibit greater effectiveness than those executed via spraying. While it has yet to be utilized in commercial contexts, deploying cold plasma technology is an innovative and promising technique, with its potential even extending to pre-packaged food items [48].

Steam pasteurization has recently emerged as a promising approach for surface decontaminating broiler carcasses [115]. This technique entails the brief application of saturated pressurised steam. A pilot-scale steam decontamination system (Deconizer) has proved efficacious in diminishing *Salmonella* and *Campylobacter* [116]. The decontamination effect ranged from a 1.22 to 3.33 log reduction for aerobic bacteria, a 1.36 to 3.05 log reduction for *Salmonella Enteritidis*, and a remarkable 0.84 to 4.32 log reduction for *C. jejuni*. Treatment for 5 s led to greater microbial reductions than 3-s treatments, observed at 95°C and 120°C. This effect was more pronounced on the breast portion than on the legs and wings [116].

Freezing poultry carcasses provides a reliable way to achieve a two-log decrease in *Campylobacter* logarithmic counts [3]. In Iceland, a requirement that broilers processed from *Campylobacter*-infected flocks be frozen had a notable effect, significantly diminishing the incidence of *Campylobacter*-induced enteritis in humans [3]. A reduction of over 90% in the risk can be attained by freezing carcasses for a period of 2–3 weeks [54]. Risk reduction ranging between 50% and 90% is possible through a combination of 2–3 days of freezing, succeeded by warm water or chemical carcass decontamination [54]. Freezing can diminish *Campylobacter* counts, but traces of viable bacteria can still be detected on poultry even after weeks of freezing at −20°C [3]. Strain variability and environmental pressures can influence survival potential and the ability to colonize hosts. Despite reducing culturable bacteria during freezing, research indicates that these organisms can maintain viability over extended periods in food stored at low or freezing temperatures [12]. The food industry's use of cold chains for product freshness can facilitate the survival of *C. jejuni* and *C. coli* in meat [12, 19]. The potency to enter a viable but non-culturable state (VBNC) may underpin this pathogen's adaptive response to environmental stress [12, 19].

Crust freezing involves rapidly freezing the carcass surface while leaving the underlying meat unfrozen. However, its effectiveness was substantially

Published by Burleigh Dodds Science Publishing Limited, 2025.

lower in real conditions before-after experimental trials compared to controlled challenge trials, suggesting that the practicality of its large-scale implementation remains to be determined [97]. Techniques such as scalding and chilling can efficiently diminish pathogens' prevalence and concentration, but their efficacy could be enhanced by incorporating processing aids to mitigate potential cross-contamination between carcasses and faecal contamination [97]. Physical decontamination methods have a more pronounced effect on reducing *Campylobacter* prevalence, decreasing relative risk by approximately ≈68.7%, as compared to a ≈30.3% reduction achieved through chemical treatment [98]. Laboratory before-after trials reported the effects of brief (up to 40 s) hot water immersion at 70–80°C [97]. The collective log reduction from these challenge trials indicated a ≈1.23 $\log_{10}$ CFU decrease. An assessment of in-plant decontamination of broiler carcasses from Norway has concluded that broiler carcass immersion for 6 s at 80°C was the most effective hot-water treatment by causing significant declines of the *E. coli* by 1.1 log CFU/mL and without causing any skin damage [117]. Consequently, while hot water immersion may show moderate efficacy in conditions of high experimental inoculation, its efficiency could be restricted under processing plant conditions with natural contamination [97].

Decontamination using immersion can be effective, resulting in the reduction of ≈ 0.9 $\log_{10}$ CFU/carcass compared to spraying, which results in a ≈0.5 $\log_{10}$ CFU/carcass reduction [97]. A meta-analysis employing a random-effects model found a significant reduction in *Campylobacter* concentration, with a pooled decrease of 0.6 $\log_{10}$ CFU/carcass and a ≈ 57.2% decline in relative risk of *Campylobacter* spp. prevalence in broiler carcasses. Using 0.1% PAA, applied through immersion and electrostatic spray methods to chicken wings, yielded an average *Campylobacter* reduction above ≈ 2.1 logs CFU/g [48].

9 Developing effective vaccines

Vaccination is still expensive, highly specific, and difficult to design against *Campylobacter*. Even though efforts have been underway for over 20 years to formulate an effective *Campylobacter* vaccine for broilers, commercial success remains elusive [24, 53, 118]. Traditional approaches towards developing a vaccine against *C. jejuni* in poultry have yet to yield a successful result that effectively stimulates immune response and combats bacterial colonization in the poultry's caeca. Given the absence of an effective vaccine and the alarming rise of antimicrobial-resistant strains, there is an urgent need to pinpoint novel intervention targets [53].

Genes essential for the growth and survival of *Campylobacter* present appealing prospects for these interventions. The diversity of genes in

Campylobacter highlights numerous indispensable factors critical to the fitness and survival of pathogen, many of which currently bear an unknown or hypothetical function [114, 119]. Deciphering these genes' roles will greatly enrich our understanding of *Campylobacter*'s fundamental biology, enabling innovative intervention strategies [114, 119]. Computational *in silico* prediction and gene expression analysis could potentially unveil promising vaccine candidates against *C. jejuni* [53]. Reverse vaccinology could pave the way for *C. jejuni* vaccine development. This novel strategy swiftly identifies promising antigenic candidates through microbial genome sequencing [53]. However, despite its ability to streamline potential vaccine candidates, the immediate prospect of yielding a successful vaccine remains modest.

In developing vaccines for broilers, several crucial factors must be taken into account [3, 53, 118]:

- the presence of multiple *C. jejuni* genotypes;
- the short lifespan of broilers;
- the immunological immaturity of the birds at vaccination;
- the potential interference of maternally derived antibodies;
- the necessity for effective antigen presentation at the gut mucosal surface;
- the lack of suitable avian mucosal adjuvants;
- the convenience of the vaccination method e.g. oral or in ovo; and
- the overall cost of the vaccine.

Three responses should be taken into account during vaccine manufacture [118, 120]:

- the influence of maternal immunity;
- initiation of innate immune responses, characterized by the involvement of Toll-like receptors, chemokines, and antimicrobial b-defensin peptides; and
- activation of adaptive immune responses, which induce the production of antibodies or cytokines.

Recent recombinant vaccines targeting *Campylobacter* colonization in broilers have shown significant reductions in Campylobacter colonization, up to ≈2 $\log_{10}$, following the administration of FlaA, FlpA, or a fused protein comprising CaDF-FlaA-FlpA [3, 24]. A ≈ 2.5 $\log_{10}$ decrease in *Campylobacter* colonization was achieved using a recombinant attenuated *Salmonella enterica* to express the Dsp antigen [121].

The immune system and microbiota experience co-evolution in the chicken gut. Gloanec et al. 2022 have recently reported a minor yet significant decrease in caecal *Campylobacter* counts in the DNA prime/protein boost flagellin vaccinated group, a finding further supported by microbiota analysis which showed a differential association of *Campylobacter* genus with the placebo group [122]. The vaccinated group exhibited temporarily heightened levels of Interleukin-10, whereas the placebo group displayed an elevated gene expression of the antimicrobial peptide avian β-defensin-10. The vaccination strategy resulted in a modest reduction in Campylobacter burden, with a mean difference of 1.3 $\log_{10}$ CFU/g noted on day 40 [118]. Moreover, the impact of vaccination on the structure and composition of the caecal microbiota was evident, with distinct augmentations in operational taxonomic units identified as *Ruminococcaceae* and *Bacillaceae* on the 40th day [122].

In-ovo vaccination of embryonated chicken eggs at 18 days with Gram-positive Enhancer Matrix (GEM) particles containing dual *Campylobacter* antigens (CjaA and CjaD) resulted in a modest reduction ≈1 $\log_{10}$ in caecal colonization by *Campylobacter* in broiler chickens aged 3 and 4 weeks post heterologous *C. jejuni* challenge [24]. Interestingly, when these antigens were encapsulated within a liposome, a greater decrease in *Campylobacter* count by 2 $\log_{10}$ was observed [24]. While the reduction in *Campylobacter* was moderately successful, these findings suggest the potential efficacy of *in-ovo* immunization, a strategy that could be augmented with booster vaccinations after hatching. Applying nanoparticle-based vaccines via oral and *in-ovo* routes has garnered interest among several research groups [123, 124]. Vaccinating broiler chickens with *C. jejuni* lysate and a Toll-Like Receptor-21 ligand (CpG ODN 2007) encapsulated within poly-lactic-co-glycolic acid nanoparticles stimulated innate mucosal responses in both the intestine and caecal tonsils, amplified the serum anti-*C. jejuni* IgY antibody titers altered the caecal microbiota composition and diminished the caecal *C. jejuni* count by ≈ 2.4 $\log_{10}$ in the vaccinated broiler chickens [123, 124].

Upcoming research should focus on discovering unique, highly conserved immunogenic proteins that can stimulate cross-protective immunity against diverse *Campylobacters*. The inventive deployment of emerging technologies, such as reverse vaccinology, could facilitate the prediction of new antigenic targets, potentially leading to the fabrication of a multi-epitope vaccine offering broad-spectrum protection against varied *Campylobacter* strains [24, 53]. Subsequently, this could drive the advancement of delivery systems designed for targeted administration of vaccine compositions to *Campylobacter* colonization sites in chickens.

10 Conclusion

Addressing the challenge of *Campylobacter* in poultry is a multifaceted endeavour that necessitates a comprehensive approach. The interventions discussed in this chapter clearly point out both the challenges and also the opportunities.

Biosecurity stands clearly as the first line of defence. A farm's biosecurity strategy is as much about the barriers it builds against external threats as the internal protocols it upholds. Cleanliness, limiting external exposure, including rigorous pest management, could help minimize the presence of *Campylobacter*.

The feed and water interventions including natural antimicrobials, organic acids, and essential oils have produced variable results, indicating that a better understanding of their mechanisms of action is required. Our journey takes us through the maze of the avian gut, a bustling metropolis of microbes, where probiotics and competitive exclusion principles come into play. This is nature's own battleground, where microbial alliances can be leveraged to keep *Campylobacter* at bay. Nevertheless, as with all natural systems, predicting outcomes becomes as much an art as a science. Bacteriophages and their role have also been described. Their precision in targeting specific bacterial strains paints a promising picture, but the storyline is still unfolding, and the results remain to be seen.

Chemical and biological interventions in a world growing increasingly wary of antibiotic resistance are also required. Physical interventions, such as steam, ultraviolet light, and chilling, are also efficient. Often seen in the post-harvest stages, these established technologies underscore the importance of integrating strategies for a holistic defence. And then there is the elusive quest for the perfect vaccine. This is a journey punctuated with highs and lows, the narrative enriched by novel methodologies, from reverse vaccinology to nanoparticle delivery mechanisms. Every twist, turn, setback, and breakthrough brings us closer to that dream of comprehensive protection.

One truth emerges in piecing together these narratives: the fight against *Campylobacter* is multifaceted, demanding many strategies. It is not about choosing one path but understanding how each can be interwoven to craft a holistic approach. Our story, thus, is not just about tackling a bacterial adversary but also about human resilience, ingenuity, and the never-ending pursuit of knowledge. It is evident that the story of *Campylobacter* interventions is far from over, but with every stride, we move closer to a safer, healthier future for livestock and humans alike.

11 References

1. Dessouky, Y.E., et al., (2022). Genomic insights into zoonotic transmission and antimicrobial resistance in Campylobacter jejuni from farm to fork: a one health perspective. *Gut Pathogens*, **14**(1): p. 44.
2. Hakeem, M.J. and Lu, X., (2021). Survival and Control of Campylobacter in Poultry Production Environment. *Frontiers in Cellular and Infection Microbiology*, **10**. https://doi.org/10.3389/fcimb.2020.615049
3. Hansson, I., et al., (2018). Knowledge gaps in control of Campylobacter for prevention of campylobacteriosis. *Transboundary and Emerging Diseases*, **65**(S1): p. 30–48.

4. King, E.O., (1957). Human infections with Vibrio fetus and a closely related vibrio. *The Journal of Infectious Diseases*, **101**(2): p. 119–128.
5. Dekeyser, P., et al., (1972). Acute enteritis due to related vibrio: first positive stool cultures. *Journal of Infectious Diseases*, **125**(4): p. 390–392.
6. Skirrow, M.B., (1994). Diseases due to Campylobacter, Helicobacter and related bacteria. *Journal of Comparative Pathology*, **111**(2): p. 113–149.
7. Skirrow, M., (1977). Campylobacter enteritis: a" new" disease. *British Medical Journal*, **2**(6078): p. 9–11.
8. Myintzaw, P., Jaiswal, A.K. and Jaiswal, S., (2023). A Review On Campylobacteriosis Associated With Poultry Meat Consumption. *Food Reviews International*, **39**(4): p. 2107–2121.
9. Gharbi, M., et al., (2022). Campylobacter spp. in eggs and laying hens in the North-East of Tunisia: high prevalence and multidrug-resistance phenotypes. *Veterinary Sciences*, **9**(3): p. 108.
10. Rowan-Nash, A.D., et al., (2019). Cross-domain and viral interactions in the microbiome. *Microbiology and Molecular Biology Reviews*, **83**(1): p. e00044-18.
11. Ghielmetti, G., et al., (2023). Whole-genome-based characterization of Campylobacter jejuni from human patients with gastroenteritis collected over an 18 year period reveals increasing prevalence of antimicrobial resistance. *Microbial Genomics*, **9**(2). https://doi.org/10.1099/mgen.0.000941
12. Vetchapitak, T. and Misawa, N., (2019). Current status of campylobacter food poisoning in Japan. *Food Safety*, **7**(3): p. 61–73.
13. Newman, K.E. and Khalid, S., (2023). Conformational dynamics and putative substrate extrusion pathways of the N-glycosylated outer membrane factor CmeC from Campylobacter jejuni. *PLOS Computational Biology*, **19**(1): p. e1010841.
14. Gitahi, N., et al., (2020). Multidrug-resistant Campylobacter jejuni, Campylobacter coli and Campylobacter lari isolated from asymptomatic school-going children in Kibera slum, Kenya [version 2; peer review: 1 approved, 1 approved with reservations]. *F1000Research*, **9**(92). https://doi.org/10.12688/f1000research.21299.2
15. Adrianza, A., et al., (2023). Campylobacter coli bacteremia associated with diarrhea. *IDCases*, **31**: p. e01734.
16. Otsuka, Y., et al., (2023). Clinical characteristics of Campylobacter bacteremia: a multicenter retrospective study. *Scientific Reports*, **13**(1): p. 647.
17. Elhadidy, M., et al., (2020). Antimicrobial resistance patterns and molecular resistance markers of Campylobacter jejuni isolates from human diarrheal cases. *PLoS One*, **15**(1): p. e0227833.
18. Bunduruș, I.A., et al., (2023). Overview of virulence and antibiotic resistance in Campylobacter spp. Livestock Isolates. *Antibiotics*, **12**(2): p. 402.
19. Balta, I., et al., (2021). The effect of natural antimicrobials on the Campylobacter coli T6SS+/− during in vitro infection assays and on their ability to adhere to chicken skin and carcasses. *International Journal of Food Microbiology*, **338**: p. 108998.
20. Chibwe, M., Odume, O.N. and Nnadozie, C.F., (2023). A review of antibiotic resistance among Campylobacter species in human, animal, and water sources in South Africa: a one health approach. *Journal of Water and Health*, **21**(1): p. 9–26.
21. Tawakol, M.M., et al., (2023). The potential role of migratory birds in the transmission of pathogenic Campylobacter species to broiler chickens in broiler poultry farms and live bird markets. *BMC Microbiology*, **23**(1): p. 66.

22. Khemnu, N., et al., (2023). Description of novel capsule biosynthesis loci of Campylobacter jejuni clinical isolates from South and South-East Asia. *PLoS One*, **18**(1): p. e0280583.
23. Dubovitskaya, O., et al., (2023). Quantitative assessment of Campylobacter spp. levels with real-time PCR methods at different stages of the broiler food chain. *Food Microbiology*, **110**: p. 104152.
24. Taha-Abdelaziz, K., et al., (2023). Intervention strategies to control campylobacter at different stages of the food chain. *Microorganisms*, **11**(1): p. 113.
25. Mota-Gutierrez, J., et al., (2022). Campylobacter spp. prevalence and mitigation strategies in the broiler production chain. *Food Microbiology*, **104**: p. 103998.
26. Morita, D., et al., (2023). Whole-genome and plasmid comparative analysis of Campylobacter jejuni from human patients in Toyama, Japan, from 2015 to 2019. *Microbiology Spectrum*, **11**(1): p. e02659-22.
27. O'Connor, L., et al., (2020). Epidemiology of Campylobacter infections in Ireland 2004–2016: what has changed? *Zoonoses and Public Health*, **67**(4): p. 362–369.
28. Šimunović, K., et al., (2022). Bacillus subtilis PS-216 antagonistic activities against Campylobacter jejuni NCTC 11168 are modulated by temperature, oxygen, and growth medium. *Microorganisms*, **10**(2): p. 289.
29. Horvat, A., et al., (2022). The impacts of biosecurity measures on Campylobacter contamination in broiler houses and slaughterhouses in the Netherlands: a simulation modelling approach. *Food Control*, **141**: p. 109151.
30. Urdaneta, S., et al., (2023). In a warm climate, ventilation, indoor temperature and outdoor relative humidity have significant effects on Campylobacter spp. colonization in chicken broiler farms which can occur in only 2 days. *Food Microbiology*, **109**: p. 104118.
31. Elmonir, W., et al., (2022). Survival capability of Campylobacter upsaliensis under environmental stresses. *BMC Research Notes*, **15**(1): p. 47.
32. Butucel, E., et al., (2022). Farm biosecurity measures and interventions with an impact on bacterial biofilms. *Agriculture*, **12**(8): p. 1251.
33. Pacífico, C., Wösten, M.M.S.M. and Hilbert, F., (2023). Low-level tetracycline resistance Gene tet(O)_3 in Campylobacter jejuni. *Antibiotics*, **12**(3): p. 426.
34. Linz, D.M., et al., (2023). Genomic characterization and wetland occurrence of a novel campylobacter isolate from Canada Geese. *Microorganisms*, **11**(3): p. 648.
35. Kobayashi, M., et al., (2022). Temporal dynamics of Campylobacter and Arcobacter in a freshwater lake that receives fecal inputs from migratory geese. *Water Research*, **217**: p. 118397.
36. Bolinger, H., et al., (2023). Campylobacter jejuni and Campylobacter coli from houseflies in commercial Turkey farms are frequently resistant to multiple antimicrobials and exhibit pronounced genotypic diversity. *Pathogens*, **12**(2): p. 230.
37. Axelsson-Olsson, D., et al., (2005). Protozoan Acanthamoeba polyphaga as a potential reservoir for Campylobacter jejuni. *Applied and Environmental Microbiology*, **71**(2): p. 987–992.
38. Royden, A., et al., (2016). A role for flies (Diptera) in the transmission of Campylobacter to broilers? *Epidemiology & Infection*, **144**(15): p. 3326–3334.
39. Niedermeyer, J.A., et al., (2018). Proximity to other commercial turkey farms affects colonization onset, genotypes, and antimicrobial resistance profiles of Campylobacter spp. in turkeys: suggestive evidence from a paired-farm model. *Applied and Environmental Microbiology*, 2018. **84**(18): p. e01212-18.

40. Smith, O.M., et al., (2023). Wind speed and landscape context mediate campylobacter risk among poultry reared in open environments. *Animals*, **13**(3): p. 492.
41. Kuhn, K.G., et al., (2020). Campylobacteriosis in the Nordic countries from 2000 to 2015: trends in time and space. *Scandinavian Journal of Public Health*, **48**(8): p. 862–869.
42. Kuhn, K.G., et al., (2020). Campylobacter infections expected to increase due to climate change in Northern Europe. *Scientific Reports*, **10**(1): p. 13874.
43. Lake, I., et al., (2019). Exploring Campylobacter seasonality across Europe using the European surveillance system (TESSy), 2008 to 2016. *Eurosurveillance*, **24**(13): p. 1800028.
44. Lindqvist, R., et al., (2022). The temporal pattern and relationship of Campylobacter prevalence in broiler slaughter batches and human campylobacteriosis cases in Sweden 2009–2019. *International Journal of Food Microbiology*, **378**: p. 109823.
45. Djennad, A., et al., (2019). Seasonality and the effects of weather on Campylobacter infections. *BMC Infectious Diseases*, **19**(1): p. 255.
46. Pedati, C., et al., (2019). Campylobacteriosis outbreak associated with contaminated municipal water supply–Nebraska, 2017. *Morbidity and Mortality Weekly Report*, **68**(7): p. 169.
47. Swanson, D., et al., (2022). A one health real-time surveillance system for nowcasting Campylobacter gastrointestinal illness outbreaks, Norway, week 30 2010 to week 11 2022. *EuroSurveillance*, **27**(43): p. 2101121.
48. Thames, H.T. and Theradiyil Sukumaran, A., (2020). A review of Salmonella and Campylobacter in broiler meat: emerging challenges and food safety measures. *Foods*, **9**(6): p. 776.
49. Gürbüz, M. and İrem Omurtag Korkmaz, B., (2022). The anti-campylobacter activity of eugenol and its potential for poultry meat safety: a review. *Food Chemistry*, **394**: p. 133519.
50. Mujahid, S., et al., (2023). Prevalence and antibiotic resistance of Salmonella and Campylobacter isolates from raw chicken breasts in retail markets in the United States and Comparison to data from the plant level. *Life*, **13**(3): p. 642.
51. Lemanske, H., (2022). *Modeling the Effects of Climate Factors on the Incidence of Future Campylobacter Infections in Connecticut*. Yale University.
52. Rousou, X., et al., (2023). Diagnostic accuracy of multiplex nucleic acid amplification tests for Campylobacter infection: a systematic review and meta-analysis. *Pathogens and Global Health*, **117**(3): p. 259–272.
53. Poudel, S., et al., (2023). In silico prediction and expression analysis of vaccine candidate genes of Campylobacter jejuni. *Poultry Science*, **102**(5): p. 102592.
54. Ma, L., et al., (2022). Campylobacter biofilms. *Microbiological Research*, **264**: p. 127149.
55. White, A.E., et al., (2019). Characteristics of Campylobacter and Salmonella infections and acute gastroenteritis in older adults in Australia, Canada, and the United States. *Clinical Infectious Diseases*, **69**(9): p. 1545–1552.
56. Wanja, D.W., et al., (2023). Antimicrobial usage, susceptibility profiles, and resistance genes in *Campylobacter* isolated from cattle, chicken, and water samples in Kajiado County, Kenya. *International Journal of Microbiology*, **2023**: p. 8394605.
57. Pogreba-Brown, K., et al., (2018). Case–case analysis of Campylobacter and Salmonella – using surveillance data for outbreak investigations and monitoring routine risk factors. *Epidemiology & Infection*, **146**(15): p. 1916–1921.

58. Varrone, L., et al., (2018). Investigating locally relevant risk factors for Campylobacter infection in Australia: protocol for a case–control study and genomic analysis. *BMJ Open*, **8**(12): p. e026630.
59. Poudel, S., et al., (2022). Prevalence, antimicrobial resistance, and molecular characterization of *Campylobacter* isolated from broilers and broiler meat raised without antibiotics. *Microbiology Spectrum*, **10**(3): p. e00251-22.
60. Williams, M.S., Ebel, E.D. and Nyirabahizi, E., (2021). Comparative history of Campylobacter contamination on chicken meat and campylobacteriosis cases in the United States: 1994–2018. *International Journal of Food Microbiology*, **342**: p. 109075.
61. Lynch, H., et al., (2022). Prevalence and levels of Campylobacter in broiler chicken batches and carcasses in Ireland in 2017–2018. *International Journal of Food Microbiology*, **372**: p. 109693.
62. Habib, I., Mohamed, M.-Y.I. and Khan, M., (2021). Current state of Salmonella, Campylobacter and Listeria in the food chain across the Arab Countries: a descriptive review. *Foods*, **10**(10): p. 2369.
63. Veltcheva, D., et al., (2022). Emerging patterns of fluoroquinolone resistance in Campylobacter jejuni in the UK [1998–2018]. *Microbial Genomics*, **8**(9). https://doi.org/10.1099/mgen.0.000875
64. Sher, A.A., et al., (2021). Epidemiological trends of foodborne Campylobacter outbreaks in the United States of America, 1998–2016. *Food Microbiology*, **97**: p. 103751.
65. Buiatte, A.B.G., et al., (2023). Virulence, antimicrobial resistance, and dissemination of Campylobacter coli isolated from chicken carcasses in Brazil. *Food Control*, **147**: p. 109613.
66. Vicary, D., et al., (2020). Hawke's Bay pharmacists' activities during a campylobacter contamination of public water supply in Havelock North during 2016. *Journal of Primary Health Care*, **12**(2): p. 122–128.
67. Amjad, M., (2023). Poultry as a source and reservoir for Campylobacteriosis. *European Journal of Veterinary Medicine*, **3**(1): p. 11–17.
68. Greenhalgh, S., et al., (2020). Reduced-crude protein diets in chicken-meat production: justification for an imperative. *World's Poultry Science Journal*, **76**(3): p. 537–548.
69. Foddai, A., Nauta, M. and Ellis-Iversen, J., (2023). A model using an inter-sectorial data integration process indicates that reducing Campylobacter cross-contamination at slaughter mitigates the risk of human campylobacteriosis effectively. *Microbial Risk Analysis*, **23**: p. 100248.
70. Ridley, A., et al., (2011). Potential sources of Campylobacter infection on chicken farms: contamination and control of broiler-harvesting equipment, vehicles and personnel. *Journal of Applied Microbiology*, **111**(1): p. 233–244.
71. Dramé, O., et al., (2020). Antimicrobial resistance of campylobacter in broiler chicken along the food chain in Canada. *Foodborne Pathogens and Disease*, **17**(8): p. 512–520.
72. Shange, N., Gouws, P. and Hoffman, L.C. (2019). Campylobacter and Arcobacter species in food-producing animals: prevalence at primary production and during slaughter. *World Journal of Microbiology and Biotechnology*, **35**(9): p. 146.
73. Gabbert, A.D., et al., (2023). The missing pieces: the role of secretion systems in Campylobacter jejuni Virulence. *Biomolecules*, **13**(1): p. 135.

74. Black, Z., et al., (2021). The fate of foodborne pathogens in manure treated soil. *Frontiers in Microbiology*, **12**. https://doi.org/10.3389/fmicb.2021.781357
75. Zhong, X., et al., (2020). Campylobacter jejuni biofilm formation under aerobic conditions and inhibition by ZnO nanoparticles. *Frontiers in Microbiology*, 2020. **11**: p. 207.
76. Butucel, E., et al., (2022). Biocides as biomedicines against foodborne pathogenic bacteria. *Biomedicines*, **10**(2): p. 379.
77. Balta, I., et al., (2022). Anti-Campylobacter probiotics: latest mechanistic insights. *Foodborne Pathogens and Disease*, **19**(10): p. 693–703.
78. Koyun, O.Y. and Callaway, T., (2019). Controlling pathogens in the poultry gut. Burleigh *Dodds Series in Agricultural Science*, p. 317–346. https://doi.org/10.19103/as.2019.0059.16
79. Koyun, O.Y., et al., (2022). Innovative treatments enhancing the functionality of gut microbiota to improve quality and microbiological safety of foods of animal origin. *Annual Review of Food Science and Technology*, **13**: p. 433–461.
80. Olson, E.G., et al., (2022). Application of bacteriophages to limit Campylobacter in poultry production. *Frontiers in Microbiology*, **12**. https://doi.org/10.3389/fmicb.2021.458721
81. Dzianach, P.A., et al., (2022). The use of interdisciplinary approaches to understand the biology of Campylobacter jejuni. *Microorganisms*, **10**(12): p. 2498.
82. Line, J.E., Seal, B.S. and Garrish, J.K., (2022). Selected antimicrobial peptides inhibit in vitro growth of Campylobacter spp. *Applied Microbiology*, **2**(4): p. 688–700.
83. Bai, Y., et al., (2022). Development of an organic acid compound disinfectant to control food-borne pathogens and its application in chicken slaughterhouses. *Poultry Science*, **101**(6): p. 101842.
84. Sohaib, M., et al., (2016). Postharvest intervention technologies for safety enhancement of meat and meat based products; a critical review. *Journal of food Science and Technology*, **53**: p. 19–30.
85. Huang, H.-H., et al., (2020). *Characterization and Utilization of Phages Specific to Campylobacter Coli*. Microbiology Society.
86. Burmeister, A.R., et al., (2020). Pleiotropy complicates a trade-off between phage resistance and antibiotic resistance. *Proceedings of the National Academy of Sciences*, **117**(21): p. 11207–11216.
87. Kittler, S., et al., (2013). Effect of bacteriophage application on Campylobacter jejuni loads in commercial broiler flocks. *Applied and Environmental Microbiology*, **79**(23): p. 7525–7533.
88. Zampara, A., et al., (2021). Developing innolysins against Campylobacter jejuni using a novel prophage receptor-binding protein. *Frontiers in Microbiology*, **12**. https://doi.org/10.3389/fmicb.2021.619028
89. Richards, P.J., Connerton, P.L. and Connerton, I.F., (2019). Phage biocontrol of Campylobacter jejuni in chickens does not produce collateral effects on the gut microbiota. *Frontiers in Microbiology*, **10**: p. 476.
90. Selle, K., et al., (2020). In vivo targeting of Clostridioides difficile using phage-delivered CRISPR-Cas3 antimicrobials. *mBio*, **11**(2): p. e00019-20.
91. Cean, A., et al., (2015). Effect of human isolated probiotic bacteria on preventing Campylobacter jejuni colonization of poultry. *Foodborne Pathogens and Disease*, **12**(2): p. 122–130.

92. Ghareeb, K., et al., (2012). Evaluating the efficacy of an avian-specific probiotic to reduce the colonization of Campylobacter jejuni in broiler chickens. *Poultry Science*, **91**(8): p. 1825–1832.
93. Smialek, M., Burchardt, S. and Koncicki, A. (2018). The influence of probiotic supplementation in broiler chickens on population and carcass contamination with Campylobacter spp. - field study. *Research in Veterinary Science*, **118**: p. 312–316.
94. Saint-Cyr, M.J., et al., (2017). Use of the potential probiotic strain Lactobacillus salivarius SMXD51 to control Campylobacter jejuni in broilers. *International Journal of Food Microbiology*, **247**: p. 9–17.
95. Helmy, Y.A., et al., (2022). Effect of probiotic E. coli Nissle 1917 supplementation on the growth performance, immune responses, intestinal morphology, and gut microbes of Campylobacter jejuni infected chickens. *Infection and Immunity*, **90**(10): p. e00337-22.
96. Khan, M., et al., (2019). Effect of newly characterized probiotic lactobacilli on weight gain, immunomodulation and gut microbiota of Campylobacter jejuni challenged broiler chicken. *Pakistan Veterinary Journal*, **39**(4): p. 473–478.
97. Dogan, O.B., et al., (2022). A systematic review and meta-analysis of the efficacy of processing stages and interventions for controlling Campylobacter contamination during broiler chicken processing. *Comprehensive Reviews in Food Science and Food Safety*, **21**(1): p. 227–271.
98. Gichure, J.N., et al., (2022). Systematic-review and meta-analysis on effect of decontamination interventions on prevalence and concentration of Campylobacter spp. during primary processing of broiler chickens. *Food Microbiology*, **102**: p. 103923.
99. USDA-FSIS, (2016). *Safe and suitable ingredients used in the production of meat, poultry, and egg products*. FSIS Directive 7120.1, revision 36.
100. Chen, S.H., et al., (2023). Proteome analysis of Campylobacter jejuni poultry strain 2704 survival during 45 min exposure to peracetic acid. *International Journal of Food Microbiology*, **385**: p. 110000.
101. Balta, I., et al., (2020). Antiviral activity of a novel mixture of natural antimicrobials, in vitro, and in a chicken infection model in vivo. *Scientific Reports*, **10**(1): p. 16631.
102. Ch Stratakos, A., et al., (2017). The in vitro and ex vivo effect of Auranta 3001 in preventing Cryptosporidium hominis and Cryptosporidium parvum infection. *Gut Pathogens*, **9**(1): p. 49.
103. Kelly, C., et al., (2017). The in vitro and in vivo effect of carvacrol in preventing campylobacter infection, colonization and in improving productivity of chicken broilers. *Foodborne Pathogens and Disease*, **14**(6): p. 341–349.
104. Pinkerton, L., et al., (2019). Attenuation of vibrio parahaemolyticus virulence factors by a mixture of natural antimicrobials. *Microorganisms*, **7**(12): p. 679.
105. Sima, F., et al., (2018). A novel natural antimicrobial can reduce the in vitro and in vivo pathogenicity of T6SS positive Campylobacter jejuni and Campylobacter coli chicken isolates. *Frontiers in Microbiology*, **9**(2139). https://doi.org/10.3389/fmicb.2018.02139
106. Stratakos, A.C., et al., (2018). The antimicrobial effect of a commercial mixture of natural antimicrobials against Escherichia coli O157:H7. *Foodborne Pathogens and Disease*, **16**(2): p. 119–129.
107. Balta, I., et al., (2022). The antioxidant effect of natural antimicrobials in shrimp primary intestinal cells infected with Nematopsis messor. *Antioxidants*, **11**(5): p. 974.

108. Butucel, E., et al., (2023). The prebiotic effect of an organic acid mixture on faecalibacterium prausnitzii metabolism and its anti-pathogenic role against vibrio parahaemolyticus in Shrimp. *Biology*, **12**(1): p. 57.
109. Silvan, J.M., et al., (2023). Olive leaf as a source of antibacterial compounds active against antibiotic-resistant strains of Campylobacter jejuni and Campylobacter coli. *Antibiotics*, **12**(1): p. 26.
110. Wagle, B.R., et al., (2023). A multiomic analysis of chicken serum revealed the modulation of host factors due to Campylobacter jejuni Colonization and in-water supplementation of eugenol nanoemulsion. *Animals*, **13**(4): p. 559.
111. El Baaboua, A., et al., (2022). Evaluation of the combined effect of antibiotics and essential oils against Campylobacter multidrug resistant strains and their biofilm formation. *South African Journal of Botany*, **150**: p. 451–465.
112. Shrestha, S., et al., (2019). Carvacrol antimicrobial wash treatments reduce Campylobacter jejuni and aerobic bacteria on broiler chicken skin. *Poultry Science*, **98**(9): p. 4073–4083.
113. Lin, L., et al., (2022). Bio-responsive composite liposomes against Campylobacter jejuni in vitro and its application in chicken preservation. *Innovative Food Science & Emerging Technologies*, **81**: p. 103122.
114. Balta, I., et al., (2021). The effect of natural antimicrobials against Campylobacter spp. and its similarities to Salmonella spp, Listeria spp., Escherichia coli, Vibrio spp., Clostridium spp. and Staphylococcus spp. *Food Control*, **121**: p. 107745.
115. Cil, G.I., et al., (2019). Effect of lactic acid and steam treatments on Campylobacter jejuni on chicken skin. *Emirates Journal of Food and Agriculture*, **31**(2): p. 143–147.
116. Kure, C.F., et al., (2020). The effects of a pilot-scale steam decontamination system on the hygiene and sensory quality of chicken carcasses. *Food Control*, **109**: p. 106948.
117. Hauge, S.J., et al., (2023). Assessment of in-plant decontamination of broiler carcasses by immersion in hot water and lactic acid in Norway. *Food Control*, **152**: p. 109873.
118. Gloanec, N., et al., (2023). Research note: analysis of immune responses in broilers after vaccination against Campylobacter jejuni. *Poultry Science*, **102**(4): p. 102510.
119. Stoakes, E., et al., (2023). Application of TraDIS to define the core essential genome of Campylobacter jejuni and Campylobacter coli. *BMC Microbiology*, **23**(1): p. 97.
120. Gahamanyi, N., et al., (2021). Antimicrobial resistance profiles, virulence genes, and genetic diversity of thermophilic campylobacter species isolated from a layer poultry farm in Korea. *Frontiers in Microbiology*, **12**. https://doi.org/10.3389/fmicb.2021.622275
121. Neal-McKinney, J.M., et al., (2014). Reducing Campylobacter jejuni colonization of poultry via vaccination. *PLoS One*, **9**(12): p. e114254.
122. Gloanec, N., et al., (2022). Impact of DNA prime/protein boost vaccination against Campylobacter jejuni on immune responses and gut microbiota in chickens. *Vaccines*, **10**(6): p. 981.
123. Taha-Abdelaziz, K., et al., (2018). Oral administration of PLGA-encapsulated CpG ODN and Campylobacter jejuni lysate reduces cecal colonization by Campylobacter jejuni in chickens. *Vaccine*, **36**(3): p. 388–394.
124. Taha-Abdelaziz, K., et al., (2018). PLGA-encapsulated CpG ODN and Campylobacter jejuni lysate modulate cecal microbiota composition in broiler chickens experimentally challenged with C. jejuni. *Scientific Reports*, **8**(1): p. 12076.

Chapter 2

Understanding and controlling *Salmonella* in poultry meat: the role of microbial ecology

Colette A. Nickodem, Steven C. Ricke and Jessica L. Hite, University of Wisconsin-Madison, USA

1 Introduction

The poultry industry is faced with combatting an ever-changing landscape of pathogens and antimicrobial resistance. The (re)-emergence of pathogens like *Campylobacter jejuni*, *Campylobacter coli*, *Clostridium perfringens*, *Staphylococcus aureus*, *Listeria monocytogenes*, and pathogenic serogroups of *Escherichia coli* continue to represent major public health concerns for the poultry industry. Each of these zoonotic pathogens has the potential to affect both poultry and human health through various direct and indirect transmission pathways (Abebe *et al.* 2020; Gomes *et al.* 2013). *Salmonella* and *Campylobacter* are the main contributors responsible for foodborne illness cases originating from poultry consumption. *Campylobacter* remains the largest contributor to all human gastroenteritis cases in the US (CDC 2022). However, in 2021, the USDA named reducing pre-harvest *Salmonella* in poultry a ***top*** priority. Thus, we focus here on non-typhoidal *Salmonella*.

http://dx.doi.org/10.19103/AS.2024.0146.02

Annually, in the US, there are 1.35 million infections, 26,000 hospitalizations, and 400 deaths related to non-typhoidal *Salmonella*, with an estimated 1 million of the cases attributed to foodborne causes. Roughly 23% of these illnesses arise from contaminated chicken and turkey (Collier *et al.* 2021). In the US, 1 in 25 packages of store-bought chicken is contaminated with non-typhoidal *Salmonella enterica*. Additionally, *S. enterica* is one of the most thoroughly studied pathogens spanning many disciplines, not just the poultry industry, and thus provides an ideal case study to identify general principles that can be applied to a wide range of pathogens and livestock systems.

Poultry and poultry products are an important reservoir for *Salmonella* and other antibiotic-resistant bacteria (ARB) (Moulana & Asgaharpour 2022). The use of therapeutic and sub-therapeutic applications of various antibiotics in poultry production systems in various regions of the world contributes to the increasing prevalence of antimicrobial-resistant bacteria in the food chain (Ager *et al.* 2023; Ayoola *et al.* 2022; Threlfall 2002; Zhou *et al.* 2023). The emergence of bacterial pathogens that are resistant to most antibiotics – even those of last resort – poses an urgent threat to both animal, human, and environmental health. Traditional cornerstone approaches to manage these challenges in poultry include a combination of stringent cleaning, antibiotics, and vaccines. Thus far, however, such approaches have been unable to sustainably limit poultry-associated *S. enterica* 'spillovers' to humans and the environment.

As such, there is renewed interest in antimicrobial-free strategies that promote alternative methods for disease prevention, improved animal health, and the production of safer poultry products, thereby contributing to the overall sustainability and health of the poultry industry. However, the widely acknowledged need to develop more integrative disease management options is hampered by large gaps in our knowledge of the basic biological processes that promote the establishment and maintenance of *Salmonella* infection (Cunrath & Palmer 2021; Ippolito *et al.* 2022; Potter 2022; Tan *et al.* 2021). Thousands of non-pharmacological alternatives ranging from pre-, pro-, and post-biotics to plant phytogenic products (e.g. garlic, oregano, and eucalyptus) have been proposed. However, for the majority of cases, the mechanisms underlying these effects remain undefined, partly due to the difficulties in identifying and manipulating interactions in complex natural communities (Cairns *et al.* 2017; Klumper *et al.* 2019). Accordingly, we currently lack a clear understanding of how, when, and why these products work or affect pathogen dynamics over longer time scales or in complex within-host environments.

Here, we aim to cement the idea that intervention strategies should aim to quantify how nonpharmacological options affect pathogen dynamics and drug resistance. We posit that any management plan for controlling *Salmonella* in poultry including, antibiotics, as well as the wide range of nutritional additives, pre-, pro-, and post-biotics, should focus not just on the traits of the focal pathogen but on the emergent properties of the microbial community in which pathogens are embedded (Bottery *et al.* 2021). Biotic feed additives and their interaction with the gut microbiome in various food animals, including poultry, are well established and have been recently covered extensively elsewhere (Callaway and Ricke, 2023). These interactions are, for the most part, obvious given that the gut microbial community is directly engaged with the biotic feed additive. However, little is known about the impact of *Salmonella* vaccines or bacteriophage applications in poultry and their potential interaction with the host gastrointestinal tract microbial ecosystem. While it is assumed that the interaction with these interventions would be much less than typically seen with feed additive biotics, there remain potential indirect impacts through host–immune interactions with the gut microbiome as a result of the introduction of a vaccine and/or peripheral unforeseen consequences to the non-*Salmonella* population after removal of *Salmonella*. In other words, regardless of the *Salmonella* intervention approach, research must shift from focusing on antimicrobial sensitivity as a trait of an individual serovar to that of the microbial community and abiotic environment in which the pathogen is embedded. This community-based approach is particularly important to understand given the wide-ranging effects that any intervention would have on the overall microbiome, whether direct or indirect. Such mechanistic insight offers the best opportunity to narrow down the overwhelming cornucopia of 'biotics' and pinpoint how they could be used to optimize microbial communities to fight pathogens – while also avoiding any unintended side effects.

Without such information, whether these products will provide reliable 'evidence-based' and sustainable management options will remain a black box waiting to be deciphered. Just as antimicrobials alter the selective pressures that drive the emergence of more virulence and drug-resistant pathogens, so too could nonpharmacological options (Hite *et al.* 2020; Hite & Cressler 2019; Wale *et al.* 2017). These products could also affect the likelihood of resistant genotypes appearing through mutation (or horizontal gene transfer) as well as the fitness of resistant and wild-type genotypes/serovars/strains once they have appeared (Abu Kwaik & Bumann 2013; Ayoola *et al.* 2022; Palomino *et al.* 2023; Plata *et al.* 2022). Nonetheless, studies examining the evolutionary consequences of these products remain rare. An important research objective, therefore, is to compare the impact of different 'biotics' on the probability and rate of resistance emergence. Such information would make it possible to identify non-pharmacological options that maximize treatment efficacy while managing resistance simultaneously. In other words, these connections are critical to moving non-pharmacological options beyond descriptive case studies.

Our goal here is to progress this enterprise by highlighting system-based approaches to understanding pathogen dynamics and antimicrobial resistance within complex microbial communities. While biotic feed amendments are certainly relevant, we chose to focus on *Salmonella* vaccines and bacteriophage therapies as examples. They also have a potential impact on the gut microbiome that is generally not considered and therefore understudied. We suggest that studies in model systems based on bacteriophage and vaccines could, in turn, more generally, serve as an invaluable guide for efforts to limit *Salmonella* and the emergence of resistance in poultry for other interventions not discussed in this review. We then identify key areas of future research focused on integrating empirical findings with functional genomics, innovative statistical approaches (e.g. structural equation models (SEMs)), and mathematical modeling. This combination of tools will (1) enable a more complete understanding of managing *Salmonella* and AMR in poultry while also improving bird health, immunity, and welfare and (2) provide valuable insights to uncover broad-scale patterns in other pathogens and livestock production systems.

2 Poultry, *Salmonella*, and antimicrobial resistance: a brief overview

The US produces approximately 22.7 million tons of poultry meat per year, placing it in the top three countries for poultry exports (AVEC 2021). Intensive poultry housing systems are integral to the high output of this industry. To protect flocks from the rapid spread of disease and infections, broilers in these intensive systems are routinely treated with preventative (metaphylaxis) antibiotics and vaccines (Roth *et al.* 2019). Consequentially, poultry production systems are among the main reservoirs of foodborne pathogens, ARB, and antibiotic-resistant genes (ARGs) that can be transmitted to humans, wildlife, and the environment via poultry products, including food and manure. Antibiotic supplementation in poultry production has a long history of commercial practice (Dibner & Richards 2005; Jones & Ricke 2003). However, mounting concerns over links between antibiotic supplementation in food animals and the emergence of antimicrobial resistance in microbial populations, particularly enteric bacteria (Roe & Pillai 2003; Van Boeckel *et al.* 2017), have drastically shifted the sole reliance on antimicrobials.

In response to this emerging public health threat, governmental regulations have attempted to limit antibiotic inclusion as animal growth promoters in both Europe and the US (Smith 2019). In the US, the Food and Drug Administration nominally eliminated clearances for antibiotic use as growth promoters, with those being used to treat disease in animals being classified as Veterinary Directive Drugs for medically important antibiotics, and those administered in the feed for disease control became classified as Veterinary Feed Directive Drugs (Smith 2019; Wythe *et al.* 2023). Ostensibly because of the reduced use of antimicrobials, on-farm antimicrobial usage from 2013 to 2017 declined markedly in major commercial broiler operations that constituted 80–90% of yearly US production (Singer *et al.* 2020).

Despite the decrease in the use of antibiotics as growth promoters in the poultry industry, antibiotic-resistant *Salmonella* remain prevalent. *Salmonella* isolated from poultry products exhibit increasing resistance to ciprofloxacin, tetracycline, streptomycin, sulfamethoxazole, and sulfisoxazole (NARMS 2019). In 2022, 53.4% of *Salmonella* isolates from chicken products and 63.3% of *Salmonella* isolates from chicken ceca in the National Antimicrobial Resistance Monitoring System were resistant to tetracycline (NARMS 2019).

Adding further complications, variation in resistance occurs across serovars, geographies, and the type of poultry production. *Salmonella* serovars are differentiated by surface proteins, which contribute to host specificity. *Salmonella* Pullorum and Gallinarum are historically host-adapted to poultry but are now considered eradicated (Foley *et al.* 2011). More recently, serovars Enteritidis, Typhimurium, Infantis, and Kentucky are most commonly identified in poultry and their products (NARMS 2019). Enteritidis and Typhimurium also contribute to the highest proportion of human cases and are commonly resistant to multiple antimicrobials (CDC 2022). Additionally, the three most common serovars isolated from chickens to harbor resistance to tetracycline and sulfamethoxazole – sulfisoxazole include *Salmonella* Infantis, Typhimurium, and Kentucky and *Salmonella* Enteriditis routinely exhibits resistance to ciprofloxacin (NARMS 2019). Although isolates of *Salmonella* Kentucky from poultry are frequently resistant to several antimicrobials, Kentucky is not often associated with human infections. However, specific strains of *Salmonella* can exhibit a high degree of antimicrobial resistance and continue to garner considerable attention from stakeholders and policymakers. For instance, in 2018, a resistant strain of *Salmonella* serovar Infantis, most commonly associated with poultry, was responsible for 25% of human infections (CDC 2019). The *S. enterica* serovar Typhimurium DT104 and new strains of *S.* Kentucky continue to illustrate resistance toward carbapenems and fluoroquinolones (Hawkey *et al.* 2019; LeHello *et al.* 2013; Threlfall *et al.* 1996).

Surveillance data from Wales, Scotland, and England revealed a greater frequency in antibiotic resistance in *Salmonella* Typhimurium isolates from broilers compared to layers (Mellor *et al.* 2019). Velasquez *et al.* (2018) surveyed southeastern US poultry flocks for *Salmonella* and isolated six serotypes (Enteritidis, Berta, Mbandaka, Typhimurium, Kentucky, and Tennessee). Of these, 7% exhibited resistance to at least one of eight antimicrobials, and a single isolate of *S.* Mbandaka was resistant to multiple antibiotics. Multiple drug resistance arises because *Salmonella* can carry multiple ARGs.

Part of what makes carriage of ARGs especially concerning for public health is the multiple of mobile genetic elements such as plasmids, integrons, and transposons. The mobile genetic elements enable the transfer and acquisition of multiple ARGs among *S. enterica*, other bacteria, and even across hosts such as from poultry to wild birds or humans (Algarni *et al.* 2022). Drug resistance in *Salmonella* involves resistance to antimicrobials that are important for human clinical use. Thus preventing cross-species transmission of ARGs is a critical public health concern. Notably, antimicrobial-resistant *Salmonella* trends in human cases resemble *Salmonella*-resistance profiles from poultry products, indicating that cross-species transmission is more prevalent than previously recognized.

3 Microbial community ecology, pathogen infectivity, and resistance

One promising avenue for reducing the emergence of drug resistance is to harness the gut microbiome to suppress drug-resistant microorganisms such as *Salmonella* (Bottery *et al.* 2021). *Salmonella*, similar to other bacterial pathogens, coexist in complex multi-strain and multi-species communities. Mounting evidence underscores that microbial dynamics including the initial density (Karslake *et al.* 2016), standing genetic variation for resistance (Wistrand-Yuen *et al.* 2018), biofilm growth (Macià *et al.* 2014; Olson *et al.* 2002), and/or species diversity at the strain, serovar, or species level (Alseth *et al.* 2019; Klumper *et al.* 2019; Sünderhauf *et al.* 2023) can strongly alter both the emergence and evolutionary trajectory of pathogens and drug resistance. Commensal gut microbes protect poultry against pathogens like *Salmonella* via effects on host immunity, the production of short-chain fatty acids, and antimicrobial peptides like bacteriocins (see Lyimu *et al.* 2023 and references therein). Additionally, the overall diversity and abundance of microbial communities can alter antimicrobial resistance through horizontal gene transfer and *de novo* resistance (Aminov 2011; Cooper *et al.* 2017; Maiden 1998). *Salmonella* infection alters the diversity of gut microbes, though the direction and magnitude of these effects vary markedly across studies (see Lyimu *et al.* 2023 and references therein). Regardless of the specific outcomes, these patterns arise because ecological interactions among microbes influence both pathogen growth and the emergence of antimicrobial resistance (Bottery *et al.* 2021). For example, competition for resources between different microbes will reduce the availability of shared resources, which can negatively affect population density, reducing the growth rates of resistant strains. Antimicrobials change the metabolic demands of the resident gut microbiome, altering nutrient requirements and microbial growth (Plata *et al.* 2022).

Additionally, the presence of other microbial species in the environment can introduce evolutionary selective pressures by sequestering antibiotics (Sabnis *et al.* 2018). Furthermore, certain bacterial species can deactivate antibiotics, such as beta-lactamases breaking down beta-lactams, which can impede the spread of resistant strains (Bottery *et al.* 2016; Yurtsev *et al.* 2013). On the other hand, some microbes engage in beneficial interactions, such as cross-feeding, where microbes rely on each other for essential nutrients, leading to increased population density and more opportunities for resistance to emerge and spread (Adamowicz *et al.* 2018; Estrela & Brown 2018). Such intra- and inter-species interactions help underscore the need to better understand antimicrobial resistance as an emergent population-level process.

Yet, microbial interactions remain largely overlooked in studies designed to test and understand the efficacy of new interventions (Bottery *et al.* 2021). This oversight is perpetuated, in part, due to logistical constraints associated with testing for the presence of antimicrobial resistance. Conventionally, conclusions about the efficacy of treatment options are drawn from measurements of single-strain cultures grown under ideal conditions. These culture-dependent laboratory methods were designed to accurately measure growth rates and identify the minimum inhibitory concentration (MIC) or, preferably, minimal selective concentrations (MSCs) and the predicted no-effect concentration (Bengtsson-Palme & Larsson 2016; Day *et al.* 2015). Such *in vitro* methods test the susceptibility of individual bacterial strains to select effective antimicrobial interventions. While this classical approach has been essential to advancing the field, it inherently focuses on antimicrobial sensitivity as a trait of an individual strain, not that of the microbial community or abiotic environment.

Together, this research demonstrates that drug resistance arises as an emergent property of the complex microbial community in which they grow, live, and transmit to new hosts. Multiple mechanisms driving these higher-order patterns have been identified, largely from simple two- or three- species communities and *in vitro* or *in silico* studies. These mechanisms, however, are often context-specific, depending on particular species, strains, or serovars of bacteria (Bottery *et al.* 2021; Spragge *et al.* 2023). As a result, we lack general principles to predict which microbial communities could be protective versus those that would allow or even facilitate pathogen colonization and the emergence of drug resistance.

Recent work with *Salmonella*-infected mice provides an experimental framework that could be used as a guide to optimize the composition of microbiomes for protection against *Salmonella* (and other pathogens) in poultry. Spragge et al. (2023) examined the ability of 100 gut symbionts to provide colonization resistance when grown with *Salmonella* individually or in diverse communities of up to 50 species. Their results suggest that more diverse microbial communities offered greater protection against pathogen colonization and growth via competition for resources. More diverse microbial communities exhibited greater niche overlap in resource use relative to less diverse microbial communities.

However, these protective properties hinged on the presence of certain microbial species even though these species offered little protection on their own. Microbes that were more closely related to the pathogens had more similar nutrient requirements and thus greater ability to competitively exclude the pathogens. The protective properties of these closely related strains were evident only when in combination with other microbes that could sufficiently starve the pathogen of key nutrients. Colonization resistance, therefore, emerged as a collective property of microbial communities.

Together, these results can help guide efforts to identify non-pharmacological approaches to control *Salmonella* in poultry. By focusing on how 'biotics' impact the diversity *and* relatedness of microbial communities in the gut (i.e. the gut microbiome), studies could begin to identify mechanistic connections to select effective treatments against a target pathogen.

4 Microbial ecology and bacteriophages

Similar approaches could be used to guide bacteriophage therapies. Bacteriophages (phages) are viruses that exclusively infect bacteria. Within poultry, *Salmonella* bacteriophages have been isolated from broiler feces, abattoirs, wastewater, litter, and fecal contents (Atterbury *et al.* 2007). Phages are then harnessed and developed into highly concentrated solutions for use in both pre-harvest and post-harvest strategies. In the pre-harvest setting, phage can be administered to animals using aerosol sprays, microencapsulated phage, intramuscular injections, or added directly to their feed or drinking water (Carvalho *et al.* 2017).

Aerosol treatments have been tested on 1-day-old chicks that were infected with *Salmonella* Enteritidis and then sprayed with a solution including phosphate-buffered saline, three different bacteriophages, and Broilact (a commercial product to promote microflora). This treatment resulted in significantly fewer *Salmonella* infections in treated chicks compared to control chicks (Borie *et al.* 2009). Additionally, an oral delivery method was investigated by inoculating broiler chickens with *Salmonella* Enteritidis (10^8 colony-forming units (CFUs)), followed by an oral phage treatment. This cocktail included phages CNPSA1, CNSPSA3, and CNSPSA4 (10^{11} plague-forming units (PFUs)) recovered from free-range layers. The treatment resulted in a 3.5 $\log_{10}$ CFU/g reduction of *Salmonella* Enteritidis recovered from broiler chicken cecal contents (Fiorentin *et al.* 2005). These studies demonstrate the effectiveness of bacteriophage in reducing *Salmonella* within poultry, especially with the use of bacteriophage cocktails. Alternatively, phage used post-harvest is applied directly to chicken breasts or eggshell surfaces by spraying, soaking, or inoculating. Directly mixing phage with ground chicken meat or phage immersion with liquid eggs are alternative methods (Khan Rahman 2022). Post-harvest applications to food products may avoid administration complications and bacterial community interference.

In general, the risk of expansive bacterial resistance is lower for narrow-range phages compared to the broad-spectrum action of antibiotics. The host-specific nature of phage allows for targeted serovar therapy while leaving space for healthy microbiota re-growth (Clavijo *et al.* 2022; Gunderson *et al.* 2023). For instance, a trial by Upadhaya *et al.* (2021) showed an increase in *Lactobacillus*, *Bifidobacteria*, and *Prevotella* relative abundance in broilers that were provided 0.05–0.10% bacteriophage supplements compared to broilers that received antibiotics during *Salmonella* infection (Upadhaya *et al.* 2021). These are beneficial bacteria that promote gut health by regulating gastrointestinal homeostasis, acting as a probiotic, and producing SCFAs, respectively. In the broilers that received antibiotics, these healthy microbial species were inhibited, and Proteobacteria proliferated, which can contribute to an unstable microbiome and lead to complicated infections. Not only did these phage treatments improve the composition of the gut microbiome, but they also led to increased body weight in the bacteriophage-treated broilers. Additionally, a trial supplementing broiler water with a phage cocktail significantly reduced *Campylobacter* while increasing other bacteria that have various effects on poultry health, and the overall resident microbiota was not impacted (Clavijo *et al.* 2022).

Relative to antimicrobials, phage therapy appears to have a lower impact on the resident gut microbial community. And like antimicrobials, the effectiveness of phage therapy is dependent on the diversity and composition of the gut microbial community. For instance, other microbes may produce protective biofilm formations that can block the adsorption of a phage or reduce the number of interactions between the phage and the bacteria target. Thus, prior knowledge of the microbial community composition in agricultural animals is proving essential for selecting effective phage therapies. While phages are an effective tool in the arsenal against *Salmonella*, resistance to phage can emerge. To ensure the long-term efficacy of phage therapy, a deeper understanding of how microbial communities impact phage–pathogen dynamics is crucial.

The mechanisms of phage resistance, including adsorption resistance, uptake resistance, restriction, and abortive infections, are well studied (Hyman & Abedon 2010). These mechanisms are not mutually exclusive and co-selection appears common. To address these concerns, bacteriophage cocktails with phages that target different bacterial surface receptors are often effective (Ormala & Jalasvuori 2013; Sulakvelidze *et al.* 2011). Bacteriophage cocktails can increase the depth of activity acted on a single bacterial strain by including multiple phages that have different adsorption targets and mechanisms (Abdeon *et al.* 2021). Selecting phages without cross-resistance (i.e. phages that use different host receptors) for bacteriophage cocktails increases the chances of sustaining bacterial sensitivity (Gill & Hyman 2010).

In addition to resistance, stressors such as temperature and pH can easily break down phages. The pH of a chicken digestive tract ranges from 3.43 in the gizzard to 6.43 in the small intestines, and in the most acidic conditions, free phage will be digested (Lorenzo-Rebenaque *et al.* 2021; Mabelebele *et al.* 2014). However, anti-acid supplementation and phage microencapsulation can lengthen phage viability in the chicken gut (Atterbury *et al.* 2007; Koo *et al.* 2001; Lorenzo-Rebenaque *et al.* 2021). Other challenges for phage therapy include selecting phage with the proper host range, determining tolerance to specific environments (i.e. complex matrices), optimizing the concentration of phage (i.e. multiplicity of infection (MOI)), and determining the necessary frequency and duration of applications (Wei *et al.* 2019). Careful selection of phage host range is particularly crucial for treatment goals. For instance, in complex matrices like agricultural soil, physical barriers to phage–bacteria interactions, as well as phage neutralization from ion interactions, can reduce the effectiveness of phage therapy. Phage therapy is also less effective when there is a low abundance of bacterial hosts for proliferation. A high phage titer (PFUs), that improves the MOI (virus-to-bacteria cell ratio) and increased application frequency improves the chances of these bacteria–phage interactions.

Together, these studies suggest that future research on phages should focus on following current *Salmonella* trends in the poultry industry to target serovars of public health concern. Although commercial phage products are available to reduce *Salmonella* in poultry, these products may not match the individual needs of each poultry house. Research that focuses on identifying the critical control point in the poultry production process and the optimal delivery method to the intended *Salmonella* infection site will improve the effectiveness of these treatments. Using phage therapy as an antibiotic alternative can preserve the chicken gut microbiome and counteract antimicrobial resistance. These phage treatments can be individualized to attack specific outbreak strains or generalized to pathogen populations commonly identified at poultry operations, providing poultry producers autonomy while developing their *Salmonella* management plan.

5 Microbial ecology and *Salmonella* vaccines

A key question for sustainably managing *Salmonella* in poultry is how vaccines shape the microbial community. At the farm level, vaccination is considered the most efficient approach to controlling *Salmonella* in poultry. Commercially available *Salmonella* vaccines for poultry come in two broad categories: Inactivated and live-attenuated vaccines. The latter are generally considered to offer more protection because they induce both cellular and humoral immune responses. Live attenuated vaccines are typically orally administered to poultry, which introduces important but poorly understood links between the attenuated pathogen strain and the gut microbiome.

Current evidence indicates that live attenuated vaccines for *Salmonella* can impact both the variability (alpha-diversity) and overall composition of the cecal microbiome. Patterns, however, are often study- and vaccine-specific, likely due to methodological differences (e.g. bird origin and vaccine administration routes). Such differences preclude a broad-scale and generalizable understanding of how vaccines impact the gut microbiome and how these changes 'scale up' to affect production and epidemiology; certain vaccines can either increase or decrease the relative abundance of Firmicutes, depending on the production system, vaccine, or route of vaccination. For instance, vaccination with the AviPro® *Salmonella* DUO vaccine increased the relative abundance of Firmicutes relative to the AviPro® *Salmonella* Vac T and AviPro® *Salmonella* Vac E vaccines. In this case, vaccination could improve overall bird health; Firmicutes carry well-known anti-inflammatory properties. However, other vaccines (e.g. the AviPro® Vac E and AviPro® Vac T) can increase the relative abundance of other harmful microbes like Proteobacteria (Lyimu *et al.* 2023). Proteobacteria include several pathogenic bacteria associated with gut dysbiosis and reduced performance metrics like poultry weight gain (Zhang *et al.* 2022).

These results join others in highlighting key links between the type of vaccine and shifts that impact the gut microbiome. Future studies that mechanistically connect these changes with pathogen dynamics and overall poultry health could provide a critical next step to sustainably managing *Salmonella*. Studies that simultaneously connect vaccination to changes in the gut microbiome, pathogens, and AMR genes, as well as bird health and production metrics, will be essential to disentangling these complicated pathways. Such studies will benefit from the groundwork laid by current work focused on piecewise comparisons between vaccination, bird health, and gut microbial communities *or* gut microbial communities, bird health, and production metrics. One promising approach to address this knowledge gap is high-throughput proximity-ligation (Hi-C). This emerging technology can link specific microbes and the antimicrobial resistance genes they harbor (Bickhart *et al.* 2019; Kent *et al.* 2020). This innovative method facilitates the direct assembly of complete genomes down to the strain level from mixed-population samples and establishes a physical connection between plasmids and their respective hosts. Such detailed mapping networks can identify both the synergistic and antagonistic effects of vaccines on the overall microbial community (Crits-Christoph *et al.* 2022).

6 Quantitative approaches to improving *Salmonella* control in complex microbial environments: structural equation models

The poultry industry is unparalleled in its ability to conduct detailed and complex experiments such as those proposed earlier. Still, making sense of the growing mountain of products and data and identifying which products will provide more long-term solutions while also avoiding unintended consequences requires a sophisticated quantitative toolkit. Luckily, quantitative tools from the fields of ecology, evolutionary biology, and statistics can be harnessed to cut through this complexity and pinpoint the most effective management tools. We provide a brief overview of this approach.

Microbial ecology can link poultry management practices (e.g. biotics vs. antimicrobials) to the emergence and spread of *Salmonella* and AMR genes via interactions among bird age, production type, nutrition, and competition or predation among microbes. However, complicated microbial interactions (i.e. trophic interactions among microbes and their impacts on the abundance and diversity of pathogens and AMR genes) often obscure the important pathways regulating pathogens and AMR genes. These complications pose major challenges to selecting which management practices (e.g. 'biotics') best reduce pathogens and AMR genes. SEMs firmly grounded in microbial ecology can provide a solution to these problems (Grace 2006). SEMs are ideal for examining the relative contributions, strength, and direction of pathways linking multi-level hypotheses, especially when both observed and unobserved 'latent' variables can be directly and indirectly connected through bidirectional and cross-system feedbacks (Grace 2006; Kendall 2015).

We posit that a two-step approach should be used to select 'biotics' by examining their impact on both *Salmonella* and AMR genes. In the first step, the goal is to identify all relevant drivers of pathogens and AMR genes and their interactions and to test all relationships with univariate patterns. In step two, any univariately significant links are used to guide the creation of SEMs. To build the SEMs iteratively (e.g. to avoid overfitting), one simply fits separate path models to first predict, for example, how a selection of 'biotics' impact the prevalence of pathogens and AMR genes and then build a second set of models to examine their effects on microbial diversity and abundance. These separate models will highlight key differences among the 'biotics' based on the strengths (regression coefficients) of the paths.

To ensure a high degree of reproducibility and open access, researchers should work with statisticians to build models in software where analyses and results are readily replicated, for example, R (R Development Core Team 2010). An example of this two-step approach detailed above using R is as follows.

6.1 Univariate analyses

Links among 'biotics' and microbial ecology (e.g. competition vs. facilitation vs. predation) will first be tested individually with univariate mixed-effect models in the package NLME. These candidate models will start with the full model with all combinations of main effects among relevant biological and methodological factors while avoiding overfitting. Specifically, to examine overall changes in the prevalence of pathogens or AMR across the sampling time frame included in the study, the full model would include 'biotic' type, *Salmonella* serovar(s), time point (e.g. 1, 3, and 5 days after administration), and their interactions as a fixed effect (Hite *et al.* 2016; Hurley 2022).

Then, to examine more fine-scale differences in the abundance and diversity of pathogens or AMR, the full model will examine the effects of each 'biotic' on the overall microbial diversity, *Salmonella* abundance (or diversity), AMR abundance (or diversity), time, and their interactions. For projects with limited repeated sampling, 'time' can be modeled as a fixed (rather than random) effect (Weisberg & Fox 2011). With this baseline model structure, the likelihood ratios can be used to test the significance of each relationship. Notably, all data should remain untransformed to preserve their natural variance structures. Most up-to-date statistical approaches can explicitly model the variance of all response variables with exponential or power functions to describe the heteroscedasticity in the data (Hite *et al.* 2016). However, an advanced understanding of the logic and underlying biology behind the selection of these statistical tools is essential, and therefore, researchers should work with trained statisticians.

The univariate analyses in step one will help identify variables that should be fit as covariances, implying correlations, and those that should be fit as regressions, implying causality. Additional covariances can then be included to account for correlations among frequencies of *Salmonella* serovars and AMR genes (since they share a common denominator). Second, models can then be fit with a maximum likelihood estimator that is robust to non-normal standard errors and uses a robust Satorra-Bentler chi-square test statistic (Satorra & Bentler 2001). After model fitting, residual covariances should be inspected to identify any potentially missing links or latent variables. Third, model fits should be assessed with multiple robust metrics, including comparative fit index (CFI), Tucker-Lewis index (TLI), root mean square error of approximation (RMSEA), and standardized root mean square residual (SRMR) test statistics (Hu & Bentler 1999). Finally, the *P* values and standardized parameter estimates (SPEs) for each relationship can be extracted. These *P* values and SPEs will be used to compare effect sizes (strength of relationships) among paths in the final models.

From this final set of models, the 'winning' model can be selected using Akaike information criteria with the *aictab* function in the R package

AICcmodavg. More specifically, candidate models can be compared using ΔAIC (the difference in AIC values for the focal model and the model with the lowest AIC, i.e., the 'winning' model) (Burnham & Anderson 2004). The Akaike weight (*w*), which further quantifies the probability that a model is the most appropriate relative to the candidate models, should also be calculated. ΔAIC less than two and a higher *w* generally indicate that a model has substantial support, while a suite of best models with low weights ($w \sim 0$) indicates that no single variable plays a substantial role in mediating infection dynamics (Burnham & Anderson 2004). Using the *Anova* function in the R package car, the significance of the effects can be assessed using Wald χ^2 statistics for the winning model (Weisberg & Fox 2011).

6.2 Key deliverables from structural equation models

The two-step approach detailed here will enable researchers and stakeholders alike to (1) quantify the relative contributions of 'biotics' on key epidemiological patterns (e.g. *Salmonella* prevalence, diversity, and abundance of AMR genes) and (2) identify the most important microbial species interactions driving the emergence and spread of *Salmonella*, other pathogens, and AMR genes and refine in management practices accordingly.

7 Improving quantitative microbial risk assessments

SEMs can help refine the list of 'biotics' that have the largest effect on the prevalence, abundance, and diversity of *Salmonella*, other pathogens, and AMR genes. Given this information, how do we design realistic and effective changes to management practices without concomitant losses in other areas of the production chain? To answer this question, quantitative microbial risk assessments can be applied (Haas 2020).

The benefit of quantitative microbial risk assessments is that they take a system-based approach that accounts for bidirectional feedback among numerous compartments. This approach has been used, for example, in an adaptive dynamics framework (Dieckmann *et al.* 2002) to examine how the common practice of feed withdrawal before transport could be modified to reduce the severity and transmission of common pathogens affecting livestock, including *Salmonella* (Hite *et al.* 2020; Hite & Cressler 2019). The models indicate that whether feed withdrawal prior to transport enhances or diminishes disease depends on whether the specific nutritional protocol (e.g. 'biotics') provides nutrients that bolster or inhibit immune function and/or pathogen growth rates. Model analyses uncovered that feedbacks driven by nutrition-mediated competition between host immune functions and pathogen production can create a unimodal relationship between feed withdrawal and parasite fitness. Subsequently, depending on the host's diet, management practices that implement feed withdrawal before transport could backfire and inadvertently select for more virulent pathogens (e.g. *Salmonella*) and larger epidemics.

A similar approach could be used to examine links between 'biotics' and the prevalence of *Salmonella*, other microbes, including beneficial and pathogenic species or serovars, and AMR across various poultry systems. These models can examine the relative costs and benefits of different combinations of practices aimed at limiting *Salmonella* (e.g. providing vaccinated broilers with a given 'biotic' that promotes microbial diversity prior to feed withdrawal) and will identify the underlying mechanisms driving the outcomes (e.g. competitive exclusion; (Bottery *et al.* 2021)). In other words, quantitative microbial risk assessments track the system-level outcomes of multiple management practices on epidemiological processes, which may act simultaneously and exert contrasting effects on transmission (Haas 2020).

7.1 Data requirements

General models can be developed based on sparse data but will be improved if model predictions can be compared to empirical data. The key empirical data needed to parameterize the model include time-series data that track the effects of management practices on changes in pathogen burden (e.g. in blood or fecal samples), metrics of virulence and drug resistance obtained from DNA sequencing, multiple metrics of immune function (e.g., selected inflammatory and regulatory cytokines and chemokines (IL-6, IL-8, IL-10, IL-18, IL-4, IL-17A, and TNFα), avian β-defensin (AvBD1), Forkhead Box P3 (FoxP3) and inducible nitric oxide synthase (iNOS)), and ideally, food intake (Bashir-Tanoli & Tinsley 2014; Hite *et al.* 2020; Pfenning-Butterworth *et al.* 2022). Food intake is particularly crucial for studies on 'biotics' because many animals may actively avoid consuming certain 'biotics' due to preference or infection-mediated anorexia. Tracking food intake helps to ensure that outcomes are, in fact, driven by changes in the offered biotic.

7.2 Key deliverables from quantitative microbial risk assessments

The chief advantage of quantitative microbial risk models is that they provide a system-level assessment of management practices and identify critical control points along the production chain that strongly influence the prevalence and diversity of pathogens and AMR genes. Together, these models can inform actionable goals for the sustainable management of *Salmonella* in poultry as well as reducing the transmission of AMR genes across the food chain.

8 Conclusion

The (re)-emergence of drug-resistant *Salmonella* serovars emphasizes the importance of infectious diseases associated with poultry products to public health. Renewed interest in alternatives to antimicrobials has spurred rapid development of pre-, pro-, and post-biotics as well as plant-based phytogenic products (e.g. garlic, oregano, and eucalyptus). While these products are certainly a step in the right direction, we are far from a consensus on what combination of non-pharmacological treatments would optimally reduce the transmission of *Salmonella* or AMR genes. This gap remains, in part, because few studies measure the suite of parameters required to track transmission or pathogen evolution over long time frames. To be fair, such data are logistically difficult to quantify, and making sense of such complex and dynamic feedbacks requires advanced quantitative tools. Nonetheless, the growing challenges of managing *Salmonella* and AMR demand new approaches.

We hope that these studies and approaches discussed herein will help facilitate the development of economical, scalable, and 'evolution-proof' solutions to better manage *Salmonella* (and other pathogens) in poultry (and other livestock).

9 References

Abdeon, S.T., Danis-Wlodarczyk, H.M. & Wozniak, D. (2021). Phage cocktail development for bacteriophage therapy: toward improving spectrum of activity breadth and depth. *Pharmaceuticals*, 14, 1019.

Abebe, E., Gugsa, G. & Ahmed, M. (2020). Review on major food-borne zoonotic bacterial pathogens. *Journal of Tropical Medicine*, 2020, 4674235.

Abu Kwaik, Y. & Bumann, D. (2013). Microbial quest for food in vivo: "nutritional virulence" as an emerging paradigm. *Cell Microbiol*, 15, 882–890.

Adamowicz, E.M., Flynn, J., Hunter, R.C. & Harcombe, W.R. (2018). Cross-feeding modulates antibiotic tolerance in bacterial communities. *The ISME Journal*, 12, 2723–2735.

Ager, E.O., Carvalho, T., Silva, E.M., Ricke, S.C. & Hite, J.L. (2023). Global trends in antimicrobial resistance on organic and conventional farms. *Sci. Rep*, 13, 22608.

Algarni, S., Ricke, S.C., Foley, S.L. & Han, J. (2022). The dynamics of the antimicrobial resistance mobilome of salmonella enterica and related enteric bacteria. *Front Microbiol*, 13, 859854.

Alseth, E.O., Pursey, E., Luján, A.M., McLeod, I., Rollie, C. & Westra, E.R. (2019). Bacterial biodiversity drives the evolution of CRISPR-based phage resistance. *Nature*, 574, 549–552.

Aminov, R.I. (2011). Horizontal gene exchange in environmental microbiota. *Front Microbiol*, 2, Article 158.

Atterbury, R.J., Van Bergen, M.A.P., Ortiz, F., Lovell, M.A., Harris, J.A., De Boer, A., *et al.* (2007). Bacteriophage therapy to reduce *Salmonella* colonization of broiler chickens. *Appl Environ Microbiol*, 73, 4543–4549.

AVEC. (2021). *Association of Poultry Processors and Poultry Trade Annual Report 2021*. AVEC, Belgium.

Ayoola, M.B., Pillai, N., Uri, B., Rothrock, M.J. & Ramkumar, M. (2022). Preharvest environmental and management drivers of multidrug resistance in major bacterial zoonoticpathogens in pastured poultry flocks. *Microorganisms*, 10, 20.

Bashir-Tanoli, S. & Tinsley, M.C. (2014). Immune response costs are associated with changes in resource acquisition and not resource reallocation. *Funct Ecol*, 28, 1011–1019.

Bengtsson-Palme, J. & Larsson, D.G.J. (2016). Concentrations of antibiotics predicted to select for resistant bacteria: proposed limits for environmental regulation. *Environ Int*, 86, 140–149.

Bickhart, D.M., Watson, M., Koren, S., Panke-Buisse, K., Cersosimo, L.M., Press, M.O., *et al.* (2019). Assignment of virus and antimicrobial resistance genes to microbial hosts in a complex microbial community by combined long-read assembly and proximity ligation. *Genome Biology*, 20, 153.

Borie, C., Sanches, M.L., Navarro, C., Ramirez, S., Morales, M.A., Retamales, J., *et al.* (2009). Aerosol spray treatment with bacteriophages and competitive exclussion reduces *Salmonella* Enteritidis infection in chickens. *Avian Dis*, 53, 250–254.

Bottery, M.J., Pitchford, J.W. & Friman, V.-P. (2021). Ecology and evolution of antimicrobial resistance in bacterial communities. *ISME J*, 15, 939–948.

Bottery, M.J., Wood, A.J. & Brockhurst, M.A. (2016). Selective conditions for a multidrug resistanceplasmid depend on the sociality of antibiotic resistance. *Antimicrob Agents Chemother*, 60, 2524–2527.

Burnham, K.P. & Anderson, D.R. (2004). Multimodel Inference: understanding AIC and BIC in Model Selection. *Sociol Method Res*, 33, 261–304.

Cairns, J., Frickel, J., Jalasvuori, M., Hiltunen, T. & Becks, L. (2017). Genomic evolution of bacterial populations under coselection by antibiotics and phage. *Mol Ecol*, 26, 1848–1859.

Callaway, T.R. & Ricke, S.C. (Eds.). 2023. *Direct Fed Microbials/Prebiotics for Animals: Science and Mechanisms of Action*, 2nd edition. Springer Science, New York, NY, 348 pp.

Carvalho, C., Costa, A.R., Silva, F. & Oliveira. (2017). Bacteriophages and their derivatives for the treatment and control of food-producing animal infections. *Crit Rev Microbiol*, 43, 583–601.

CDC. (2019). *Antibiotic resistance threats in the United States*. Centers for Disease Control and Prevention, Atlanta, Georgia.

CDC. (2022). *FoodNet Fast-Salmonella*. Centers for Disease Control and Prevention, Atlanta, Georgia.

Clavijo, V., Morales, T., Vives-Flores, M.J. & Munoz, A.R. (2022). The gut microbiota of chickens in a commercial farm treated with a Salmonella phage cocktail. *Sci Rep*, 12.

Collier, S.A., Deng, L., Adam, E.A., Benedict, K.M., Beshearse, E.M., Blackstock, A.J., *et al.* (2021). Estimate of burden and direct healthcare cost of infectious waterborne disease in the United States. *Emerg Infect Dis*, 27, 140.

Cooper, R.M., Tsimring, L. & Hasty, J. (2017). Inter-species population dynamics enhance microbial horizontal gene transfer and spread of antibiotic resistance. *Elife*, 1, e25950.

Crits-Christoph, A., Hallowell, H.A., Koutouvalis, K. & Suez, J. (2022). Good microbes, bad genes? The dissemination of antimicrobial resistance in the human microbiome. *Gut Microbes*, 14, 2055944.

Cunrath, O. & Palmer, J.D. (2021). An overview of Salmonella enterica metal homeostasis pathways during infection. *Microlife*, 2, uqab001.

Day, T., Huijben, S. & Read, A.F. (2015). Is selection relevant in the evolutionary emergence of drug resistance? *Trends Microbiol*, 23, 126–133.

Dibner, J.J. & Richards, J.D. (2005). Antibiotic growth promoters in agriculture: history and mode of action. *Poult Sci*, 84, 634–643.

Dieckmann, U., Metz, J.A., Sabelis, M.W. & Sigmund, K. (2002). Adaptive dynamics of infectious diseases. In Dieckmann, U., Metz, J. A. J., Sabelis, M. W., & Sigmund, K., eds. *Pursuit of virulence management*, 460–463. Cambridge University Press, Cambridge.

Estrela, S. & Brown, S.P. (2018). Community interactions and spatial structure shape selection on antibiotic resistant lineages. *PLOS Comput Biol*, 14, e1006179.

Fiorentin, L., Vieira, N.D. & Barioni, W. (2005). Oral treatment with bacteriophages reduces the concentration of *Salmonella* Enteritidis PT4 in caecal contents of broilers. *Avian Pathol*, 34, 258–263.

Foley, S.L., Nayak, R., Hanning, I.B., Johnson, T.J., Han, J. & Ricke, S.C. (2011). Population dynamics of Salmonella enterica serotypes in commercial egg and poultry production. *Appl Environ Microbiol*, 77, 4273–4279.

Gill, J.J. & Hyman, P. (2010). Phage choice, isolation, and preparation for phage therapy. *Curr Pharm Biotechnol*, 11, 2–14.

Gomes, B.C., Franco, B.D.G. de M. & De Martinis, E.C.P. (2013). Microbiological food safety issues in Brazil: bacterial pathogens. *Foodborne Pathogens and Disease*, 10, 197–205.

Grace, J.B. (2006). *Structural equation modeling and natural systems*. Cambridge University Press, Cambridge.

Gunderson, M.S., Fielder, A.W., Bakke, I. & Vadstein, O. (2023). The impact of phage treatment on bacterial community structure is minor compared to antibiotics. *Sci Rep*, 13.

Haas, C.N. (2020). Quantitative microbial risk assessment and molecular biology: paths to integration. *Environ Sci Technol*, 54, 8539–8546.

Hawkey, J., Le Hello, S., Doublet, B., Granier, S.A., Hendriksen, R.S., Fricke, W.F., *et al.* (2019). Global phylogenomics of multidrug-resistant Salmonella enterica serotype Kentucky ST198. *Microb Genom*, 5. https://doi.org/10.1099/mgen.0.000269

Hite, J.L., Bosch, J., Fernández-Beaskoetxea, S., Medina, D. & Hall, S.R. (2016). Joint effects of habitat, zooplankton, host stage structure and diversity on amphibian chytrid. *Proc Biol Sci*, 283, 20160832.

Hite, J.L. & Cressler, C.E. (2019). Parasite-mediated anorexia and nutrition modulate virulence evolution. *Integrative and comparative biology*, 59, 1264–1274.

Hite, J.L., Pfenning, A.C. & Cressler, C.E. (2020). Starving the enemy? Feeding behavior shapes host-parasite interactions. *Trends Ecol Evol*, 35, 68–80.

Hu, L. & Bentler, P.M. (1999). Cutoff criteria for fit indexes in covariance structure analysis: conventional criteria versus new alternatives. *Struct Equ Model Multidis J*, 6, 1–55.

Hurley, J.C. (2022). How to apply structural equation modelling to infectious diseases concepts. *Clinic Microbiol Infect*, 28, 1567–1571.

Hyman, P. & Abedon, S.T. (2010). Bacteriophage host range and bacterial resistance. *Adv Appl Microbiol*, 70, 217–248.

Ippolito, J.R., Piccolo, B.D., Robeson, M.S., Barney, D.E.J., Ali, J., Singh, P., *et al.* (2022). Iron deficient diets modify the gut microbiome and reduce the severity of enteric infection in a mouse model of S. Typhimurium-induced enterocolitis. *J Nutr Biochem*, 107, 109065.

Jones, F.T. & Ricke, S.C. (2003). Observations on the history of the development of antimicrobials and their use in poultry feeds. *Poult Sci*, 82, 613–617.

Karslake, J., Maltas, J., Brumm, P. & Wood, K.B. (2016). Population density modulates drug inhibition and Ggves rise to potential Bbstability of treatment outcomes for bacterial infections. *PLOS Comput Biol*, 12, e1005098.

Kendall, B.E. (2015). 149A statistical symphony: instrumental variables reveal causality and control measurement error. In Fox, G.A., Negrete-Yankelevich, S., & Sosa, V.J., eds. *Ecological Statistics: Contemporary Theory and application*. Oxford University Press, Oxford.

Kent, A.G., Vill, A.C., Shi, Q., Satlin, M.J. & Brito, I.L. (2020). Widespread transfer of mobile antibiotic resistance genes within individual gut microbiomes revealed through bacterial Hi-C. *Nat Commun*, 11, 4379.

Khan, M.A.S. & Rahman, S.R. (2022). Use of phages to treat antimicrobial-resistant *Salmonella* infections in poultry. *Vet Sci*, 9, 438.

Klumper, U., Recker, M., Zhang, L., Yin, X., Zhang, T., Buckling, A., *et al.* (2019). Selection for antimicrobial resistance is reduced when embedded in a natural microbial community. *ISME J*, 13, 2927–2937.

Koo, J., Marshal, D.L. & DePaola, A. (2001). Antacid increases survival of virbio vulnificus phage in a gastrointestinal model. *Appl Environ Microbiol*, 67, 2895–2902.

LeHello, S., Harrois, D., Bouchrif, B., Sontag, L., Elhani, D., Guibert, V., *et al.* (2013). Highly drug-resistant *Salmonella enterica* serotype Kentucky ST198-X1: a microbiological study. *Lancet Infect Dis*, 13, 672–679.

Lorenzo-Rebenaque, L., Malik, D.J., Catala-Gregori, P., Marin, C. & Sevilla-Navarro, S. (2021). In vitro and in vivo gastrointestinal survival of non-encapsulated and microencapsulated *Salmonella* bacteriophages: implications for bacteriophage therapy in poultry. *Pharm*, 14. https://doi.org/10.3390/ph14050434

Lyimu, W.M., Leta, S., Everaert, N. & Paeshuyse, J. (2023). Influence of live attenuated *Salmonella* vaccines on cecal microbiome composition and microbiota abundances in young broiler chickens. *Vaccines (Basel)*, 11, 1116.

Mabelebele, M., Alabi, O.J., Ngambi, J.W., Norris, D. & Ginindza, M.M. (2014). Comparison of gastrointestinal tracts and pH values of digestive organs of ross 308 broiler and indigenous venda chickens fed the same diet. *Asia J Anim Vet Adv*, 1, 71–76.

Macià, M.D., Rojo-Molinero, E. & Oliver, A. (2014). Antimicrobial susceptibility testing in biofilm-growing bacteria. *Clin Microbiol Infect*, 20, 981–990.

Maiden, M.C. (1998). Horizontal genetic exchange, evolution, and spread of antibiotic resistance in bacteria. *Clin Infect Dis*, 27, S12–S20.

Mellor, K., Petrovska, L., Thomson, N., Harris, K. Reid, S. & Mather, A (2019). Antimicrobial Resistance Diversity Suggestive of Distinct *Salmonella* Typhimurium Sources or Selective Pressures in Food-Production Animals. *Front Microbiol 10*.

Moulana, Z. & Asgaharpour, F. (2022). Prevalence and antimicrobial resistance of *Salmonella enterica* serovar Infantis isolates from poultry: ā review. *Poult Sci J*, 10, 13–26.

NARMS. (2019). *2019 NARMS Update: Integrated Report*. The National Antimicrobial Resistance Monitoring System.

Olson, M.E., Ceri, H., Morck, D.W., Buret, A.G. & Read, R.R. (2002). Biofilm bacteria: formation and comparative susceptibility to antibiotics. *Can J Vet Res*, 66, 86–92.

Ormala, A. & Jalasvuori, M. (2013). Should bacterial resistance to phages be a concern, even in the long run? *Bacteriophage*, 3, e24219.

Palomino, A., Gewurz, D., DeVine, L., Zajmi, U., Moralez, J., Abu-Rumman, F., *et al.* (2023). Metabolic genes on conjugative plasmids are highly prevalent in Escherichia coli and can protect against antibiotic treatment. *ISME J*, 17, 151–162.

Pfenning-Butterworth, A.C., Vetter, R.E. & Hite, J.L. (2022). Natural variation in host defense strategies impacts both host and pathogen fitness. *bioRxiv*. 2022.09.22.509093.

Plata, G., Baxter, N.T., Susanti, D., Volland-Munson, A., Gangaiah, D., Nagireddy, A., *et al.* (2022). Growth promotion and antibiotic induced metabolic shifts in the chicken gut microbiome. *Comm Biol*, 5. https://doi.org/10.1038/s42003-022-03239-6

Potter, B. (2022). Reducing *Salmonella* in pre-harvest poultry. *Food Qual Saf*, 29, 20–21.

Roe, M.T. & Pillai, S.D. (2003). Monitoring and identifying antibiotic resistance mechanisms in bacteria. *Poult Sci*, 82, 622–626.

Roth, N., Kasbohrer, A., Mayrhofer, S., Zitz, U., Hofacre, C. & Domig, K.J. (2019). The application of antibiotics in broiler production and the resulting antibiotic resistance in *Escherichia coli*: a global overview. *Poult Sci*, 4. https://doi.org/10.3382/ps/pey539

Sabnis, A., Ledger, E.V.K., Pader, V. & Edwards, A.M. (2018). Antibiotic interceptors: creating safe spaces for bacteria. *PLOS Pathog*, 14, e1006924.

Satorra, A. & Bentler, P.M. (2001). A scaled difference chi-square test statistic for moment structure analysis. *Psychometrika*, 66, 507–514.

Singer, R.S., Porter, L.J., Schrag, N.F.D., Davies, P.R., Apley, M.D. & Bjork, K. (2020). Estimates of on-farm antimicrobial usage in broiler chicken production in the United States, 2013–2017. *Zoonoses Public Health*, 67(Suppl 1), 22–35.

Smith, J.A. (2019). Broiler production without antibiotics: United States field perspectives. *Anim Feed Sci Technol*, 250, 93–98.

Spragge, F., Bakkeren, E., Jahn, M.T., Araujo, E.B.N., Pearson, C.F., Wang, X., *et al.* (2023). Microbiome diversity protects against pathogens by nutrient blocking. *Science*, 382. https://doi.org/10.1126/science.adj3502

Sulakvelidze, A., Alavidze, Z. & Morris, J.G. (2011). Bacteriophage therapy. *Antimicrob Agents Chemother*, 45, 649–659.

Sünderhauf, D., Klümper, U., Gaze, W.H., Westra, E.R. & van Houte, S. (2023). Interspecific competition can drive plasmid loss from a focal species in a microbial community. *ISME J*, 17, 1765–1773.

Tan, Z., Lu, P., Adewole, D., Diarra, M.S., Gong, J. & Yang, C. (2021). Iron requirement in the infection of *Salmonella* and its relevance to poultry health. *J Appl Poult Res*, 30, 100101.

Threlfall, E.J. (2002). Antimicrobial drug resistance in *Salmonella:* problems and perspectives in food- and water-borne infections. *FEMS Microbiol*, 26, 141–148.

Threlfall, E.J., Frost, J.A., Ward, L.R. & Rowe, B. (1996). Increasing spectrum of resistance in multiresistant Salmonella typhimurium. *Lancet*, 347, 1053–1054.

Upadhaya, S., Ahn, J.M., Cho, J.H., Kim, J.Y., Kang, D.K., Kim, S.W., *et al.* (2021). Bacteriophage cocktail supplementation improves growth performance, gut microbiome and production traits in broiler chickens. *J Anim Sci Biotechnol*, 12. https://doi.org/10.1186/s40104-021-00570-6

Van Boeckel, T.P., Glennon, E.E., Chen, D., Gilbert, M., Robinson, T.P., Grenfell, B.T., *et al.* (2017). Reducing antimicrobial use in food animals. *Science*, 357, 1350–1352.

Velasquez, C.G., Macklin, K.S., Kumar, S., Bailey, M., Ebner, P.E., Oliver, H.F., *et al.* (2018). Prevalence and antimicrobial resistance patterns of Salmonella isolated from poultry farms in southeastern United States. *Poult Sci*, 97, 2144–2152.

Wale, N., Sim, D.G., Jones, M.J., Salathe, R., Day, T. & Read, A.F. (2017). Resource limitation prevents the emergence of drug resistance by intensifying within-host competition. *Proc Natl Acad Sci*, 114, 13774–13779.

Wei, S., Chelliah, R., Rubab, M., Oh, D.-H., Uddin MJ & Ahn, J. (2019). Bacteriophages as potential tools for detection and control of *Salmonella* spp. in food systems. *Microorganisms*, 7, 570.

Weisberg, S. & Fox, J. (2011). *An R companion to applied regression*. Sage, Thousand Oaks.

Wistrand-Yuen, E., Knopp, M., Hjort, K., Koskiniemi, S., Berg, O.G. & Andersson, D.I. (2018). Evolution of high-level resistance during low-level antibiotic exposure. *Nat Commun*, 9, 1599.

Wythe, L.A., D.K. Dittoe, and S.C. Ricke. (2023). *In Direct Fed Microbials/Prebiotics for Animals: Science and Mechanisms of Action Advancements in Poultry Nutrition and Genetics, The Role of Antibiotic Growth Promoters, and the Introduction of Feed Additive Alternatives.* Chapter 4. 2nd edn. Springer Science, New York, NY, USA.

Yurtsev, E.A., Chao, H.X., Datta, M.S., Artemova, T. & Gore, J. (2013). Bacterial cheating drives the population dynamics of cooperative antibiotic resistance plasmids. *Mol Syst Biol*, 9, 683.

Zhang, X., Akhtar, M., Chen, Y., Ma, Z., Liang, Y., Shi, D., *et al.* (2022). Chicken jejunal microbiota improves growth performance by mitigating intestinal inflammation. *Microbiome*, 10, 107.

Zhou, K., Sun, L., Zhang, X., Xu, X., Mi, K., Ma, W., *et al.* (2023). Salmonella antimicrobials inherited and the non-inherited resistance: mechanisms and alternative therapeutic strategies. *Front Microbiol*, 14, 1176317.

Chapter 3

Escherichia coli in poultry meat

Hunter G. Perez, University of Georgia, USA; S. C. Ricke, University of Wisconsin-Madison, USA; and Todd R. Callaway, University of Georgia, USA

1 Introduction

Each year in the United States, approximately 48 million people contract foodborne illnesses, and there are at least 128 000 hospitalizations (Scallan et al., 2011; Tack, 2020). The aggregated cost of foodborne illness to the U.S. economy is in excess of US$75 billion each year (Scharff, 2012). The leading foodborne bacterial pathogens responsible for the majority of these human illnesses include *Salmonella* spp., *Campylobacter* (*jejuni* and *coli*), and pathogenic *Escherichia coli*, which can affect both humans and poultry (Scallan et al., 2011; Delahoy et al., 2023). The poultry production industry predominantly focuses on addressing *Campylobacter* and *Salmonella* contamination because these are the pathogens most commonly associated with poultry products (Jorgensen et al., 2002; Golden et al., 2021)

However, *E. coli* generally remains a major concern due to its widespread environmental prevalence and impact on both human and avian health (Dziva and Stevens, 2008; Kabir, 2010; Nolan et al., 2020; Gross, 1994; Karmali et al., 2010). *E. coli* infection (colibacillosis) can impact the respiratory system, digestive system, and immune system activity of birds, leading to decreased production efficiency and potential mortality (Usman et al., 2022; Barnes

http://dx.doi.org/10.19103/AS.2024.0146.03

et al., 2003; Nolan et al., 2020). Therefore, understanding and managing *E. coli* contamination during poultry production are crucial steps in ensuring the availability, affordability, and safety of poultry products for consumers.

Escherichia coli is an easily grown, Gram-negative facultative anaerobic bacterium that can be found in various environments and is considered a typical inhabitant of the gastrointestinal tract (GIT) and mucosal surfaces of poultry (Johnson et al., 2008b; Oakley et al., 2014; Fancher Courtney et al., 2021). *E. coli*, being a diverse and adaptable pathogen, has evolved to thrive in various environments and ecosystems, particularly in harsh environments such as the GIT. Within *E. coli*, there are non-pathogenic and pathogenic strains, distinguished based on clinical symptoms and pathogenicity mechanisms (Köhler and Dobrindt, 2011). The pathogenic strains of *E. coli* can opportunistically impact humans, animals, and birds (Stein and Katz, 2017; El-Sawah et al., 2018; Efsa Biohaz Panel et al., 2020). Although it has a symbiotic relationship with gut microorganisms, *E. coli* can also be an opportunistic foodborne pathogen in humans (Duffy et al., 2000; Kaper et al., 2004; Gyles, 2007). Given its vast ability to adapt and thrive in various environments, poultry production in relation to *E. coli* emphasizes the need to develop effective strategies for mitigating its impact on both animal and public health (Callaway and Oliver, 2009).

2 Pathotypes of *Escherichia coli* which impact poultry

Escherichia coli serves as a model microorganism because it is very diverse, genetically (ranging between 3.98 and 5.8 Mb), biochemically, and ecologically. Pathogenicity in this species is also diverse and dependent on strain- and host-specific colonization, virulence, and fitness factors (Kaper et al., 2004). Some strains have acquired specific virulence traits which provide them with a competitive advantage and enable them to cause a broad spectrum of disease (Kaper et al., 2004). Only the most ecologically advantageous combinations of virulence factors have led to the development of *E. coli* 'pathotypes', which can lead to enteric disease, urinary tract infections, and sepsis (Table 1) (Kaper et al., 2004).

There are several specifically that cause diarrhea:

- enteropathogenic *E. coli* (EPEC);
- enterohemorrhagic *E. coli* (EHEC; also known as Shiga oxin-producing/ Verotoxin-producing *E. coli* [STEC/VTEC]);
- enterotoxigenic *E. coli* (ETEC);
- enteroaggregative *E. coli* (EAEC);
- enteroinvasive *E. coli* (EIEC); and
- diffusely adherent *E. coli* (DAEC).

Table 1 *Escherichia coli* pathotypes of importance

Pathogenic *E. coli* location	Pathotype	Abbreviation
Diarrheagenic	Enteropathogenic	EPEC
	Enreroinvasive	EIEC
	Enterotoxigenic	ETEC
	Diffusely adherent	DAEC
	Enteroaggregative	EAEC
	Enterohemorrhagic (includes Shiga toxin-producing and Verotoxin-producing)	EHEC (STEC, VTEC)
Urinary tract	Uropathogenic	UPEC
Extraintestinal	Extraintestinal pathogenic	ExPEC
	Avian pathogenic	APEC

Source: Adapted from Kaper et al. (2004).

Urinary tract infections are associated with uropathogenic (UPEC) (Kaper et al., 2004). Several types of pathogenic *E. coli* groups include extraintestinal pathogenic *E. coli* strains (ExPEC), with the latter typically part of the normal intestinal microbiota but capable of causing systemic diseases affecting animals and humans (Johnson et al., 2008b; Köhler and Dobrindt, 2011).

While many of these pathotypes have been isolated from birds sporadically, it should be noted that some of these *E. coli* pathotypes present specific threats

- to poultry health (e.g. avian pathogenic *E. coli*) and
- to human health via poultry products (e.g. STEC/VTEC; e.g. serotypes O157:H7).

Extraintestinal pathogenic *E. coli* (ExPEC) infections challenge the poultry industry as these infections result in significant economic losses (Barnes and Calnek, 1997).

Avian pathogenic *E. coli* (APEC) is a subgroup of ExPEC (and is typically considered to be serotypes O1, O2, and O78) which poses a significant threat to poultry production, causing both localized and systemic infections in avian species that can impact growth and mortality (Barbieri et al., 2021; Nolan et al., 2020; Kaper et al., 2004; Dho-Moulin and Fairbrother, 1999). Disease states in poultry caused by APEC include air sacculitis, polyserositis, septicemia, and other extraintestinal diseases in birds (Dho-Moulin and Fairbrother, 1999). Unlike other pathogenic *E. coli* groupings, APEC lacks a singular defining trait, contributing to the complexity of its pathogenicity and transmission dynamics (Dziva and Stevens, 2008). Avian colibacillosis is caused by APEC, and it presents a significant challenge to the poultry industry, impacting poultry health and productivity (Kabir, 2010; Dziva and Stevens, 2008; Guabiraba and

Schouler, 2015; Kathayat et al., 2021). Some APEC (e.g. *E. coli* O78 serotype) have been successfully used in *in vivo* models of infection, showing some indications of colibacillosis (Tarabees et al., 2020). Depending on the strain of the APEC infection, localized inflammation and lesions can lead to yolk sac infections, respiratory tract infections, swollen head syndrome, septicemia, pericarditis, perihepatitis, salpingitis, coligranuloma, omphalitis, and cellulitis (Table 2) (Gross, 1994; Dziva and Stevens, 2008; Kabir, 2010; Kathayat et al., 2021; Barnes et al., 2003).

More than half (53.5%) of yolk sac infections impacting young chicks are due to *E. coli* infections, which lead to septicemia, causing high mortality within the first week of life (Olsen et al., 2012). Respiratory infections, primarily air sacculitis, also contribute to production losses, as this is one of the most common forms of colibacillosis (Stearns et al., 1987; Pourbakhsh et al., 1997; Mellata, 2013). Beyond its implications for avian health, APEC causes financial losses due to the cost of treatment, mortality, lost production, containment, and carcass condemnations (Barnes et al., 2003). Moreover, the zoonotic transmission potential of APEC raises concerns regarding its transmission to humans through contaminated poultry products in the context of foodborne

Table 2 Systemic diseases affecting poultry related to *Escherichia coli* infection

Disease	Location
Colibacillosis	General infection by *E. coli*, losses in the range of millions of USD/year. Some cases are worsened by predisposing conditions (e.g. mycoplasmosis, Newcastle Disease).
Omphalitis	Infected yolk sacs, or poorly healed navels (due to poor incubation conditions). Listless and anorexic.
Peritonitis	Most common isolate from peritonitis cases (inflammation of the lining of abdominal cavity).
Salpingitis (layers)	Inflammation of the oviduct and uterus late in life in laying hens, reducing egg production.
Cellulitis	Most common cause of cellulitis, which is a discoloration and thickening of subcutaneous tissues. Growing issue in the U.S. poultry flock.
Septicemia	Most common disease of farmed poultry, blood stream infection often subsequent to an upper respiratory disease. Lack of lymph nodes leads to pathogen spread in the blood stream. Systemic infection and rapid deterioration due to multiplerelated conditions.
Airsacculitis	Bacterial infection of air sacs with *E. coli* is one of the most common causes. Broad range of clinical signs, including pneumonia and pulmonary edema.
Pericarditis	Often associated with septicemia, it is the inflammation of the heart sac. It can lead to whole carcass rejection.
Perihepatitis	Inflammation of the capsule around the liver. Related to septicemia as well.

pathogens (Markland et al., 2015; Ali et al., 2020). Avian pathogenic *E. coli* clearly presents a significant concern for public health due to the potential zoonotic transmission risk associated with undercooked poultry consumption, but it is also impactful to bird health and productivity.

Extraintestinal pathogenic *E. coli* strains are a zoonotic pathogen that can be transmitted through the consumption of undercooked chicken meat (Bergeron et al., 2012). Emerging research suggests that poultry may indeed act as a reservoir for ExPEC infections in humans. Studies have found ExPEC in the intestines of healthy poultry and in poultry meat sold at retail stores (Manges and Johnson, 2012; Johnson et al., 2005). In humans, *E. coli* infections can result in a range of gastrointestinal symptoms, including diarrhea, abdominal cramps, and vomiting (O'Brien et al., 1992). Although, certain ExPEC strains are shared between chicken products and humans products which have been linked to colibacillosis in chickens and urinary tract infections and other extraintestinal infections in humans (Zhao et al., 2009). Poultry products have also been identified as potential sources of ExPEC strains responsible for sepsis infections in humans (Manges and Johnson, 2012).

3 Shiga toxin-producing/verotoxigenic *Escherichia coli* as a foodborne pathogen in poultry

Shiga toxin-producing/verotoxigenic *E. coli* strains, such as O157:H7, are often most closely associated with cattle and can be found throughout the GIT (Chapman et al., 1993; Karmali et al., 2010). The severe impacts on humans of this potent foodborne pathogen have led to its categorization as an adulterant in ground beef by USDA/FSIS, effectively prohibiting the sale of any meat product found to be contaminated with this strain (FSIS, 1999). This emphasizes the importance of strain-level identification of *E. coli*, as not all strains are characterized as adulterants. Within the first 10 years following the recognition of *E. coli* O157:H7, the beef industry invested in excess of US$2.3 billion (Kay, 2003) to reduce VTEC/STEC spread to human consumers of beef. The economic impact of STEC/VTEC infection in humans, in terms of the basic cost of human illnesses, has been estimated at greater than US$2 billion per year (Scharff, 2012).

While the foodborne threat of *E. coli* is typically associated with the cattle industry (Karmali et al., 2010), STEC/VTEC (e.g. serotypes O157, O145, O121, O111, O103, O26, and O45) can also be isolated sporadically from wild and domestic birds (Mellor et al., 2016; Stanford et al., 2016; Noll et al., 2018; Cernicchiaro et al., 2020; Schneider et al., 2018; Alharbi et al., 2022). While VTEC/STEC are primarily found in beef cattle, contamination with these potent pathogens can impact not only consumer health but also the economic sustainability of the poultry industry, impacting production metrics as well as morbidity and mortality (Usman et al., 2022; Nolan et al., 2020; Guabiraba and Schouler, 2015; Kabir, 2010). Unfortunately, wild and domestic birds (including broilers and layers) in the United States and European Union potentially serve as a reservoir and vector/vehicle for transmission of several VTEC serotypes, including O157:H7 (Chapman et al., 1997; Dipineto et al., 2006; Santaniello et al., 2007; Kalin et al., 2012; Kauffman and LeJeune, 2011).

Several studies have shown that poultry are also susceptible to colonization by O157:H7 and seven other STEC/VTEC strains (La Ragione et al., 2005) and this carriage has led to human illnesses (Ejidokun et al., 2006; Ferens and Hovde, 2011). Young birds (as early as one day) can be colonized by *E. coli* O157 and can persist in the cecum for up to 3 months (Chapman et al., 1997; Stavric et al., 1993; La Ragione et al., 2005). Further studies with 6-week-old broilers demonstrated that some strains could persist for more than 100 days (La Ragione et al., 2005). Migratory and local birds (e.g. seagulls, pigeons, cowbirds, egrets, grackles, and starlings) can carry STEC/VTEC and disseminate these organisms over long distances (thousands of kilometers), as well as spread them from farm to farm locally (Gaukler et al., 2009; Kauffman and LeJeune, 2011; Wallace et al., 1997; Makino et al., 2000; Großmann et al., 2005; Dutta et al., 2013; Callaway et al., 2014). Wild and domestic parrots have been demonstrated to contain VTEC/STEC and other pathogenic strains of *E. coli* (Schremmer et al., 1999; Hubálek, 2004; Marietto-Gonçalves et al., 2011).

Moreover, analyses have revealed occasional detection of low concentrations of STEC/VTEC in chicken meat products, suggesting the potential for transmission through the food chain (Heuvelink et al., 1999; Stromberg et al., 2017; Hussain et al., 2017; Momtaz and Jamshidi, 2013; Dishan et al., 2023). Further evidence found that EPEC and STEC could be carried in poultry (Alonso et al., 2012), and could be simultaneously isolated from chicken meat samples (Zarei et al., 2021). Given the increasing consumption of poultry products globally, the spread of STEC/VTEC to human consumers is a growing concern (Ferens and Hovde, 2011). While little is known surrounding VTEC presence and pathogenicity in poultry, the pathogenicity of *E. coli* O157:H7 in chicks remains unclear, though it does appear that colonization with *E. coli* O157:H7 can negatively impact broiler growth performance and efficiency (El-Sawah et al., 2018). An *E. coli* O157:H7 challenge in broilers reduced villi height/crypt depth ratio in the ileum, along with reduced microbial diversity, and led to an increase in TLR-4, NF-KB, and IL-8 expression in the cecum (Xiang et al., 2022). When different routes of *E. coli* O157:H7 infection were examined (enema, oral gavage, and intraperitoneal injection), they all had different types of physiological impacts (Guo et al., 2024). Intraperitoneal injection caused the most damage to chick physiology and resulted in liver, spleen, and intestinal damage, whereas enema inoculation resulted in more colonic damage, and oral gavage resulted in more typical colibacillosis and caused liver, spleen, and colon damage (Guo et al., 2024).

4 Sources of *E.coli* contamination

The flexible and adaptable physiology and metabolic versatility of *E. coli* enables it to thrive as an environmental pathogen within livestock and poultry environments on farms, including water, soil, and litter (Blaak et al., 2015; Nolan et al., 2020), posing persistent risks for contamination and multiple potential transmission pathways (Jang et al., 2017; Papouskova et al., 2023). The prevalence of *E. coli* is common in both poultry production and processing environments, carrying risks of compromised poultry health, economic losses, and public health concerns (Cortés et al., 2010; Mitchell et al., 2015; Usman et al., 2022). For example, avian colibacillosis can be caused by environmental contamination with *E. coli* and can manifest in a variety of avian illnesses, including yolk sac infections, respiratory tract infections, swollen head syndrome, and a spectrum of localized infections (Gross, 1994; Dziva and Stevens, 2008; Kabir, 2010; Kathayat et al., 2021). Colibacillosis infections have a tremendous economic impact on the poultry industry in the United States alone, estimated as high as US$40 million annually, particularly from cellulitis lesions, with many isolates being antimicrobial resistant, which leads to downgrading and condemnation losses (de Brito et al., 2003; Norton et al., 1997; Barnes et al., 2003; Dziva and Stevens, 2008; Müller et al., 2024).

Poultry farms can contain high levels of environmental *E. coli* contamination, ranging between 61% and 81% in sampled hens and broilers from multiple farms, with significant contamination detected in diverse farm elements such as wastewater, sediments, dust, and soil (Blaak et al., 2015) and is linked with poor hygiene in hatcheries and on farms (Kim et al., 2011). Moreover, elevated concentrations of *E. coli* have been documented at concentrations up to 4 $\log_{10}$ CFU/m^3 in the air (Zucker et al., 2000), 3 $\log_{10}$ CFU/g in feeds (Munoz et al., 2021), and 7 $\log_{10}$ CFU/g in poultry litter (Martin et al., 1998), further emphasizing the pervasive nature of this pathogen on farms.

Despite the primary mode of transmission of *E. coli* being fecal–oral, concerns persist regarding contamination during poultry harvesting processes (Kabir, 2010; Linton et al.; 1977, Swelum et al., 2021). During harvest and processing, the risk of cross-contamination is high as bacteria from the poultry gut can easily contaminate carcasses from birds colonized with EAEC and STEC (Trung et al., 2016), thereby leading to *E. coli* contamination in poultry meat (Chaslus-Dancla and Lafont, 1985; Bensink and Botham, 1983). *E. coli* strains such as O2:K1 have been isolated from both human urinary tract infections and septicemic chickens, which have phenotypical similarities, indicating potential transmission routes between human and avian populations (Kabir, 2010). When over 2400 meat samples were examined, *E. coli* was isolated in over 79% of them, with the emergence of the *E. coli* ST131-H22 lineage, now prevalent in poultry flocks and identified as a foodborne uropathogen (UPEC) in humans (Liu et al., 2018). A link has been proposed between *E. coli* urinary tract infections in humans and the consumption of poultry meat (Jakobsen et al., 2010; Ewers et al., 2009; Rodriguez-Siek et al., 2005).

Across the meat production industry, reducing cross-contamination remains a primary focus for pathogen mitigation at harvesting facilities. The risk of cross-contamination extends beyond direct contact of gastrointestinal contents with poultry carcasses and includes processing equipment and surfaces (Golden et al., 2021; Dishan et al., 2023). During poultry processing steps, there are several moving parts (e.g. belts and shackles) and large tanks (e.g. chill and rinse tanks) that come into carcass contact, each presenting its own specific set of risks for contamination and cross-contamination, which are addressed systematically by the implementation of Hazard Analysis and Critical Control Points (HACCP) procedures along with pathogen (e.g. *Salmonella*, *Campylobacter*, and *E. coli*) reduction hurdles in each processing facility (Mukhopadhyay and Gorris, 2014; Bucher et al., 2012; Guo et al., 2011; Fries, 2002).

All steps during poultry processing can play a role in reducing pathogenic microbial contamination levels of the carcasses (Mulder et al., 1978; Cason and Hinton Jr, 2006; Bucher et al., 2012) and these steps can also alter the microbiome of the carcass (Ricke et al., 2022). Reducing the organic material load in rinse and chill water decreases pathogens, *E. coli*, and other coliform numbers (Incili and Çalicioğlu, 2018; Fries, 2002), but during the course of a processing day, the organic material in these tanks naturally increases. A study utilizing a 'generic' *E. coli* K12 marker found that cross-contamination can occur in broilers during defeathering, primarily through aerosols, large droplets, and feathers (Allen et al., 2003). The effect of chilling and desiccation when combined does reduce *E. coli* culturability (Mellefont et al., 2015).

Introduction of an experimentally inoculated broiler carcass using fluorescent markers into the processing line led to both vertical and horizontal contamination of subsequent chicken carcasses from ruptured crops or intestinal tracts (Byrd et al., 2002). Equipment contamination becomes a concern when the GIT of a bird is inadvertently ruptured during the evisceration process, leading to the leakage of fecal material (Abu-Ruwaida et al., 1994; Lillard et al., 1984; Warriss et al., 2004). This is a risk due to fecal matter potentially contaminating the poultry's skin during evisceration. It is important to point out that failure to properly clean and sanitize surfaces between processing batches poses additional risks of *E. coli* transmission and increased concerns regarding food safety (Mead et al., 1995). Therefore, it is vital to ensure proper protocols are being utilized throughout the entirety of poultry production and processing and undergo a minimum of procedural drift.

5 Control of *E.coli*: the problem of antibiotic resistance

The demand for poultry meat and other animal-based meats continues to rise due to increasing population growth and changing dietary preferences. However, with this increased demand comes the need to ensure a continuous, plentiful, sustainable, and efficient clean supply of poultry products, necessitating effective management of broiler farms, broiler breeders, and processors to meet market demands (Leeson and Summers, 2010). As these competing demands on producers and processors grow, it is becoming imperative that we focus on creating a safer and more efficient food supply. We know *E. coli* contamination poses significant health risks to the production of poultry and consumers of poultry products, and that a variety of pathotypes can impact bird health and productivity and be spread to human consumers, potentially leading to severe illnesses in human consumers (including the spread of antibiotic-resistant organisms and genetic material) via contamination during processing.

Resistance to antimicrobials has become one of the leading public health challenges of the twenty-first century (Lhermie et al., 2016; Williams-Nguyen et al., 2016; McEvoy, 2016; O'Neill, 2016), and the risks posed by antibiotic-resistant *E. coli* strains pose a significant threat to both public health and food safety (Stanford et al., 2012; Karaiskos et al., 2019; Kathayat et al., 2021). Antibiotics have long been one of the most widely used additives within the poultry industry to improve feed conversion, growth rate, and overall health (Lourenco et al., 2019; Gadde et al., 2017). Numerous antibiotics, including penicillin, tetracycline, macrolide, aminoglycoside, and amphenicol, were utilized in this role (Diarra and Malouin, 2014; O'Brien, 2002). However, the use of antimicrobials as growth promotants led to more widespread antibiotic-resistant organisms (including *E. coli*) thriving within poultry production scenarios (Engberg et al., 2000; Diarra et al., 2007; Forgetta et al., 2012; Bensink and Botham, 1983; Nadeau et al., 2000).

Poultry meat exhibits relatively high overall levels of *E. coli* contamination (greater than 79% in some studies), and *E. coli* has a high prevalence of multidrug-resistant strains compared to other meat animals (Manges and Johnson, 2012; Liu et al., 2018). Due to the high poultry consumption in the United States, the risk of antibiotic-resistant ExPEC and the subsequent emergence of ExPEC infections in humans has increased (Pitout, 2012). Antimicrobial-resistant *E. coli* strains exhibit various phenotypes, each associated with resistance to specific classes of antimicrobial agents (Yates et al., 2004; Šmídková and Čížek, 2017; Barbieri et al., 2021; Dishan et al., 2023).

Poultry production farms and facilities are a source of extended-spectrum beta-lactamase (ESBL)-producing bacteria (which confer resistance to penicillins, cephalosporins, and aztreonam) that can cause serious human infection (Falgenhauer et al., 2019). *Escherichia coli* strains with AmpC-resistance phenotypes were found in 84.9% of cage swabs and 77.8% of retail ground meat samples, whereas 22.6% of cage swabs and 11.1% of meat samples contained strains with ESBL resistance phenotypes (Mollenkopf et al., 2018). Prevalence rates of ESBL-producing *E. coli* in broiler chicken meat in India was 46% (Hussain et al., 2017). Canadian results concluded that not all isolates possess the ESBL types most commonly associated with human infections (Ghosh et al., 2019). In 2020, the tet(X3) resistance gene was found in *E. coli* isolated from retail chicken meat, which is known to have high levels of tigecycline resistance (Lv et al., 2020). The presence of these potent antibiotic-resistant *E. coli* in poultry products not only poses a risk to animal health but also can be transmitted to humans through poultry consumption, resulting in foodborne illnesses that cannot be managed using conventional antibiotic treatments.

6 Strategies to reduce *Escherichia coli* in poultry on the farm

Consumer concerns about antibiotic use in poultry production have increased; as a result, researchers have shifted their focus to developing novel alternatives to antibiotics (ATA) (Taylor, 2001; Seal et al., 2013; Salaheen et al., 2017; Callaway et al., 2021). Antibiotic alternatives ideally need to provide the same production benefits as antibiotics (e.g. lowering flock mortality rates, maximizing animal productivity) all while meeting changing consumer, sustainability, and regulatory demands (Koyun and Callaway, 2019). A variety of non-antibiotic strategies (e.g. ATAs) have been examined in research and production situations, such as probiotics, prebiotics, postbiotics, organic acids, bacteriophages, along with others (Shi et al., 2019; Callaway et al., 2013; Grilli et al., 2015; Callaway et al., 2017). As the use of these antibiotic alternatives becomes increasingly accepted in animal agriculture, their role in reducing the load of *E. coli* in poultry warrants further investigation and validation.

Probiotic-type approaches (including postbiotics, prebiotics, organic acids, phytochemicals, and bacteriophages) hold the potential to replace the benefits of antibiotic usage on farms (El Jeni et al., 2024), and several of these approaches have been used in the past few years to reduce pathogen populations in poultry (Huff et al., 2015; Zhang et al., 2016; Guo et al., 2017; Vase-Khavari et al., 2019; Tarabees et al., 2020; Papouskova et al., 2023).

Traditionally, a probiotic is defined as 'a live microbial feed supplement which beneficially affects the host animal by improving its intestinal balance'

(Fuller, 1991). Numerous studies have demonstrated beneficial impacts of probiotics on various aspects of poultry production (Rodjan et al., 2018; He et al., 2019; Wu et al., 2019; Callaway et al., 2012), including impacts on nutrient absorption (Rotz, 2004; Applegate et al., 2010), bodyweight gain (Askelson et al., 2014), and immune response (Rinttilä and Apajalahti, 2013). The specific impacts of any individual probiotic culture may vary based on probiotic culture species, dosage, production status, animal health, and environmental/production factors (El Jeni et al., 2024). It is thought that probiotics enhance gut epithelial integrity/health by strengthening the tight junctions between epithelial cells, thus contributing to improved digestion and absorption of nutrients, a more effective separation of self from non-self, reducing non-growth energy expenditure by the immune system, ultimately leading to increased broiler performance (Koyun and Callaway, 2019; Koyun et al., 2022). During times of dietary change or health stress, such as changes in diet or high animal density, probiotics reduced the frequency of morbidity and mortality in poultry, thereby improving welfare and production efficiency (Yeo and Kim, 1997; Callaway et al., 2012; Palamidi et al., 2016; Yazhini et al., 2018; Vase-Khavari et al., 2019). Probiotic treatment has demonstrated the ability to reduce cholesterol and fat content in poultry meat, which can result in healthier meat products for consumers (Hossain et al., 2012). Furthermore, probiotics have demonstrated promise in combating the deleterious impact of mycotoxins in feeds such as aflatoxin, which causes mortality and morbidity throughout poultry production (Ma et al., 2012; Motawe et al., 2014).

Prebiotic approaches are the inclusion of specific nutrients to the host that are utilized by intestinal microbes but are unavailable to the host's digestion (Walker and Duffy, 1998) and are often known as 'colonic food.' Prebiotic compounds typically include oligosaccharides, including (Patterson and Burkholder, 2003; Murate et al., 2015; Jung et al., 2008; Janardhana et al., 2009):

- inulin;
- fructooligosaccharides (FOS);
- galactooligosaccharides (GOS);
- mannanoligosaccharides (MOS);
- xylo-oligosaccharides (XOS);
- pyrodextrins;
- isomaltooligosaccharides; and
- lactulose.

In poultry, FOS decreases populations of harmful bacteria such as *Clostridium perfringens* and *E. coli* while increasing beneficial native *Lactobacillus* populations (Kim et al., 2011). Prebiotics support the growth of native microbiota in the GIT which can help exclude pathogen penetration (Hume, 2011). Supplementing poultry diets with yeast extracts (postbiotics) rich in MOS increased heterophil numbers and provided resistance against *E. coli*-related diseases in poultry (Huff et al., 2010). Similarly, MOS supplementation in broiler chickens undergoing suboptimal environmental conditions improved the growth of cecal bacteria (e.g. *Lactobacillus* and *Bifidobacterium*) (Pourabedin et al., 2014). By altering the gut microbial population, prebiotics are thought to increase the microbial production of vitamins (especially B vitamins), short-chain fatty acids, and antioxidants from bacterial fermentation, thereby improving poultry and host health (Callaway et al., 2014). However, further research is needed to better understand prebiotic modes of action and their potential benefits for animal health and food safety.

Bacteriophages, or phages, are viruses that infect bacteria, hijacking the bacterial cell to replicate and multiply (Clavijo and Flórez, 2018; Callaway et al., 2003a; Young, 1992). Phages are thought of as a 'smart bomb' because they tend to be active against a relatively narrow spectrum of bacterial strains, which allows them to be deployed widely in the gut without concerns about unintended impacts throughout the gastrointestinal microbial population (Lederberg, 1996). Phages are classified into two main types based on their life cycles: virulent phages, which have a lytic cycle involving bacterial lysis, and temperate phages, which have a lysogenic cycle where they integrate into the bacterial genome and can later switch to a lytic cycle when conditions are favorable for phage propagation (Clavijo and Flórez, 2018, Young, 1992).

Although phages were discovered in the early twentieth century (d'Herelle, 1919), they are now being promoted as a potential alternative to combat antibiotic-resistant bacteria due to advantages such as *in situ* effect, non-toxicity, high specificity, and the ability to target biofilms (Koyun and Callaway, 2019). Phages specific to VTEC/STEC have been used in ruminant animals to reduce foodborne pathogen carriage in cattle (Bach et al., 2002; Callaway et al., 2003b; Bach et al., 2003; Raya et al., 2006; Callaway et al., 2006). Lytic bacteriophage have been used successfully to reduce *E. coli* populations and control colibacillosis and airsaculitis in poultry (Johnson et al., 2008a; Oliveira et al., 2010; López-Cuevas et al., 2011; Atterbury et al., 2011; Huff et al., 2002). However, despite their potential, further research is needed to fully establish phage therapy as a sustainable strategy for controlling animal health impacts and foodborne pathogens in poultry and other food animal species.

Organic acids and phytochemicals have been used for millennia to control pathogens and spoilage organisms in human foods (Cowan, 1999; Nascimento et al., 2000; Lillehoj et al., 2018; Dittoe et al., 2018). These compounds exhibit antimicrobial action against a wide variety of microorganisms, including *E. coli*. Organic acids have been used for many years as part of carcass rinses/sprays to reduce pathogens entering the food chain (Berry and Cutter, 2000;

McWilliam Leitch and Stewart, 2002; Carlson et al., 2008), but recently interest has increased in using them to reduce *E. coli* in live birds on the farm of origin (Menconi et al., 2014; Grilli et al., 2015; Rodjan et al., 2018; Mortada et al., 2020). Phytochemicals exhibit a number of modes of action specific to each compound, but the surge in interest in utilization of ATA has focused researchers on exploring the use of phytochemicals in live animals to improve animal health and human food safety (Kim et al., 2015; Gadde et al., 2017; Lillehoj et al., 2018; Kurt et al., 2019). Benefits of these compounds include the fact that many of them are Generally Recognized as Safe, which means they can be used in animals without further regulatory approval.

While these new methods show distinct promise in reducing *E. coli* in poultry, other traditional means of combating food-borne pathogens have not been abandoned. Traditionally, poultry producers have relied on a combination of hygiene measures, antibiotics, and vaccination strategies to prevent and control APEC infections in poultry (Dho-Moulin and Fairbrother, 1999; Guabiraba and Schouler, 2015). Hygiene barriers, such as strict biosecurity protocols and regular facility cleaning with disinfectants, serve to minimize naïve bird exposure to pathogenic bacteria in the farm environment (Dame-Korevaar et al., 2020). Vaccination strategies have remained effective against most infections, but only recently has vaccination emerged as another key strategy to combat avian pathogenic *E. coli*, with two commercially available vaccines targeting specific APEC strains. These vaccines include the live-attenuated APEC O78 ΔaroA Poulvac® *E. coli* vaccine and the inactivated Nobilis® *E. coli* vaccine, which offer targeted protection against APEC infections in chickens (Kathayat et al., 2021). As the poultry industry seeks ATAs, vaccination is increasingly recognized as a valuable tool in the arsenal against APEC, complementing traditional hygiene practices and contributing to the overall health and welfare of poultry flocks.

7 Conclusion

Escherichia coli is a common gastrointestinal bacterium that can be a commensal member of the gut population or can act as a transient pathogen. The high level of diversity within this bacterium provides it the ability to thrive in a variety of environments and can exhibit a number of pathogenic/virulence traits that can impact poultry producers and consumers. Pathotypes of *E. coli* range from those that impact poultry performance and gut physiology to those that cause colibacillosis and include strains that can cause severe disease in human consumers. We have long been dependent on the use of antimicrobials to increase poultry performance, efficiency, and reduce health issues. In spite of the benefits provided by antibiotic use, concerns surrounding the dissemination of antimicrobial resistance mean that we must find ATAs that we can use in poultry production. The use of probiotics, prebiotics, postbiotics, bacteriophage, and organic acids has the potential to alter the gastrointestinal microbial population and to further impact the carriage and shedding of *E. coli* strains, including pathogenic groups of *E. coli*. These alternatives offer promising avenues for maintaining poultry health while reducing reliance on antibiotics and mitigating the risks associated with antimicrobial resistance.

8 Where to look for further information

For further information on food safety in poultry production, the impact of Escherichia coli and other foodborne pathogens, and strategies for mitigation, you can vist the following resources:

- Centers for Disease Control and Prevention (CDC): The CDC provides guidelines on poultry handling, cooking recommendations, and foodborne illness prevention. Their resources cover key bacterial pathogens such as Salmonella, Campylobacter, and E. coli, along with strategies to reduce contamination risks in poultry production. Available at: https://www.cdc.gov/food-safety/foods/chicken.html#:~:text=When%20cooked%2C%20chicken%20can%20be,illness%2C%20also%20called%20food%20poisoning.
- Springer Handbook of Foodborne Diseases: This reference details the major bacterial, viral, and parasitic pathogens associated with foodborne illnesses, including their epidemiology, detection methods, and control measures. It is a useful resource for understanding the broader context of food safety and pathogen mitigation strategies. Available at: https://doi.org/10.1007/978-3-030-05011-5.
- U.S. Department of Agriculture (USDA) Food Safety and Inspection Service (FSIS): The USDA FSIS regulates meat, poultry, and egg products in the U.S. Their website provides extensive resources on poultry processing regulations, pathogen control measures, and national food safety policies. Available at: https://www.fsis.usda.gov/
- World Health Organization (WHO) – Food Safety: WHO provides international guidelines on foodborne disease prevention, including risk assessment strategies for poultry production. Their reports and publications offer a global perspective on food safety standards and best practices. Available at: https://www.who.int/health-topics/food-safety

9 References

Abu-Ruwaida, A. S., et al. 1994. Microbiological quality of broilers during processing in a modern commercial slaughterhouse in Kuwait. *Journal of Food Protection*, 57, 887–892.

Alharbi, M. G., et al. 2022. The "Big Six": hidden emerging foodborne bacterial pathogens. *Tropical Medicine and Infectious Disease*, 7, 356.

Ali, A., et al. 2020. Novel avian pathogenic escherichia coli genes responsible for adhesion to chicken and human cell lines. *Applied and Environmental Microbiology*, 86, 1–13.

Allen, V. M., et al. 2003. Microbial cross-contamination by airborne dispersion and contagion during defeathering of poultry. *British Poultry Science*, 44, 567–576.

Alonso, M. Z., et al. 2012. Enteropathogenic (EPEC) and Shigatoxigenic *Escherichia coli* (STEC) in broiler chickens and derived products at different retail stores. *Food Control*, 23, 351–355.

Applegate, T. J., et al. 2010. Probiotics and phytogenics for poultry: myth or reality?. *Journal of Applied Poultry Research*, 19, 194–210.

Askelson, T. E., et al. 2014. Evaluation of phytate-degrading *Lactobacillus* culture administration to broiler chickens. *Applied and Environmental Microbiology*, 80, 943–950.

Atterbury, R. J., et al. 2011. Effects of orally administered *Bdellovibrio bacteriovorus* on the well-being and *Salmonella* colonization of young chicks. *Applied and Environmental Microbiology*, 77, 5794–5803.

Bach, S. J., et al. 2002. Evaluation of bacteriophage DC22 for control of *Escherichia coli* O157:H7. *Journal of Animal Science*, 80(Suppl. 1), 263 (Abstract).

Bach, S. J., et al. 2003. Effect of bacteriophage DC22 on *Escherichia coli* O157:H7 in an artificial rumen system (RUSITEC) and inoculated sheep. *Animal Research*, 52, 89–101.

Barbieri, N. L., et al. 2021. mcr-1 Identified in Fecal Escherichia coli and Avian Pathogenic E. coli (APEC) From Brazil. *Frontiers in Microbiology*, 12, 659613.

Barnes, H., et al. 2003. *Colibacillosis. Diseases of Poultry*. Saif, Y., ed. Ames, IA: Iowa State University Press.

Barnes, H. J. & Calnek, B. 1997. *Diseases of Poultry*. London: Mosby-Wolfe.

Bensink, J. C. & Botham, F. P. 1983. Antibiotic resistant coliform bacilli, isolated from freshly slaughtered poultry and from chilled poultry at retail outlets. *Australian Veterinary Journal*, 60, 80–83.

Bergeron, C. R., et al. 2012. Chicken as reservoir for extraintestinal pathogenic Escherichia coli in humans, Canada. *Emerging Infectious Diseases*, 18, 415.

Berry, E. D. & Cutter, C. N. 2000. Effects of acid adaptation of *Escherichia coli* O157:H7 on efficacy of acetic acid spray washes to decontaminate beef carcass tissue. *Applied and Environmental Microbiology*, 66, 1493–1498.

Blaak, H., et al. 2015. Distribution, numbers, and diversity of ESBL-producing E. coli in the poultry farm environment. *PLoS One*, 10, e0135402.

Bucher, O., et al. 2012. Evaluating interventions against *Salmonella* in broiler chickens: applying synthesis research in support of quantitative exposure assessment. *Epidemiology and Infection*, 140, 925–945.

Byrd, J. A., et al. 2002. Fluorescent marker for the detection of crop and upper gastrointestinal leakage in poultry processing plants. *Poultry Science*, 81, 70–74.

Callaway, T. R., et al. 2003a. Preslaughter intervention strategies to reduce food-borne pathogens in food animals12. *Journal of Animal Science*, 81, E17–E23.

Callaway, T. R., et al. 2003b. Isolation of naturally-occuring bacteriophage from sheep that reduce populations of *E. coli* O157:H7 *in vitro* and *in vivo*. Proc. 5th Int. Symp. on Shiga Toxin-Producing *Escherichia coli* Infections, Edinburgh, UK. 25.

Callaway, T., et al. 2006. Fecal prevalence of escherichia coli O157, salmonella, listeria, and bacteriophage infecting E-coli O157: H7 in feedlot cattle in the southern plains region of the United States. *Foodborne Pathogens and Disease*, 3, 234–244.

Callaway, T., et al. 2012. Current status of practical applications: probiotics in dairy cattle. *In*: Callaway, T. R. & Ricke, S. C. (eds.) *Direct-Fed Microbials and Prebiotics for Animals: Science and Mechanisms of Action*, 121–135. New York, NY: Springer.

Callaway, T., et al. 2013. Novel methods for pathogen control in livestock pre-harvest: an update. *Advances in Microbial Food Safety*, 259, 275–304.

Callaway, T., et al. 2014. Isolation of escherichia coli O157:H7 and salmonella from migratory brown-headed cowbirds (Molothrus ater), common grackles (Quiscalus quiscula), and cattle egrets (Bubulcus ibis). *Foodborne Pathogens and Disease*, 11, 791–794.

Callaway, T., et al. 2017. Use of direct-fed microbials in layer hen production-performance response and salmonella control. *In*: Ricke, S. & Gast, R. K. (eds.) *Producing Safe Eggs: Microbial Ecology of Salmonella*. London, UK: Elsevier Science Publishing.

Callaway, T. R., et al. 2021. Alternatives to antibiotics: a symposium on the challenges and solutions for nimal health and production. *Antibiotics (Basel)*, 10, 471.

Callaway, T. R. & Oliver, S. P. 2009. On-farm strategies to reduce foodborne pathogen contamination. *Foodborne Pathogens and Disease*, 6, 753.

Carlson, B. A., et al. 2008. Comparison of antimicrobial efficacy of multiple beef hide decontamination strategies to reduce levels of Escherichia coli O157:H7 and Salmonella. *Journal of Food Protection*, 71, 2223–2237.

Cason, J. & Hinton Jr, A. 2006. Coliforms, Escherichia coli, Campylobacter, and Salmonella in a counterflow poultry scalder with a dip tank. *International Journal of Poultry Science*, 5, 846–849.

Cernicchiaro, N., et al. 2020. Associations between season, processing plant, and hide cleanliness scores with prevalence and concentration of Major Shiga Toxin-producing escherichia coli on beef cattle hides. *Foodborne Pathogens and Disease*, 17, 611–619.

Chapman, P., et al. 1997. A 1-year study of Escherichia coli O157 in cattle, sheep, pigs and poultry. *Epidemiology & Infection*, 119, 245–250.

Chapman, P. A., et al. 1993. Cattle as a possible source of verocytotoxin-producing Escherichia coli O157 infections in man. *Epidemiology and Infection*, 111, 439–447.

Chaslus-Dancla, E. & Lafont, J. P. 1985. IncH plasmids in Escherichia coli strains isolated from broiler chicken carcasses. *Applied and Environmental Microbiology*, 49, 1016–1018.

Clavijo, V. & Flórez, M. J. V. 2018. The gastrointestinal microbiome and its association with the control of pathogens in broiler chicken production: a review. *Poultry Science*, 97, 1006–1021.

CortéS, P., et al. 2010. Isolation and characterization of potentially pathogenic antimicrobial-resistant escherichia coli strains from chicken and pig farms in Spain. *Applied and Environmental Microbiology*, 76, 2799–2805.

Cowan, M. M. 1999. Plant products as antimicrobial agents. *Clinical Microbiology Reviews*, 12, 564–582.

D'herelle, F. 1919. Sur le role du microbe bacteriophage dans la typhose aviare. *Comptes rendus de l'Académie des Sciences*, 169, 932–934.

Dame-Korevaar, A., et al. 2020. Competitive exclusion prevents colonization and compartmentalization reduces transmission of ESBL-producing escherichia coli in broilers. *Frontiers in Microbiology*, 11. https://doi.org/10.3389/fmicb.2020.566619

De Brito, B. G., et al. 2003. Virulence factors and clonal relationships among Escherichia coli strains isolated from broiler chickens with cellulitis. *Infection and Immunity*, 71, 4175–4177.

Delahoy, M. J., et al. 2023. Preliminary incidence and trends of infections caused by pathogens transmitted commonly through food – foodborne diseases active surveillance network, 10 U.S. Sites, 2022. *Morbidity and Mortality Weekly Report*, 72, 701–706.

Dho-Moulin, M. & Fairbrother, J. M. 1999. Avian pathogenic *Escherichia coli* (APEC). *Veterinary Research*, 30, 299–316.

Diarra, M. S. & Malouin, F. 2014. Antibiotics in Canadian poultry productions and anticipated alternatives. *Frontiers in Microbiology*, 5, 87153.

Diarra, M. S., et al. 2007. Impact of feed supplementation with antimicrobial agents on growth performance of broiler chickens, Clostridium perfringens and enterococcus counts, and antibiotic resistance phenotypes and distribution of antimicrobial resistance determinants in Escherichia coli isolates. *Applied and Environmental Microbiology*, 73, 6566–6576.

Dipineto, L., et al. 2006. Presence of Shiga toxin-producing Escherichia coli O157: H7 in living layer hens. *Letters in Applied Microbiology*, 43, 293–295.

Dishan, A., et al. 2023. Biofilm formation, antibiotic resistance and genotyping of Shiga toxin-producing *Escherichia coli* isolated from retail chicken meats. *British Poultry Science*, 64, 63–73.

Dittoe, D. K., et al. 2018. Organic acids and potential for modifying the avian gastrointestinal tract and reducing pathogens and disease. *Frontiers in Veterinary Science*, 5, 216.

Duffy, L. L., et al. 2000. Acid resistance of enterohaemorrhagic and generic Escherichia coli associated with foodborne disease and meat. *International Journal of Food Microbiology*, 60, 83–89.

Dutta, P., et al. 2013. Isolation, histopathology and antibiogram of *Escherichia coli* from pigeons (*Columba livia*). *Veterinary World*, 6, 91–94.

Dziva, F. & Stevens, M. P. 2008. Colibacillosis in poultry: unravelling the molecular basis of virulence of avian pathogenic Escherichia coli in their natural hosts. *Avian Pathology*, 37, 355–366.

Efsa Biohaz Panel, K., et al. 2020. Pathogenicity assessment of Shiga toxin-producing Escherichia coli (STEC) and the public health risk posed by contamination of food with Stec. *Efsa Journal*, 18, e05967.

Ejidokun, O. O., et al. 2006. Human Vero cytotoxigenic *Escherichia coli* (VTEC) O157 infection linked to birds. *Epidemiology and Infection*, 134, 421–423.

El Jeni, R., et al. 2024. Invited review: "probiotic" approaches to improving dairy production: reassessing "magic foo-foo dust". *Journal of Dairy Science*, 107, 1832–1856.

El-Sawah, A. A., et al. 2018. Pathogenicity of Escherichia coli O157 in commercial broiler chickens. *Beni-Suef University Journal of Basic and Applied Sciences*, 7, 620–625.

Engberg, R. M., et al. 2000. Effect of zinc bacitracin and salinomycin on intestinal microflora and performance of broilers. *Poultry Science*, 79, 1311–1319.

Ewers, C., et al. 2009. Intestine and environment of the chicken as reservoirs for extraintestinal pathogenic Escherichia coli strains with zoonotic potential. *Applied and Environmental Microbiology*, 75, 184–192.

Falgenhauer, L., et al. 2019. Detection and characterization of ESBL-producing escherichia coli from humans and poultry in Ghana. *Frontiers in Microbiology*, 9. https://doi.org/10.3389/fmicb.2018.03358

Fancher Courtney, A., et al. 2021. Prevalence and molecular characteristics of avian pathogenic escherichia coli in "no antibiotics ever" broiler farms. *Microbiology Spectrum*, 9, e00834-21.

Ferens, W. A. & Hovde, C. J. 2011. *Escherichia coli* O157:H7: animal reservoir and sources of human infection. *Foodborne Pathogens and Disease*, 8, 465–487.

Forgetta, V., et al. 2012. Pathogenic and multidrug-resistant Escherichia fergusonii from broiler chicken. *Poultry Science*, 91, 512–525.

Fries, R. 2002. Reducing *Salmonella* transfer during industrial poultry meat production. *World's Poultry Science Journal*, 58, 527–540.

Fsis 1999. USDA-Food Safety Inspection Service. Policy on non-intact raw beef products contaminated with *E. coli* O157:H7. *In:* Agriculture, U. S. D. O. (ed.). 64, 803–2805.

Fuller, R. 1991. Probiotics in human medicine. *Gut*, 32, 439–442.

Gadde, U., et al. 2017. Alternatives to antibiotics for maximizing growth performance and feed efficiency in poultry: a review. *Animal Health Research Reviews*, 18, 26–45.

Gaukler, S. M., et al. 2009. *Escherichia coli*, *Salmonella*, and *Mycobacterium avium* subsp. *paratuberculosis* in Wild European starlings at a Kansas cattle feedlot. *Avian Diseases*, 53, 544–551.

Ghosh, K. K., et al. 2019. Extended-spectrum β-lactamase and AmpC β-lactamase-producing escherichia coli isolates from chickens raised in small flocks in Ontario, Canada. *Microbial Drug Resistance*, 25, 1250–1256.

Golden, C. E., et al. 2021. Mapping foodborne pathogen contamination throughout the conventional and alternative poultry supply chains. *Poultry Science*, 100, 101157.

Grilli, E., et al. 2015. Organic acid blend with pure botanical product treatment reduces escherichia coli and salmonella populations in pure culture and in in vitro mixed ruminal microorganism fermentations. *Foodborne Pathogens and Disease*, 12, 56–61.

Gross, W. 1994. Diseases due to Escherichia coli in poultry. *In:* Gyles C. L. (ed.) Escherichia Coli in Domestic Animals and Humans. Wallingford, UK: CAB International.

Großmann, K., et al. 2005. Racing, ornamental and city pigeons carry Shiga toxin producing Escherichia coli (STEC) with different Shiga toxin subtypes, urging further analysis of their epidemiological role in the spread of STEC. *Berliner und Munchener Tierarztliche Wochenschrift*, 118, 456–463.

Guabiraba, R. & Schouler, C. 2015. Avian colibacillosis: still many black holes. *FEMS Microbiology Letters*, 362, fnv118.

Guo, C., et al. 2011. Application of Bayesian techniques to model the burden of human salmonellosis attributable to U.S. food commodities at the point of processing: adaptation of a Danish model. *Foodborne Pathogens and Disease*, 8, 509–516.

Guo, J. R., et al. 2017. Effects of long-term *Bacillus subtilis* CGMCC 1.921 supplementation on performance, egg quality, and fecal and cecal microbiota of laying hens. *Poultry Science*, 96, 1280–1289.

Guo, Y. N., et al. 2024. Three different routes of EHEC O157:H7 infection were used to establish EHEC broiler model. *Poultry Science*, 103, 103561.

Gyles, C. L. 2007. Shiga toxin-producing *Escherichia coli*: an overview. *Journal of Animal Science*, 85, E45–E62.

He, T., et al. 2019. Effects of probiotics as antibiotics substitutes on growth performance, serum biochemical parameters, intestinal morphology, and barrier function of broilers. *Animals*, 9, 985–995.

Heuvelink, A. E., et al. 1999. Isolation and characterization of verocytotoxin-producing Escherichia coli O157 from slaughter pigs and poultry. *International Journal of Food Microbiology*, 52, 67–75.

Hossain, M. E., et al. 2012. Growth performance, meat yield, oxidative stability, and Fatty Acid composition of meat from broilers fed diets supplemented with a medicinal plant and probiotics. *Asian-Australasian Journal of Animal Sciences*, 25, 1159–1168.

Hubálek, Z. 2004. An annotated checklist of pathogenic microorganisms associated with migratory birds. *Journal of Wildlife Diseases*, 40, 639–659.

Huff, G. R., et al. 2010. Bacterial clearance, heterophil function, and hematological parameters of transport-stressed turkey poults supplemented with dietary yeast extract. *Poultry Science*, 89, 447–456.

Huff, G. R., et al. 2015. Efficacy of a novel prebiotic and a commercial probiotic in reducing mortality and production losses due to cold stress and Escherichia coli challenge of broiler chicks. *Poultry Science*, 94, 918–926.

Huff, W. E., et al. 2002. Prevention of *Escherichia coli* respiratory infection in broiler chickens with bacteriophage (SPR02). *Poultry Science*, 81, 437–441.

Hume, M. E. 2011. Historic perspective: prebiotics, probiotics, and other alternatives to antibiotics. *Poultry Science*, 90, 2663–2669.

Hussain, A., et al. 2017. Risk of transmission of antimicrobial resistant escherichia coli from commercial broiler and free-range retail chicken in India. *Frontiers in Microbiology*, 8, 2120.

Incili, G. K. & Çalicioğlu, M. 2018. Change in scalding fluids by time in poultry slaughterhouse and its effect on microbiological quality of carcasses. *Journal of Food Safety*, 38, e12485.

Jakobsen, L., et al. 2010. Broiler chickens, broiler chicken meat, pigs and pork as sources of ExPEC related virulence genes and resistance in Escherichia coli isolates from community-dwelling humans and Uti patients. *International Journal of Food Microbiology*, 142, 264–272.

Janardhana, V., et al. 2009. Prebiotics modulate immune responses in the gut-associated lymphoid tissue of chickens. *Journal of Nutrition*, 139, 1404–1409.

Jang, J., et al. 2017. Environmental Escherichia coli: ecology and public health implications–a review. *Journal of Applied Microbiology*, 123, 570–581.

Johnson, J. R., et al. 2005. Antimicrobial-resistant and extraintestinal pathogenic Escherichia coli in retail foods. *The Journal of Infectious Diseases*, 191, 1040–1049.

Johnson, R. P., et al. 2008a. Bacteriophages for prophylaxis and therapy in cattle, poultry and pigs. *Animal Health Research Reviews*, 9, 201–215.

Johnson, T. J., et al. 2008b. Identification of minimal predictors of avian pathogenic Escherichia coli virulence for use as a rapid diagnostic tool. *Journal of Clinical Microbiology*, 46, 3987–3996.

Jorgensen, F., et al. 2002. Prevalence and numbers of *Salmonella* and *Campylobacter* spp. on raw, whole chickens in relation to sampling methods. *International Journal of Food Microbiology*, 76, 151–164.

Jung, S. J., et al. 2008. Effects of galacto-oligosaccharides and a *Bifidobacteria lactis*-based probiotic strain on the growth performance and fecal microflora of broiler chickens. *Poultry Science*, 87, 1694–1699.

Kabir, S. M. L. 2010. Avian colibacillosis and salmonellosis: a closer look at epidemiology, pathogenesis, diagnosis, control and public health concerns. *International Journal of Environmental Research and Public Health* [Online], 7.

Kalin, R., et al. 2012. Isolation and molecular characterization of Escherichia coli O157 from broiler and human samples. *Foodborne Pathogens and Disease*, 9, 313–318.

Kaper, J. B., et al. 2004. Pathogenic *Escherichia coli*. *Nature Reviews Microbiology*, 2, 123–140.

Karaiskos, I., et al. 2019. The "old" and the "new" antibiotics for MDR gram-negative pathogens: for whom, when, and how. *Frontiers in Public Health*, 7, 151.

Karmali, M. A., et al. 2010. Verocytotoxin-producing Escherichia coli (VTEC). *Veterinary Microbiology*, 140, 360–370.

Kathayat, D., et al. 2021. Avian Pathogenic Escherichia Coli (APEC): an overview of virulence and pathogenesis factors, zoonotic potential, and control strategies. *Pathogens* [Online], 10. https://doi.org/10.3390/pathogens10040467

Kauffman, M. D. & Lejeune, J. 2011. European starlings (*Sturnus vulgaris*) challenged with *Escherichia coli* O157 can carry and transmit the human pathogen to cattle. *Letters in Applied Microbiology*, 53, 596–601.

Kay, S. 2003. *E. coli* O157:H7: the costs during the past 10 years. *Meat and Poultry News*.

Kim, G. B., et al. 2011. Effect of dietary prebiotic supplementation on the performance, intestinal microflora, and immune response of broilers. *Poultry Science*, 90, 75–82.

Kim, J. E., et al. 2015. Dietary capsicum and curcuma longa oleoresins increase intestinal microbiome and necrotic enteritis in three commercial broiler breeds. *Research in Veterinary Science*, 102, 150–158.

Köhler, C.-D. & Dobrindt, U. 2011. What defines extraintestinal pathogenic Escherichia coli? *International Journal of Medical Microbiology*, 301, 642–647.

Koyun, O. Y. & Callaway, T. 2019. Controlling pathogens in the poultry gut. *In*: Ricke, S. C. (ed.). *Improving Gut Health in Poultry*. Cambridge: Burleigh Dodd Publishing.

Koyun, O. Y., et al. 2022. Innovative treatments enhancing the functionality of gut microbiota to improve quality and microbiological safety of foods of animal origin. *Annual Review of Food Science and Technology*, 13, 433–461.

Kurt, T., et al. 2019. Strategic priorities for research on antibiotic alternatives in animal agriculture–results from an expert workshop. *Frontiers in Veterinary Science*, 6, 429.

La Ragione, R. M., et al. 2005. Variable and strain dependent colonisation of chickens by *Escherichia coli* O157. *Veterinary Microbiology*, 107, 103–113.

Lederberg, J. 1996. Smaller Fleas...ad infinitum: therapeutic bacteriophage redux. *Proceedings of the National Academy of Sciences of the United States of America*, 93, 3167–3168.

Leeson, S. & Summers, J. D. 2010. *Broiler Breeder Production*. Nottingham: Nottingham University Press.

Lhermie, G., et al. 2016. Addressing antimicrobial resistance: an overview of priority actions to prevent suboptimal antimicrobial use in food-animal production. *Frontiers in Microbiology*, 7, 2114.

Lillard, H., et al. 1984. Effect of reduced processing on recovery of foodborne pathogens from hot-boned broiler meat and skin. *Journal of Food Protection*, 47, 209–213.

Lillehoj, H. S., et al. 2018. Phytochemicals as antibiotic alternatives to promote growth and enhance host health. *Veterinary Research*, 49, 76.

Linton, A. H., et al. 1977. Antibiotic Resistance among *Escherichia coli* O-serotypes from the gut and carcasses of commercially slaughtered broiler chickens: a potential public health hazard. *Journal of Applied Bacteriology*, 42, 365–378.

Liu, C. M., et al. 2018. Escherichia coli ST131-H22 as a foodborne uropathogen. mBio, 9. e00470-18.

López-Cuevas, O., et al. 2011. Characterization of bacteriophages with a lytic effect on various *Salmonella* serotypes and *Escherichia coli* O157:H7. *Canadian Journal of Microbiology*, 57, 1042–1051.

Lourenco, J. M., et al. 2019. The effects of feeding a soybean-based or a soy-free diet on the gut microbiome of pasture-raised chickens throughout their lifecycle. *Frontiers in Sustainable Food Systems*, 3, 25–31.

Lv, H., et al. 2020. Identification of novel plasmids containing the tigecycline resistance gene tet(X4) in escherichia coli isolated from retail chicken meat. *Foodborne Pathogens and Disease*, 17, 792–794.

Ma, Q. G., et al. 2012. Protective effect of Bacillus subtilis ANSB060 on egg quality, biochemical and histopathological changes in layers exposed to aflatoxin B1. *Poultry Science*, 91, 2852–2857.

Makino, S., et al. 2000. Detection and characterization of Shiga toxin-producing *Escherichia coli* from seagulls. *Epidemiology and Infection*, 125, 55–61.

Manges, A. R. & Johnson, J. R. 2012. Food-borne origins of escherichia coli causing extraintestinal infections. *Clinical Infectious Diseases*, 55, 712–719.

Marietto-Gonçalves, G. A., et al. 2011. Presence of a human diarrheagenic *Escherichia coli* clone in captivity kept psittacidaes. *The Open Microbiology Journal*, 5, 72–75.

Markland, S. M., et al. 2015. Old friends in new places: exploring the role of extraintestinal e. coli in intestinal disease and foodborne illness. *Zoonoses and Public Health*, 62, 491–496.

Martin, S. A., et al. 1998. Microbiological survey of georgia poultry litter. *Journal of Applied Poultry Research*, 7, 90–98.

Mcevoy, J. D. 2016. Emerging food safety issues: an EU perspective. *Drug Testing and Analysis*, 8, 511–520.

Mcwilliam Leitch, E. C. & Stewart, C. S. 2002. Susceptibility of *Escherichia coli* O157 and non-O157 isolates to lactate. *Letters in Applied Microbiology*, 35, 176–180.

Mead, G., et al. 1995. Effect of changes in processing to improve hygiene control on contamination of poultry carcasses with Campylobacter. *Epidemiology & Infection*, 115, 495–500.

Mellata, M. 2013. Human and avian extraintestinal pathogenic Escherichia coli: infections, zoonotic risks, and antibiotic resistance trends. *Foodborne Pathogens and Disease*, 10, 916–932.

Mellefont, L. A., et al. 2015. Combined effect of chilling and desiccation on survival of Escherichia coli suggests a transient loss of culturability. *International Journal of Food Microbiology*, 208, 1–10.

Mellor, G. E., et al. 2016. National survey of Shiga toxin-producing *Escherichia coli* serotypes O26, O45, O103, O111, O121, O145, and O157 in Australian beef cattle feces. *Journal of Food Protection*, 79, 1868–1874.

Menconi, A., et al. 2014. Evaluation of a commercially available organic acid product on body weight loss, carcass yield, and meat quality during preslaughter feed withdrawal in broiler chickens: a poultry welfare and economic perspective. *Poultry Science*, 93, 448–455.

Mitchell, N. M., et al. 2015. Zoonotic potential of Escherichia coli isolates from retail chicken meat products and eggs. *Applied and Environmental Microbiology*, 81, 1177–1187.

Mollenkopf, D. F., et al. 2018. Salmonella spp. and extended-spectrum cephalosporin-resistant escherichia coli frequently contaminate broiler chicken transport cages of an organic production company. *Foodborne Pathogens and Disease*, 15, 583–588.

Momtaz, H. & Jamshidi, A. 2013. Shiga toxin-producing *Escherichia coli* isolated from chicken meat in Iran: Serogroups, virulence factors, and antimicrobial resistance properties. *Poultry Science*, 92, 1305–1313.

Mortada, M., et al. 2020. In vivo and in vitro assessment of commercial probiotic and organic acid feed additives in broilers challenged with Campylobacter coli. *Journal of Applied Poultry Research*, in press.

Motawe, H. F. A., et al. 2014. Reducing the toxicity of aflatoxin in broiler chickens' diet by using probiotic and yeast. *International Journal of Poultry Science*, 13, 397–407.

Mukhopadhyay, S. & Gorris, L. G. M. 2014. Hurdle technology*. *In:* Batt, C. A. & Tortorello, M. L. (eds.) *Encyclopedia of Food Microbiology* (2nd ed.). Oxford: Academic Press.

Mulder, R. W. A. W., et al. 1978. Cross-contamination during the scalding and plucking of broilers. *British Poultry Science*, 19, 61–70.

Müller, A., et al. 2024. Molecular characterization of Escherichia coli isolates recovered from broilers with cellulitis. *Poultry Science*, 103, 103704.

Munoz, L. R., et al. 2021. Evaluation of commercially manufactured animal feeds to determine presence of Salmonella, Escherichia coli, and Clostridium perfringens. *Journal of Applied Poultry Research*, 30, 100142.

Murate, L. S., et al. 2015. Efficacy of prebiotics, probiotics, and synbiotics on laying hens and broilers challenged with Salmonella enteritidis. *Journal of Poultry Science*, 52, 53–56.

Nadeau, M., et al. 2000. Surveillance of antibiotic resistance bacteria originating from poultry and swine from 1993 to 1999 in Quebec. *Veterinary Medicine Quebec*, 30, 195–199.

Nascimento, G. G. F., et al. 2000. Antibacterial activity of plant extracts and phytochemicals on antibiotic-resistant bacteria. *Brazilian Journal of Microbiology*, 31, 247–256.

Nolan, L., et al. 2020. Colibacillosis. *In:* Swayne, D., Boulianne, M., Logue, C., Mcdougald, L., Venugopal, N. & Suarez, D. (eds.) *Diseases of Poultry*. Ames, IW: Wiley and Sons.

Noll, L. W., et al. 2018. Detection and quantification of seven major serogroups of Shiga toxin–producing *Escherichia coli* on hides of cull Dairy, cull beef, and fed beef cattle at slaughter. *Journal of Food Protection*, 81, 1236–1244.

Norton, R. A., et al. 1997. A reproducible model for the induction of avian cellulitis in broiler chickens. *Avian Diseases*, 41, 422–428.

Oakley, B. B., et al. 2014. The chicken gastrointestinal microbiome. *FEMS Microbiology Letters*, 360, 100–112.

O'Brien, A. D., et al. 1992. Shiga toxin: biochemistry, genetics, mode of action, and role in pathogenesis. *Current Topics in Microbiology and Immunology*, 180, 65–94.

O'Brien, T. F. 2002. Emergence, spread, and environmental effect of antimicrobial resistance: how use of an antimicrobial anywhere can increase resistance to any antimicrobial anywhere else. *Clinical Infectious Diseases*, 34, S78–S84.

Oliveira, A., et al. 2010. In vivo efficiency evaluation of a phage cocktail in controlling severe colibacillosis in confined conditions and experimental poultry houses. *Veterinary Microbiology*, 146, 303–308.

O'Neill, J. M. 2016. *Tackling Drug-resistant Infections Globally: Final Report and Recommendations*. London: HM Government and Wellcome Trust.

Olsen, R., et al. 2012. An investigation on first-week mortality in layers. *Avian Diseases*, 56, 51–57.

Palamidi, I., et al. 2016. Probiotic form effects on growth performance, digestive function, and immune related biomarkers in broilers. *Poultry Science*, 95, 1598–1608.

Papouskova, A., et al. 2023. Research note: a mixture of Bacteroides spp. and other probiotic intestinal anaerobes reduces colonization by pathogenic E. coli strain O78:H4-ST117 in newly hatched chickens. *Poultry Science*, 102, 102529.

Patterson, J. A. & Burkholder, K. M. 2003. Application of prebiotics and probiotics in poultry production. *Poultry Science*, 82, 627–631.

Pitout, J. 2012. Extraintestinal pathogenic Escherichia coli: a combination of virulence with antibiotic resistance. *Frontiers in Microbiology*, 3, 9. https://doi.org/10.3389/fmicb.2012.00009

Pourabedin, M., et al. 2014. Effects of mannan oligosaccharide and virginiamycin on the cecal microbial community and intestinal morphology of chickens raised under suboptimal conditions. *Canadian Journal of Microbiology*, 60, 255–266.

Pourbakhsh, S. A., et al. 1997. Dynamics of Escherichia coli infection in experimentally inoculated chickens. *Avian Diseases*, 41, 221–233.

Raya, R. R., et al. 2006. Isolation and characterization of a new T-even bacteriophage, CEV1, and determination of its potential to reduce *Escherichia coli* O157:H7 levels in sheep. *Applied and Environmental Microbiology*, 72, 6405–6410.

Ricke, S. C., et al. 2022. Practical opportunities for microbiome analyses and bioinformatics in poultry processing. *Poultry Science*, 101, 101787.

Rinttilä, T. & Apajalahti, J. 2013. Intestinal microbiota and metabolites–implications for broiler chicken health and performance. *Journal of Applied Poultry Research*, 22, 647–658.

Rodjan, P., et al. 2018. Effect of organic acids or probiotics alone or in combination on growth performance, nutrient digestibility, enzyme activities, intestinal morphology and gut microflora in broiler chickens. *Journal of Animal Physiology and Animal Nutrition*, 102, e931–e940.

Rodriguez-Siek, K. E., et al. 2005. Comparison of Escherichia coli isolates implicated in human urinary tract infection and avian colibacillosis. *Microbiology*, 151, 2097–2110.

Rotz, C. A. 2004. Management to reduce nitrogen losses in animal production. *Journal of Animal Science*, 82(E-Suppl), E119–E137.

Salaheen, S., et al. 2017. Alternative growth promoters modulate broiler gut microbiome and enhance body weight gain. *Frontiers in Microbiology*, 8, 2088.

Santaniello, A., et al. 2007. Survey of Shiga toxin-producing *Escherichia coli* O157:H7 in urban pigeons (*Columba livia*) in the city of Napoli, Italy. *Italian Journal of Animal Science*, 6, 313–316.

Scallan, E., et al. 2011. Foodborne illness acquired in the United States–major pathogens. *Emerging Infectious Diseases*, 17, 7–15.

Scharff, R. L. 2012. Economic burden from health losses due to foodborne illness in the United States. *Journal of Food Protection*, 75, 123–131.

Schneider, L. G., et al. 2018. Cross-sectional study to estimate the prevalence of enterohaemorrhagic Escherichia coli on hides of market beef cows at harvest. *Zoonoses Public Health*, 65, 625–636.

Schremmer, C., et al. 1999. Enteropathogenic Escherichia coli in psittaciformes. *Avian Pathology*, 28, 349–354.

Seal, B. S., et al. 2013. Alternatives to antibiotics: a symposium on the challenges and solutions for animal production. *Animal Health Research Reviews*, 14, 78–87.

Shi, Z., et al. 2019. Applications of microbiome analyses in alternative poultry broiler production systems. *Frontiers in Veterinary Science*, 6, 157–163.

Šmídková, J. & Čížek, A. 2017. The effect of Enterococcus faecium M74 feed additive on the extended-spectrum beta-lactamases/AmpC-positive Escherichia coli faecal counts in pre-weaned dairy calves. *Acta Veterinaria Brno*, 86, 333–338.

Stanford, K., et al. 2012. Genetic diversity and antimicrobial resistance among isolates of Escherichia coli O157: H7 from feces and hides of super-shedders and low-shedding pen-mates in two commercial beef feedlots [electronic resource]. *BMC Veterinary Research*, 8, 586–586.

Stanford, K., et al. 2016. Influence of season and feedlot location on prevalence and virulence factors of seven serogroups of Escherichia coli in feces of western-Canadian slaughter cattle. *PLoS One*, 11. https://doi.org/10.1371/journal.pone.0159866

Stavric, S., et al. 1993. Intestinal colonization of young chicks with Escherichia coli O157: H7 and other verotoxin-producing serotypes. *The Journal of Applied Bacteriology*, 74, 557–563.

Stearns, R. C., et al. 1987. Deposition and phagocytosis of inhaled particles in the gas exchange region of the duck, Anas platyrhynchos. *Respiration Physiology*, 67, 23–36.

Stein, R. A. & Katz, D. E. 2017. Escherichia coli, cattle and the propagation of disease. *FEMS Microbiology Letters*, 364.https://doi.org/10.1093/femsle/fnx050

Stromberg, Z. R., et al. 2017. Evaluation of Escherichia coli isolates from healthy chickens to determine their potential risk to poultry and human health. *PLoS One*, 12, e0180599.

Swelum, A. A., et al. 2021. Ways to minimize bacterial infections, with special reference to Escherichia coli, to cope with the first-week mortality in chicks: an updated overview. *Poultry Science*, 100, 101039.

Tack, D. M. 2020. Preliminary incidence and trends of infections with pathogens transmitted commonly through food–Foodborne Diseases Active Surveillance Network, 10 Us Sites, 2016–2019. MMWR. Morbidity and mortality weekly report, 69.

Tarabees, R., et al. 2020. Effects of the probiotic candidate E. faecalis-1, the Poulvac E. coli vaccine, and their combination on growth performance, caecal microbial composition, immune response, and protection against E. coli O78 challenge in broiler chickens. *Probiotics Antimicrob Proteins*, 12, 860–872.

Taylor, D. J. 2001. Effects of antimicrobials and their alternatives. *British Poultry Science*, 42(Suppl.), S67–S68.

Trung, N. V., et al. 2016. Colonization of enteroaggregative *Escherichia coli* and Shiga toxin-producing *Escherichia coli* in chickens and humans in southern Vietnam. *BMC Microbiology*, 16, 208.

Usman, S., et al. 2022. Antibiotic resistance pattern and pathological features of avian pathogenic Escherichia coli O78:K80 in chickens. *Brazilian Journal of Biology*, 84, e257179.

Vase-Khavari, K., et al. 2019. The effect of three tropical medicinal plants and superzist probiotic on growth performance, carcass characteristics, blood constitutes, immune response, and gut microflora of broiler. *Tropical Animal Health and Production*, 51, 33–42.

Walker, W. A. & Duffy, L. C. 1998. Diet and bacterial colonization: role of probiotics and prebiotics. *The Journal of Nutritional Biochemistry*, 9, 668–675.

Wallace, J. S., et al. 1997. Isolation of vero cytotoxin-producing *Escherichia coli* O157 from wild birds. *Journal of Applied Microbiology*, 82, 399–404.

Warriss, P., et al. 2004. Defaecation and weight of the gastrointestinal tract contents after feed and water withdrawal in broilers. *British Poultry Science*, 45, 61–66.

Williams-Nguyen, J., et al. 2016. Antibiotics and antibiotic resistance in agroecosystems: state of the science. *Journal of Environmental Quality*, 45, 394–406.

Wu, Y., et al. 2019. Pretreatment with probiotic *Enterococcus faecium* Ncimb 11181 ameliorates necrotic enteritis-induced intestinal barrier injury in broiler chickens. *Scientific Reports*, 9, 10256–1061.

Xiang, L., et al. 2022. A novel Lactobacillus bulgaricus isolate can maintain the intestinal health, improve the growth performance and reduce the colonization of E. coli O157:H7 in broilers. *British Poultry Science*, 63, 621–632.

Yates, C. M., et al. 2004. High frequency transfer and horizontal spread of apramycin resistance in calf faecal *Escherichia coli*. *Journal of Antimicrobial Chemotherapy*, 54, 534–537.

Yazhini, P., et al. 2018. Dietary encapsulated probiotic effect on broiler serum biochemical parameters. *Veterinary World*, 11, 1344–1348.

Yeo, J. & Kim, K. I. 1997. Effect of feeding diets containing an antibiotic, a probiotic, or yucca extract on growth and intestinal urease activity in broiler chicks. *Poultry Science*, 76, 381–385.

Young, R. 1992. Bacteriophage lysis: mechanism and regulation. *Microbiological Reviews*, 56, 430–481.

Zarei, O., et al. 2021. The prevalence of shiga toxin-producing E*scherichia coli* and enteropathogenic *Escherichia coli* isolated from raw chicken meat samples. *International Journal of Microbiology*, 2021, 3333240.

Zhang, L., et al. 2016. Effects of dietary supplementation of probiotic, Clostridium butyricum, on growth performance, immune response, intestinal barrier function, and digestive enzyme activity in broiler chickens challenged with Escherichia coli K88. *Journal of Animal Science and Biotechnology*, 7, 3.

Zhao, L., et al. 2009. Comparison of virulence factors and expression of specific genes between uropathogenic Escherichia coli and avian pathogenic E. coli in a murine urinary tract infection model and a chicken challenge model. *Microbiology*, 155, 1634–1644.

Zucker, B. A., et al. 2000. Airborne gram-negative bacterial flora in animal houses. *Journal of Veterinary Medicine Series B-Infectious Diseases and Veterinary Public Health*, 47, 37–46.

Part 2

Pre-harvest safety management on the farm

Chapter 4

Advances in rapid detection of zoonotic pathogens in poultry production and processing environments

Jeanne Malet-Villemagne, Francesco Rizzotto, Vincent Léguillier and Jasmina Vidic, Paris-Saclay University/AgroParisTech, France; Vanessa Escobar and Yanxia Hou, Grenoble Alpes University/CNRS/CEA/Grenoble INP/IRIG/SyMMES, France; and Nicole Jaffrezic-Renault, University of Lyon, France

1 Introduction

In recent years, poultry consumption has seen an exponential increase worldwide, especially in developed western countries, most notably the United States, with an average consumption of 50.90 kg per person in 2022. Likewise, an increase has also been observed in the European Union (OECD – FAO *Agricultural Outlook* (Edition 2021), 2021). According to the AVEC (Association of Poultry Processors and Poultry Trade in the EU countries), poultry meat is the second most important meat species in the European Union today with an annual production of approximately 11 million tons and an average consumption of about 23 kilos per capita per year. Importantly, recent Food

http://dx.doi.org/10.19103/AS.2024.0146.06

and Agriculture Organization (FAO) figures show that the poultry sector, which accounts for 35% of global livestock production, contributes only 7% to total livestock emissions, making poultry one of the most sustainable meats globally. Globally, it has been estimated that poultry meat consumption has surpassed pork consumption. Two-thirds of future growth are predicted to occur in Asia. Southeast Asian countries (e.g. India, Indonesia, Malaysia and Thailand) will contribute to production gains with possibilities of oversupply that might impact prices and profitability in these countries. Considerable dynamic growth is expected in Argentina in becoming the world's fifth largest poultry producer. Poultry production therefore represents clear environmental advantages but also is a more healthy and economical option for consumers. The European poultry industry, e.g. employs more than 300 000 staff across Europe and has an annual turnover of 30 billion Euros and is thus of significant economic importance.

Projections from FAO suggest that global egg consumption will rise from 6.5 kg per person per year in 1997/99 to 8.9 kg per person per year in 2030 in developing countries. More than 67% of the demand for eggs in 2017 came from Asia. In China, where egg consumption is more than double the average for developing countries, a rise in consumption from 15 kg per person per year in 1997/99 to 20 kg per person per year in 2030 is likely, based on FAO data. Additional data from FAO suggest that the world's production of eggs reached 70.4 million tons in 2015 and will reach 89.9 million tons in 2030, at a growth rate of 1.6% per year from 2015 through 2030. Poultry will therefore be one of the most important livestock species for future global meat and animal-derived products.

Much poultry production is highly integrated and industry driven. This structure can make it easier to implement new integrated technology to improve efficiency. In Europe, e.g. there are more than 70 000 broiler farms and 10 000 laying farms. In many countries, raising productivity will involve a shift from extensive to more intensive food production animal management, with higher stocking densities, more targeted animal feed and growing conditions. However, higher stocking densities can increase the risk of the spread of animal diseases. The animal health sector can therefore play an important role in this trend towards greater productivity.

There is therefore an increasing demand to ensure the microbiological safety of poultry products, which are subject to contamination by bacterial pathogens and responsible for gastrointestinal infections among consumers. These pathogens, naturally present in the animals and the slaughtering and transformation processes, can easily contaminate meat or egg products (Rouger et al., 2017). Circulating avian influenza viruses pose serious threats to the poultry industry and humans. In 2016, e.g. 89 outbreaks of the highly pathogenic H5N8 virus were reported in France, requiring 800 000 birds to be killed in January

2017. South Korea experienced an H5N6/H5N8 outbreak which paralyzed its poultry industry with 20 million birds killed in January 2017. Such pandemic highly pathogenic avian influenza viruses (HPAIV) are difficult to control because no tool for in-field diagnostics is available. Samples collected by veterinarians at farms are initially transported to authorized laboratories where the diagnosis of an HPAIV strain takes at least a whole day. Influenza viruses are not treated in this chapter, only bacterial pathogens. In contrast to influenza viruses whose transmission requires close contact with birds, bacterial pathogens cause illnesses mainly through the consumption of contaminated food (Marin et al., 2021).

Systemic reviews of several studies conducted in Europe to assess the prevalence of pathogenic bacteria in poultry meat and eggs indicate that the main contaminating microorganisms are:

- *Staphylococcus aureus*;
- *Campylobacter* spp.;
- *Listeria monocytogenes*; and
- *Salmonella* spp.

All four enteric bacterial pathogens have the potential for zoonotic transmission from poultry to humans. They have been found in poultry and poultry products, at both the end-processing step and retail level, along with in-packed and unpacked products (Gonçalves-Tenório et al., 2018). In this chapter, we will present the main advances in the rapid detection of these more prevalent foodborne pathogens in poultry. The medical issues faced following the consumption of contaminated poultry products with the four main bacterial pathogens: i.e. *S. aureus*, *Campylobacter jejuni*, *L. monocytogenes* and *Salmonella Enteritidis* (Fig. 1), and the infections caused by the ingestion of contaminated products will be briefly outlined, together with the standard methods for their detection. Finally, we discuss future developments of detection methods.

2 Zoonotic pathogens in poultry: *Salmonella* spp., *Campylobacter* spp., *Listeria monocytogenes* and *Staphylococcus aureus*

Salmonella and *Campylobacter* are among the most frequently occurring foodborne bacteria worldwide and are the most important poultry foodborne biological hazards to human health (Cheung et al., 2021; European Centre for Disease Prevention and Control, 2022).

2.1 Salmonella *spp.*

Salmonella spp. are Gram-negative, motile, non-spore-forming facultatively anaerobic rods. *Salmonella* comprises the two species *Salmonella enterica*

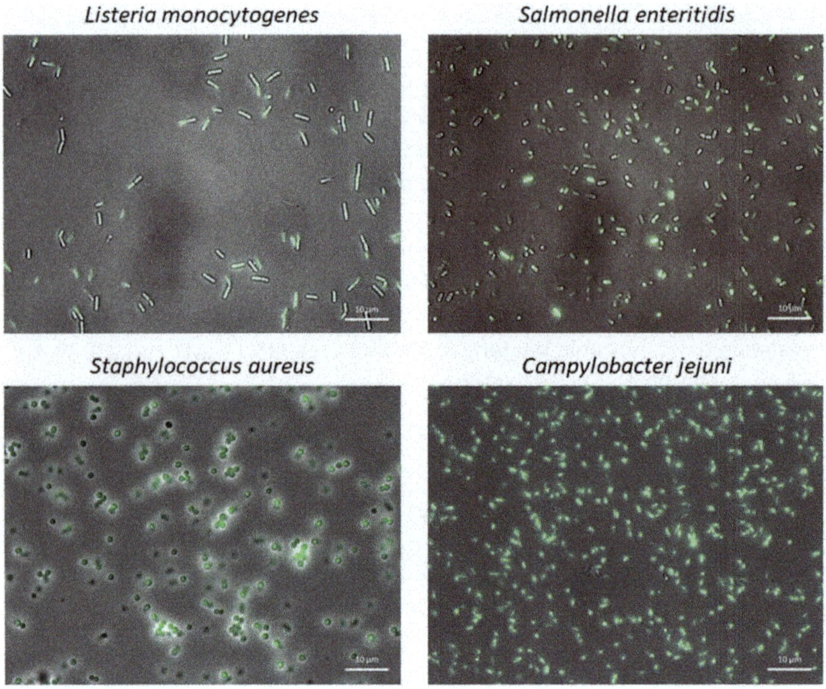

Figure 1 Microscopic view of the four main poultry pathogens: *Listeria monocytogenes*, *Salmonella Enteritidis*, *Staphylococcus aureus* and *Campylobacter jejuni*. Bacteria were observed with an AxioObserver.Z1 Zeiss inverted fluorescence microscope with a Zeiss AxioCam MRm digital camera and with Zeiss fluorescence filters. LIVE/DEAD BacLight Bacterial Viability Kit was used to color alive and dead bacteria. Images were processed using the ZEN software package (Zeiss). Scale = 10 μm.

and *Salmonella bongori*. In the agri-food industry, *S. enterica* is the primary species and can cause different clinical symptoms depending on the serotype, gastroenteritis, bacteremia and enteric fever or can be carried asymptomatically (Falkow, 2004). Among the most common causal agents of outbreaks are the following *Salmonella* serovars: *S. enterica* serovar Typhimurium and *S. enterica* serovar Enteritidis. Infections with *S. enterica* occur worldwide (Fàbrega & Vila, 2023). However, non-typhoid salmonellosis is more common in industrialized countries, whereas enteric fever is more frequent in developing countries (especially Asia).

Poultry are considered one of the most important *Salmonella* reservoirs (Gast & Porter, 2020). Furthermore, increased human infections with some less frequent serovars, such as *S. enterica* serovar Infantis, are correlated with an elevated frequency of serovar Infantis from poultry (Gal-Mor et al., 2010). Affected birds are anorectic, weak, have diarrhea and may die shortly after hatching. Layer chicks who survive remain small in size and frequently exhibit localized infection of the ovary. Surviving hens may therefore transmit the infection to eggs and produce infected chicks. In humans, *Salmonella* spp. leads to salmonellosis that causes diarrhea, abdominal cramps and fever within 8–72 h after eating contaminated food. The infectious dose is different from one serotype to the other. For non-typhoidal salmonellosis, it is approximately 1000 bacilli whereas for typhoidal salmonellosis, it is approximately 10 000 by ingestion (Falkow, 2004). Interestingly, the infectious dose may also vary depending on the acidity of the host stomach.

2.2 Campylobacter *spp.*

The World Health Organization (WHO) recommended improved control and surveillance of *Campylobacter* in poultry to reduce the risk for humans since this bacterium has a significant impact on public health. *Campylobacter* spp. are Gram-negative spiral-shaped rods. They are motile, non-spore-forming and microaerobic pathogens. The presence of a polar flagella at one or both ends leads to a characteristic corkscrew-like motion (Borriello et al., 2005). *Campylobacter* spp. is responsible for a form of gastroenteritis known as campylobacteriosis which is primarily associated with the consumption of raw meat, especially poultry. Even if *Campylobacter* spp. infections are mostly asymptomatic or mild, several complications have been reported such as the Guillain-Barré syndrome, a rare neurologic disorder affecting the nerves, whose symptoms are diarrhea with abdominal cramps, vomiting and fever.

European Food Safety Authority (EFSA) in its 2022 Report confirmed that campylobacteriosis is the most reported zoonosis in the European Union and has been so since 2005 (Authority & European Centre for Disease Prevention and Control, 2022). They also report evidence of cross-contamination from broiler meat to ready-to-eat food (EFSA Panel on Biological Hazards (BIOHAZ), 2010). In fact, contaminated, undercooked poultry is responsible for 50–80% of cases of campylobacteriosis investigated. *Campylobacters* grow slowly and have an optimal growth temperature of 42°C. *C. jejuni*, along with *Campylobacter coli*, represents the major cause of gastroenteritis and acute enterocolitis in humans in developed countries (Fitzgerald & Nachamkin, 2015). In addition, the prevalence of *Campylobacter* spp. on chicken carcasses is high and cross-contamination occurs easily during food preparation. The infectious dose of this pathogen by ingestion is very low, approximately 500 organisms (Robinson, 1981). This is why the detection of this pathogen is mandatory in food products in the European Union and the detection process is defined in the ISO norm 10272-1:2017 (Table 1).

Table 1 List of the ISO standards for poultry meat pathogenic bacteria detection

Staphylococcus aureus	ISO 6888-1:2021
Campylobacter spp.	ISO 10272-1:2017
Listeria spp.	ISO 11290-1:2017
Salmonella spp.	ISO 6579-1:2017

2.3 Listeria monocytogenes

L. monocytogenes, the causative organism of foodborne listeriosis, is commonly found in poultry products, but it is rarely isolated from live poultry (Rodriguez et al., 2021). *L. monocytogenes* is responsible for listeriosis, mainly associated with ready-to-eat products. Listeriosis, whose symptoms range from headache, fever, loss of balance, and confusion, to convulsions and meningitis, is a foodborne disease characterized by a high number of hospitalized cases and a high mortality rate (Krauss et al., 2003). It is also particularly associated with pregnancy cases, with severe impacts on the fetus (Sibanda & Buys, 2022).

Listeria spp. are Gram-positive rod-shaped coccobacillus. They are motile, facultatively anaerobic and non-spore-forming bacteria. They express diverse virulence factors implicated in adhesion, internalization, survival in the host cells and dissemination. These virulence factors include some bacteriocins, phospholipases, internalins and the listeriolysin O. They are found scattered across the genome and clustered in the pathogenicity islands LIP-A, LIP-3 and LIP-4 (Quereda et al., 2021). The major transmission mode of *L. monocytogenes* is ingestion of contaminated food and the infectious dose is estimated to be 10–100 million colony-forming units (CFU) in healthy hosts but 0.1–10 million in high-risk individuals (Farber et al., 1996). Listeriosis is observed worldwide but is more frequent in industrialized countries (Krauss et al., 2005). This is why the detection of this pathogen is mandatory in Europe for food products at risk for contamination by this pathogen (as governed by the ISO norm 11290-1:2017, as mentioned in Table 1).

2.4 Staphylococcus aureus

S. aureus are Gram-positive and catalase positive cocci bacteria. They are non-motile and non-spore-forming facultative anaerobes (with the exception of *S. aureus* subsp. *anaerobius*). Many strains produce staphylococcal toxins such as the superantigen toxic shock syndrome toxin (TSST-1), some enterotoxins or exfoliative toxins. As part of the human microbiota, high amounts of *S. aureus* are found in the nose and skin bacterial communities. These bacteria are opportunistic pathogens responsible for diverse diseases in humans, from self-limiting to life-threatening illnesses (Murray et al., 2003). *S. aureus* is a leading cause of food poisoning after consumption of contaminated food containing staphylococcal enterotoxins (SEs). SEs are highly stable and can resist thermal treatments used by the food industry. The intoxication results in rapid on set of nausea, vomiting, diarrhea and abdominal cramps. The symptoms usually become resolved after 24 h. The infectious dose of *S. aureus* in humans is estimated to be at least 100 000 organisms (Schmid-Hempel & Frank, 2007).

S. aureus is present in the poultry gut microbiota and can contaminate the egg surface and its contents. The most probable route is that the egg shell becomes contaminated is when going across the vent or immediately after laying due to attachment with infected surfaces. After laying, *S. aureus* deposited on the egg surface, similar to *Salmonella*, can cross the shell and subsequently infect egg contents (Jalali et al., 2012). Considering the pathogenicity and the worldwide distribution of this pathogen, its detection in high-risk food products such as poultry meat and eggs is very important. The detection process is described in the ISO norm 6888-1:2021 (Table 1).

3 Processing challenges: safe limits and sampling

To ensure food safe levels for consumers, health agencies (notably the EFSA) have fixed maximum bacterial concentrations acceptable in food products for foodborne pathogens. These bacterial criteria were developed for each pathogen depending on its infectious dose in humans and taking into account its participation in healthy microbiota or not. For *C. jejuni*, which can be found in the microbiota, a limit of 1000 CFU/g on carcasses of broilers after chilling has been set (European Union, 2017). On the other hand, regarding *S. Enteritidis* and *L. monocytogenes*, there is a zero tolerance in 25 g for products placed on the market during shelf life. Finally, for *S. aureus*, the target value is 5×10^5 CFU/g and the critical value is 5×10^6 CFU/g in minced meat products (European Union, 2005).

During poultry processing, bacteria are exposed to numerous stresses during the farm-to-fork chain: heat (immersion into a hot water bath for scalding) and cold stress (refrigeration for chilling step), oxidative stress (all along the process) and sometimes acid stress. In response to these stressful conditions, bacteria adapt and set up various defense mechanisms such as the production of heat shock proteins like DnaK, GroEL and ClpB (Reid et al., 2008) and detoxification enzymes. All these mechanisms allow foodborne pathogens to survive in slaughterhouse environments and poultry processing (Duqué et al., 2021). Thus, detection of food-poisoning of food products is essential for food producers to verify their compliance with the microbiologic criteria.

Several sampling techniques are commonly utilized, depending on the stage of the poultry production chain being analyzed. Early on, boot sampling can be performed, wherein pathogens can be detected from feces collected from boots. However, this process can be labor-intensive and suffer from low sensitivity (Søndergaard et al., 2014). Usually, a swabbing process is employed to check various surfaces. Some studies have therefore focused on the functionalization of cotton swabs for in-line foodborne pathogen detection (Alamer et al., 2018b; Qu et al., 2016).

Conventionally, a Whole Carcass Rinse (WCR) is used as a sampling method to detect and quantify microorganisms present in processing facilities; however, this process can be labor-intensive. Alternatively, the cumulative drip sampling method can be employed at various stages of food processing, in which random drops of rinsing water are sampled at selected locations throughout the process with the purpose of obtaining a more representative sample and minimizing bird-to-bird variability (Line et al., 2013).

Another innovative strategy for pathogen surveillance throughout the processing chain is air sampling. Some examples of this technology are gelatin membrane filters for microbial capture, or the AeroCollect, consisting of air-sampling equipment and a disposable sample chamber, in which microorganisms are collected at a set flow rate and height off the ground. Captured microorganisms can then be analyzed by PCR for identification (Andersen et al., 2022; Søndergaard et al., 2014). Notably, the various stress conditions of the food processing chain may result in bacteria entering a viable but non-culturable state, rendering this sampling technique inefficient, which further highlights the need to develop sensing systems capable of detecting cells in this state (Cossettini et al., 2022).

4 Detecting bacteria: from culture-based techniques to molecular methods and biosensors

To address the dangers of food contamination, efficient methods for bacterial detection at the various stages of the production chain are indispensable. A list of the ISO standards for the detection of the primary poultry pathogenic bacteria is presented in Table 1. The standard procedure used to detect bacteria in food samples is the use of culture-based methods, which are highly sensitive when appropriate media are used and able to recognize viable from dead cells in samples. However, culture-based methods have the disadvantage of being laborious, time and resource intensive, with a high risk of contamination and the need for highly qualified operators (Zhao et al., 2014). In addition, further analysis is often required for confirmation. For instance, the recommended procedure for the detection of *Campylobacter* requires 4 days to provide an answer on its presence or absence, while up to 7 more days are required for the identification of a particular *Campylobacter* strain.

To overcome the drawbacks of culturing, molecular methods, such as polymerase chain reaction (PCR), real-time PCR (RT-PCR) and immuno-assays, have been developed. Molecular methods are characterized by high specificity, reproducibility and relatively modest skill level required. However, they are still laborious and time and resource intensive, with results that are sometimes difficult to interpret. These methods are time-consuming and frequently unable to distinguish the DNA of live and dead bacteria. Furthermore, DNA-based methodologies need isolated genetic materials and sophisticated instruments, that are not suitable for in-field analysis, and can be inefficient in some food matrices because of interference from matrix components (Leonard et al., 2003). The demand for developing more rapid, accurate, easy-to-use, cost-effective and preferably portable analytical methods, capable of conducting point-of-care analyses, is continuously increasing.

In recent years, most research efforts have focused on the development of high-throughput, generic, rapid and low-cost methods, a number of which have been validated on an intra-laboratory basis or through limited trials. Before they are used outside research laboratories, these methods often have to undergo additional testing, e.g. further ring tests, development of reference materials, harmonization or adaptation for implementation in-field conditions (sampling methods, multitargeting, pen-side tests, mobile analysis). A diagnostic test for poultry infection agents is crucial for effective clinical management and for selecting appropriate disease control activities. Many variables such as genetic variation of the pathogen and/or host, test methodology and its physical format may influence the diagnostic performance of an assay (Manhas et al., 2021).

In an effort to address the shortcomings of the gold standard methods of pathogen detection, the development of biosensors has exponentially increased in the last decades (Escobar et al., 2023; Rizzotto et al., 2023). A biosensor is an analytical device containing a biorecognition element, also known as bioreceptor (e.g. antibodies, enzymes, aptamers and nucleic acids), designed to interact specifically with the analyte, and a transducer capable of translating a biometric recognition event into a measurable signal (Rotariu et al., 2016). Depending on the nature of the transducer, biosensors can be classified into optical, electrochemical, piezoelectric, acoustic and calorimetric types. Biosensors are also classified according to the bioreceptor, e.g. immuno-sensors (antibodies), genosensors (nucleic acids) and aptasensors (aptamers). Biosensors offer wide possibilities to diagnose bacteria in complex food samples with the specificity and selectivity of a laboratory-based analytical methods. Here, we will focus on electrochemical and optical biosensors, developed to detect the main pathogenic bacteria in poultry samples.

5 Optical biosensors: fluorescence, colorimetry, lateral flow, surface plasmon resonance and surface-enhanced Raman spectroscopy

Optical biosensors detect analytes by measuring changes in the characteristics of light waves, either in the visible light, UV or NIR regions. The four most common approaches are:

- fluorescence;
- colorimetry;
- surface plasmon resonance (SPR); and
- surface-enhanced Raman spectroscopy (SERS).

These are described below, together with lateral flow biosensors.

5.1 Fluorescence

Fluorometric biosensors are the most widely used in the field of optical biosensors. Recent years have seen significant improvements in pathogen detection for *Escherichia coli* (Jenie et al., 2021; Qiao et al., 2020; Xue et al., 2018), *Salmonella* (Ding et al., 2022; Wang et al., 2019), *Campylobacter* (Cheng et al., 2022; Dehghani et al., 2020; He et al., 2018; Liu et al., 2024), *Listeria* (Hossein-Nejad-Ariani et al., 2018; Servarayan et al., 2023; Zhao et al., 2019) and *S. aureus* (Jiao et al., 2023; Xu et al., 2023).

Some of the lowest limits of detection have been achieved by coupling fluorophores with highly selective recognition biomolecules, such as antibodies and aptamers. In order to achieve outstanding sensitivity, the amplification of the fluorescent signal is paramount, be it by the use of optical filters or nanomaterials. Recently, Ding et al. (2022) used a highly specific monoclonal antibody to capture *Salmonella* in milk samples on 96-well plates. They then exploited the streptavidin–biotin interaction, one of the strongest in nature, to effectively form a complex with amplified fluorescence by using Streptavidin-coated Quantum Dots (QDs) and biotinylated poly-L-lysine. This system's limit of detection (LOD) was 4.9×10^3 CFU/mL under optimal conditions.

In another example, Liu et al. designed a dual-fluorescence channel biosensor able to differentiate between *C. jejuni* and *Arcobacter butzleri* (*A. butzleri*) in human stool and chicken broiler samples (Fig. 2). In the first module, magnetic nanoparticles (MNPs) are employed to separate the bacteria from the complex sample matrix. In the second module, by coupling *C. jejuni*-specific aptamers to red-emitting carbon dots (CDs) and *A. butzleri*-specific aptamers to green-emitting CDs, bifunctional recognition is achieved. The fluorescence readouts are then transformed into red–green–blue (RGB) values, which

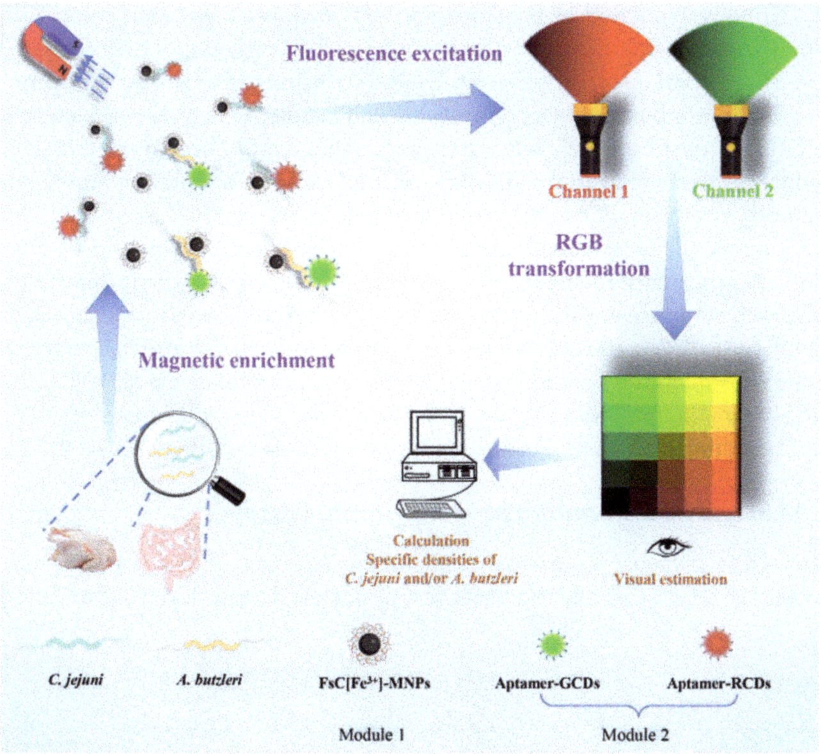

Figure 2 Schematic illustration of the principle designed by Liu et al. for the detection and discrimination of *Campylobacter* and *Aliarcobacter*. Source: Adapted from Liu et al. (2024) with permission.

represent the proportions of each bacteria present in the sample. This signal can also be treated to calculate each of the fluorescence intensities to obtain the specific bacterial densities. The lowest LOD obtained was 1 CFU/mL for each bacterium (Liu et al., 2024).

5.2 Colorimetry

The main advantage of colorimetric platforms is their ease of use and versatility. Upon the interaction of a chromophore with an analyte, absorbance or reflectance events in the UV–Vis spectrum are measured and used to quantify an analyte of interest. Various colorimetric platforms have been developed recently for the detection of *E. coli* (Gupta et al., 2021; He et al., 2022b; Peng & Chen, 2019; Xie et al., 2021; Yuan et al., 2023), *Salmonella* (Su et al., 2023; Y. Wang et al., 2023; Yuan et al., 2023), *Campylobacter* (Alamer et al., 2018b; Dehghani et al., 2018; Shams et al., 2019), *Listeria* (Chen et al., 2018; Yan et al., 2019) and *Staphylococcus* (Alamer et al., 2018a, 2018b).

Notably, Alamer et al. have developed a low-cost, ultra-rapid colorimetric biosensor for the detection and identification of *S. Typhimurium*, *S. Enteritidis*, *S. aureus* and *C. jejuni* on surfaces (Fig. 3). They constructed a sandwich lactoferrin immunosensor by coating a cotton swab with lactoferrin, a glycoprotein known to bind to most bacterial cells. After performing a swab on the contaminated surface with this instrument, they dipped it into a solution containing antibody-coated nanobeads of different colors. The antibodies specific to *S. Typhimurium*, *S. Enteritidis*, *S. aureus* and *C. jejuni* were immobilized on black, blue, orange and green beads, respectively. This allowed for the formation of a sandwich complex, consisting of lactoferrin, bacteria and antibody-conjugated nanobeads. Finally, the color of the cotton swab indicated the specific bacterial strain present in the contaminated surfaces. LODs were as low as 10 CFU/mL for *S. Typhimurium* and *C. jejuni*, 100 CFU/mL for *S. Enteritidis* and 100 CFU/mL for *S. aureus* on contaminated chicken meat surfaces (Alameret al., 2018a).

5.3 Lateral flow immunochromatographic assay

Lateral flow biosensors are widely used for the detection of various proteins and nucleic acids. Historically the detection of nucleic acids using this technology

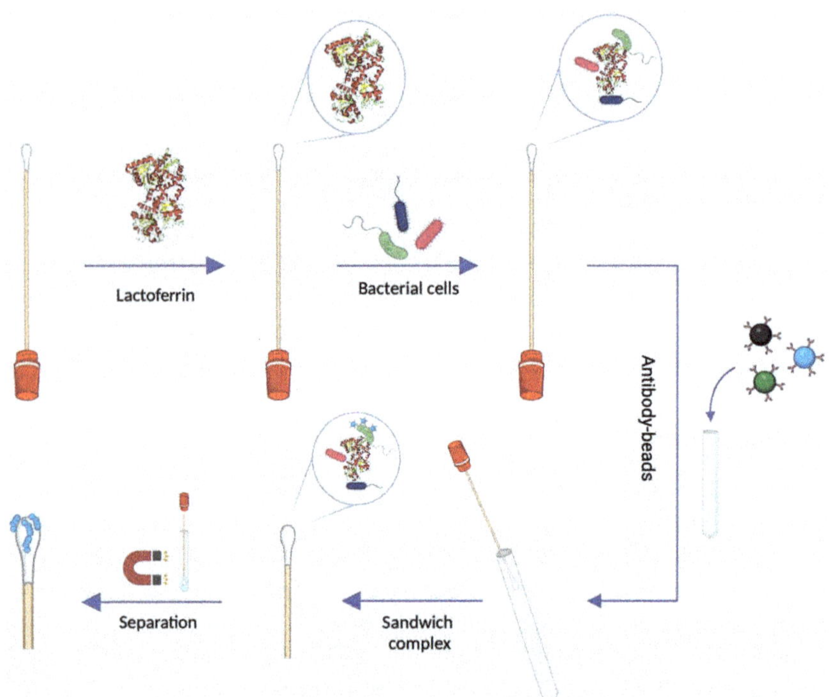

Figure 3 Schematic illustration of sandwich lactoferrin immunosensor onto cotton swab for screening pathogenic bacteria.Source: Adapted from Alamer et al. (2018a).

was coupled with amplification techniques. He et al. (2022a) designed a label-free dual signal LFIA by employing methylene blue-embedded zirconium-based organic framework (UIO@MB) as a capture carrier and signal indicator for the sensitive detection of *E. coli* O157:H7 (Fig. 4). For a colorimetric application, UIO@MB adhered to *E. coli* O157:H7 to form a visible conjugate as a blue band. For a fluorescent application, the organic framework was excited with Al^{3+} through electrostatic affinity, generating a green fluorescent band. They also developed a smartphone-based portable reading system capable of providing results immediately. Both the colorimetric and fluorescent modes were achieved over 90 min when using optimal conditions with a LOD of 10^3 CFU/mL. Finally, the method was further confirmed by detecting bacteria in spiked drinking water and cabbage samples (He et al., 2022b).

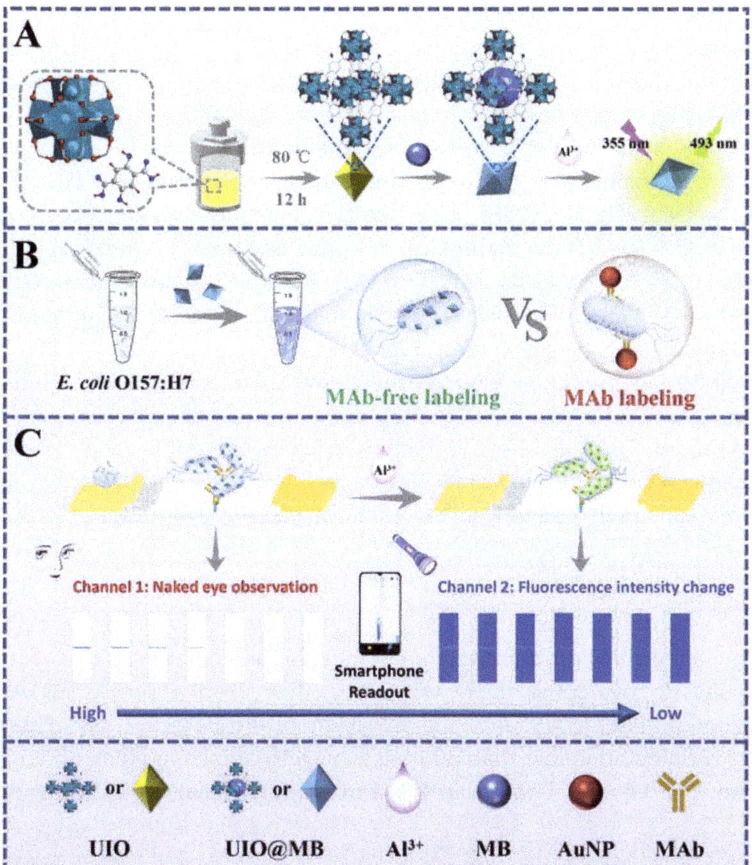

Figure 4 Schematic illustration of the working principle of the dual-signal LFIA platform designed by He et al. Source: Adapted from He et al. (2022a) with permission.

5.4 Surface plasmon resonance

SPR-based biosensors measure the excitation of surface plasmons of a thin gold layer, which is extremely sensitive to variations in the refractive index of a dielectric medium that occurs upon binding events between the analyte and surface-functionalized biosensors. These instruments permit the label-free, real-time monitoring of binding events between biological elements while also providing kinetic information (El Kazzy et al., 2021). Due to its many advantages, various SPR-based systems have been developed for the detection of *E. coli* (Bulard et al., 2015; Mondani et al., 2014; Wang et al., 2016; Wang et al., 2013), *Salmonella* (Bhandari et al., 2022; Fathi et al., 2023; Hyeon et al., 2021), *Campylobacter* (Masdor et al., 2017, 2019; Shams et al., 2019) and *Staphylococcus* (Eissa & Zourob, 2020; Jiao et al., 2023; Templier & Roupioz, 2017; Wang et al., 2022).

Bulard et al. (2015) have designed a SPR biochip capable of differentiating five closely related *E. coli* strains, including *E. coli* O157:H7, by grafting simple carbohydrates onto a gold surface and using SPR imaging to monitor the interactions during a bacterial culture. Each strain reacted differently with the carbohydrate chip, allowing for their sensitive detection and discrimination in less than 10 h. This chip was easily developed and regenerated and was able to successfully analyze a sample containing bacteria at 10^2 CFU/mL. Similarly, localized SPR (LSPR) refers to the SPR phenomenon occurring on gold nanoparticles or nanostructures. It is also exploited for the development of biosensors for bacterial detection. For example, Yaghubi et al. recently achieved a 10 CFU LOD using spherical gold nanoparticles conjugated to an anti-*E. coli* O157:H7 antibody (Yaghubi et al., 2020).

Marin et al. (2022) have successfully developed a colorimetric detection test for *S. aureus* based on a DNA aptamer and gold nanoparticles (AuNPs) in food samples. This naked-eye detection relies on the aggregation of AuNPs, resulting in a color change from wine-red to violet in the presence of *S. aureus* contamination in milk samples. This test takes place in two stages:

- Incubating the aptamer to the sample; and
- Adding the collected supernatants to the AuNPs.

Unfortunately, dairy products are rarely suitable for a colorimetric visualization of results. To overcome these difficulties, they used a simple EDTA milk treatment strategy to enhance the sensitivity of the detection test. Finally, *S. aureus* is detected in less than 30 min, achieving detection limits of 7.5×10^4 CFU/mL and 8.4×10^4 CFU/mL in milk and infant formula, respectively.

5.5 Surface-enhanced Raman spectroscopy

Raman spectroscopy relies on using a monochromatic light to excite and detect inelastic photon scattering, which permits the creation of a unique fingerprint showing the molecular vibration frequencies present in the sample. In SERS, the sample is adsorbed onto a nanostructured gold surface (Balbinot et al., 2021). This allows for the creation of a localized plasmonic light field, which amplifies the signal by up to 11 orders of magnitude. This is one of the most popular strategies for the sensitive detection of *E. coli* (Ye et al., 2022; You et al., 2020; Zhu et al., 2017), *Salmonella* (Jia et al., 2023; Li et al., 2021) and *Campylobacter* (He et al., 2022a).

Jia et al. (2023) coupled the high sensitivity of this transduction technique with the excellent specificity of Clustered Regularly Interspaced Short Palindromic Repeats-Cas12a (CRISPR-Cas12a) technology to design a CRISPR-SERS biosensor for the rapid detection of *S. Typhimurium* through targeting its invasin A (*invA*) gene. The team introduced an Single-stranded DNA (ssDNA) labeled with a Raman active molecule as a signal report probe to track the changes during target identification and detection (Fig. 5). After optimization,

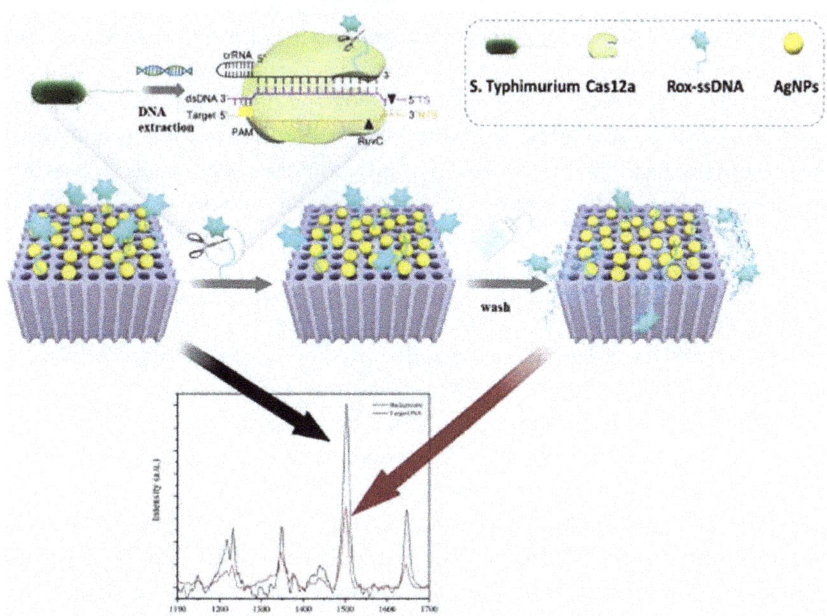

Figure 5 Schematic illustration of the CRISPR-SERS biosensor platform for the detection of *Salmonella Typhimurium* designed by Jia et al. Source: Adapted from Jia et al. (2023) with permission).

they were able to detect 110 CFU/mL in spiked poultry samples within 2 h. Furthermore, selectivity was successfully confirmed by testing against *E. coli* O157:H7, *Vibrio parahemolyticus*, *L. monocytogenes* and *S. aureus* (Jia et al., 2023).

6 Electrochemical methods: immunosensors, aptasensors, genosensors and imprinted polymers

An electrochemical biosensor is a device based on an electrochemical transducer. This provides selective quantitative analytical information on the modification of the electrochemical properties due to interaction between the analyte and the bioreceptor fixed onto an electrode, converting the biological information recorded into electrical current or voltage (Singh et al., 2021). In biosensors, several electrochemical techniques are employed: potentiometry, voltammetry, impedance and amperometry. Potentiometry measures the potential variation of the system containing the analyte, holding a constant current. Voltammetry techniques include cyclic voltammetry (CV), square wave voltammetry, different potential voltammetry or linear sweep voltammetry. In this technique, a controlled variation in potential is used to analyze the change in current of the system. Electrochemical impedance spectroscopy (EIS) measures the relationship between the alternative current and an applied sinusoidal potential in a given frequency, offering insights into reaction kinetics. In amperometry, the monitoring of the current is carried over holding constant the potential of the system (Rizzotto et al., 2023; Vidic & Manzano, 2021). The significant interest in electrochemical methods is due to the possibility of performing an accurate, rapid, real-time and label-free analysis, as well as their versatile detection, disposability and cost-effectiveness (Kaya et al., 2021).

6.1 Electrochemical immunosensors

Several electrochemical biosensors for the detection of *Salmonella* and *Campylobacter* have been developed in recent years, exploring different strategies to offer a rapid, costless, sensitive and selective detection of such pathogens in food samples. Mahari and Gandhi (2022) proposed an electrochemical immunosensor for the detection of *Salmonella gallinarum* and *Salmonella pullorum* in poultry meat (Mahari & Gandhi, 2022).

A reduced graphene oxide-based (rGO) fluorine-doped tin oxide (FTO)-electrode was functionalized with polyclonal antibodies (Fig. 6). FTO was chosen as an electrode, due to its resistance to corrosion and its ability to enhance electrical conductivity. rGO was used as a low-cost electroactive nanomaterial, due to its high surface area and ability for enabling the direct transfer of electrons from the redox probe to the electrode.

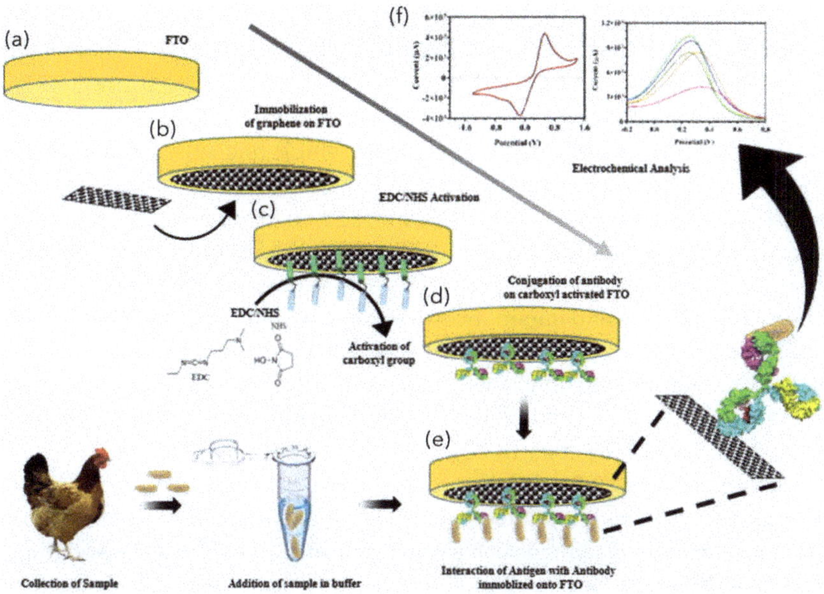

Figure 6 Design and principle of the fabricated FTO/rGO/*Salmonella gallinarum* or *Salmonella pullorum* Ab sensor. (a) Bare FTO electrode as a platform for development of electrochemical immunosensor; (b) fabrication of FTO electrode with rGO; (c) activation of rGO using EDC and NHS; (d) immobilization of *Salmonella gallinarum/Salmonella pullorum*-Ab via amide crosslinking between activated rGO and antibody; (e) addition of *Salmonella gallinarum/Salmonella pullorum*-Ag; (f) electrochemical detection. Source: Adapted from Mahari and Gandhi (2022) with permission.

The FTO/rGO/Ab electrochemical sensor was initially tested using *Salmonella* cultures, obtaining a LOD of 37 cells for *S. gallinarum* and 25 cells for *S. pullorum*. In order to evaluate the specificity of the sensor, negative control samples containing *S. Typhimurium* and *S. Enteritidis* cells were tested. Similarly, to the blank signal, no change in current was observed, concluding that the developed system presented a high specificity for the targets it was designed for. The ability of the biosensor to detect target strains in poultry meat samples was tested too, observing a LOD of 218 and 173 cells for *S. gallinarum* and *S. pullorum*, respectively (Mahari & Gandhi, 2022).

Che et al. (2001) developed an electrochemical biosensor coupled with the immunomagnetic separation technique for the detection of *C. jejuni*. Magnetic beads containing an anti-*Campylobacter* antibody were added to rinse water of inoculated chicken carcass. A secondary antibody labeled with alkaline phosphatase was subsequently linked. Beads were subsequently collected with an external magnet and were then incubated in phenylphosphate substrate to produce phenol. A carbon paste enzymatic electrode (tyrosinase) was used for the amperometric detection of produced phenol. This system was applied to *C. jejuni* pure culture and to poultry samples inoculated with *C. jejuni*. This detection method could be completed within 2.5 h and had a detection limit of 2.1×10^4 CFU/mL.

Roushani et al. (2020) developed a sensitive and label-free electrochemical immunosensor *S. aureus* detection system employing chicken anti-protein A antibody IgY as a bioreceptor. This antibody is selective and specific to *S. aureus*, targeting the protein A (spA) expressed on its surface. The researchers utilized glassy carbon electrode (GCE) modified with AuNPs to immobilize the IgY antibody, taking advantage of its amine group. *S. aureus* detection is carried out using CV and impedance spectroscopy with the $[Fe(CN)_6]^{4-/3-}$ redox probe in 60 min. This immunosensor shows a linear dynamic range from 10 to 10^7 CFU/mL with a detection limit of 3.3 CFU/mL. This method was applied also for the detection of *S. aureus* in milk and human blood samples using chicken antibody IgY (Roushani et al., 2020).

Soares et al. (2020) developed an electrochemical immunosensor for the rapid and label-free detection of the food-borne *S. enterica* serovar Typhimurium in food sample such as chicken broth. They used a laser-induced graphene (LIG) electrode functionalized with anti-*Salmonella* antibodies on its surface. Food samples contaminated with *Salmonella* could specifically bind to the antibody immobilized on the LIB electrode surface and be quantified using the EIS method. This biosensor detected the pathogen in 22 min in complex media without any pretreatment. It exhibited a broad linear range from 10 to 10^5 CFU/mL with a LOD of 13 CFU/mL. Moreover, this laser-induced graphene (LIG) immunosensor assured a high selectivity to *Salmonella* and a nonresponse to other bacteria tested in this study (Soares et al., 2020).

6.2 Electrochemical aptasensors and genosensors

Appaturi et al. (2020) have developed a rGO-carbon nanotube (CNT)-based electrochemical aptasensor for the detection of *S. Typhimurium* in chicken meat CNT acts as one-dimensional spacer between rGO sheets hybridized on it. Furthermore, it increases the surface area and electrical conductivity of the sensor. Amino-modified aptamers, selected for their affinity to *S. Typhimurium* cells, were covalently bound to the carboxyl group of the rGO–CNT complex. The system was constructed by a simple drop-casting onto a GCE. The performance of the aptasensor was tested under optimal conditions and in real chicken samples, showing fast, sensitive and selective detection, with a LOD of 10^1 CFU/mL. However, the biosensor also demonstrated some limitations related to the viscosity, pH and ionic strength of the samples analyzed, suggesting the need for further studies (Appaturi et al., 2020).

Morant-Miñana et al. (2015) developed an electrochemical genosensor based on microscale electrodes integrated on Cyclo Olefin Polymer (COP),

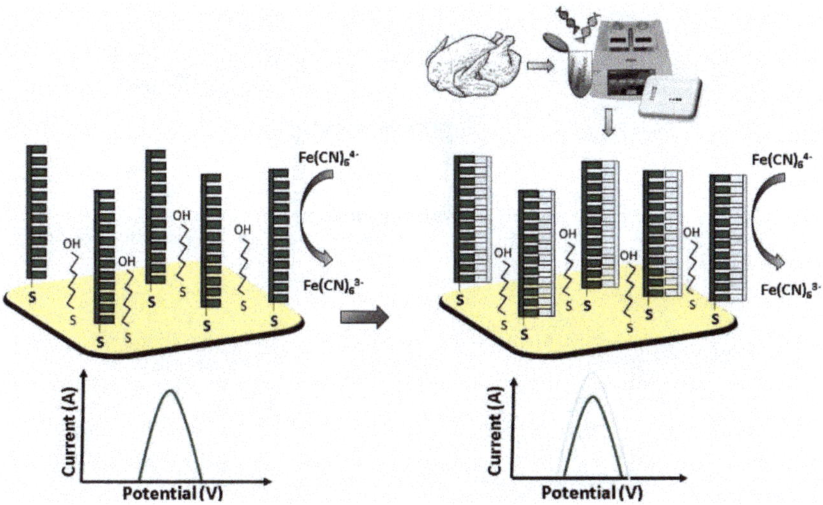

Figure 7 Schematic representation of *Campylobacter* spp. DNA detection by COP-based electrochemical genosensor. Source: Adapted from Vizzi et al. (2019) with permission.

for the detection of *Campylobacter* spp. in chicken meat, illustrated in Fig. 7 (Vizzini et al., 2019). COP was chosen for its chemical compatibility with organic solvents and low water absorption, high resistance to acids and bases and good optical properties. A thin gold film was then deposited onto the polymer and subsequently functionalized by immobilizing thiolated DNA probes designed to hybridize *Campylobacter* spp. flagellin gene sequences. The system obtained presented good sensitivity and selectivity, with a LOD of 90 pM using DNA target from PCR and of 50 nM using DNA target extracted from inoculated chicken meat samples (Morant-Miñana & Elizalde, 2015).

6.3 Imprinted polymers

Although antibodies are the most frequently used bioreceptors due to their high selectivity and binding affinity, they generally face several inherent restrictions such as complex preparation requiring animal experiments, high cost and limited stability during and after immobilization on the electrode surface. Moreover, although other biological molecules used as bioreceptors, such as peptides, DNA probes or aptamers, are more resistant than antibodies, they can be unstable in complex matrices.

Molecularly imprinted polymers (MIPs) are tailor-made synthetic receptors possessing specific recognition sites complementary to the template both in shape and chemical functionality. They have a unique combination of properties such as simple preparation, high physical and chemical stability, and low-cost production, which make them promising alternatives to natural bioreceptors.

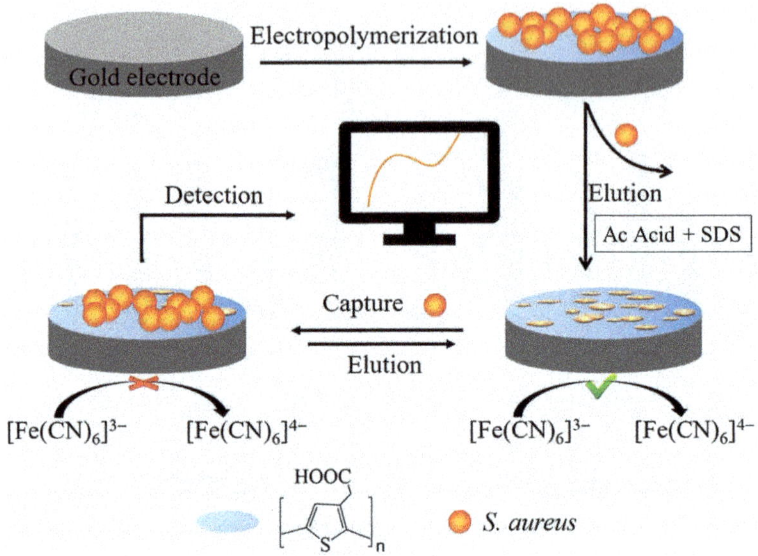

Figure 8 Schematic illustration of the fabrication of the bacteria-imprinted conductive poly(3-thiopheneacetic acid) film-based impedimetric sensor for *S. aureus* detection. Source: Adapted from Wang et al. (2021).

Over the past few years, MIPs have demonstrated great potential in the isolation, purification and sensing of small organic molecules. However, the imprinting of large structured species such as whole bacterial cells is still very challenging due to the increased difficulty in the complete removal of micrometer-sized templates.

Electropolymerization represents an advantageous method for bacteria imprinting as it allows the direct and fast deposition of a polymer film on the electrode surface with controllable film thickness. Wang et al. (2021) electropolymerized 3-substituted thiophene bearing a carboxyl group, 3-thiopheneacetic acid (TAA) as a functional monomer for bacteria imprinting, because the carboxyl group of TAA is not involved in polymerization and would be available for non-covalent interaction with groups such as carboxyl, amino, hydroxyl and phosphate on the outer surface of the bacterial cell wall, *S. aureus*. After the lysis of the entrapped bacteria in a mixture of acetic acid and SDS surfactant, the capture of *S. aureus* was detected after 10 min, using EIS (Fig. 8). The obtained detection limit was 2 CFU/mL and the dynamic range is 10–10^8 CFU/mL.

Recent years have seen a significant increase of MIP-based electrochemical sensors in pathogen detection, such as, for *S. aureus* in contaminated milk samples with the LOD of only 2 CFU/mL within 10 min (Wang et al., 2021), *L. monocytogenes* with the LOD of 70 CFU/mL in pure culture, *C. coli* and *C. jejuni* with the LOD of 1.1×10^3 CFU/mL for and 2.7×10^4 CFU/mL, respectively, directly in chicken cecal droppings within a 60-min analysis (Givanoudi et al., 2021).

7 Conclusion and future trends

PCR, RT-PCR, and immunoassays now are replacing culturing methods for microbial control in poultry production. Research has focused on the development of high-throughput and low-cost methods through analytical microdevices based on biosensors with short response time for on-site use and low detection limits. This chapter presents recently published optical and electrochemical biosensors for foodborne pathogen detection. Among the optical techniques, the four most common approaches are described in this chapter: fluorescence, colorimetry, SPR and SERS. Electrochemical approaches are based on different bioreceptors: antibodies, aptamers or imprinted polymers. Different electrochemical methods are used: voltammetry, amperometry or EIS. The main criteria for the choice of detection methods are sample preparation, response time and limit of detection. For the extraction of the specific bacteria from the sample, the immunomagnetic method is frequently suggested. For the same response time, and the same detection limit, the instrumentation cost will be the criteria of choice. Portable devices allowing on-site detection will be important for the delocalization of microbial control.

8 References

Alamer, S., et al. (2018a). Rapid colorimetric lactoferrin-based sandwich immunoassay on cotton swabs for the detection of foodborne pathogenic bacteria. *Talanta*, *185*, 275–280. https://doi.org/10.1016/j.talanta.2018.03.072

Alamer, S., et al. (2018b). A rapid colorimetric immunoassay for the detection of pathogenic bacteria on poultry processing plants using cotton swabs and nanobeads. *Microchimica Acta*, *185*(3), 164. https://doi.org/10.1007/s00604-018-2696-7

Andersen, K. B., et al. (2022). A new tool for air sample-based surveillance of *Campylobacter* and *Salmonella* in poultry flocks. *Journal of Applied Poultry Research*, *31*(2), 100236. https://doi.org/10.1016/j.japr.2022.100236

Appaturi, J. N., et al. (2020). Rapid and sensitive detection of *Salmonella* with reduced graphene oxide-carbon nanotube based electrochemical aptasensor. *Analytical Biochemistry*, *589*, 113489. https://doi.org/10.1016/j.ab.2019.113489

Authority, E. F. S. & European Centre for Disease Prevention and Control. (2022). The European Union One Health 2021 Zoonoses Report. *EFSA Journal*, *20*(12), e07666. https://doi.org/10.2903/j.efsa.2022.7666

Balbinot, S., et al. (2021). Plasmonic biosensors for food control. *Trends in Food Science & Technology*, *111*, 128–140. https://doi.org/10.1016/j.tifs.2021.02.057

Bhandari, D., et al. (2022). Magnetic nanoparticles enhanced surface plasmon resonance biosensor for rapid detection of *Salmonella Typhimurium* in Romaine Lettuce. *Sensors*, *22*(2), Article 2. https://doi.org/10.3390/s22020475

Borriello, S. P., et al. (2005). Enteroccoci. In W. W. C. Topley & G. S. Wilson (Eds.), *Bacteriology* (10th ed, Vol. 1, pp. 883–902). Hodder Arnold.

Bulard, E., et al. (2015). Carbohydrates as new probes for the identification of closely related *Escherichia coli* strains using surface plasmon resonance imaging. *Analytical Chemistry*, *87*(3), 1804–1811. https://doi.org/10.1021/ac5037704

Che, Y., et al. (2001). Detection of *Campylobacter jejuni* in poultry samples using an enzyme-linked immunoassay coupled with an enzyme electrode. *Biosensors & Bioelectronics*, *16*(9–12), 791–797. https://doi.org/10.1016/s0956-5663(01)00222-6

Chen, Q., et al. (2018). An optical biosensor using immunomagnetic separation, urease catalysis and pH indication for rapid and sensitive detection of *Listeria monocytogenes*. *Sensors and Actuators B: Chemical*, *258*, 447–453. https://doi.org/10.1016/j.snb.2017.11.087

Cheng, X., et al. (2022). Improved triple-module fluorescent biosensor for the rapid and ultrasensitive detection of *Campylobacter jejuni* in livestock and dairy. *Food Control*, *137*, 108905. https://doi.org/10.1016/j.foodcont.2022.108905

Cheung, G. Y. C., et al. (2021). Pathogenicity and virulence of *Staphylococcus aureus*. *Virulence*, *12*(1), 547–569. https://doi.org/10.1080/21505594.2021.1878688

Cossettini, A., et al. (2022). Rapid detection of *Listeria monocytogenes*, *Salmonella*, *Campylobacter* spp., and *Escherichia coli* in food using biosensors. *Food Control*, *137*, 108962. https://doi.org/10.1016/j.foodcont.2022.108962

Dehghani, Z., et al. (2018). Colorimetric aptasensor for *Campylobacter jejuni* cells by exploiting the peroxidase like activity of Au@Pd nanoparticles. *Mikrochimica Acta*, *185*(10), 448. https://doi.org/10.1007/s00604-018-2976-2

Dehghani, Z., et al. (2020). Whole cell FRET immunosensor based on graphene oxide and graphene dot for *Campylobacter jejuni* detection. *Food Chemistry*, *309*, 125690. https://doi.org/10.1016/j.foodchem.2019.125690

Ding, S., et al. (2022). A fluorescent biosensor based on quantum dot–labeled streptavidin and poly-l-lysine for the rapid detection of *Salmonella* in milk. *Journal of Dairy Science*, *105*(4), 2895–2907. https://doi.org/10.3168/jds.2021-21229

Duqué, B., et al. (2021). Quantification of *Campylobacter jejuni* gene expression after successive stresses mimicking poultry slaughtering steps. *Food Microbiology*, *98*, 103795. https://doi.org/10.1016/j.fm.2021.103795

EFSA Panel on Biological Hazards (BIOHAZ). (2010). Scientific Opinion on Quantification of the risk posed by broiler meat to human *Campylobacter*iosis in the EU. *EFSA Journal*, *8*(1), 1437. https://doi.org/10.2903/j.efsa.2010.1437

Eissa, S. & Zourob, M. (2020). Ultrasensitive peptide-based multiplexed electrochemical biosensor for the simultaneous detection of *Listeria monocytogenes* and *Staphylococcus aureus*. *Microchimica Acta*, *187*(9), 486. https://doi.org/10.1007/s00604-020-04423-3

El Kazzy, M., et al. (2021). An overview of artificial olfaction systems with a focus on surface plasmon resonance for the analysis of volatile organic compounds. *Biosensors*, *11*(8), Article 8. https://doi.org/10.3390/bios11080244

Escobar, V., et al. (2023). Recent advances on peptide-based biosensors and electronic eoses for eoodborne pathogen detection. *Biosensors*, *13*(2), 258. https://doi.org/10.3390/bios13020258

European Centre for Disease Prevention and Control. (2022). The European Union One Health 2021 Zoonoses Report. *EFSA Journal*, *20*(12), e07666. https://doi.org/10.2903/j.efsa.2022.7666

European Union, P. O. of the E. (2005, November 15). *Commission Regulation (EU) 2005/2073 of 15 November 2005 on microbiological criteria for foodstuffs (Text with EEA relevance)* [Website]. Publications Office of the EU; Publications Office of the European Union. https://eur-lex.europa.eu/legal-content/EN/TXT/PDF/?uri=CELEX:02005R2073-20200308&rid=1

European Union, P. O. of the E. (2017, August 23). *Commission Regulation (EU) 2017/1495 of 23 August 2017 amending Regulation (EC) No 2073/2005 as regards Campylobacter in broiler carcases (Text with EEA relevance), C/2017/5746* [Website]. Publications Office of the EU; Publications Office of the European Union. https://op.europa.eu/en/publication-detail/-/publication/218f322f-8890-11e7-b5c6-01aa75ed71a1

Fàbrega, A. & Vila, J. (2013). Salmonella enterica serovar Typhimurium skills to succeed in the host: virulence and regulation. *Clinical Microbiology Reviews*, *26*(2), 308–341.

Falkow, S. (2004). Chapter 10: Host--pathogen interactions. In K. J. Ryan, C. G. Ray, & J. C. Sherris (Eds.), *Sherris medical microbiology: An introduction to infectious diseases: Vol. Part III* (4th ed). McGraw-Hill.

Farber, J. M., et al. (1996). Health risk assessment of *Listeria monocytogenes* in Canada. *International Journal of Food Microbiology*, *30*(1–2), 145–156. https://doi.org/10.1016/0168-1605(96)01107-5

Fathi, S., et al. (2023). Surface plasmon resonance-based oligonucleotide biosensor for *Salmonella Typhi* detection. *Analytical Biochemistry*, *677*, 115250. https://doi.org/10.1016/j.ab.2023.115250

Fitzgerald, C. & Nachamkin, I. (2015). *Campylobacter* and *Arcobacter*. In J. H. Jorgensen, K. C. Carroll, G. Funke, M. A. Pfaller, M. L. Landry, S. S. Richter, & D. W. Warnock (Eds.), *Manual of clinical microbiology* (11th ed., pp. 998–1012). John Wiley & Sons, Ltd. https://doi.org/10.1128/9781555817381.ch56

Gal-Mor, O., et al. (2010). Multidrug-resistant *Salmonella enterica* serovar Infantis, Israel. *Emerging Infectious Diseases*, *16*(11), 1754. https://doi.org/10.1128/CMR.00066-12. https://doi.org/10.3201/eid1611.100100

Gast, R. K. & Porter Jr., R. E. (2020). *Salmonella* infections. In D. E. Swayne (Ed.), *Diseases of poultry* (14th ed., pp. 717–753). John Wiley and Sons. https://doi.org/10.1002/9781119371199.ch16

Givanoudi, S., et al. (2021). Selective *Campylobacter* detection and quantification in poultry: a sensor tool for detecting the cause of a common zoonosis at its source. *Sensor Actuator B Chem*, 332, 129484. https://doi.org/10.1016/j.snb.2021.129484

Gonçalves-Tenório, A., et al. (2018). Prevalence of pathogens in poultry meat: a meta-analysis of european published surveys. *Foods*, *7*(5), Article 5. https://doi.org/10.3390/foods7050069

Gupta, R., et al. (2021). Naked eye colorimetric detection of *Escherichia coli* using aptamer conjugated graphene oxide enclosed Gold nanoparticles. *Sensors and Actuators B: Chemical*, *329*, 129100. https://doi.org/10.1016/j.snb.2020.129100

He, K., et al. (2018). Sensitive detection of *Campylobacter jejuni* using one-step strategy based on functional nanospheres of immunomagnetic capture and quantum dots. *Ecotoxicology and Environmental Safety*, *163*, 274–278. https://doi.org/10.1016/j.ecoenv.2018.07.050

He, D., et al. (2022a). Quantitative detection of *Campylobacter jejuni* with a core-satellite assemblies-based dual-modular aptasensor. *Food Control*, *135*, 108828. https://doi.org/10.1016/j.foodcont.2022.108828

He, K., et al. (2022b). "Lighting-up" methylene blue-embedded zirconium based organic framework triggered by Al3+ for advancing the sensitivity of E. coli O157:H7 analysis in dual-signal lateral flow immunochromatographic assay. *Journal of Hazardous Materials*, *425*, 128034. https://doi.org/10.1016/j.jhazmat.2021.128034

Hossein-Nejad-Ariani, H., et al. (2018). Peptide-based biosensor utilizing fluorescent gold nanoclusters for detection of *Listeria monocytogenes*. *ACS Applied Nano Materials*, *1*(7), 3389–3397. https://doi.org/10.1021/acsanm.8b00600

Hyeon, S. H., et al. (2021). Novel surface plasmon resonance biosensor that uses full-length Det7 phage tail protein for rapid and selective detection of *Salmonella Enterica* serovar *Typhimurium*. *Biotechnology and Applied Biochemistry*, *68*(1), 5–12. https://doi.org/10.1002/bab.1883

Jalali, M., et al. (2012). Microbial quality and prevalence of *Salmonella* and *Listeria* in eggs. *International Journal of Environmental Health Engineering*, *1*(1), 48. https://doi.org/10.4103/2277-9183.105347

Jenie, S. N. A., et al. (2021). Rapid fluorescence quenching detection of *Escherichia coli* Using Natural Silica-Based Nanoparticles. *Sensors*, *21*(3), Article 3. https://doi.org/10.3390/s21030881

Jia, F., et al. (2023). An amplification-free CRISPR-SERS biosensor for specific, sensitive and rapid detection of *Salmonella typhimurium* in poultry. *LWT*, *189*, 115476. https://doi.org/10.1016/j.lwt.2023.115476

Jiao, J., et al. (2023). Integrated multifunctional nanoplatform for fluorescence detection and inactivation of *Staphylococcus aureus*. *Food Chemistry*, *428*, 136780. https://doi.org/10.1016/j.foodchem.2023.136780

Kaya, H. O., et al. (2021). Pathogen detection with electrochemical biosensors: advantages, challenges and future perspectives. *Journal of Electroanalytical Chemistry*, *882*, 114989. https://doi.org/10.1016/j.jelechem.2021.114989

Krauss, H., et al. (2003). Zoonoses: infectious diseases transmissible from animals to humans. In S. Higgs (Trans.), *Vector-borne and zoonotic diseases* (3rd ed., Vol. 6, pp. 113–113). ASM Press.

Krauss, H., et al. (2005). Zoonoses: infectious diseases transmissible from animals to humans (W. Preiser, Trans.). *Medical Microbiology and Immunology*, *194*(4), 219–220. https://doi.org/10.1007/s00430-004-0232-3

Leonard, P., et al. (2003). Advances in biosensors for detection of pathogens in food and water. *Enzyme and Microbial Technology*, *32*(1), 3–13. https://doi.org/10.1016/S0141-0229(02)00232-6

Li, A., et al. (2021). An aptamer biosensor based dual signal amplification system for the detection of *Salmonella Typhimurium*. *Analytical Biochemistry*, *615*, 114050. https://doi.org/10.1016/j.ab.2020.114050

Line, J. E., et al. (2013). Comparison of cumulative drip sampling with whole carcass rinses for estimation of *Campylobacter* species and quality indicator organisms associated with processed broiler chickens. *Poultry Science*, *92*(1), 218–224. https://doi.org/10.3382/ps.2012-02217

Liu, W., et al. (2024). A biosensor encompassing fusarinine C-magnetic nanoparticles and aptamer-red/green carbon dots for dual-channel fluorescent and RGB discrimination of *Campylobacter* and *Aliarcobacter*. *Talanta*, *266*, 125085. https://doi.org/10.1016/j.talanta.2023.125085

Mahari, S. & Gandhi, S. (2022). Electrochemical immunosensor for detection of avian salmonellosis based on electroactive reduced graphene oxide (rGO) modified electrode. *Bioelectrochemistry*, *144*, 108036. https://doi.org/10.1016/j.bioelechem.2021.108036

Manhas, P. K., et al. (2021). Enhanced detection of major pathogens and toxins in poultry and livestock with zoonotic risks using nanomaterials-based diagnostics. *Frontiers in Veterinary Science*, *8*, 673718. https://doi.org/10.3389/fvets.2021.673718

Marin, M., et al. (2021). Rapid point-of-need detection of bacteria and their toxins in food using gold nanoparticles. *Comprehensive Reviews in Food Science and Food Safety*, *20*(6), 5880–5900. https://doi.org/10.1111/1541-4337.12839

Marin, M., et al. (2022). Naked-eye detection of *Staphylococcus aureus* in powdered milk and infant formula using gold nanoparticles. *Journal of Microbiological Methods*, *201*, 106578. https://doi.org/10.1016/j.mimet.2022.106578

Masdor, N. A., et al. (2019). Subtractive inhibition assay for the detection of *Campylobacter jejuni* in chicken samples using surface plasmon resonance. *Scientific Reports*, *9*(1), Article 1. https://doi.org/10.1038/s41598-019-49672-2

Masdor, N. A., et al. (2017). Surface plasmon resonance immunosensor for the detection of *Campylobacter jejuni*. *Chemosensors*, *5*(2), Article 2. https://doi.org/10.3390/chemosensors5020016

Mondani, L., et al-. (2014). Simultaneous enrichment and optical detection of low levels of stressed *Escherichia coli* O157:H7 in food matrices. *Journal of Applied Microbiology*, *117*(2), 537–546. https://doi.org/10.1111/jam.12522

Morant-Miñana, M. C. & Elizalde, J. (2015). Microscale electrodes integrated on COP for real sample *Campylobacter* spp. Detection. *Biosensors & Bioelectronics*, *70*, 491–497. https://doi.org/10.1016/j.bios.2015.03.063

Murray, P. R., et al. (Eds.). (2003). Manual of clinical microbiology (8th edition). In *Diagnostic Microbiology and Infectious Disease* (Albert Balows, 1–2, pp. 625–626). ASM Press. https://www.ncbi.nlm.nih.gov/pmc/articles/PMC7135312/

OECD - FAO Agricultural Outlook (Edition 2021). (2021). OECD. http://data.oecd.org/agroutput/meat-consumption.htm

Peng, H. & Chen, I. A. (2019). Rapid colorimetric detection of bacterial species through the capture of gold nanoparticles by chimeric phages. *ACS Nano*, *13*(2), 1244–1252. https://doi.org/10.1021/acsnano.8b06395

Qiao, Y., et al. (2020). Polymyxin B–modified upconversion nanoparticles for selective detection of Gram-negative bacteria such as *Escherichia coli*. *Journal of Chemical Research*, *44*(11–12), 756–761. https://doi.org/10.1177/1747519820911266

Qu, L.-L., et al. (2016). Silver nanoparticles on cotton swabs for improved surface-enhanced Raman scattering, and its application to the detection of carbaryl. *Microchimica Acta*, *183*(4), 1307–1313. https://doi.org/10.1007/s00604-016-1760-4

Quereda, J. J., et al. (2021). Pathogenicity and virulence of *Listeria monocytogenes*: a trip from environmental to medical microbiology. *Virulence*, *12*(1), 2509–2545. https://doi.org/10.1080/21505594.2021.1975526

Reid, A. N., et al. (2008). Identification of *Campylobacter jejuni* genes contributing to acid adaptation by transcriptional profiling and genome-wide mutagenesis. *Applied and Environmental Microbiology*, *74*(5), 1598–1612. https://doi.org/10.1128/AEM.01508-07

Rizzotto, F., et al. (2023). Recent advances in electrochemical biosensors for food control. *Micromachines*, *14*(7), Article 7. https://doi.org/10.3390/mi14071412

Robinson, D. A. (1981). Infective dose of *Campylobacter jejuni* in milk. *British Medical Journal (Clinical Research Ed.)*, *282*(6276), 1584. https://doi.org/10.1136/bmj.282.6276.1584

Rodriguez, C., et al. (2021). *Listeria monocytogenes* dissemination in farming and primary production: sources, shedding and control measures. *Food Control*, *120*, 107540. https://doi.org/10.1016/j.foodcont.2020.107540

Rotariu, L., et al. (2016). Electrochemical biosensors for fast detection of food contaminants – trends and perspective. *TrAC Trends in Analytical Chemistry*, *79*, 80–87. https://doi.org/10.1016/j.trac.2015.12.017

Rouger, A., et al. (2017). Bacterial contaminants of poultry meat: sources, species, and dynamics. *Microorganisms*, *5*(3), Article 3. https://doi.org/10.3390/microorganisms5030050

Roushani, M., et al. (2020). Electrochemical immunosensor for determination of *Staphylococcus aureus* bacteria by IgY immobilized on glassy carbon electrode with electrodeposited gold nanoparticles. *Mikrochimica Acta*, *187*(10), 567. https://doi.org/10.1007/s00604-020-04547-6

Schmid-Hempel, P. & Frank, S. A. (2007). Pathogenesis, virulence, and infective dose. *PLoS Pathogens*, *3*(10), e147. https://doi.org/10.1371/journal.ppat.0030147

Servarayan, K. L., et al. (2023). Optical immunosensor for the detection of *Listeria monocytogenes* in Food Matrixes. *ACS Omega*, *8*(18), 15979–15989. https://doi.org/10.1021/acsomega.2c07848

Shams, S., et al. (2019). A sensitive gold-nanorods-based nanobiosensor for specific detection of *Campylobacter jejuni* and *Campylobacter coli*. *Journal of Nanobiotechnology*, *17*(1), 43. https://doi.org/10.1186/s12951-019-0476-0

Sibanda, T. & Buys, E. M. (2022). *Listeria monocytogenes* pathogenesis: the role of stress adaptation. *Microorganisms*, *10*(8), 1522. https://doi.org/10.3390/microorganisms10081522

Singh, A., et al. (2021). Recent advances in electrochemical biosensors: applications, challenges, and future scope. *Biosensors*, *11*(9), 336. https://doi.org/10.3390/bios11090336

Soares, R. R. A., et al. (2020). Laser-induced graphene electrochemical immunosensors for rapid and label-free monitoring of *Salmonella* enterica in Chicken Broth. *ACS Sensors*, *5*(7), 1900–1911. https://doi.org/10.1021/acssensors.9b02345

Søndergaard, M. S. R., et al. (2014). Low-cost monitoring of *Campylobacter* in poultry houses by air sampling and quantitative PCR. *Journal of Food Protection*, *77*(2), 325–330. https://doi.org/10.4315/0362-028X.JFP-13-268

Su, Z., et al. (2023). Smartphone-assisted colorimetric detection of *Salmonella* Typhimurium based on the catalytic reduction of 4-nitrophenol by β-cyclodextrin-capped gold nanoparticles. *Analytica Chimica Acta*, *1239*, 340672. https://doi.org/10.1016/j.aca.2022.340672

Templier, V. & Roupioz, Y. (2017). On the challenges of detecting whole *Staphylococcus aureus* cells with biosensors. *Journal of Applied Microbiology*, *123*(5), 1056–1067. https://doi.org/10.1111/jam.13510

Vidic, J. & Manzano, M. (2021). Electrochemical biosensors for rapid pathogen detection. *Current Opinion in Electrochemistry*, *29*, 100750. https://doi.org/10.1016/j.coelec.2021.100750

Vizzini, P., et al. (2019). Electrochemical and optical biosensors for the detection of *Campylobacter* and *Listeria*: An update look. *Micromachines*, 10(8), 500. https://doi.org/10.3390/mi10080500

Wang, S., et al. (2016). The development of a portable SPR bioanalyzer for sensitive detection of *Escherichia coli* O157:H7. *Sensors (Basel, Switzerland)*, *16*(11), 1856. https://doi.org/10.3390/s16111856

Wang, S., et al. (2019). A microfluidic biosensor for online and sensitive detection of *Salmonella Typhimurium* using fluorescence labeling and smartphone video processing. *Biosensors and Bioelectronics*, *140*, 111333. https://doi.org/10.1016/j.bios.2019.111333

Wang, R., et al. (2021). Rapid sensitive and label-free detection of pathogenic bacteria using a bacteria-imprinted conducting polymer film-based electrochemical sensor, *Talanta* 26, 122135. https://doi.org/10.1016/j.talanta.2021.122135

Wang, Y., et al. (2023). A visual colorimetric assay based on phage T156 and gold nanoparticles for the sensitive detection of *Salmonella* in lettuce. *Analytica Chimica Acta*, *1272*, 341501. https://doi.org/10.1016/j.aca.2023.341501

Wang, Y., et al. (2013). Monitoring of *Escherichia coli* O157:H7 in food samples using lectin based surface plasmon resonance biosensor. *Food Chemistry*, *136*(3–4), 1303–1308. https://doi.org/10.1016/j.foodchem.2012.09.069

Wang, Z., et al. (2022). A novel PEG-mediated boric acid functionalized magnetic nanomaterials based fluorescence biosensor for the detection of *Staphylococcus aureus*. *Microchemical Journal*, *178*, 107379. https://doi.org/10.1016/j.microc.2022.107379

Xie, Y., et al. (2021). A trigger-based aggregation of aptamer-functionalized gold nanoparticles for colorimetry: an example on detection of *Escherichia coli* O157:H7. *Sensors and Actuators B: Chemical*, *339*, 129865. https://doi.org/10.1016/j.snb.2021.129865

Xu, Y., et al. (2023). Ultrasensitive hairpin mediated upconversion fluorescence biosensor for *Staphylococcus aureus* detection in foods and waters exploiting g-C3N4-assisted catalysis. *Analytica Chimica Acta*, *1239*, 340738. https://doi.org/10.1016/j.aca.2022.340738

Xue, L., et al. (2018). An ultrasensitive fluorescent biosensor using high gradient magnetic separation and quantum dots for fast detection of foodborne pathogenic bacteria. *Sensors and Actuators B: Chemical*, *265*, 318–325. https://doi.org/10.1016/j.snb.2018.03.014

Yaghubi, F., et al. (2020). Design of localized surface plasmon resonance (LSPR) biosensor for immunodiagnostic of *E. coli* O157:H7 Using Gold Nanoparticles Conjugated to the Chicken Antibody. *Plasmonics*, *15*(5), 1481–1487. https://doi.org/10.1007/s11468-020-01162-2

Yan, P., et al. (2019). Colorimetric sensor array based on Wulff-type boronate functionalized AgNPs at various pH for bacteria identification. *Analytical Chemistry*, *91*(19), 12134–12137. https://doi.org/10.1021/acs.analchem.9b03172

Ye, Y., et al. (2022). A surface-enhanced Raman scattering aptasensor for *Escherichia coli* detection based on high-performance 3D substrate and hot spot effect. *Analytica Chimica Acta*, *1221*, 340141. https://doi.org/10.1016/j.aca.2022.340141

You, S.-M., et al. (2020). Gold nanoparticle-coated starch magnetic beads for the separation, concentration, and SERS-based detection of *E. coli* O157:H7. *ACS Applied Materials & Interfaces*, *12*(16), 18292–18300. https://doi.org/10.1021/acsami.0c00418

Yuan, P., et al. (2023). Bimetallic metal–organic framework nanorods with peroxidase mimicking activity for selective colorimetric detection of *Salmonella Typhimurium* in food. *Food Control*, *144*, 109357. https://doi.org/10.1016/j.foodcont.2022.109357

Zhao, X., et al. (2014). Advances in rapid detection methods for foodborne pathogens. *Journal of Microbiology and Biotechnology*, *24*(3), 297–312. https://doi.org/10.4014/jmb.1310.10013

Zhao, X., et al. (2019). Preparation of fluorescent molecularly imprinted polymers via pickering emulsion interfaces and the application for visual sensing analysis of *Listeria monocytogenes*. *Polymers*, *11*(6), Article 6. https://doi.org/10.3390/polym11060984

Zhu, T., et al. (2017). A novel SERS nanoprobe based on the use of core-shell nanoparticles with embedded reporter molecule to detect *E. coli* O157:H7 with high sensitivity. *Microchimica Acta*, *185*(1), 30. https://doi.org/10.1007/s00604-017-2573-9

Chapter 5

Biosecurity measures to prevent zoonotic and other diseases in poultry farm environments

Dita Julianingsih, Chuan-Wei Tung and Debabrata Biswas, University of Maryland, USA

1 Introduction

The safety and biosecurity of poultry farms in the USA and around the world are critically important issues. The prevention of both poultry-borne zoonotic diseases and diseases affecting poultry is a vital aspect of maintaining the health and well-being of both poultry, the individuals involved in the production of poultry products at pre-harvest and post-harvest stages, as well as consumers of poultry meat. Farm biosecurity serves as a comprehensive framework encompassing all the necessary measures and actions taken to prevent the introduction and spread of diseases in poultry farm environments. Its primary objective is to keep causative agents of diseases, including viruses, bacteria, fungi, parasites, and other microorganisms, away from birds, farm environments, and workers at various levels (Gržinić et al., 2023; Hedman et al., 2020; Hafez and Attia, 2020). Farm biosecurity incorporates a wide range of measures involving a combination of physical, operational, and management

http://dx.doi.org/10.19103/AS.2024.0146.07

practices that work in unison to establish a multi-layered defense system (Huber et al., 2022; Alarcón et al., 2021).

Zoonotic diseases, which can be transmitted between animals and humans, pose a considerable threat to public health (Rahman et al., 2020). The close interaction between humans and poultry in farm settings is a major concern in terms of transmission of zoonotic diseases within the poultry industry (Bissong et al., 2022). Moreover, by reducing the potential for disease outbreaks, biosecurity measures minimize economic losses associated with decreased production, increased mortality rates, and the need for costly treatment and control measures (Butucel et al., 2022; Kruger et al., 2009; Butterworth and Weeks, 2009). It is therefore imperative to implement stringent biosecurity measures to mitigate this risk and protect both poultry and the individuals working on the farm (MacMahon et al., 2008).

Farm biosecurity is a comprehensive approach that encompasses various measures and protocols (Cunningham and Fairchild, 2020). Farm biosecurity measures can be broadly categorized into two main types: structural biosecurity measures and operational biosecurity measures (USDA, 2021). Other management practices also play a crucial role in farm biosecurity, including disease surveillance, regular health checks, and effective maintaining records (Gehrett, 2022). Early detection of disease signs and prompt reporting to veterinary authorities are essential for timely intervention and control measures. In addition, providing comprehensive training and education to farm personnel on biosecurity practices, disease recognition, and proper handling of animals and waste ensures that everyone involved in the farm understands and follows the necessary protocols (Dhaka et al., 2023; Tyler et al., 2022). In this chapter, we will discuss a combination of biosecurity measures that can be established by poultry farms to create a safe and disease-free environment for birds and people involved in the industry (MacMahon et al., 2008).

2 Assessing the effectiveness of biosecurity: transmission routes into poultry farms

In order to provide a comprehensive understanding of the effectiveness of biosecurity measures, it is essential to investigate the main transmission routes of zoonotic diseases in poultry farms. This and following sections will briefly explore the primary transmission routes, examine existing research on their significance, and review trends in zoonotic disease outbreaks originating from poultry farms.

According to the Center for Food Security and Public Health (CFSPH, 2021), zoonotic diseases can infiltrate poultry farms through diverse routes, including but not limited to:

1 Direct contact with infected animals: the proximity of poultry to other animals, both domestic and wild, poses a direct risk of disease transmission. This includes contact with mammals, birds, and insects carrying zoonotic pathogens.
2 Contaminated feed and water: the introduction of contaminated feed and water into poultry farms serves as a potential avenue for zoonotic diseases. The origin of feed ingredients and water sources plays a pivotal role in farm biosecurity.
3 Vector-borne transmission: insects, such as mosquitoes and ticks, can act as vectors for zoonotic diseases. Understanding the prevalence and significance of insect-mediated transmission is crucial for effective preventive measures.
4 Airborne transmission: some pathogens can be transmitted through the air, emphasizing the importance of ventilation systems and air-quality management in preventing the spread of zoonoses within poultry facilities.
5 Human-mediated transmission: human activity and movement, including those of farm workers, visitors, and service providers, can inadvertently introduce and spread zoonotic diseases. Strict biosecurity protocols are essential for minimizing this risk.

The transmission of many zoonotic diseases into poultry farms, particularly avian influenza, is intricately linked to interactions between domesticated birds and wild bird populations. The Centers for Disease Control and Prevention (CDC) has highlighted the risk of avian influenza A viruses, which naturally occur among wild aquatic birds worldwide. Wild birds, especially species of dabbling duck (which feed mainly on vegetative matter on the surface of water), are considered reservoirs for these viruses, capable of infecting domestic poultry and other bird species. The mode of transmission involves infected birds shedding the virus in saliva, nasal secretions, and feces. Direct contact with infected wild birds is a significant transmission route for zoonotic diseases affecting poultry flocks. Infected birds can transmit avian influenza A viruses to susceptible birds through contact or exposure to contaminated surfaces. This pathway poses a substantial risk to domesticated birds, including chickens, ducks, and turkeys (CDC, 2022c).

The CDC recommendations for poultry workers and bird outbreak responders emphasize the importance of worker protection and personal protective equipment (PPE) to mitigate the risk of zoonotic transmission. Biosecurity measures, including surveillance measures, play a crucial role in monitoring and controlling the spread of avian influenza viruses between wild birds and domesticated poultry (CDC, 2022c).

3 Assessing the effectiveness of biosecurity: trends in zoonotic disease outbreaks

An in-depth analysis of the latest annual EU One Health Zoonosis Report, jointly released by the European Food Safety Authority (EFSA) and the European Centre for Disease Prevention and Control (ECDC) (2022), reveals critical insights into the trends of zoonotic disease outbreaks originating from poultry farms. This comprehensive report, covering the year 2021, offers a nuanced perspective on the status of zoonosis monitoring and surveillance activities across 27 Member States, the United Kingdom (Northern Ireland), and nine non-Member States.

Overall zoonotic disease trends in this report indicate an overall increase in reported cases of zoonotic diseases and foodborne outbreaks in 2021 compared to the previous year. However, it is noteworthy that these levels remain well below those observed in pre-pandemic years. The continued impact of the COVID-19 pandemic and associated control measures, including lockdowns, is evident in the challenges faced during data collection and analysis at the EU level. Campylobacteriosis and Salmonellosis were reported as the first and second most significant zoonoses in humans related to poultry production in 2021. While cases increased compared to 2020, they showed a longer-term decrease compared with pre-pandemic years. *Salmonella* Enteritidisremained the most frequently reported causative agent for foodborne outbreaks. *Salmonella* in 'eggs and egg products' and in 'mixed foods' were highlighted as the agent/food pairs of most concern. A positive note is the achievement of official targets for reducing *Salmonella* prevalence in poultry populations by 16 Member States and the UK. This suggests progress in poultry biosecurity measures (EFSA and ECDC, 2022).

In the USA, zoonotic disease outbreaks from poultry farms, particularly *Salmonella*, show a worrying trend. Despite efforts to improve the situation, the incidence of *Salmonella*-related foodborne illnesses in the U.S. has not changed significantly since 2000, affecting a million Americans annually (Nazir et al., 2023). Poultry products, especially chicken, remain a notable source of outbreaks. While some progress in reducing reported contamination rates has been made, the number of *Salmonella* outbreaks linked to chicken increased from 2000 to 2018 (Collins et al., 2022). This reveals a gap between goals and actual outcomes, suggesting the need for stronger regulation. The Food Safety and Inspection Service (FSIS) of the United States Department of Agriculture (USDA) has outlined plans for an improved risk-based approach, targeted specific *Salmonella* serotypes and specific on-farm interventions to effectively control hazards. The proposed new framework includes measures such as testing incoming poultry flocks for *Salmonella*, enhancing process control monitoring and verification, and implementing an enforceable final product safety standard (Talkington, 2021; FSIS-USDA, 2023).

A more recent issue of concern is Highly Pathogenic Avian influenza (HPAI). A(H5N1) virus infections in humans were initially documented in 1997 during a poultry outbreak in the Hong Kong Special Administrative Region (SAR), China. This virus was isolated in 2003 and has since expanded to become endemic among bird populations (including poultry) throughout Asia, Europe, Africa, and the Americas by 2021. Millions of chicken infections, several hundred human cases, and numerous human fatalities have been caused by outbreaks. The current bird flu situation in both the U.S. domestic and global poultry, as outlined by the CDC, reveals a dynamic landscape (CDC, 2022d, 2023c). Since February 2022, there have been HPAI A(H5N1) virus outbreaks in commercial poultry facilities and backyard flocks across multiple states in the U.S. The genetic sequencing of the viruses suggests a clade 2.3.4.4b classification. Globally, from late 2021 to 2022, the predominant HPAI A(H5) virus subtype causing outbreaks is A(H5N1), with reported incidents in Africa, Europe, and Asia. The World Organization for Animal Health data indicates that, between 2005 and 2020, HPAI A viruses led to significant poultry deaths worldwide, with peaks in 2006 and 2016. While the detection of A(H5N1) bird flu in poultry and wild birds in the U.S. is seen to pose a low risk to public health currently, human infections are possible. More than 58.7 million poultry and over 7100 wild birds were impacted in the U.S. as of August 9, 2023 (WHO, 2023; CDC, 2022d, 2023c).

In the context of recorded outbreaks, identifying sources of infection is essential for devising effective prevention strategies. The 2020 CDC Annual Report on Animal Contact Outbreaks highlights that direct contact with infected animals in poultry, especially baby chicks and ducklings, stands out as a significant contributor to outbreaks, constituting 15 outbreaks and causing the highest number of illnesses (1149 illnesses, 84%) (CDC, 2020). As noted earlier, *Salmonella* from poultry is a major concern, accounting for 12 outbreaks and resulting in the highest number of outbreak-associated illnesses (1138), along with hospitalizations (250) and deaths (2) among outbreaks with a confirmed single etiology. This data underscores the central role of poultry-related sources in recorded outbreaks, highlighting the need for targeted interventions and heightened biosecurity measures within the poultry farming sector (CDC, 2020).

4 Assessing the effectiveness of biosecurity: current biosecurity practices on poultry farms

Understanding the contemporary landscape of biosecurity in poultry farms is imperative for effective disease prevention and control. Recent studies from a range of countries shed light on common biosecurity practices adopted by commercial poultry farms, how well they are implemented and potential areas for improvement.

A recent study of 192 large-scale poultry farms across the EU found a high overall level of compliance with 62 identified biosecurity measures (BMs) (Souillard et al., 2024). The most frequently implemented practices were daily surveillance of birds, pest control measures (e.g. for rodents), and secure storage of feed. Areas of weakness included disinfection of rendering tanks and feed silos after each collection, monitoring for bacteria after house cleaning and disinfection between cycles, controlling entry of vehicles and individuals into poultry farms, and failure to consistently follow through 'all-in/all-out' practices (where birds move through each stage as a group, with no new animals added). Reasons for not implementing practices included low awareness of the benefits of a particular BM, lack of training, lack of time, and perceived cost.

A survey of 259 poultry farms in Northeast Italy between 2018 and 2019 showed a high level of biosecurity compliance. Turkey farms exhibited the highest compliance, particularly in farm hygiene, infrastructure (bird housing), and cleaning and disinfection tools. However, some deficiencies were identified, highlighting areas for improvement, such as the cleanliness of hygiene lock areas (areas such as changing rooms and disinfection areas controlling access to birds) and dealing with the buildup of litter in poultry barns. The study underscores the need for continuous vigilance and improvement, even in regions with stringent national regulations (Tilli et al., 2022).

In a study conducted in the Rajshahi district of Bangladesh from June to November 2021, researchers assessed the biosecurity status of 204 poultry farms. The findings indicate that, overall, biosecurity measures are well-practiced and implemented in most broiler and layer farms in the region. Noteworthy practices include restricting visitor access, using company-supplied feed, and prompt cleaning of farm litter. However, the study emphasizes the importance of stricter enforcement and monitoring of biosecurity measures in farms that do not adhere to essential practices (Islam et al., 2023).

Similarly, a study in two regions in Ethiopia explored the biosecurity practices and status in small- and medium-scale commercial poultry farms. The results, based on a structured questionnaire, revealed a need for interventions, with 40.7% of the farms having a biosecurity status below 50%. Poor biosecurity implementation, particularly in the area of employee training, contributed to weaknesses in disease prevention (Tsegaye et al., 2023). Assessment of 44 commercial poultry farms in Bishoftu town, Ethiopia, between December 2017 and April 2018 found that, while some farms exhibited good biosecurity practices, such as the presence of fences and footbaths, others lacked essential facilities such as isolation rooms for diseased chickens and proper record-keeping. The study highlights the association between farm biosecurity status and factors such as ownership, training, source of premises, and farm capacity, emphasizing the need for tailored interventions (Ismael et al., 2021).

A study conducted in Paraná State, Brazil, focused on the biosecurity and sustainability practices of 70 broiler farms. Hierarchical cluster analysis revealed two distinct groups – low- and high-biosecurity farms. Factor analysis identified three key factors: farmer knowledge and training, mandatory environmental safety practices, and non-mandatory environmental safety practices. Despite similarities in production variables and operating conditions, there were significant differences in biosecurity levels. High biosecurity farms exhibited greater compliance with both mandatory and non-mandatory safety requirements, with operators possessing higher education levels compared to those running low biosecurity farms (de Oliveira et al., 2021).

This sample of studies across different regions show a spectrum of biosecurity scenarios, spanning from well-established measures to regions in need of improvement. These insights provide a foundation for our investigation of critical biosecurity measures, which are an important part of disease prevention and control in poultry farming. Let us now turn our attention to the critical practices and protocols that, when combined, constitute the foundation for successful biosecurity, playing a critical role in protecting both poultry and public health.

5 Key components of biosecurity: structural measures

Key biosecurity measures encompass a diverse range of practices and protocols used to mitigate the risk of disease on poultry farms (FAO, 2010). Implementing these measures helps poultry farms avoid disease outbreaks and ensure the well-being and welfare of poultry (Fagrach et al., 2023). These measures include physical barriers, operational protocols, and sound management practices, all working together to prevent the transmission of diseases (Tyler et al., 2022). Among biosecurity measures, structural and operational biosecurity measures are two main components of biosecurity, crucial in establishing a secure poultry farm environment (USDA, 2021).

Structural biosecurity measures focus on the physical environment, including construction and maintenance of farm infrastructure, which can minimize the risk of disease transmission (USDA, 2021). Some examples of structural biosecurity measures include facility design, secure fencing and enclosures, ventilation systems, and waste management (Drayer and Russell, 2004; Oliveira et al., 2017).

5.1 Facility design

The design, construction, and the layout of facilities play a significant role in preventing the entry and spread of diseases within the poultry farm environment (USDA, 2019a; Butucel et al., 2022; Alarcón et al., 2021). Layout of facilities involves careful planning of the placement of poultry houses, pens, and other structures to ensure adequate segregation and control of different groups of birds. The risk of disease transmission between groups can be decreased by designating distinct locations for certain age groups or production phases, as well as planning the layout to reduce contact with outside sources of infections, such as wild birds or rats (Samanta et al., 2018; NAMI, 2022).

The construction of the facilities should focus on creating a secure and bio-secure environment. This includes using appropriate building materials that are durable, easy to clean, and resistant to the entry of pests and disease vectors. Proper ventilation systems should be in place to maintain optimal air quality and minimize the accumulation of harmful pathogens. Adequate waste management systems, including the proper disposal of manure and other waste materials, are essential to prevent the build-up of disease-causing organisms (USDA, 2016; HHS, 2015; Lyman, 2020).

Implementing biosecurity features such as controlled access points and secure fencing is crucial in facility design. Controlled access points ensure that only authorized personnel and vehicles enter the farm premises, reducing the risk of disease introduction. Secure fencing acts as a physical barrier to prevent the entry of wildlife, stray animals, and unauthorized individuals who may inadvertently bring pathogens onto the farm. Implementing these measures not only protects poultry but also helps maintain the integrity of the farm and the surrounding environment (Alarcón et al., 2021; Towers, 2014; USDA, 2019a).

5.2 Secure fencing and enclosures

Secure fencing and enclosures are essential components of structural biosecurity measures in poultry farming. These steps also provide a physical barrier that prevents the entry of disease carriers and unauthorized individuals onto the farm premises, thereby reducing the risk of disease transmission within the poultry flock (CEVA, 2005; Martin et al., 2009; Hayward Kerley, 2009).

One of the functions of secure fencing is to establish a perimeter around the poultry farm, effectively restricting access to the facility. The fencing should be designed to prevent entry of wildlife, stray animals, and other potential disease vectors, such as wild birds or rodents, which can carry and spread pathogens to the poultry flock (Alarcón et al., 2021; Hayward and Kerley, 2009).

In addition to perimeter fencing, enclosures within the farm premises are used to segregate different areas or groups of flock based on their age, health status, or production stage. These enclosures act as an extra layer of protection by preventing direct contact between different flock groups. It is possible to reduce the risk of disease transmission and prevent subsequent outbreaks within the flock by isolating diseased or vulnerable birds from the healthy population (ACFM, 2020; Breukelman et al., 2006; Jagmeet and Leishangthem, 2022).

Secure fencing and enclosures also serve as a deterrent to unauthorized access. Unauthorized individuals may inadvertently bring pathogens onto the farm through contaminated clothing or equipment, posing a serious threat to the biosecurity of the poultry flock. The risk of potential disease introduction is decreased by creating controlled entry points and adopting steps to ensure that only authorized individuals and vehicles can enter the farm premises. These measures also help enforce strict biosecurity protocols and maintain the integrity of the farm environment (Reiss and Woods, 2011; FAO and OIE, 2010).

5.3 Ventilation system

Ventilation systems play a crucial role in maintaining a healthy and disease-free environment for the poultry flock (David et al., 2015; Ritz et al., 2017; Gržinić et al., 2023). Proper ventilation can regulate temperature, humidity, and air quality within poultry housing facilities, creating optimal conditions for the well-being of the flock (Abdisa, 2023; USDA, 2021). One of the primary benefits of ventilation systems as a biosecurity measure is their ability to control airborne pathogens. Ventilation systems help to dilute and remove airborne microorganisms that may carry and spread diseases by ensuring appropriate air circulation and exchange, minimizing the risk of respiratory infections and other airborne diseases among poultry (Zhao et al., 2014; Jacob et al., 2013).

In addition, ventilation systems contribute to reducing moisture levels in the poultry house. Excess moisture can create a favorable environment for the growth of bacteria and fungi, increasing the risk of transmission. Proper ventilation helps to decrease moisture, keeping the litter and surfaces dry, essential to maintain good hygiene and prevent proliferation of pathogens (Ritz et al., 2017; Menges, 2013).

Ventilation systems also reduce the accumulation of odors and pollutants, such as ammonia, in chicken houses. The presence of these chemicals can adversely affect the respiratory health of both birds and the facility staff. Through the removal of these contaminants, effective ventilation contributes to the establishment of a more comfortable and healthier environment for both farm personnel and the chickens, reducing susceptibility to infection (Gržinić et al., 2023; David et al., 2015).

5.4 Waste management

Waste management focuses on the proper handling and disposal of waste materials generated in the production process (Gržinić et al., 2023). Effective waste management serves to limit transmission of diseases and maintain a clean and healthy environment for the poultry flock (Rahman et al., 2022).

Poultry farms produce a variety of waste, including manure, bedding materials, and feed residues (Rahman et al., 2022). If not managed properly, these wastes can harbor pathogens and serve as potential sources of disease transmission (Gržinić et al., 2023). Implementing appropriate waste management practices ensures that these materials are handled, stored, and disposed of in a way that minimizes the risk of disease transmission (Begum et al., 2023).

A key aspect of waste management involves the proper storage and containment of manure and other waste products (Bollwahn, 2014). Poultry farms should designate areas or structures, such as manure storage facilities or composting systems, to safely store and process waste materials (LPELC, 2017). These structures reduce the risk of pathogens spreading to the surrounding environment or contaminating water sources (EPA, 2001).

Proper waste management also involves regular removal of accumulated waste from the poultry facilities. Timely removal prevents the build-up of organic matter and reduces the presence of disease-causing microorganisms. Additionally, waste materials should be transported and disposed of in compliance with local regulations and guidelines to minimize environmental contamination (USDA, 2009; EPA, 2001).

Moreover, effective waste management practices encompass the incorporation of biosecurity measures during waste handling. This may involve the use of dedicated equipment and vehicles for waste removal, maintaining hygiene protocols for personnel engaged in waste management activities, and implementing measures to prevent cross-contamination between different areas of the farm (Abdul-Rahman, 2014; Schneider et al., 2010).

6 Key components of biosecurity: operational measures

Operational biosecurity measures encompass a range of practices and protocols implemented on a day-to-day basis to minimize the risk of disease introduction and spread within the farm environment (USDA, 2021). These practices include keeping visitors to a minimum and implementing visitor guidelines, washing hands before and after interacting with live poultry, providing disposable boot covers (preferred) and/or disinfectant footbaths for anyone having contact with your flock, changing clothes before entering poultry areas and before leaving the property, cleaning and disinfecting tools or equipment before moving them to a new poultry facility, limiting the borrowing of equipment from other poultry/livestock keepers, inspecting your flock daily, looking for signs of illness and segregating any sick birds immediately, and keeping other animals out of the poultry house (USDA, 2021). Implementing these measures consistently and diligently enables farms to establish a robust biosecurity framework, reducing the risk of disease outbreaks, protecting the poultry flock, and upholding the integrity of the entire poultry farming operation (Jones, 2002).

6.1 Keeping visitors to a minimum and implementing visitor guidelines

Keeping visitors to a minimum and ensuring they adhere to visitor guidelines is a key process to effectively reduce the risk of disease introduction and spread within the farm (Cunningham and Fairchild, 2020). Visitors, including farm suppliers, veterinarians, and service providers, should be restricted to those who have a legitimate need to access the farm premises. This reduces the likelihood of introducing pathogens from external sources and helps maintain a more secure and controlled environment (Robertson, 2020; Renault et al., 2021). Allowing only essential visitors and minimizing non-essential personnel on the farm helps to control the entry of potential disease carriers.

Implementing visitor guidelines is crucial to ensure that any visitor entering the farm complies with biosecurity protocols. Visitors should be informed about the farm's biosecurity requirements, including the use of PPE, proper hygiene practices, and designated areas they are allowed to access. These measures include wearing disposable coveralls boots and following handwashing procedures before entering poultry houses or other sensitive areas (USDA, 2021; Robertson, 2020).

Recording visitor information is another important aspect of operational biosecurity (Cunningham and Fairchild, 2020). Farms should maintain a visitor log that includes details such as date, name, purpose of visit, and contact information (AFIA, 2019). This recordkeeping allows for traceability in case of any disease outbreak and facilitates contact tracing if necessary (Wagner, 2019).

Regular communication with visitors is essential to ensure their understanding and compliance with biosecurity measures (CFSPH, 2016). Farm personnel should provide clear instructions, offer training if needed, and regularly remind visitors about the importance of following the guidelines. This helps to create awareness and shared responsibility for biosecurity among all individuals who enter the farm premises (Liu et al., 2022).

6.2 Washing hands

Hand washing thoroughly is one important procedure that should be undertaken before and after interacting with live poultry (USDA, 2021; CDC, 2022a, 2023b). Hand washing removes dirt, bacteria, or viruses, especially when handling live poultry, which may involve contact with feathers, droppings, or other bodily fluids. Using soap and water, the hands are lathered, ensuring all surfaces, including the palms, back of hands, between fingers, and under the nails are scrubbed. The hands are rinsed well under running water to remove any soap residue and potential contaminants (Domel, 2023; CDSA, 2009; CDC, 2022a, 2023b; OVMA, 2023).

Proper hand hygiene significantly reduces the risk of disease transmission within poultry flocks (Racicot et al., 2013). It also acts as a barrier to protect human handlers and other individuals who may encounter birds, preventing the transfer of harmful microorganisms from poultry to humans and vice versa (EHS, 2018; Didier et al., 2021). It is a simple and cost-effective measure that can have a significant impact on maintaining biosecurity on poultry farms (Racicot, 2013).

Education and promotion of handwashing practices among individuals working with live poultry is essential. Farm workers, poultry handlers, and visitors should be made aware of the importance of hand hygiene. Clear instructions and reminders about washing hands before and after handling poultry should be provided to ensure that this biosecurity measure is consistently followed. Facilities such as handwashing stations with access to soap and clean water should be available in convenient locations to encourage compliance (CFSPH, 2023; Gorgo et al., 2019).

6.3 Providing disposable boot covers (preferred) and/or disinfectant footbaths for anyone having contact with your flock

To ensure the biosecurity of your poultry flock, it is recommended to provide disposable boot covers and/or disinfectant footbaths for anyone who has contact with birds (USDA, 2021). This practice helps prevent the introduction and spread of diseases by reducing the risk of contamination from external sources (Dee et al., 2004; Rashid et al., 2016).

Disposable boot covers are the preferred option as they provide a single-use barrier that can be easily disposed of after each visit to the poultry area. These covers should be worn over regular footwear and should be put on before entering the poultry premises. They serve as an effective protective layer, preventing any potential pathogens from being carried into the area (Hersom et al., 2008; MacMahon, 2008).

In addition to disposable boot covers, disinfectant footbaths can be used as an extra precautionary measure. A footbath is a container filled with a disinfectant solution that individuals can step into before and after entering the poultry area (Seaman et al., 2023). The disinfectant solution helps kill any potential pathogens on the footwear, further reducing the risk of disease transmission (Harrison et al., 2022).

When using footbaths, it is important to properly maintain and replenish the disinfectant solution regularly to ensure its effectiveness (Kustritz, 2022). The footbath should be deep enough for to ensure adequate immersion of footwear and should be placed at the entrance of the poultry area. Users should follow the recommended contact time in the footbath to allow sufficient disinfection (McCrea and Bradley, 2008).

Educating and training all individuals who have contact with the flock on the proper use of disposable boot covers and footbaths is crucial. They should be instructed on how to properly put on and dispose of boot covers, as well as the correct procedure for using footbaths. Emphasize the importance of adhering to these biosecurity measures to protect the health and well-being of the poultry flock (McCrea and Bradley, 2008).

6.4 Changing clothes before entering poultry areas and before leaving

Changing clothes before entering poultry areas and before leaving areas is important for operational biosecurity measures (USDA, 2021). It involves having dedicated and designated clothing solely for poultry areas to minimize the risk of introducing or spreading diseases through contaminated attire (Jean et al., 2020).

Similarly, changing out of poultry clothing before leaving the farm is crucial in preventing the spread of pathogens to other locations. This measure serves as a protective barrier, minimizing the chances of carrying any potential contaminants with individuals, thereby reducing the risk of cross-contamination between poultry premises and other environments (USDA, 2021).

Dedicated poultry clothing should be stored separately and regularly laundered using detergent and hot water to eliminate any potential pathogens. Providing designated changing areas equipped with handwashing facilities encourages individuals to adhere to these biosecurity practices. This, in turn, safeguards the health of the poultry flock and prevents the spread of diseases (Jean et al., 2020).

6.5 Cleaning and disinfecting tools or equipment before moving them

A crucial operational biosecurity precaution is to clean and disinfect tools or equipment before relocating them to a different part of the farm or to a new poultry facility (USDA, 2021). Following this procedure prevents pathogens that might be on contaminated equipment from entering the system and spreading (FSIS, 2021). The risk of disease transmission across poultry facilities is reduced by meticulously cleaning and disinfecting tools and equipment (USDA, 2019b).

Proper cleaning involves removing any organic matter, such as dirt, manure, or debris, from the surfaces of tools and equipment (USDA, 2021). This can be done by scrubbing them with a brush and water or using appropriate cleaning agents (Naeve. 2016). After cleaning, disinfection should be carried out using approved disinfectants that are effective against targeted pathogens (CDC, 2022b).

It is crucial to adhere to cleaning and sanitization recommendations made specifically for poultry facilities. This includes paying attention to high-risk areas, such as feeding and watering equipment, as well as tools used for bird handling (USDA, 2019b). Poultry farmers can protect flock health by employing stringent cleaning and disinfection procedures that reduce the possibility of introducing or transmitting diseases through contaminated tools or equipment (USDA, 2018; Rossi et al., 2017).

6.6 Looking for signs of illness

Regularly observing and looking for signs of illness in the poultry flock can mean early detection of any signs of disease that may then be crucial in preventing the spread of infections (Henning, 2004; Levis and Baker, 2011; Balsamo et al., 2017; USDA, 2021). Rapidly spotting potential problems requires familiarity with the typical look and behavior of birds (Hess, 2023).

Signs of illness in poultry can vary depending on the specific disease, but common indicators include changes in appetite, water consumption, egg production, and behavior (e.g. reduced movement). Abnormal respiratory sounds, sneezing, coughing, diarrhea, or lethargy can be warning signs of potential health issues (Jacob et al., 1998). It is important to promptly report any suspicious or unusual signs to a veterinarian or local animal health authority for further evaluation and guidance, and to consider segregation of birds with potential signs of illness (Levis and Baker, 2011; Balsamo et al., 2017; ZAHP, 2023).

6.7 Limiting borrowing of equipment from other poultry/livestock keepers

Borrowing equipment from other poultry or animal keepers should be avoided or minimized for operational biosecurity (USDA, 2021). Each poultry or livestock operation has its unique set of biosecurity protocols and practices (AFIA, 2019). Sharing equipment can introduce unfamiliar pathogens, parasites, or residues that can compromise the health of your flock and increase the risk of introducing diseases or contaminants. Even if the borrowed equipment appears clean and well-maintained, it may still carry pathogens that have survived insufficient cleaning and disinfection (CDC, 2023a). It is best to use equipment that is dedicated solely to your own operation and has undergone proper cleaning and disinfection procedures (CDC, 2023a; Rossi et al., 2017). If borrowing equipment is unavoidable, it is essential to thoroughly clean and disinfect it before use (Smith and Adam, 2023). This includes washing the equipment with appropriate detergents and disinfectants, followed by proper drying and storage (CFSPH, 2022a; Smith and Adam, 2023).

6.8 Segregating any sick birds immediately

It is essential to inspect your flock daily and segregate any sick birds immediately in a separate area when you identify any signs of illness or abnormal behavior in the flock. As noted earlier, report the issue to a veterinarian or the local animal health authority for further evaluation and guidance to prevent the spread of potential diseases to the rest of the flock (USDA, 2019b; Hess and Griffler, 2023; NCAGR, 2019; ZAHP, 2023).

Segregating sick birds serves multiple purposes. First, separating sick birds prevents them from coming into direct contact with healthy individuals, minimizing the potential for the spread of infectious agents (CFSPH, 2022b). Second, it allows you to provide appropriate care and treatment to the affected birds in a controlled environment. This isolation facilitates closer monitoring and targeted interventions to improve their chances of recovery (Hotaling, 2017).

When segregating sick birds, ensure that they have access to clean water, appropriate nutrition, and proper ventilation (O'Malley, 2011). Maintain strict hygiene practices when handling and caring for isolated birds, including frequent handwashing and disinfection of equipment. If possible, consider consulting a veterinarian to diagnose and treat sick birds effectively. Regular monitoring of the progress of isolated birds is essential, including consulting a veterinarian if the condition worsens or if additional measures are necessary to safeguard the health of the flock (CDC, 2023a).

6.9 Keeping other animals out of the poultry house

Animals such as rodents, wild birds, and other domestic animals can carry pathogens that are harmful to poultry. These pathogens can be transmitted through direct contact or through contaminated feces, saliva or respiratory secretions (Gržinić et al., 2023). To keep other animals out of the poultry house, ensure the structure is secure and properly maintained. Repair any holes or gaps in walls, doors, and windows that may serve as entry points for animals. Install screens or wire mesh on windows and vents. Regularly inspect the perimeter of the poultry house and fence off the area to discourage the entry of other animals (Bissong et al., 2022).

Additionally, practice good waste management to minimize attractants for animals. Keep feed stored in secure containers to prevent access by rodents or other animals. Properly dispose of any spilled feed or food waste to avoid attracting wildlife or stray animals. Clear any vegetation or debris around the poultry house that may provide shelter or hiding places for animals (Schneider et al., 2010). Regular monitoring and diligence in enforcing these measures will contribute to a healthier and safer environment for poultry (EPA, 2001).

7 Assessing effectiveness and compliance

Assessing the effectiveness and compliance of safety and biosecurity measures allows farm owners and managers to identify potential gaps or weaknesses in their existing protocols and make necessary improvements (Racicot et al., 2013; Tilli et al., 2022). Here are five key aspects to consider when assessing the effectiveness and compliance of safety and biosecurity measures on poultry farms.

First, the implementation of standard operating procedures (SOPs) related to biosecurity and safety is evaluated. This includes assessing whether employees are following established protocols such as handwashing, wearing PPE, and maintaining proper hygiene practices. Regular inspections can help identify any deviations from SOPs and provide an opportunity for retraining or reinforcing good practices (USDA, 2010).

Secondly, the physical infrastructure and facilities on the farm are monitored. Any structural issues that may compromise biosecurity, such as gaps in walls, damaged fences, or malfunctioning ventilation systems are checked. Whether biosecurity measures, such as controlled access points and barriers, are in place and functioning effectively are ensured. Regular maintenance and repairs are essential to uphold the integrity of the infrastructure of the farm (PIC, 2021).

Thirdly, the disease surveillance and monitoring program of the farms are reviewed. This involves assessing the frequency and comprehensiveness of disease testing and monitoring activities, including regular veterinary inspections and diagnostic testing. The effectiveness of early warning systems for detecting any signs of disease outbreaks and the promptness of response measures implemented to mitigate risks are evaluated (Vourc'h et al., 2006).

Fourth, the training and education programs provided to farm personnel are evaluated. The adequacy and quality of training materials, workshops, and educational sessions related to safety and biosecurity practices are assessed. Whether all employees have received the necessary training and understood their roles and responsibilities in preventing disease transmission are verified (Delabbio, 2006).

Lastly, the record-keeping practices of the farms are considered. Accurate and comprehensive documentation is crucial for assessing compliance and identifying trends or patterns are related to safety and biosecurity. Review records of biosecurity audits, employee training, disease outbreaks, and veterinary visits. This information can provide valuable insights into the overall performance of the farms and areas that require improvement (MDA, 2021).

Regular assessment and monitoring of safety and biosecurity measures on poultry farms are essential to ensure the protection of both animal and human health. Evaluating the effectiveness and compliance of these measures allows farm owners and managers to make informed decisions, enhancing their protocols and mitigating the risks associated with zoonotic and other diseases (Tanquilut et al., 2020).

8 Conclusion

Implementing robust safety and biosecurity measures is crucial in preventing zoonotic and other diseases in poultry farm environments. Strict adherence to biosecurity measures can significantly reduce the risk of disease transmission. Isolation and quarantine procedures should be followed for new or returning birds while minimizing contact with wild birds. Education and training programs for farmers and workers help increase awareness and understanding of biosecurity practices. Collaboration with veterinary professionals and government agencies ensures access to expert guidance and support in disease prevention and control. Regular monitoring of the health of the flocks, early detection of any signs of illness, and prompt reporting to veterinary authorities are vital in preventing disease outbreaks. Poultry farmers may maintain the health and well-being of their flocks, protect human health, and contribute to sustainable and safe poultry production by applying these complete procedures. Additionally, our exploration of transmission routes, trends, and outbreak sources emphasizes the need for dynamic and responsive biosecurity measures on poultry farms. As we move forward, it is essential to acknowledge ongoing challenges, anticipate future developments, and identify key research needs. Advancements in technology, genetic resistance, and vaccine development hold promise, but a collaborative effort involving farmers, government authorities, and researchers is paramount for the sustained success of poultry farm biosecurity.

9 References

Abdisa, T. (2023). Review of diseases and health management of poultry and swine. *International Journal of Avian & Wildlife Biology*, 7(1), 27–38. https://doi.org/10.15406/ijawb.2023.07.00187.

Abdul-Rahman, F. (2014). *Reduce, Reuse, Recycle:Alternatives for Waste Management*. Extension Family Resource Management Specialist, Department of Extension Family and Consumer Sciences, New Mexico State University. Retrieved from https://pubs.nmsu.edu/_g/G314/index.html

ACMF (Australian Chicken Meat Federation). (2020). *National Farm Biosecurity Manual for Chicken Growers*. Retrieved from https://www.farmbiosecurity.com.au/wp-content/uploads/2022/10/National-Farm-Biosecurity-Manual-for-Chicken-Growers_Oct2022.pdf

AFIA (American Feed Industry Association). (2019). *Developing Biosecurity Practices for Feed & Ingredient Manufacturing*. Retrieved from AFIA website: https://www.afia.org/pub/?id=E348BF9F-98ED-09DB-A45D-504737FE7AE2

Alarcón, L. V., Allepuz, A. and Mateu, E. (2021). Biosecurity in pig farms: a review. *Porcine Health Management*, 7, 5. https://doi.org/10.1186/s40813-020-00181-z

Balsamo, G., Maxted, A. M., Midla, J. W., Murphy, J. M., Wohrle, R., Edling, T. M., Fish, P. H., Flammer, K., Hyde, D., Kutty, P. K., Kobayashi, M., Helm, B., Oiulfstad, B., Ritchie, B. W., Stobierski, M. G., Ehnert, K. and Tully Jr., T. N. (2017). Compendium of measures to control chlamydia psittaci infection among humans (Psittacosis) and pet birds (Avian chlamydiosis). *Journal of Avian Medicine and Surgery*, 31(3), 262–282.

Bissong, M. E. A., Lyombe, J. C. N., Asongalem, E., Ngamsha, R. B. and Tendongfor, N. (2022). Zoonotic diseases risk perception and infection prevention and control practices among poultry farmers in the Buea Health District, Cameroon: a one health perspective. *Veterinary World,* 15(11), 2744–2753. https://doi.org/10.14202/vetworld.2022.2744-2753. Epub 2022 Nov 30. PMID: 36590116; PMCID: PMC9798056.

Begum, J. A., Nooruzzaman, M., Modak, M., Khanam, D., Hossain, I., Hasnat, A., Kabiraj, C. K. and Chowdhury, E. H. (2023). Poultry waste management practices in Bangladesh: -farmer's perceptions, and food and environmental hazards. *Journal of Advanced Veterinary and Animal Research*, 10(1), 72–79. https://doi.org/10.5455/javar.2023.j654. PMID: 37155533; PMCID: PMC10122936.

Bollwahn, S. (2014). *Storing Manure on Small Farms – Deciding on a Storage Option*. Michigan State University Extension. Retrieved from https://www.canr.msu.edu/news/storing_manure_on_small_farms_deciding_on_a_storage_option

Breukelman, C., James, A., Falk, K., Thiessen, R., Wiebe, D., Heppel, S., Janzen, D., den Hertog, C., Vaandrager, J., Burm, L., Legendre, E., Schwartz, D., Dyble, D., Wetzstein, M., Paulson, S., Kilmury, R., McKee, A. and Ambrose, N. (2006). *British Columbia On-Farm Poultry Biosecurity Program: Reference Guide* (1st ed.). British Columbia. ISBN 0-9738261-2-6.

Butterworth, A. and Weeks, C. (2009). The impact of disease on welfare. *The Welfare of Domestic Fowl and Other Captive Birds,* 9, 189–218. https://doi.org/10.1007/978-90-481-3650-6_8. PMCID: PMC7121082.

Butucel, E., Balta, I., McCleery, D., Morariu, F., Pet, I., Popescu, C. A., Stef, L. and Corcionivoschi, N. (2022). Farm biosecurity measures and interventions with an impact on bacterial biofilms. *Agriculture,* 12(8), 1251. https://doi.org/10.3390/agriculture12081251

CDC (Centers for Disease Control and Prevention). (2020). *Annual Report on Animal Contact Outbreaks*. Retrieved from https://www.cdc.gov/acoss/annual-reports/index.html

CDC (Centers for Disease Control and Prevention). (2022a). *When and How to Wash Your Hands*. Retrieved from https://www.cdc.gov/handwashing/when-how-handwashing.html

CDC (Centers for Disease Control and Prevention). (2022b). *When and How to Clean and Disinfect a Facility*. Retrieved from https://www.cdc.gov/hygiene/cleaning/facility.html

CDC (Centers for Disease Control and Prevention). (2022c). *Avian Influenza in Birds*. Retrieved from https://www.cdc.gov/flu/avianflu/avian-in-birds.htm

CDC (Centers for Disease Control and Prevention). (2022d). *Current Bird Flu Situation in Poultry*. Retrieved from https://www.cdc.gov/flu/avianflu/poultry.htm

CDC (Centers for Disease Control and Prevention). (2023a). *Backyard Poultry*. Retrieved from https://www.cdc.gov/healthypets/pets/farm-animals/backyard-poultry.html

CDC (Centers for Disease Control and Prevention). (2023b). *Show Me the Science – Why Wash Your Hands?* Retrieved from https://www.cdc.gov/handwashing/why-handwashing.html#:~:text=Germs%20from%20unwashed%20hands%20can%20be%20transferred%20to,may%20even%20help%20prevent%20skin%20and%20eye%20infections

CDC (Centers for Disease Control and Prevention). (2023c). *Reports of Avian Influenza (AI) "Bird Flu" Outbreaks Increased Globally from 2013–2022*. Retrieved from https://www.cdc.gov/flu/avianflu/spotlights/2022-2023/avian-influenza-outbreaks-increased-globally.htm#:~:text=While%20the%20recent%20detections%20of,affected%20in%20the%20United%20States

CDSA. (2009). *Australian Certification Program – Level 1: Food Safety Handbook*. Retrieved from http://www.clubdemoaustralia.com/wp-content/uploads/2013/03/FoodSafetyHandbook_CDSA.pdf

CEVA Santé Animale. (2005). *Vaccines. Chapter 1: Biosecurity*. Retrieved from https://www.sapoultry.co.za/pdf-training/Vaccines-biosecurity.pdf

CFSPH (Center for Food Security and Public Health, Iowa State University of Science and Technology). (2016). *NAHEMS Guidelines: Biosecurity*. Retrieved from https://www.cfsph.iastate.edu/pdf/fad-prep-nahems-guidelines-biosecurity

CFSPH (Center for Food Security and Public Health). (2021). *Routes of Disease Transmission in Poultry*. https://www.cfsph.iastate.edu/Assets/routes-disease-list-poultry.pdf

CFSPH (Center for Food Security and Public Health). (2022a). *Control of Direct Contact Disease Transmission for Beef and Dairy Producers*. Retrieved from https://www.cfsph.iastate.edu/Infection_Control/Routes/English/direct_contact_control.pdf

CFSPH (Center for Food Security and Public Health). (2022b). *Safely Introducing New or Returning Animals and Separating Sick Animals from Your Herd on Organic and Alternative Farms*. Retrieved from https://www.cfsph.iastate.edu/thelivestockproject/safely-introducing-new-or-returning-animals-and-separating-sick-animals-from-your-herd-on-organic-and-alternative-farms/

CFSPH (Center for Food Security and Public Health, Iowa State University of Science and Technology). (2023). *Protecting People & Animals When You Have Visitors to Your Organic/Alternative Farm*. The Livestock Project. Retrieved from https://www.cfsph.iastate.edu/thelivestockproject/protecting-people-animals-visitors/

Collins, J. P. et al. (2022). Preliminary incidence and trends of infection caused by pathogens transmitted through food. *Morbidity and Mortality Weekly Report*, 71, 1260–1284.

Cunningham, D. L. and Fairchild, B. D. (2020). *Biosecurity Basics for Poultry Growers*. Retrieved from https://secure.caes.uga.edu/extension/publications/files/pdf/B%201306_6.PDF

David, B., Mejdell, C., Michel, V., Lund, V. and Moe, R. O. (2015). Air quality in alternative housing systems may have an impact on laying hen welfare. *Part II-Ammonia. Animals (Basel)*, 5(3), 886–896. https://doi.org/10.3390/ani5030389. PMID: 26479391; PMCID: PMC4598711.

de Oliveira, M. E. A., Sidinei, S., Marcato, S. M., Perez, H. L. and Bánkuti, F. I. (2021). Biosecurity, environmental sustainability, and typological characteristics of broiler farms in Paraná State, Brazil. *Preventive Veterinary Medicine*, 194, 105426. https://doi.org/10.1016/j.prevetmed.2021.105426

Dee, S., Deen, J. and Pijoan, C. (2004). Evaluation of 4 intervention strategies to prevent the mechanical transmission of porcine reproductive and respiratory syndrome virus. *Canadian Journal of Veterinary Research*, 68(1), 19–26. PMID: 14979431; PMCID: PMC1142125.

Delabbio, J. (2006). How farm workers learn to use and practice biosecurity. *Journal of Extension*, 44(6). Retrieved from https://archives.joe.org/joe/2006december/a1.php

Dhaka, P., Chantziaras, I., Vijay, D., Bedi, J. S., Makovska, I., Biebaut, E. and Dewulf, J. (2023). Can improved farm biosecurity reduce the need for antimicrobials in food animals? A scoping review. *Antibiotics (Basel),* 12(5), 893. https://doi.org/10.3390/antibiotics12050893. PMID: 37237795; PMCID: PMC10215210.

Didier, P., Nguyen-The, C., Martens, L., Foden, M., Dumitrascu, L., Mihalache, A. O., Nicolau, A. I., Skuland, S. E., Truninger, M., Junqueira, L. and Maitre, I. (2021). Washing hands and risk of cross-contamination during chicken preparation among domestic practitioners in five European countries. *Food Control*, 127, 108062. ISSN 0956-7135. https://doi.org/10.1016/j.foodcont.2021.108062.

Domel, J. (2023). *Tracking, Preventing the Spread of HPAI*. Texas Farm Bureau. Retrieved from https://texasfarmbureau.org/tracking-preventing-the-spread-of-hpai/

Drayer, D. D., & Russell, J. L. (2004). *Physical Security of Carcass Disposal Sites. In Carcass disposal: A comprehensive review*. National Agricultural Biosecurity Center, Kansas State University. http://hdl.handle.net/2097/662

EFSA and ECDC (European Food Safety Authority and European Centre for Disease Prevention and Control). (2022). The European Union One Health 2021 Zoonoses Report. *EFSA Journal,* 20(12), 7666, 273 pp. https://doi.org/10.2903/j.efsa.2022.7666 ISSN: 1831-4732

EHS (Environmental Health and Safety, University of Washington). (2018). *Working with Birds and Poultry*. EH&S Research and Occupational Safety. Retrieved from https://www.ehs.washington.edu/system/files/resources/working-with-birds.pdf

EPA (United States Environmental Protection Agency). (2001). *Source Water Protection Practices Bulletin: Managing Livestock, Poultry, and Horse Waste to Prevent Contamination of Drinking Water*. Office of Water EPA: 916-F-01-026. Retrieved from https://www.marionutilities.com/uploads/1/2/6/5/126583560/epa_livestock_bmp.pdf

Fagrach, A., Arbani, O., Karroute, O., El-Ftouhy, F. Z., Kichou, F., Bouslikhane, M., & Fellahi, S. (2023). *Prevalence of major infectious diseases in backyard chickens from rural markets in Morocco. Veterinary World*, 1897–1906. https://doi.org/10.14202/vetworld.2023.1897-1906

FAO (Food and Agriculture Organization of the United Nations) and OIE (World Organisation for Animal Health). (2010). *Good Practices for Biosecurity in the Pig Sector – Issues and Options in Developing and Transition Countries*. FAO Animal Production and Health Paper No. 169. FAO: Rome.

FSIS (Food Safety and Inspection Service). (2021). FSIS Guideline for Controlling *Salmonella* in Raw Poultry. Retrieved from https://www.fsis.usda.gov/sites/default/files/media_file/2021-07/FSIS-GD-2021-0005.pdf

FSIS-USDA. (2023). *Proposed Regulatory Framework to Reduce Salmonella Illnesses Attributable to Poultry*. Washington, DC: Food Safety and Inspection Service (FSIS) – United States Department of Agriculture (USDA).

Gehrett, S. (2022). Bio-Security on Your Farm Continues to be Important. Retrieved from https://extension.psu.edu/bio-security-on-your-farm-continues-to-be-important

Gorgo-Gourovitch, M., Chawner, M., Seifrit, D., Stoltzfus, J., LaBorde, L. and Baugher, T. (2019). *Best Practices for Fresh Produce Food Safety*. Penn State Extension. Retrieved from https://extension.psu.edu/best-practices-for-fresh-produce-food-safety

Gržinić, G., Piotrowicz-Cieślak, A., Klimkowicz-Pawlas, A., Górny, R. L., Ławniczek-Wałczyk, A., Piechowicz, L., Olkowska, E., Potrykus, M., Tankiewicz, M., Krupka, M., Siebielec, G. and Wolska, L. (2023). Intensive poultry farming: a review of the impact on the environment and human health. *Science of The Total Environment*, 858(Part 3), 160014. ISSN 0048-9697. https://doi.org/10.1016/j.scitotenv.2022.160014.

Hafez, M. and Attia Youssef, A. (2020). Challenges to the poultry industry: current perspectives and strategic future after the COVID-19 outbreak. *Frontiers in Veterinary Science,* 7. https://doi.org/10.3389/fvets.2020.00516

Harrison, O., Houston, G., Blomme, A., Ottot, H., Bai, J., Porter, E., Woodworth, J., Paulk, C., Gebhardt, J. and Jones, C. (2022). Evaluating dry vs. wet disinfection in boot baths on detection of porcine epidemic diarrhea virus and porcine reproductive and respiratory virus RNA. *Translational Animal Science*, 6. https://doi.org/10.1093/tas/txac150

Hayward, M. W. and Kerley, G. I. H. (2009). Fencing for conservation: restriction of evolutionary potential or a riposte to threatening processes? *Biological Conservation*, 142, 1–13.

Hedman, H. D., Vasco, K. A. and Zhang, L. (2020). A review of antimicrobial resistance in poultry farming within low-resource settings. *Animals (Basel),* 10(8), 1264. https://doi.org/10.3390/ani10081264. PMID: 32722312; PMCID: PMC7460429.

Henning, K. J. (2004). Overview of syndromic surveillance: what is syndromic surveillance? *MMWR Supplements*, 53(Suppl), 5–11. Retrieved from https://www.cdc.gov/mmwr/preview/mmwrhtml/su5301a3.htm

Hersom, M., Irsik, M. and Thrift, T. (2008). *Biosecurity and Biological Risk Management for Livestock Enterprises*. IFAS Extension Service, University of Florida. Retrieved from https://www.nifa.usda.gov/sites/default/files/resource/Biosecurity-and-biological-risk-management-for-livestock.pdf

HHS (U.S. Department of Health & Human Services). (2015). Physical Security. Retrieved from Public Health Emergency website: https://www.phe.gov/s3/BioriskManagement/biosecurity/Pages/Physical-Security.aspx

Hess, T. (2023). *Daily Observation for Chicken Health and Well-Being*. The Open Sanctuary Project. Retrieved from https://opensanctuary.org/daily-observation-for-chicken-health-and-well-being/

Hess, T. and Griffler, M. (2023). *How to Conduct a Chicken Health Check*. The Open Sanctuary Project. Retrieved from https://opensanctuary.org/chicken-health-exam/

Hotaling, A. (2017). *Having Patients: How To Help A Sick Chicken*. Hobby Farms. Retrieved from https://www.hobbyfarms.com/sick-chicken-chickens-help/

Huber, N., Andraud, M., Sassu, E. L., Prigge, C., Zoche-Golob, V., Käsbohrer, A., D'Angelantonio, D., Viltrop, A., Żmudzki, J., Jones, H., Smith, R. P., Tobias, T. and Burow, E. (2022). What is a biosecurity measure? A definition proposal for animal production and linked processing operations. *One Health,* 15:100433. https://doi.org/10.1016/j.onehlt.2022.100433. PMID: 36277103; PMCID: PMC9582555.

Islam, Z., Hsan, K., Ripon, R. K., Madhu, J., Hossain, S., Al Masud, A., Islam, S., Kundu, L. R. and Hossain, M. (2023). Assessment of biosecurity measures in commercial poultry farms of Rajshahi district in Bangladesh. *Preventive Veterinary Medicine*, 219, 106027. https://doi.org/10.1016/j.prevetmed.2023.106027

Ismael, A., Abdella, A., Shimelis, S., Tesfaye, A. and Muktar, Y. (2021). Assessment of biosecurity status in commercial chicken farms found in Bishoftu Town, Oromia Regional State, Ethiopia. *Veterinary Medicine International*, 2021. Retrieved from https://api.semanticscholar.org/CorpusID:237289930

Jacob, J. P., Wilson, H. R., Miles, R. D., Butcher, G. D. and Mather, F. B. (1998). *Factors Affecting Egg Production in Backyard Chicken Flocks*. UF/IFAS Extension. Retrieved from https://edis.ifas.ufl.edu/publication/PS029 (Original publication date April 1998. Revised March 2009 and December 2017)

Jacob, J. T., Kasali, A., Steinberg, J. P., Zimring, C. and Denham, M. E. (2013). The role of the hospital environment in preventing healthcare-associated infections caused by pathogens transmitted through the air. *HERD: Health Environments Research & Design Journal*, 7(1_suppl), 74–98. https://doi.org/10.1177/193758671300701S07

Jagmeet and Leishangthem, G. (2022). Biosecurity in poultry farms. *Just Agriculture*, 2(11). Retrieved from https://justagriculture.in/files/newsletter/2022/july/10.pdf

Jean, M., Tocco, P., Gross, P., Zangaro, C., Oliveri, A. and Millerick-May, M. (2020). *Farm Biosecurity Basics: Keeping Employees Healthy*. Michigan State University Extension. Retrieved from https://www.canr.msu.edu/news/farm-biosecurity-basics-keeping-employees-healthy

Jones, F. T. (2002). Economic consequences of a major poultry disease break. *Avian Advice*, 4(4). The Arkansas Cooperative Extension Service. Retrieved from https://poultry-science.uark.edu/_resources/pdf/avian_advice_Wi2002.pdf

Kustritz, M. R. (2022). *Veterinary Preventive Medicine. Chapter 13: Biosecurity*. University of Minnesota Libraries Publishing. Retrieved from https://pressbooks.umn.edu/vetprevmed/chapter/chapter-13-biosecurity/

Kruger, H., Thompson, L., Clarke, R., Stenekes, N. and Carr, A. (2009). *Engaging in Biosecurity: Gap Analysis*. Retrieved from https://www.agriculture.gov.au/sites/default/files/sitecollectiondocuments/abares/publications/biosecuritygapsanalysis.pdf

Levis, D. G. and Baker, R. "Butch". (2011). *Biosecurity of Pigs and Farm Security*. Pork Gateway. Retrieved from https://porkgateway.org/resource/biosecurity-of-pigs-and-farm-security/

Liu, T.-L., Merrill, S. C., O'Keefe, A., Clark, E. M., Langle-Chimal, O. D., Trinity, L., Shrum, T. R., Koliba, C., Zia, A., Sellnow, T. L., Sellnow, D. D. and Smith, J. M. (2022). Effects of message delivery on cross-cultural biosecurity compliance: Insights from experimental simulations. *Frontiers in Veterinary Science*, 9, Article 984945. https://doi.org/10.3389/fvets.2022.984945

LPELC (LPE Learn Center). (2017). *Layer Hen Housing and Manure Management*. Retrieved from https://lpelc.exposure.co/layer-chicken-housing-and-manure-management

Lyman, J. (2020). *Biosecurity Basics for Facilities*. Retrieved from Neogen website: https://www.neogen.com/neocenter/blog/biosecurity-basics-for-facilities/

MacMahon, K. L., Delaney, L. J., Kullman, G., Gibbins, J. D., Decker, J. and Kiefer, M. J. (2008). Protecting poultry workers from exposure to avian influenza viruses. *Public Health Reports*, 123(3), 316–322. https://doi.org/10.1177/003335490812300311. PMID: 19006973; PMCID: PMC2289984.

Martin, V., Forman, A. and Lubroth, J. (2009). *FAO Animal Production and Health Manual: Preparing for Highly Pathogenic Avian Influenza*. Rome: FAO. ISBN 978-92-5-106249-4.

McCrea, B. A. and Bradley, F. A. (2008). *Footbaths for Animal Facilities: Easier -Than You Think*. University of California-Division of Agriculture and Natural Resources. Publication 8281. Retrieved from https://anrcatalog.ucanr.edu/pdf/8281.pdf

MDA (Maryland Department of Agriculture). (2021). *Poultry Operation Record Keeping Guide & Quick Reference Booklet*. Retrieved from https://mda.maryland.gov/resource_conservation/counties/Poultry%20Operator%27s%20Guide.pdf

Menges, J. (2013). *Managing Cold Weather Ventilation*. BVS Poultry. Retrieved from https://www.bestvetsolutions.com/media/documents/BVS_Poultry_Talk_Fall_2013_8EFD1A5EAC0AA.pdf

Naeve, L. (2016). *Clean and Maintain Garden Tools*. Iowa State University Extension and Outreach. Retrieved from https://www.extension.iastate.edu/smallfarms/clean-and-maintain-garden-tools

NAMI. 2022. "The United States Meat Industry at a Glance." *North American Meat Institute*. Accessed October 28, 2022. https://www.meatinstitute.org/index.php?ht=d/sp/i/47465/pid/47465.

Nazir, A. et al. (2023). Rising trends of foodborne illnesses in the United States: short communication. *Annals of Medicine and Surgery*, 85(5), 2280–2281.

NCAGR (North Carolina Department of Agriculture & Consumer Services). (2019). *NC PREP Newsletter: Biosecurity Basics-Practices to Keep Your Flock Healthy*. Retrieved from https://www.ncagr.gov/NCPrep/EducationalContent.htm

Oliveira, V. H. S., Sørensen, J. T., & Thomsen, P. T. (2017). *Associations between biosecurity practices and bovine digital dermatitis in Danish dairy herds. Journal of Dairy Science, 100*(10), 8398–8408. https://doi.org/10.3168/jds.2017-12815

O'Malley, B. (2011). *Avian Emergencies*. World Small Animal Veterinary Association World Congress Proceedings. Retrieved from https://www.vin.com/apputil/content/defaultadv1.aspx?id=5124282&pid=11343&print=1

OVMA (Oregon Veterinary Medical Association). (2023). *Zoonotic Diseases & Birds/Poultry*. Retrieved from https://www.oregonvma.org/care-health/companion-animals/zoonotic-diseases/zoonotic-diseases-birdspoultry

PIC. (2021). *Biosecurity Standards for PIC Multiplication Units and Gene Transfer Centers*. Retrieved from https://www.pic.com/wp-content/uploads/sites/3/2019/09/PIC-Biosecurity-Standards.pdf

Racicot, M., Kocher, A., Beauchamp, G., Letellier, A. and Vaillancourt, J.-P. (2013). Assessing most practical and effective protocols to sanitize hands of poultry catching crew members. *Preventive Veterinary Medicine*, 111(1–2), 92–99. https://doi.org/10.1016/j.prevetmed.2013.03.014. ISSN 0167-5877.

Rahman, M. M., Hassan, A., Hossain, I., Jahangir, M. M. R., Chowdhury, E. H. and Parvin, R. (2022). Current state of poultry waste management practices in Bangladesh, environmental concerns, and future recommendations. *Journal of Advanced Veterinary and Animal Research*, 9(3), 490–500. https://doi.org/10.5455/javar.2022.i618. PMID: 36382043; PMCID: PMC9597914.

Rahman, M. T., Sobur, M. A., Islam, M. S., Ievy, S., Hossain, M. J., El Zowalaty, M. E., Rahman, A. T. and Ashour, H. M. (2020). Zoonotic diseases: etiology, impact, and control. *Microorganisms*, 8(9), 1405. https://doi.org/10.3390/microorganisms8091405. PMID: 32932606; PMCID: PMC7563794.

Rashid, T., VonVille, H. M., Hasan, I. and Garey, K. W. (2016). Shoe soles as a potential vector for pathogen transmission: a systematic review. *Journal of Applied Microbiology*, 121, 1223–1231. https://doi.org/10.1111/jam.13250. -ISSN 1364-5072.

Reiss, A. E. and Woods, R. W. (2011). *National Zoo Biosecurity Manual: A Cooperative Initiative between the Zoo and Aquarium Association, the Australian Wildlife Health Network, the Commonwealth Department of Agriculture, Fisheries and Forestry, and the Australian Zoo Industry* (1st ed.). ISBN 978-1-921575-16-7. Commonwealth of Australia.

Renault, V., Humblet, M. F., Pham, P. N. and Saegerman, C. (2021). Biosecurity at cattle farms: strengths, weaknesses, opportunities and threats. *Pathogens*, 10(10), 1315. https://doi.org/10.3390/pathogens10101315. PMID: 34684268; PMCID: PMC8538770.

Ritz, C. W., Fairchild, B. D. and Lacy, M. P. (2017). *Litter Quality and Broiler Performance*. UGA Extension Bulletin 1267. Retrieved from https://secure.caes.uga.edu/extension/publications/files/pdf/B%201267_5.PDF

Robertson, I. D. (2020). Disease control, prevention and on-farm biosecurity: the role of veterinary epidemiology. *Engineering*, 6(1), 20–25. ISSN 2095-8099. https://doi.org/10.1016/j.eng.2019.10.004

Rossi, G., Smith, R. L., Pongolini, S. and Bolzoni, L. (2017). Modelling farm-to-farm disease transmission through personnel movements: from visits to contacts, and back. *Scientific Reports*, 7(1), 2375. https://doi.org/10.1038/s41598-017-02567-6. PMID: 28539663; PMCID: PMC5443770.

Samanta, I., Joardar, S.N., and Das, P.K. (2018) *Biosecurity Strategies for Backyard Poultry: A Controlled Way for Safe Food Production*. In *Food Control and Biosecurity*. Elsevier, pp. 481–517. https://doi.org/10.1016/B978-0-12-811445-2.00014-3

Schneider, K. R., Goodrich-Schneider, R. M. and Chang, A. (2010). *Food Safety on the Farm: Good Agricultural Practices and Good Handling Practices–Manure and Municipal Biosolids*. IFAS Extension Service, University of Florida. Retrieved from https://www.nifa.usda.gov/sites/default/files/resource/Food-safety-on-the-farm-GAP-manure-municipal-biosolids.pdf

Seaman, J. S., Fangman, T. J. and Carlson Shannon, M. (2023). *Biosecurity for Today's Swine Operation*. University of Missouri Extension. Retrieved from https://extension.missouri.edu/publications/g2340

Smith, J. and Adam, E. (2023). *Science Sleuths: the Science that Shapes Diagnostic Tests: Infectious or Contagious – Which Is It? Equine Disease Quarterly, Science Sleuths*. Retrieved from https://equine.ca.uky.edu/content/science-sleuths-science-shapes-diagnostic-tests-infectious-or-contagious-%E2%80%93-which-it

Souillard, R. et al. (2024). Biosecurity implementation on large-scale poultry farms in Europe: a qualitative interview study with farmers. *Preventive Veterinary Medicine*, 224, 106119.

Talkington, K. (2021, December 10). *12 Charts Explore America's Salmonella Problem–and Steps to Solve It*. Pew Trusts. Retrieved from https://www.pewtrusts.org/en/research-and-analysis/data-visualizations/2021/12-charts-explore-americas-salmonella-problem-and-steps-to-solve-it

Tanquilut, N. C., Espaldon, M. V. O., Eslava, D. F., Ancog, R. C., Medina, C. D. R., Paraso, M. G. V., Domingo, R. D. and Dewulf, J. (2020). Quantitative assessment of biosecurity in broiler farms using Biocheck UGent in Central Luzon, Philippines. *Poultry Science*, 99(6), 3047–3059. https://doi.org/10.1016/j.psj.2020.02.004. PMID: 32475441; PMCID: PMC7597688.

Tilli, G., Laconi, A., Galuppo, F., Mughini-Gras, L. and Piccirillo, A. (2022). Assessing biosecurity compliance in poultry farms: a survey in a densely populated poultry area in Northeast Italy. *Animals (Basel)*, 12(11), 1409. https://doi.org/10.3390/ani12111409. PMID: 35681871; PMCID: PMC9179503.

Towers, L. (2014). *Biosecurity on the Farm: Guidelines & Resources for Developing a Biosecurity Plan*. Retrieved from The Fish Site: https://thefishsite.com/articles/biosecurity-on-the-farm-guidelines-resources-for-developing-a-biosecurity-plan

Tsegaye, D., Tamir, B. and Gebru, G. (2023). Assessment of biosecurity practices and its status in small- and medium-scale commercial poultry farms in Arsi and East Showa Zones, Oromia, Ethiopia. *Poultry*, 2, 334–348. https://doi.org/10.3390/poultry2020025

Tyler, S., Pearson, B. and Hill, E. (2022). *Kansas Cattle Feedyard Biosecurity Guide*. Retrieved from https://agriculture.ks.gov/docs/default-source/ah---disease-control/master-ks-cattle-feedyard-biosecurity-guide-template-march-22.docx?sfvrsn=34adb5c1_14

USDA (United States Department of Agriculture). (2009). *Part 651 Agricultural Waste Management Field Handbook: Chapter 10 Agricultural Waste Management System Component Design*. Natural Resources Conservation Service. Retrieved from https://directives.sc.egov.usda.gov/OpenNonWebContent.aspx?content=31529.wba

USDA (United States Department of Agriculture). (2010). *Foot-and-mouth Disease Standard Operating Procedures: 9*. Biosecurity. Retrieved from https://cms.agr.wa.gov/WSDAKentico/Imported/FMD_SOP_09_Biosecurity_Dec2010.pdf

USDA (United States Department of Agriculture) (2016). *Biological Safety Manual*. Agricultural Research Service. Retrieved from https://www.ars.usda.gov/northeast-area/docs/she/biological-safety-manual/#wastesum

USDA (United States Department of Agriculture). (2018). *Standard Operating Procedures: 15. Cleaning and Disinfection*. The Foreign Animal Disease Preparedness and Response Plan (FAD PReP). Retrieved from https://www.aphis.usda.gov/animal_health/emergency_management/downloads/sop/sop_cd.pdf

USDA (United States Department of Agriculture). (2019a). *Information Manual for Implementing Poultry Biosecurity*. Retrieved from https://poultrybiosecurity.org/files/Poultry-Biosecurity-Info-Manual.pdf

USDA (United States Department of Agriculture). (2019b). *Checklist for Cleaning and Disinfecting Poultry Enclosures*. Retrieved from https://www.aphis.usda.gov/publications/animal_health/fsc-birds-checklist-english.pd

USDA (United States Department of Agriculture). (2021). *Defend the Flock - Biosecurity 101*. Retrieved from https://www.aphis.usda.gov/aphis/ourfocus/animalhealth/animal-disease-information/avian/defend-the-flock-program/dtf-biosecurity#:~:text=Structural%20biosecurity%3A%20measures%20used%20in,are%20consistently%20followed%20by%20people

Vourc'h, G., Bridges, V. E., Gibbens, J., De Groot, B. D., McIntyre, L., Poland, R. and Barnouin, J. (2006). Detecting emerging diseases in farm animals through clinical observations. *Emerging Infectious Diseases*, 12(2), 204–210. https://doi.org/10.3201/eid1202.050498. [Erratum in: *Emerging Infectious Diseases*, 12(4), 714]. PMID: 16494743; PMCID: PMC3293432.

Wagner, N. (2019). *Visitor Biosecurity in Poultry Production*. Retrieved from Neogen Blog: https://www.neogen.com/neocenter/blog/visitor-biosecurity-in-poultry-production/

WHI (World Health Organization). (2023). *Influenza (Avian and Other Zoonotic)*. https://www.who.int/news-room/fact-sheets/detail/influenza-(avian-and-other-zoonotic)#:

~:text=In%201997%2C%20human%20infections%20with,poultry%20populations%20in%20many%20countries

ZAHP (Zoo and Aquarium All Hazards Partnership). (2023). *Highly Pathogenic Avian Influenza (HPAI) Resources*. Retrieved from https://zahp.org/highly-pathogenic-avian-influenza-hpai-2022-resources/

Zhao, Y., Aarnink, A. J. A., De Jong, M. C. M. and Groot Koerkamp, P. W. G. (2014). Airborne microorganisms from livestock production systems and their relation to dust. *Critical Reviews in Environmental Science and Technology*, 44(10), 1071–1128. https://doi.org/10.1080/10643389

Chapter 6

Understanding and dealing with antibiotic resistance in poultry production

Poonam Vinayamohan, Leya Susan Viju, Divya Joseph, and Kumar Venkitanarayanan, University of Connecticut, USA

1 Introduction

Poultry farming is a critical component of global agriculture, providing a vital source of protein to consumers, employment to those working in the industry and contributing to economic activity worldwide. The consumption of poultry meat and eggs continues to rise steadily, driven by population growth, changing dietary habits and increased consumer awareness of its nutritional benefits. In addition to its significance in food security, poultry farming plays a significant role in rural development, particularly in developing countries, offering

http://dx.doi.org/10.19103/AS.2024.0146.08

opportunities to millions of smallholder farmers. The economic contributions of the poultry industry are substantial, with billions of dollars invested annually across the various stages in the supply chain, from production to processing, distribution and retail. Poultry production thus represents a significant share of the agricultural sector in many countries, contributing to national gross domestic product (GDP) and international trade. Moreover, the industry's adaptability and efficiency make it a resilient component of farming economies, capable of withstanding fluctuations in demand and market conditions (National Chicken Council, https://www.nationalchickencouncil.org, Accessed 1st March 2024).

The history of poultry production is intertwined with the widespread use of antibiotics, which revolutionized farming practices in the mid-twentieth century. Since the 1950s, the US poultry industry has been using antibiotic growth promoters (AGPs) at sub-therapeutic levels as feed additives to improve performance metrics such as feed conversion ratio and weight gain. For example, in 1965, the industry standard was a 112-day rearing period for a 2.5-lb chicken with a 4.7 feed conversion ratio. In contrast, modern practices have reduced the rearing period to 42 days for chickens to reach a weight of 6 pounds, with a significantly improved feed conversion ratio of 1.8 (National Chicken Council, https://www.nationalchickencouncil.org/, Accessed 22nd March 2024). This substantial leap in efficiency is partly credited to the use of AGPs alongside other techniques, particularly advances in genetic selection.

Initially celebrated for their ability to enhance animal health and overall productivity, antibiotics soon became a staple in poultry farming utilized not just for disease treatment but also for disease prevention as well as growth promotion. However, this use has now been called into question by the growing global concern over the rise of antimicrobial-resistant (AMR) bacteria which are resistant to the range of antibiotics used to treat human diseases. According to the Centers for Disease Control and Prevention (CDC), the global challenge of AMR bacteria results in approximately 2.8 million infections and 35 000 fatalities each year in the US, incurring billions in economic costs due to increased healthcare costs and decreased productivity (CDC, 2022, https://www.cdc.gov/drugresistance/about.html/, Accessed March 24, 2024). Studies have shown that poultry operations can be significant reservoirs for AMR bacteria, highlighting the critical role of agricultural practices in the proliferation of AMR (Kumar, 2019; Cookey and Otokunefor, 2016). In response to consumer concerns, there has been a global trend toward imposing restrictions or prohibiting antibiotics as growth promoters in livestock feeds. However, despite implementing these restrictions and adopting alternative methods aimed at reducing AMR within poultry farming, there continue to be numerous AMR bacteria linked to poultry.

Several factors contribute to the ongoing AMR challenge, even given widespread antibiotic bans and restrictions. These include incomplete compliance, cross-contamination, the role of global trade in spreading AMR bacteria and the natural evolution of bacterial resistance mechanisms. This evolving scenario has spurred countries, including the US, to further tighten antibiotic use policies, balancing therapeutic needs against the risks of sub-therapeutic applications in promoting AMR. Consequently, this has led to the rise of a distinct market segment for 'antibiotic-free,' 'organic' and 'all-natural' poultry products, aligning with growing consumer demands for safer, healthier and more sustainable food products. Contrary to common assumptions, AMR bacteria have also been detected in organic farming environments (which prohibits the use of antibiotics as growth promoters), suggesting a complex interplay between environmental factors and agricultural practices.

AMR is a naturally occurring phenomenon that has evolved over millions of years as microorganisms compete for survival, primarily through genetic mutations and horizontal gene transfer (HGT). This natural resistance is prevalent even in environments where antibiotics are not used, with resistance genes existing long before the clinical use of antibiotics, highlighting its deep evolutionary roots (Davies and Davies, 2010). These resistance genes are believed to have evolved as a defense mechanism against naturally occurring antibiotics produced by soil microorganisms. These findings emphasize the critical need for a One Health approach, an integrated strategy acknowledging the interconnectedness of human, animal and environmental health in addressing AMR. Understanding how AMR bacteria circulate between environments, animals and humans can help develop strategies that encompass the entire ecosystem, reducing the overall spread of resistance. This chapter explores patterns of antimicrobial use (AMU) in poultry production, reviews recent modifications in use and assesses their impact on trends in AMR. The chapter also examines the common AMR bacteria associated with poultry and the impact of specific agricultural practices on AMR-related environmental and human health. Finally, the chapter outlines and evaluates strategies for managing and reducing AMR in poultry production.

2 Trends in antimicrobial use in poultry production

In establishing different levels of risk, the World Health Organization (WHO) distinguishes between AMU in humans only, in both humans and animals and for use only in animals. WHO also makes a distinction between medically and non-medically important antimicrobial drugs (WHO, 2024). The latter are characterized as antimicrobials authorized only for use in animals for which there is no substantial evidence that their use could result in resistance to medically important antimicrobials (MIAs) for use in humans.

Following earlier bans around the world (e.g. by the European Union (EU) in 2006), the use of antibiotics as growth promotors in feed in the US was banned following the Food and Drug Administration (FDA)'s 2017 regulations against over-the-counter sales of medically important antimicrobials (https://www.fda.gov/animal-veterinary/development-approval-process/veterinary-feed-directive-vfd/, Accessed: March 21, 2024). This regulatory change signified a major move toward reducing AMR and improving health standards.

In poultry production, antibiotics are currently primarily used for the treatment of disease, prevention of disease and metaphylaxis, which involves administering treatment to a group of healthy animals that are in close proximity to other animals showing signs of infectious disease (Roth *et al.*, 2019). Drug regulations vary between broilers and layers. In the US, for example, ionophores are classified as non-medically important antimicrobial drugs. Two ionophores, monensin and salinomycin, are approved for use in pullets, though neither is allowed for use in laying hens (Bello *et al.,* 2023; Singer 2023). In both the US and the European Union (EU), broilers may receive ionophores to prevent coccidiosis, whereas their use is prohibited in layers due to concerns about residue transfer to eggs and resulting consumer health risks. Certain medications, such as enrofloxacin, are disallowed in both broilers and layers (Diaz-Sanchez *et al.*, 2015).

AMU practices in poultry production also vary significantly across different stages of growth and according to specific disease conditions. The use of extended-spectrum cephalosporin (ESC) antimicrobials aims to reduce mortality rates associated with *E. coli* during hatching and addressing *Enterococcus* spp. infections in breeding populations. Injectable antimicrobials are used at one day of age or in-ovo to prevent omphalitis. However, not all drugs (e.g. cephalosporins) are approved for in-ovo use due to resistance concerns. Exceptions include certain drugs, including oxytetracycline, florfenicol and penicillin, that are used for treating specific diseases such as fowl cholera and erysipelas under strict extra-label drug use regulations (AAAP Guidelines, AVMA https://www.avma.org/resources-tools/avma-policies/aaap-guidelines-judicious-therapeutic-use-antimicrobials-poultry/, Accessed: March 21, 2024).

The application of antimicrobial medications to eggs or chicks in hatcheries is a notable factor in the rise and spread of AMR organisms in these environments. The problem is exacerbated by factors such as environmental contamination, hatchery equipment, egg cross-contamination, and the direct spread of bacteria among hatchlings. In a Canadian study, researchers examined fecal *E. coli* from 371 late-rear broiler flocks for the presence of *bla*CMY-2, a gene responsible for resistance to the antibiotic ceftiofur (Caffrey *et al.*, 2017). Their findings showed a correlation between in-ovo administration of ceftiofur at the originating hatchery and a higher occurrence of this resistant phenotype in broilers. A comprehensive eight-year study of avian extra-intestinal *E. coli* (APEC-type) isolates in Ontario has also demonstrated a direct correlation between specific AMUs in hatcheries (particularly ceftiofur, gentamicin, and a lincomycin-spectinomycin combination) and increased resistance levels to these drugs (Kadykalo *et al.*, 2019).

Given these issues, there has been a marked reduction in the use of some antimicrobials such as gentamicin and penicillin within hatcheries. A Japanese study analyzing fecal *E. coli* samples from 362 broiler farms between 2010 and 2013 observed a significant decline in ESC resistance – from over 16% of isolates in 2010 to 4.6% in 2013 – after ceftiofur use was halted in Japanese hatcheries in 2012 (Hiki *et al.*, 2015). Likewise, there has also been a substantial decrease in the use of feed-based antimicrobials, such as ionophores like lasalocid, and medically important antibiotics such as tetracycline and virginiamycin in grow-out birds, signaling a trend toward limiting antibiotic exposure in poultry feed (Singer *et al.*, 2023).

Between 2018 and 2021 water-based antimicrobial treatments were widely used in broiler chickens for treating respiratory conditions like airsacculitis as well as necrotic enteritis (NE) and gangrenous dermatitis (GD). During this period, GD was commonly addressed using penicillin and lincomycin, whereas NE was treated with bacitracin and tylosin along with penicillin and lincomycin. *E. coli* and other respiratory infections were treated with tetracyclines, typically for 3–7 days (Singer *et al.*, 2023). However, after 2021 a noticeable subsequent decline in these treatments, especially with penicillin, lincomycin, tetracycline, sulfonamides, and tylosin, was observed with a shift toward more selected AMU in line with updated global health guidelines (Singer *et al.*, 2023).

Collecting accurate data on actual levels of antibiotic use in the US poultry industry remains challenging, in part due to the absence of a standardized national reporting system. A key source for regulatory bodies is antimicrobial sales data as an indirect measure. However, this data lacks specifics on usage, doses or administration details, highlighting the complexity of enforcing guidelines (Wallinga *et al.*, 2022). However, despite challenges in gathering farm-level data, the available evidence points a significant overall decrease in the use of hatchery, in-feed and water-soluble antimicrobials (Singer *et al.*, 2023).

Between 2013 and 2021, substantial reductions in AMU were noted in the US broiler industry, driven by factors including increased veterinary oversight and a shift toward improved animal management practices to prevent disease. The industry underwent a significant increase in the production of birds raised without antimicrobials, with current estimates indicating that 50–60% of broiler chickens are raised in such environments (Feeks, https://poultryhealthtoday.com/mobile/article/?id=7963/, Accessed March 21, 2024) (Table 1).

However, there has been a recent shift back to incorporating certain non-medically important antimicrobials due to increased disease prevalence and

Table 1 Timeline of antimicrobial use and resistance studies in poultry

Year	Study	Antibiotic use	Impact on AMR
2010	Hiki *et al.*, 2015	High levels of ceftiofur use in Japanese hatcheries.	Over 16% of fecal *E. coli* isolates from broiler farms showed ESC resistance.
2012	Hiki *et al.*, 2015	Japanese hatcheries ceased the use of ceftiofur.	Initiation of a decline in ESC resistance among broiler farms.
2013	Hiki *et al.*, 2015	Cessation of ceftiofur use in Japanese hatcheries.	ESC resistance in fecal *E. coli* isolates from broiler farms dropped to 4.6%.
2017	Caffrey *et al.*, 2017	In-ovo administration of ceftiofur in the Canadian hatcheries.	Higher occurrence of *blaCMY-2* gene in broilers.
2019	Kadykalo *et al.*, 2019	High usage of ceftiofur, gentamicin and lincomycin-spectinomycin in Ontario hatcheries.	Increased resistance levels to ceftiofur, gentamicin, and lincomycin-spectinomycin.
2020	Kadykalo *et al.*, 2019	Complete cessation of gentamicin and penicillin use in Ontario hatcheries.	Expected reduction in resistance to these antibiotics in broiler flocks.
2018-2021	Singer *et al.*, 2023	Increased use of water-based antimicrobials in broiler chickens for treating respiratory conditions.	Use of penicillin, lincomycin, bacitracin, tylosin, and tetracyclines in water treatments.
2021	Singer *et al.*, 2023	Decline in the use of water-based treatments with penicillin, lincomycin, tetracycline, sulfonamides, and tylosin.	Shift toward more prudent antimicrobial use in line with global health guidelines, likely contributing to reduced resistance levels.

animal welfare concerns. Diseases such as NE and GD, primarily caused by *Clostridium* species, continue to pose significant health challenges to broiler production. Specifically, while the use of medically important in-feed antibiotics such as tetracycline was eliminated by 2020 and the use of virginiamycin declined by 97%, the use of non-medically important antimicrobials, such as ionophores, avilamycin and bambermycins, has increased to manage NE, coccidiosis and other diseases (Singer *et al.*, 2023; US Poultry & Egg Association, https://www.uspoultry.org/poultry-antibiotic-use-report/docs/USPOULTRYAntibioticReport-2023.pdf/, Accessed July 1, 2024). The shift from medically important to non-medically important antimicrobials in the poultry industry signifies a commitment to more sustainable health management practices. This change, part of a broader global movement, aims to ensure responsible antimicrobial usage. Concurrently, there is an emphasis on developing accurate on-farm tracking systems to improve antimicrobial stewardship, prioritizing more targeted and effective use over simple reduction.

3 Antimicrobial resistance in particular poultry-associated bacteria: *Salmonella* spp. *and Escherichia coli*

3.1 Salmonella *spp.*

Salmonella is one of the most prevalent AMR pathogens found in poultry. The long-term use of antibiotics in healthy birds is known to promote AMR in this genus (Venkitanarayanan, Thakur and Ricke, 2019). A review by Castro-Vargas *et al.* (2020) demonstrated that, out of 7033 chicken meat samples collected across the world, about 40.5% had *Salmonella* resistant to antibiotics. *Salmonella* Enteritidis and *Salmonella* Typhimurium were the most commonly identified serotypes, along with *Salmonella* Heidelberg and *Salmonella* Infantis.

Another study from Canada examined 193 *Salmonella* isolates from commercial broiler farms, finding the presence of resistant *Salmonella* Kentucky (29%), *S.* Typhimurium (23.8%), *S.* Enteritidis (13.5%) and *S.* Hadar (11.9%) (Diarra *et al.*, 2014). About 58.9% of the isolates of *S.* Kentucky were resistant to multiple antibiotics, and some of them carried *bla* (beta-lactamases) genes (the largest class of antibiotic resistance genes) that produced extended-spectrum β-lactamases (*blaCMY*-2, *blaSHV*, and *blaTEM*). Of the isolates of *S.* Typhimurium, *S.* Enteritidis and *S.* Heidelberg, resistance to β-lactams was found in 32.6%, 23.0% and 100% of the isolates, respectively. Additionally, certain isolates demonstrated resistance to sulfonamides, specifically sulfisoxazole, including strains of *S.* Kentucky, *S.* Typhimurium and *Salmonella* Hadar. These strains were found to carry the *sul1* gene, which is responsible for producing the enzyme dihydropteroate synthase which contributes to resistance (Diarra *et al.*, 2014).

In the US, information gathered from 46 937 samples of poultry meat between 2008 and 2017, as part of the FDA's National Antimicrobial Resistance Monitoring System (NARMS) retail meat program, indicated that conventional poultry samples had a higher prevalence of resistant *Salmonella* at 8.6%, compared to 5.1% in poultry farm samples where no antibiotics were used. According to this report, poultry meat from conventionally grown animals had a higher risk of containing multidrug-resistant (MDR) *Salmonella* and is more likely to harbor AMR genes (Yin *et al.*, 2021). Similarly, the fosfomycin-resistant gene, *fosA7*, was identified in *S.* Heidelberg isolated from conventional broilers (Rehman *et al.*, 2017). MDR *S.* Infantis has also been recovered from chicken samples. The most common AMR patterns of MDR *Salmonella* include resistance to penicillins, tetracyclines, cephalosporins, and fluoroquinolones (de Mesquita Souza Saraiva *et al.*, 2022). Studies conducted in diverse geographical locations, from Nigeria (Igbinosa *et al.*, 2022) to Romania and Brazil (Rau *et al.*, 2021), have consistently demonstrated high resistance levels in *Salmonella* strains isolated from poultry meat.

In recent decades, the US has seen several resistant *Salmonella* outbreaks associated with poultry, drawing attention to the ongoing problem of AMR (Punchihewage-Don *et al.*, 2022). In 2011, a *S.* Heidelberg outbreak related to ground turkey showed resistance to multiple antibiotics, such as ampicillin and tetracycline, leading to widespread illness and numerous hospitalizations. In 2013, another outbreak involving the same bacterial serotype exhibited resistance to a broad spectrum of antibiotics, with many isolates being multiple drug resistant (MDR). Particularly concerning was the presence of the IncI1 plasmid carrying the *blaCMY-2* gene, which led to resistance against antibiotics such as amoxicillin/clavulanic acid and ceftriaxone. A 2018 outbreak, driven by MDR *Salmonella* strains originating from raw turkey, along with a significant 2020 outbreak connected to backyard poultry, confirmed widespread AMR across different *Salmonella* strains (Punchihewage-Don *et al.*, 2022).

3.2 *Escherichia coli*

Findings from multiple studies indicate a widespread issue of AMR *E. coli* within poultry, with a significant percentage showing resistance to several antibiotics even in the absence of antibiotic administration (Nhung, Chansiripornchai and Carrique-Mas, 2017; Chuppava *et al.*, 2019). These studies observed antibiotic resistance to tetracycline, ampicillin and piperacillin, with resistance rates of 97%, 51% and 31%, respectively. In another study, tetracycline was the most frequently resistant antibiotic (≥40%), followed by streptomycin, sulfisoxazole and ampicillin (15–39% of isolates), and trimethoprim-sulfamethoxazole and gentamicin (5–14% of isolates). The most prevalent resistance pattern in chicken *E. coli* isolates was found to be streptomycin-tetracycline-sulfisoxazole-trimethoprim-sulfamethoxazole (Varga *et al.*, 2019).

The detection of *E. coli* resistant to multiple antibiotics in very young chicks, previously unexposed to antibiotics, suggests vertical transmission from parent flocks as the source (Moreno *et al.*, 2019; Chuppava *et al.*, 2019). The resistance might originate from conditions within hatcheries or during transport, leading to colonization by resistant bacteria in newly hatched chicks, thereby becoming a part of their normal gut microflora (Projahn *et al.*, 2018). Broiler chicken manure has been identified as another significant source of AMR *E. coli*, with intensive poultry farming highlighted as a key contributing factor (Fatoba et al., 2022).

Advanced surveillance techniques, such as machine learning and metagenomics, have been applied to chicken production in China, revealing a core subset of the chicken gut resistome and microbiome correlating with AMR profiles in *E. coli* strains (Baker *et al.*, 2023). The study demonstrated that AMR profiles were consistent across multiple chicken farms and abattoirs, suggesting that resistance can spread through shared environmental reservoirs and HGT, rather than solely through direct antibiotic treatment. Antimicrobial resistance arises from complex interactions across AMR bacteria, microbial communities, geographical niches, environments, evolutionary forces, climate and human practices, highlighting the complexity of addressing AMR in agricultural settings.

Third-generation cephalosporin resistance induced by extended-spectrum beta-lactamase (ESBL) production is another major public health concern, as ESBL-producing bacteria are resistant to almost all beta-lactam antibiotics. ESBL *E. coli* is commonly encountered in poultry production and has a high prevalence in chicken meat (Blaak *et al.*, 2015). The most common ESBL-type from poultry include *CTX-M-1*, *TEM-52* and *SHV-12* (Saliu, Vahjen and Zentek, 2017). Further investigations into the source of ESBL *E. coli* have identified their presence in both broiler and layer farms. High prevalence observed in environments close to poultry barns further contributes to an increase in infection pressure with these *E. coli* strains (Saliu, Vahjen and Zentek, 2017).

A recent study focusing on *E. coli* strains from treating poultry respiratory infections found that all isolates exhibited complete resistance to several antibiotics. The analysis also identified multiple genes associated with AMR, including *strA*, *ere*, *tetA*, *aac*-(3)-(IV) and *blaTEM* (Hassan, 2014). Multiple resistance to amoxicillin–clavulanic acid, ceftiofur, ceftriaxone, cefoxitin, gentamicin, sulfonamide and tetracycline in commensal and avian-pathogenic *E. coli* (APEC) isolates have been documented in poultry. For example, an analysis of 116 APEC isolates from Egyptian broiler farms revealed a high number of strains (58.6%) that produced ESBL. The most common genes in these strains were *blaTEM* and *blaCTX−M−1*, as observed in the majority of the poultry ESBL *E. coli* isolates identified in other studies. Antibiogram results indicated the highest levels of resistance to ampicillin, tetracycline, nalidixic acid and chloramphenicol. Further investigation showed the existence of gene cassettes encoding resistance to streptomycin/spectinomycin (*aadA1*, *aadA2*, *aadA5*, aadA23), trimethoprim (*dfrA1*, *dfrA5*, *dfrA7*, *dfrA12*) and streptothricin (*sat2*) (Awad, Arafat and Elhadidy, 2016). Eight isolates from China (out of 31 tested) and four from Egypt (out of 20 tested) had the plasmid-mediated *mcr*-1 colistin resistance gene according to a recent investigation on a sizable collection (980) of APEC isolates from various nations. The majority of these isolates were resistant to ten or more antimicrobial agents (Barbieri *et al.*, 2017).

4 Antimicrobial resistance in particular poultry-associated bacteria: *Campylobacter* spp. and *Staphylococcus* spp.

4.1 Campylobacter *spp.*

Campylobacter is the primary source of human campylobacteriosis and is often associated with the consumption of contaminated chicken meat (Sahin,

Morishita and Zhang, 2002). Contamination typically occurs at several points during the poultry slaughtering process, particularly during the evisceration and chilling phases.

The incidence of AMR in *Campylobacter* strains from both animals and humans has escalated. This increase is particularly evident in resistance to tetracyclines, quinolones and aminoglycosides. High frequencies of resistance to nalidixic acid, ampicillin, cephalexin, ciprofloxacin, erythromycin, gentamicin and tetracycline have also been reported in *Campylobacter* spp. isolated from broilers, laying hens, chicken carcasses and chicken meat (de Mesquita Souza Saraiva *et al.*, 2022). In conventionally raised broiler chickens, *Campylobacter* strains exhibited very high resistance rates, spanning from 88.6% to 100%, against medically important antibiotics such as macrolides, chloramphenicol, tetracyclines and quinolones (Gharbi *et al.*, 2018).

There have been reports of extremely high tetracycline resistance in *Campylobacter jejuni* (99.5%) and *Campylobacter coli* (96.3%) (Son *et al.*, 2007). The widespread resistance to tetracycline in *Campylobacter* is largely due to the *tet*(O) gene, which promotes the transfer of resistance traits between *C. jejuni* and *C. coli* strains (Kim *et al.*, 2010). This is of significant concern as tetracycline resistance is typically present on a conjugative plasmid containing the *tet*(O) gene (Son *et al.*, 2007). The previous over-the-counter availability of tetracycline in the US likely contributed to a significant prevalence of resistance in commercial chicken flocks (Food and Drug Administrations, https://www.fda.gov/animal-veterinary/antimicrobial-resistance/gfi-263-frequently-asked-questions-farmers-and-ranchers, Accessed July 1, 2024).

AMR in *Campylobacter* is exacerbated by selection pressure from oral fluoroquinolone use in both humans and animals, leading to increased fluoroquinolone resistance in *C. jejuni* (van Boven *et al.*, 2003). The persistent spread of fluoroquinolone resistance from breeding flocks to their progeny highlights the deep-rooted nature of AMR within the poultry industry (Idris *et al.*, 2006). The most common cause of *C. jejuni* and *C. coli* resistance to fluoroquinolones is mutations in the quinalone resistance-determining regions (QRDR) regions of *gyr*A. Specifically, high levels of ciprofloxacin resistance are linked to the Thre-86-Ile mutation, whereas the Asp-90-Asn and Thr-86-Lys mutations result in moderate levels of resistance.

Erythromycin, an antibiotic for the treatment of human campylobacteriosis, has seen a significant rise in resistance among *C. jejuni* and *C. coli* strains in poultry. The use of tylosin, a macrolide-like erythromycin, added at subtherapeutic levels to poultry feed in the USA and Canada for growth promotion, has inadvertently contributed to increased resistance to erythromycin. While 11.4% of *Campylobacter* isolates from chickens receiving a therapeutic dose exhibited resistance to erythromycin, the number soared to 63% among isolates from chickens given subtherapeutic levels of tylosin (Gyles, 2008).

4.2 Staphylococcus spp.

Staphylococci is frequently isolated from chickens and has demonstrated antibiotic resistance. Biofilm-forming *Staphylococci* isolated from air samples in broiler production facilities have been found to harbor genes conferring resistance to a range of antibiotics (Vela *et al.*, 2012). Isolates of *S. aureus* resistant to fluoroquinolones have been detected in chicken products, probably linked to the use of fluoroquinolones in the US broiler industry from 1995 to 2005 (Nelson *et al.*, 2007).

Staphylococcal isolates in the US have been found to be resistant to multiple drugs, including gentamycin, vancomycin, cotrimoxazole, rifampin and chloramphenicol, with over half of isolates (52%) identified as MDR (Waters *et al.*, 2011). MDR *S. aureus* has been isolated from raw poultry products, demonstrating prevalent resistance to critical antibiotic classes such as β-lactams, macrolides, quinolones and fluoroquinolones (Buzón-Durán, Capita and Alonso-Calleja, 2017). As a result, there are currently very few medications available for treating human *Staphylococcal* infections. This situation is exacerbated by the emergence of methicillin-resistant *S. aureus* (MRSA) isolates, showing high resistance to multiple antibiotics, including co-trimoxazole, ciprofloxacin, gentamicin and tetracycline (Kali *et al.*, 2013). In India, *Staphylococcal* isolates were shown to harbor the *mecA* gene conferring colistin resistance (Bhedi *et al.*, 2018). The trend of increasing AMR in *Staphylococcus* strains indicates a growing public health challenge that demands urgent action to mitigate its spread and impact.

5 Antimicrobial resistance in particular poultry-associated bacteria: *Clostridium perfringens* and other Enterobacteriaceae

5.1 Clostridium perfringens

Clostridium perfringens, a bacterium associated with NE in poultry, poses significant challenges in the broiler industry due to its varied AMR patterns and contribution to poultry morbidity and mortality. The variability in AMR among *Clostridium perfringens* strains in poultry farms highlights their genetic diversity and the presence of specific resistance genes. The *tet*(M) gene, for example, confers resistance to oxytetracycline (Martel et al., 2004).

Over 95% of *C. perfringens* samples from broiler chickens with NE in Egypt were found to be resistant to multiple antibiotics, including colistin, doxycycline, perfloxacin, sulfamethoxazole-trimethoprim and neomycin, though they remained largely susceptible to cephradine, florfenicol, fosfomycin, ampicillin and amoxicillin (Osman *et al.*, 2012). Comparable research in Korea found 30 *C. perfringens* strains from chickens with NE which displayed resistance to gentamycin, neomycin, streptomycin, apramycin and colistin while showing susceptibility to other antibiotics such as ampicillin, amoxicillin/clavulanic acid, cephalothin, cefepime, chloramphenicol, cefoxitin, ceftiofur and penicillin (Park *et al.*, 2015). *C. perfringens* strains isolated from the intestines of broiler chickens showed high susceptibility to penicillin, avilamycin, monensin and narasin (Silva *et al.*, 2009). Results of AMR gene profiling indicated that the tetracycline resistance gene (41.33%) was more prevalent than erythromycin (34.66%) and bacitracin (17.33%) resistance (Anju *et al.*, 2021). In a study in Pakistan, 34 *C. perfringens* strains were isolated from 134 samples (an overall incidence of 25.37%) and demonstrated resistance to multiple drugs, with the highest levels of resistance against tetracycline, neomycin, trimethoprim and lincomycin (Haider *et al.*, 2022).

5.2 Other Enterobacteriaceae

Enterobacteriaceae comprise Gram-negative bacteria, encompassing over 30 genera and more than 100 species. The family includes many pathogens, such as *Enterococci*, *Enterobacter*, *Pseudomonas* and *Shigella* as well as *Salmonella* and *E. coli*. Isolates of colistin-resistant *Enterobacteriaceae* obtained from poultry exhibited resistance to multiple drugs. These isolates were shown to carry the *mcr* gene providing colistin resistance along with multiple other AMR genes (Karim *et al.*, 2023).

AMR has been observed in *Enterococcus* spp. and *E. coli* isolated from the gastrointestinal microbiota of broiler chickens (Ribeiro *et al.*, 2023). Identification of high levels of resistance against aminoglycosides (streptomycin), tetracyclines (doxycycline and tetracycline) and quinolones (ciprofloxacin and enrofloxacin) in *Enterococci* isolated from poultry have been reported (de Mesquita Souza Saraiva *et al.*, 2022). Elevated resistance from a number of pathogens to key antibiotics such as tetracycline, ampicillin and fluoroquinolones is particularly alarming.

Pseudomonas aeruginosa isolates from poultry litter have shown some susceptibility to aztreonam and levofloxacin but resistance to aminoglycoside, monobactam, penicillins, carbapenems, quinolones and cephalosporin antibiotics. About 89.4% of the MDR strains had the class 1 integron, which is known to contain multiple AMR genes. The *blaVIM* gene in *P. aeruginosa* has been linked to resistance in a majority of β-lactam antibiotics such as meropenem, imipenem, aztreonam and ceftazidime (Zhang *et al.*, 2017). A

Nigerian study reported that the *P. aeruginosa* isolates were highly resistant to β-lactams, tetracycline, tobramycin, nitrofurantoin and sulfamethoxazole-trimethoprim (Iroha *et al.*, 2016).

6 Mechanisms of action of antimicrobial resistance

Bacteria coexisting in environments where antimicrobials are present have evolved enduring strategies to withstand the effects of these antimicrobials. Two primary pathways are linked to the evolution and emergence of AMR:

- Resistance mediated by pre-existing phenotypes
- Acquired resistance

The first pathway involves resistance mediated by pre-existing phenotypes within natural bacterial populations. Over time, bacteria develop genetic mutations in existing genes (found in the bacterial chromosome or plasmids) and transfer these genetic determinants responsible for inherent or intrinsic resistance to subsequent generations of cells through vertical gene transfer. The second scenario relates to acquired resistance, which occurs through a direct pathway involving gene mutations or an indirect pathway involving the acquisition of DNA fragments encoding resistance (Holmes *et al.*, 2016).

Research indicates that the development of resistance within bacterial populations on poultry farms occurs primarily through point mutations within the bacterial DNA or by the acquisition of resistance genes from another bacterium (Kumar, Pornsukarom and Thakur, 2019). Intrinsic resistance in bacteria involves mechanisms such as the degradation of the drug by bacterial enzymes or the modification of drug targets within the bacteria, making the antibiotics less effective. Environmental factors, such as changes in radiation, light or pH levels, can trigger these resistance mechanisms. Consequently, these bacteria can resist the effects of antibiotics without prior exposure, showcasing a natural form of resistance that is critical in the context of antimicrobial exposure (Ali, Rafiq and Ratcliffe, 2018).

Acquired resistance is primarily facilitated by two genetic strategies:

- gene mutations; and
- HGT.

Mutations can alter drug targets within the bacteria or reduce the drug's uptake, rendering antibiotics ineffective. Additionally, bacterial enzymes can inactivate antibiotics by adding chemical groups to them, preventing their binding to target sites (Munita and Arias, 2016).

HGT can occur through various mechanisms, including:

- conjugation;
- transduction; and
- transformation.

Conjugation, the direct transfer of genetic material, often involves plasmids and occurs through cell-to-cell contact. Transformation involves the uptake of naked DNA fragments from the environment, while transduction involves the transfer of genetic material between bacteria via bacteriophages. These processes collectively facilitate the rapid dissemination of resistance traits among bacterial populations (Munita and Arias, 2016).

The role of mobile genetic elements such as transposons, integrons and plasmids is central to the dissemination of AMR genes. Transposons, or 'jumping genes,' transfer resistance traits between different DNA molecules within a cell, while integrons, which are common in Gram-negative bacteria, aggregate and express various resistance genes, facilitating the spread of MDR. Class 1 integrons are notable for carrying genes conferring resistance to a broad spectrum of antibiotics, contributing significantly to the AMR problem (Wellington *et al.*, 2013; Holmes *et al.*, 2016). While clonal spread and HGT are recognized as primary routes of AMR distribution among *Salmonella*, AMR also arises through other mechanisms, including enzymatic inactivation of antimicrobials, modification or protection of target sites, restricting antimicrobial entry into microbial cells and active efflux of antimicrobial agents (Walsh, 2010).

7 Factors contributing to colonization of antimicrobial-resistant bacteria in poultry

In addition to concerns about the spread of AMR bacteria associated with poultry farming, there are significant human health concerns regarding the potential presence of antimicrobial residues in meat and eggs (Reig and Toldrá, 2008; Goetting, Lee and Tetl, 2011). Several factors, such as cross-contaminated feed, litter recirculation, and direct contamination contribute to drug residues in poultry products. Additionally, eggs may contain residues in both yolk and albumen. Poultry meat and offal such as liver and kidneys have been found to contain high antibiotic residue levels, often surpassing maximum residue limits (MRLs) due to non-compliance with withdrawal periods and poor biosafety measures (Mund *et al.*, 2017). Consumption of poultry contaminated with antibiotic residues, such as tetracycline, can lead to health issues, including gastrointestinal problems, allergic reactions and the development of AMR bacterial strains in humans (Morshdy *et al.*, 2013).

In the US, the United States Department of Agriculture (USDA) Food Safety and Inspection Service (FSIS) conducts regular monitoring of antibiotic residues in meat and egg products. The results from these tests help inform regulatory strategies and ensure compliance with food safety standards (2018-red-book.pdf https://www.fsis.usda.gov/sites/default/files/media_file/2020-07/2018-red-book.pdf/, Accessed March 22, 2024). Federal monitoring indicates that antibiotic residues in poultry are significantly lower than public perception, with only a small fraction of samples from 1997 to 2000 above permitted levels. These results confirm that US poultry meat and eggs generally maintain safety standards and are unlikely to pose serious health risks from such residues (Donoghue, 2003). In a study carried out in EU countries, 3 out of 20 poultry meat samples from markets in Germany and Lithuania were found to contain antibiotic residues. However, these residues did not surpass MRLs (Bartkiene *et al.*, 2020). In developing countries, the widespread and often unregulated use of antimicrobial drugs in often smallholder poultry farming remains a significant concern (Wang *et al.*, 2019; Al-Ghamdi *et al.*, 1999). A study in Egypt, for example, detected residues above MRLs of oxytetracycline, gentamicin and tilmicosin in chicken meat (Hussein, Ahmed and Morshedy, 2016).

The use of antimicrobials in poultry both promotes the selection and persistence of AMR in human populations and facilitates colonization of human microbiota by avian-origin-resistant bacteria (Levy, Fitzgerald and Macone, 1976; Ben *et al.*, 2019). AMU in veterinary medicine has been shown to contribute to the emergence of drug-resistant bacteria such as *Enterococcus*, *Campylobacter*, *Salmonella*, and *E. coli* (Bager *et al.*, 1997; Dutil *et al.*, 2010; Nelson *et al.*, 2007). The application of fluoroquinolones in broiler chickens, e.g., correlates with rising instances of quinolone-resistant *Salmonella* strains (Cui *et al.*, 2019). Similarly, contamination of retail chicken with ceftiofur-resistant *Salmonella* Heidelberg strains has been associated with ceftiofur-resistant *S.* Heidelberg infections in humans (Dutil *et al.*, 2010).

Humans primarily acquire AMR microorganisms through contaminated food or contact with resistant non-pathogenic isolates (Laxminarayan *et al.*, 2013). Exposure to environments contaminated with resistant bacteria (such as water, air, soil or manure) can also lead to the acquisition of AMR in humans (Ma et al., 2021; Mak *et al.*, 2022).

A significant portion of the antibiotics used in livestock farming can be found in the gastrointestinal tract of animals at sub-lethal concentrations, suppressing the growth of susceptible bacterial populations. This scenario creates a selective environment that promotes gut bacteria to acquire AMR genes, resulting in populations that are resistant (Gullberg *et al.*, 2011). The chicken gut microbiota has been identified as a significant reservoir for AMR genes (Yang *et al.*, 2022). Additionally, up to 90% of antimicrobials administered in the poultry are excreted, which accumulates in poultry litter. Fecal waste from poultry facilitates the spread of AMR genes (Colomer-Lluch et al., 2011; Dhanarani et al., 2009). The number of residues left behind by chicken feed additives, including bacitracin, chlortetracycline, monensin, narasin, nicarbazin, penicillin, salinomycin and virginiamycin, might vary between 0.07 and 66 mg/L (Furtula et al., 2010). The highest concentration (11.0 mg/kg) of chlorotetracycline was detected in livestock manure and amended soils in Northern China (Huang *et al.*, 2013).

Antibiotic residues in the poultry environment led to environmental reservoirs of AMR bacteria and AMR genes by exerting selection pressure on the environmental microbiome (Ben *et al.*, 2019). The dissemination of antibiotic residues, along with AMR bacteria and AMR genes, into the surrounding environments from poultry farms contributes significantly to environmental pollution and the spread of AMR contaminants. Research has consistently identified the presence of resistant strains such as *Staphylococcus*, *E. coli* and *Salmonella* in various poultry farm-related samples, including litter, water and air, indicating that the impact of poultry farming on AMR extends beyond the farms to affect broader ecosystems (Dhanarani *et al.*, 2009; Furtula *et al.*, 2013; Laube *et al.*, 2014; Brower *et al.*, 2017). For example, when AMR bacteria in the livestock gastrointestinal environment are excreted, AMR genes are disseminated into environments such as soil and water.

AMR gene transmission into soil primarily occurs via fecal water contamination or the use of organic manure-based fertilizers, leading to the lateral movement of resistance genes across different environments and their incorporation into pathogenic bacteria genomes. Studies have also documented the transfer of specific resistance genes, such as the *aadA* gene, from poultry manure to soil, highlighting the interconnectedness of soil and gut microbiomes in sharing significant quantities of these genes (Binh *et al.*, 2009). The shared resistome between soil and gut microbiome indicates that a substantial number of resistance genes are commonly found across both environments (Forsberg *et al.*, 2012; Nesme *et al.*, 2014). AMR *E. coli* strains have been shown to survive in soil for months, carrying genes resistant to major antibiotics, including β-lactams, tetracyclines and streptomycin. Soil microbiota have been shown to be a reservoir of resistance genes for clinical pathogens which have been transmitted from soil bacteria to human pathogens (Forsberg *et al.*, 2012).

Antibiotic residues have been found in wastewater and surface water on a regular basis, typically at values of 0.01–1.0 μg/L (Le Page *et al.*, 2017). Quinolones, chloramphenicol, sulfonamides and macrolides are the major antibiotics detected in drinking water, and ciprofloxacin exhibits the highest detected concentration, at up to 679.7 ng/L (Ben *et al.*, 2019). Wastewater samples from a German poultry slaughterhouse show the presence of ESKAPE bacteria such as ESBL-producing *E. coli*, AMR genes and antibiotic residues (Savin *et al.*, 2021).

Environmental factors play an important role in the dissemination and proliferation of AMR pathogens within poultry production systems. The predominant pathway for the introduction of these pathogens is through HGT from environmental sources, with contaminated feed and water serving as potential reservoirs that facilitate the transfer of AMR pathogens to the flock (Krauland *et al.*, 2009; Pérez-Boto *et al.*, 2010). The presence of residual antimicrobial agents in poultry litter and manure also significantly contributes to the enrichment of AMR genes in soil and water systems, perpetuating the cycle of resistance (Ghosh and LaPara, 2007). The typical commercial poultry production setting, where birds are housed on littered concrete floors, promotes accumulation of and contact with excreta and manure, promoting AMR and other bacteria (Furtula *et al.*, 2010).

Bioaerosols originating from litter and manure within poultry facilities also play a significant role in the transmission of AMR pathogens (Just *et al.*, 2012). Bioaerosols have been found to harbor erythromycin-resistant methylase (*erm*) genes such as *ermA* and *ermC*. Airborne dissemination of AMR bacteria poses a risk not only to the immediate flock but also to neighboring farms and surrounding areas. The detection of AMR genes such as erythromycin (*erm*), tetracycline (*tet*) and zinc bacitracin (*bcr*) in poultry bioaerosols also suggests that poultry farmers could serve as carriers of AMR bacteria in their nasal passages, providing another transmission route (Just *et al.*, 2012). A study involving shotgun metagenome sequencing in Italian poultry farms revealed significant similarities between the microbiomes of poultry, farm workers and their families, highlighting a direct pathway for AMR gene transmission from animals to humans (Scicchitano *et al.*, 2024). Similarly, findings from Zambia and Bangladesh have demonstrated the widespread presence of resistant bacteria, including ESBL-producing *E. coli* among poultry farm workers, as well as a high prevalence of MDR pathogens in food products (Mwansa *et al.*, 2023; Rafiq *et al.*, 2022).

Hatcheries have been identified as an important source of AMR in poultry production (Wales and Davies, 2020). Projahn *et al.* (2018), exploring the prevalence of *E. coli* across two antimicrobial-free broiler production chains, showed that plasmid-borne ESC resistance (*blacmy-2*) found in samples from a flock and its transport truck had probably transmitted these bacteria from parent to production flocks via the hatchery environment (Projahn *et al.*, 2018). Another Japanese study of pathogen prevalence in chick delivery box liners detected transmissible ESBL (blaCTX-M-25) genes within *Enterobacter* and *Klebsiella* species (Yossapol, Sugiyama and Asai, 2017). Genetic analysis and resistance profiles of those isolates revealed unique strains from each of the three hatcheries studied. Similar findings were observed in research conducted in Germany and Denmark, where unrelated broiler flocks sharing common hatcheries exhibited colonization by similar AMR *Enterobacteriaceae* strains such as MDR avian pathogenic *E. coli* (APEC), further highlighting hatchery-based cross-contamination and the persistence of AMR organisms without direct selective pressure from antibiotic use (Daehre *et al.*, 2018; Petersen *et al.*, 2006). Environmental assessments in hatcheries across Egypt and the UK revealed resident strains of *E. coli* with distinct AMR patterns and strong biofilm-forming capabilities (Osman *et al.*, 2018).

Another critical factor contributing to AMR in poultry production is inadequate implementation of biosecurity measures and hygiene practices. Poor biosecurity protocols can facilitate the movement of AMR pathogens into poultry facilities, increasing the risk of colonization and subsequent transmission among birds (Poudel *et al.*, 2024). Tawakol *et al.* (2023) have demonstrated that migratory birds are key vectors in transmitting AMR and pathogenic *Campylobacter* strains to broiler chickens, emphasizing the critical need for stringent biosecurity practices. Additionally, suboptimal hygiene conditions, such as inadequate cleaning and disinfection of equipment and facilities, can promote the persistence of AMR bacteria, creating reservoirs of contamination that perpetuate the spread of resistance within the production environment (van Dijk and Verbrugh, 2022).

Organic and conventional farms employ differing farming practices, impacting the spread and prevalence of AMR bacteria in agricultural settings. Research e.g. has suggested lower rates of *Campylobacter* infection in organic farms (Rothrock *et al.*, 2016). However, studies conducted in various parts of the US have reported a slightly higher overall prevalence of AMR in poultry from organic farms compared to those from conventional ones (Ager *et al.*, 2023). This trend has been found in research from other parts of the world, including the UK and Portugal. For example, systems raising poultry organically or without antibiotics displayed higher frequencies of *E. coli* strains resistant to tigecycline than those found in conventional farming; organic samples also showed higher resistance levels to azithromycin and gentamicin compared to those from antibiotic-free and conventional systems (Musa *et al.*, 2021).

Conversely, conventional poultry systems were found to exhibit higher resistance to ceftazidime and cefotaxime than their organic counterparts, suggesting that organic practices might offer some protection against certain AMR strains (Musa *et al.*, 2021). Similarly, antibiotic-free and organic poultry systems have demonstrated decreased occurrences of ESBL *E. coli*, indicating potential protective benefits against this type of resistance compared to conventional methods (Musa *et al.*, 2021). Conventional poultry samples in Italy revealed the greatest levels of resistance to chloramphenicol-resistant

E. coli despite a ban on chloramphenicol in European livestock since 1997. The detected resistance might be attributed to the use of amoxicillin in treating broilers. These findings suggest co-selection for resistance to these antibiotics (Burow *et al.*, 2020).

Other studies show varying results, with some indicating higher MDR in conventional farming, while others find no significant difference between conventional, organic, and antibiotic-free practices (Miranda *et al.*, 2008; Musa *et al.*, 2021). Furthermore, although organic broiler production initially exhibits lower *Campylobacter* prevalence compared to conventional production, post-processing levels are similar, suggesting minimal overall impact from antibiotic-free rearing (Bailey *et al.*, 2019). AMR genes have been found in poultry litter from both conventional and antibiotic-free farms, with tetracycline resistance genes being more prevalent in antibiotic-free flocks (Farooq *et al.*, 2022). The variability in organic farming regulation and practices across different regions may contribute to these inconsistent findings. For instance, organic poultry's access to outdoor areas may inadvertently expose them to contaminated soil, which could contain antimicrobials, as well as microbes or insects acting as reservoirs for drug-resistant bacteria (Bailey *et al.*, 2019).

Backyard poultry farming, a practice that ranges from small-scale to more extensive free-range systems, has become increasingly popular worldwide. However, during the period from 2014 to 2023, the US Centers for Disease Control (CDC) documented a series of outbreaks with multiple resistant *Salmonella* serotypes, including Enteritidis, Infantis, Typhimurium, and Hadar that were linked to backyard poultry production. Investigations revealed that most of the isolates were resistant to antibiotics, including tetracycline, ampicillin, ciprofloxacin, nalidixic acid and sulfisoxazole, which resulted in heightened hospitalization rates across multiple states (CDC, 2023; US Outbreaks of Zoonotic Diseases Spread between Animals & People | Healthy Pets, Healthy People Available at: https://www.cdc.gov/healthypets/outbreaks .html/, Accessed: April 1, 2024). Similarly, a study from India isolated *Salmonella* from backyard poultry showing resistance to key antibiotics, including chloramphenicol and ciprofloxacin. The presence of the same isolates in the feed, drinking water and eggs highlights the extensive transmission pathways of AMR from poultry to their immediate environment (Samanta *et al.*, 2014). However, research in Pakistan revealed that backyard poultry harbored fewer drug-resistant enteric bacteria from the Enterobacteriaceae family compared to commercial broilers, with the exception of higher resistance to chloramphenicol and oxytetracycline. This trend was attributed to the easy availability and common use of these antibiotics by backyard farmers in Pakistan, despite the ban on use of chloramphenicol in food animals in developed countries (Kamboh *et al.*, 2018).

8 Control strategies to reduce antimicrobial resistance in poultry production: flock management practices and gastrointestinal colonization control

The most cost-effective and practical way to control MDR foodborne pathogens is by implementing interventions upstream in the food supply chain, particularly at the farm level. This approach reduces the occurrence and transmission of these pathogens throughout the food chain. As a response to earlier overuse of antimicrobial drugs in poultry production, the US Food and Drug Administration (FDA) has devised strategies for 'The Judicious Use of Medically Important Antimicrobial Drugs in Food-Producing Animals' in guidance document 209 (*CVM-GFI--209-The-Judicious-Use-of-Medically-Important-Antimicrobial-Drugs-in-Food-ProducingAnimals.Pdf*, https://www.fda.gov/regulatory-information/search-fda-guidance-documents/cvm-gfi-209-judicious-use-medically-important-antimicrobial-drugs-food-producing-animals/, Accessed April 1, 2024). This guideline is based on two fundamental principles:

- firstly, the use of medicall -important antimicrobials in food-producing animals should be limited to those uses that are considered necessary for assuring animal health; and
- secondly, their use should be limited to those uses that include veterinary oversight or consultation.

Adhering to these guidelines creates a roadmap for reducing the impact of AMR, while simultaneously safeguarding the health and productivity of food-producing animals. This approach is consistent with the broader One Health perspective, which integrates efforts across human, animal and environmental health sectors to tackle AMR (De Mesquita Souza Saraiva *et al.*, 2022). Emphasizing primary control strategies, such as improved flock management and control of gastrointestinal colonization, further align with the FDA's guidelines, enhancing overall efforts to minimize the development and spread of AMR across ecosystems.

8.1 Flock management practices

Implementing proper management practices is vital for reducing AMR development. This includes vaccination, sample testing and surveillance, as well as biosecurity and sanitation measures. Vaccinations can trigger immune responses that help reduce both the severity and duration of *Salmonella* infections in chickens while also providing some level of protection against reinfection. However, it is important to note that vaccination alone cannot completely eradicate infections. Factors such as poor sanitation practices and environmental stressors affecting chickens can all negatively impact vaccine performance. Furthermore, vaccines may offer limited defence against field strains or new serotypes that differ antigenically from the strains included in the vaccine (Van Immerseel *et al.*, 2009).

In response to the significant increase in human salmonellosis linked to *S.* Enteritidis transmitted through eggs, various testing and quality assurance programs have been proposed by government agencies and others for commercial laying flocks. In the US, a combination of environmental sampling, serologic testing and bacteriological culturing of birds has been implemented. The National Poultry Improvement Plan (NPIP) was established as a cooperative effort involving industry, state, and federal partners to combat poultry diseases like *Salmonella* Pullorum and *Salmonella* Enteritidis. However, despite these testing efforts, there remains uncertainty about their effectiveness in identifying or addressing emerging reservoirs of *Salmonella* infection or new serotypes.

The National Antimicrobial Resistance Monitoring System (NARMS) is a surveillance program in the US that tracks AMR in foodborne pathogens, including those found in poultry and chicken products. It was established in 1996 following a recommendation by an expert FDA panel. NARMS involves collaboration between three federal agencies – the CDC, FDA and USDA, along with state and local health departments nationwide. Its purpose is to monitor trends in AMR among various bacteria, disseminate timely information to stakeholders, encourage research and assist the FDA in making decisions related to antimicrobial drug approval for animals (Karp *et al.*, 2017).

While programs like NARMS and NPIP are critical in addressing *Salmonella* and AMR challenges in poultry, proper implementation of biosecurity and hygiene protocols in flock management is equally essential. These practices encompass a range of actions, such as disinfection procedures, pest management, monitoring movements of animals, people and vehicles, as well as ensuring the quality of feed, water and of poultry flocks. Scientifically validated studies are essential to understand the impact of these practices on AMR and to develop good industry practices (Yang *et al.*, 2019).

8.2 Gastrointestinal colonization control

The Reduce, Refine, Replace (RRR) framework offers a strategic approach to managing AMU, particularly in addressing the challenge of AMR in sectors such as poultry farming. This approach encourages the reduction of unnecessary antibiotic use, refinement of existing treatment protocols to minimize the impact on beneficial microbiota and the replacement of traditional antibiotics with alternative treatments. Alternative methods for combating AMR in poultry production include the use of probiotics, prebiotics and plant-derived antimicrobial compounds (Mak *et al.*, 2022).

Probiotics are live microorganisms which, when added to feed, aid in maintaining gut microbial balance and promoting host health. Commonly utilized probiotics in poultry include *Enterococcus faecium* and *Enterococcus faecalis*, *Streptococcus thermophilus*, *Bacillus subtilis* and various *Lactobacillus* spp. (*L. bulgaricus, L. plantarum, L. acidophilus, L. helveticus, L. lactis, L. salivarius, L. casei* and *L. reuteri*) (Patterson and Burkholder, 2003). These probiotics, administered within a range of 10^4–10^9 CFU/ml, enhance broiler growth performance, support intestinal equilibrium and hinder pathogenic bacteria colonization (Patterson and Burkholder, 2003). Major mechanisms of action of these probiotics include competitive exclusion, barrier function enhancement, immunomodulation, and metabolic effects such as inhibition of quorum sensing and virulence modulation in pathogens (Mak *et al.*, 2022). For example, *L. bulgaricus*, *L. rhamnosus* and *L. paracasei* were effective in attenuating the colonization of *S.* Enteritidis, *S.* Heidelberg and *S.* Typhimurium *in vitro* (Muyyarikkandy and Amalaradjou, 2017).

Prebiotics refer to beneficial substances that promote microbial growth and include fiber, polyphenols, polyunsaturated fatty acids and conjugated linoleic acid (Peng *et al.*, 2020; Dong *et al.*, 2024). Examples of prebiotics such as fructooligosaccharides (FOSs), galactooligosaccharides and mannanoligosaccharides (MOS) are administered at 1–10 g/kg feed (Ricke, 2015). Fructans and galactans, widely recognized prebiotics, have been studied for their ability to enhance the population of beneficial gut microbiota such as *Lactobacillus* and/or *Bifidobacterium* spp. (Yang *et al.*, 2019). The prebiotic FOS indirectly influences the gut microbiota by increasing the production of short-chain fatty acids (SCFA) and also includes some that promote fermentation. These SCFAs play essential roles in gastrointestinal and immune functions while also exerting acid stress on pathogenic bacteria (Patterson and Burkholder, 2003). They inhibit pathogenic bacteria colonization, improve digestibility and promote the growth of healthy gut bacteria, indirectly altering gut microbiota (Patterson and Burkholder, 2003; Ricke, 2015; Ricke, 2018). When combined with probiotics, prebiotics have been shown to lower the *Campylobacter* load more effectively than when prebiotics are used alone (Arsi *et al.*, 2015). They also have been shown to eradicate *C. perfringens* from the intestines of broilers (Murshed, Abudabos and Qaid, 2024).

Plant-derived antimicrobials (PDAs) are antimicrobial compounds produced by plants as a defense mechanism against plant pathogens. Historically, these compounds have been extensively employed for food preservation, health maintenance and disease treatment, owing to their cost-effectiveness and non-toxic attributes. These phytochemicals when administered as supplements in feed or water (nanoemulsions), possess antioxidant, antifungal, antimicrobial and anti-inflammatory properties (Gracia *et al.*, 2016; Islam *et al.*, 2019; Das *et al.*, 2020, 2021). Certain PDAs are known to influence the chicken microbiome, prevent the formation of biofilms and protect broilers from gastrointestinal diseases. Importantly, they achieve these benefits without causing harm to the birds or promoting AMR in bacteria (Xue *et al.*, 2017; Hu *et al.*, 2019; Yadav and Jha, 2019; Lemos *et al.*, 2020).

A study by Johny *et al.* (2010) found that PDAs such as trans-cinnamaldehyde, β-resorcylic acid, carvacrol, thymol and eugenol individually and synergistically increased the sensitivity of MDR *S.* Typhimurium DT104 to five antibiotics, namely, ampicillin, chloramphenicol, streptomycin, sulfamethoxazole and tetracycline (Johny, Hoagland and Venkitanarayanan, 2010). Caprylic acid, a medium-chain fatty acid mainly obtained from palm oil and coconut oil, effectively reduced the colonization of *S.* Enteritidis in broiler chickens when supplemented with poultry feed (Kollanoor-Johny *et al.*, 2012a). Studies by Kollanoor Johny *et al.* (2010) have documented the ability of thymol and carvacrol in effectively inhibiting *S.* Enteritidis in the contents of chicken ceca *in vitro* (Kollanoor Johny *et al.*, 2010). They also showed that supplementing feed transcinnamaldehyde and eugenol decreased colonization of *S.* Enteritidis in market-age chickens (Kollanoor-Johny *et al.*, 2012b). Another study has shown that using an antimicrobial solution containing either eugenol or carvacrol rapidly inactivated *S.* Enteritidis on eggs, reducing its levels to undetectable limits (Upadhyaya *et al.*, 2013). Wagle et al. (2017) found that supplementing poultry feed with β-resorcylic acid for 14 days at 0.5% and 1% reduced *Campylobacter* populations in cecal contents by approximately 2.5 and 1.7 Log CFU/g, respectively (Wagle *et al.*, 2017).

Apart from prebiotics, probiotics and PDAs, bacteriophages, which are living viruses targeting bacteria, have been shown to be effective in reducing the colonization of many bacteria. Bacteriophages have been shown to decrease *S.* Enteritidis colonization *in vitro* and *in vivo* (Andreatti Filho *et al.*, 2007). Selected bacteriophages were shown to possess a high lysis capability against ciprofloxacin-resistant *Salmonella* isolates, indicating their potential as a biocontrol strategy within the broiler production chain (Pelyuntha *et al.*, 2022). Further research has demonstrated that phage therapy can be as effective as traditional antibiotics such as enrofloxacin or colistin in treating *S.* Typhimurium infections in chickens but with the advantage of being less invasive and preserving the microbiome integrity (Kosznik-Kwaśnicka *et al.*, 2022). Research also highlighted the efficacy of a phage cocktail (consisting of GRNsp6, GRNsp8, and GRNsp51) as a potent antimicrobial against MDR *S.* Enteritidis (Li *et al.*, 2022). Studies have also identified phage GN06 and phage flora as potent agents against APEC and *E. coli*, showcasing significant inhibition of bacterial growth and biofilm formation (Wang *et al.*, 2022). Interestingly, phage flora, in particular, demonstrated a broader lytic spectrum and superior antibiofilm effects compared to the kanamycin sulfate (Jiang *et al.*, 2022). Moreover, in-ovo administration of phage cocktails has shown promising results in preventing colibacillosis in chicks, suggesting an early intervention strategy (Nicolas *et al.*, 2023). Likewise, phage therapy achieved a significant reduction in AMR *C. jejuni* loads in broilers, highlighting its effectiveness as a pre-slaughter intervention (D'Angelantonio *et al.*, 2021).

9 Conclusion

The emergence, spread and persistence of AMR remain a public health hazard worldwide. Animal agriculture, including poultry production, accounts for a significant portion of global AMU. Bacteria are capable of sharing genetic information, resulting in the potential spread of antibiotic resistance determinants in diverse environmental niches, including soil, chicken droppings and bird gut. HGT is the major route for AMR spread in farm settings. With poultry being documented to serve as a reservoir for AMR, surveillance and control measures are critical for reducing AMR persistence in chickens and the environment, improving food safety and protecting human health. Future research is also critical to understanding the epidemiology and ecology of AMR throughout the poultry production continuum.

10 Where to look for further information

For those interested in exploring AMR in poultry production further, the following resources provide excellent insights, from foundational overviews to cutting-edge research:

For readers interested in the historical context and fundamental mechanisms behind AMR, the article "Origins and Evolution of Antibiotic Resistance" offers an in-depth analysis. This seminal work provides a comprehensive overview of the genetic, biochemical and evolutionary aspects that have shaped AMR over time (https://journals.asm.org/doi/full/10.1128/mmbr.00016-10).

For an in-depth exploration of the factors influencing AMR in animal production, readers can refer to the FAO publication "Drivers, Dynamics and Epidemiology of Antimicrobial resistance in Animal Production" (https://openknowledge.fao.org/server/api/core/bitstreams/95efae06-4da2-4689-b12a-1b33fffc2244/content).

Leading Research Centers: National Antimicrobial Resistance Monitoring System (NARMS) in the USA tracks AMR trends in foodborne bacteria from poultry and other animals, providing data critical to understanding AMR spread (https://www.cdc.gov/narms/about/index.html).

For practical guidance on AMR prevention in animal agriculture, the CDC's resource on Antimicrobial Resistance and Animals provides valuable insights into best practices and preventive measures. This page offers accessible information on how to limit the spread of AMR in agricultural settings (https://www.cdc.gov/antimicrobial-resistance/prevention/animals.html).

USDA provides a comprehensive overview of AMR which is accessible here. This resource covers essential aspects of AMR as it related to the USDA's One Health approach, focusing on the interconnectedness of human, animal and environmental health (https://www.usda.gov/topics/animals/one-health/antimicrobial-resistance-overview-amr).

For a detailed perspective on the environmental implications of veterinary antimicrobial use in Europe, as well as EU's goal to reduce antimicrobial use by 2030, readers can explore the publication by the European Environment Agency (https://www.eea.europa.eu/publications/veterinary-antimicrobials-in-europes-environment#:~:text=In%20the%20EU%2D27%2C%20the,antimicrobials%20by%2050%25%20by%202030).

11 References

2018-red-book.pdf. Available at: https://www.fsis.usda.gov/sites/default/files/media_file/2020-07/2018-red-book.pdf (Accessed: 22 March 2024).

AAAP guidelines, AVMA. Available at: https://www.avma.org/resources-tools/avma-policies/aaap-guidelines-judicious-therapeutic-use-antimicrobials-poultry (Accessed: 21 March 2024).

Ager, E.O. *et al.* (2023) 'Global trends in antimicrobial resistance on organic and conventional farms', *Scientific Reports*, 13(1), p. 22608. Available at: https://doi.org/10.1038/s41598-023-47862-7.

Al-Ghamdi, M.S. *et al.* (1999) 'Antibiotic resistance of Escherichia coli isolated from poultry workers, patients and chicken in the eastern province of Saudi Arabia', *Tropical Medicine & International Health*, 4(4), pp. 278–283. Available at: https://doi.org/10.1046/j.1365-3156.1999.00392.x.

Ali, J., Rafiq, Q.A. and Ratcliffe, E. (2018) 'Antimicrobial resistance mechanisms and potential synthetic treatments', *Future Science OA*, 4(4), p. FSO290. Available at: https://doi.org/10.4155/fsoa-2017-0109.

Andreatti Filho, R.L. *et al.* (2007) 'Ability of bacteriophages isolated from different sources to reduce salmonella enterica serovar enteritidis in vitro and in vivo', *Poultry Science*, 86(9), pp. 1904–1909. Available at: https://doi.org/10.1093/ps/86.9.1904.

Anju, K. *et al.* (2021) 'Toxinotyping and molecular characterization of antimicrobial resistance in *Clostridium perfringens* isolated from different sources of livestock and poultry', *Anaerobe*, 67, p. 102298. Available at: https://doi.org/10.1016/j.anaerobe.2020.102298.

Arsi, K. *et al.* (2015) 'The efficacy of selected probiotic and prebiotic combinations in reducing Campylobacter colonization in broiler chickens', *Journal of Applied Poultry Research*, 24(3), pp. 327–334. Available at: https://doi.org/10.3382/japr/pfv032.

Awad, A., Arafat, N. and Elhadidy, M. (2016) 'Genetic elements associated with antimicrobial resistance among avian pathogenic Escherichia coli', *Annals of Clinical Microbiology and Antimicrobials*, 15(1), p. 59. Available at: https://doi.org/10.1186/s12941-016-0174-9.

Bager, F. *et al.* (1997) 'Avoparcin used as a growth promoter is associated with the occurrence of vancomycin-resistant *Enterococcus faecium* on Danish poultry and pig farms', *Preventive Veterinary Medicine*, 31(1), pp. 95–112. Available at: https://doi.org/10.1016/S0167-5877(96)01119-1.

Bailey, M.A. *et al.* (2019) 'Prevalence and antimicrobial resistance of *Campylobacter*from antibiotic-free broilers during organic and conventional processing', *Poultry Science*, 98(3), pp. 1447–1454. Available at: https://doi.org/10.3382/ps/pey486.

Baker, M. *et al.* (2023) 'Machine learning and metagenomics reveal shared antimicrobial resistance profiles across multiple chicken farms and abattoirs in China', *Nature Food*, 4(8), pp. 707–720. Available at: https://doi.org/10.1038/s43016-023-00814-w.

Barbieri, N.L. *et al.* (2017) 'mcr-1 identified in Avian Pathogenic Escherichia coli (APEC)', *PLoS One*, 12(3), p. e0172997. Available at: https://doi.org/10.1371/journal.pone.0172997.

Bartkiene, E. *et al.* (2020) 'Study of the antibiotic residues in poultry meat in some of the EU countries and selection of the best compositions of lactic acid bacteria and essential oils against *Salmonella enterica*', *Poultry Science*, 99(8), pp. 4065–4076. Available at: https://doi.org/10.1016/j.psj.2020.05.002.

Bello, Abubakar, et al. (2023) 'Ionophore coccidiostats–disposition kinetics in laying hens and residues transfer to eggs', *Poultry Science* 102(1), p. 102280.

Ben, Y. *et al.* (2019) 'Human health risk assessment of antibiotic resistance associated with antibiotic residues in the environment: a review', *Environmental Research*, 169, pp. 483–493. Available at: https://doi.org/10.1016/j.envres.2018.11.040.

Bhedi, K.R. *et al.* (2018) 'Detection and molecular characterization of methicillin-resistant staphylococcus aureus obtained from poultry and poultry house environment of Anand District, Gujarat, India', *International Journal of Current Microbiology and Applied Sciences*, 7(2), pp. 867–872. Available at: https://doi.org/10.20546/ijcmas.2018.702.109.

Binh, C.T.T. *et al.* (2009) 'Diverse *aadA* gene cassettes on class 1 integrons introduced into soil via spread manure', *Research in Microbiology*, 160(6), pp. 427–433. Available at: https://doi.org/10.1016/j.resmic.2009.06.005.

Blaak, H. *et al.* (2015) 'Distribution, Numbers, and Diversity of ESBL-Producing E. coli in the Poultry Farm Environment', *PLoS One*, 10(8), p. e0135402. Available at: https://doi.org/10.1371/journal.pone.0135402.

Brower, C.H. *et al.* (2017) 'The prevalence of extended-spectrum beta-lactamase-producing multidrug-resistant escherichia coli in poultry chickens and variation according to farming practices in Punjab, India', *Environmental Health Perspectives*, 125(7), p. 077015. Available at: https://doi.org/10.1289/EHP292.

Burow, E. *et al.* (2020) 'Antibiotic resistance in escherichia coli from broiler chickens after amoxicillin treatment in an experimental environment', *Microbial Drug Resistance*, 26(9), pp. 1098–1107. Available at: https://doi.org/10.1089/mdr.2019.0442.

Buzón-Durán, L., Capita, R. and Alonso-Calleja, C. (2017) 'Microbial loads and antibiotic resistance patterns of *Staphylococcus aureus* in different types of raw poultry-based meat preparations', *Poultry Science*, 96(11), pp. 4046–4052. Available at: https://doi.org/10.3382/ps/pex200.

Caffrey, N. *et al.* (2017) 'Risk factors associated with the A2C resistance pattern among *E. coli* isolates from broiler flocks in Canada', *Preventive Veterinary Medicine*, 148, pp. 115–120. Available at: https://doi.org/10.1016/j.prevetmed.2017.11.001.

Castro-Vargas, R.E. *et al.* (2020) 'Antibiotic resistance in Salmonella spp. isolated from poultry: a global overview', *Veterinary World*, 13(10), pp. 2070–2084. Available at: https://doi.org/10.14202/vetworld.2020.2070-2084.

CDC (2022) *What Exactly is Antibiotic Resistance? Centers for Disease Control and Prevention*. Available at: https://www.cdc.gov/drugresistance/about.html (Accessed: 24 March 2024).

Chuppava, B. *et al.* (2019) 'Resistance Reservoirs and Multi-Drug Resistance of Commensal Escherichia coli From Excreta and Manure Isolated in Broiler Houses With Different Flooring Designs', *Frontiers in Microbiology*, 10. Available at: https://doi.org/10.3389/fmicb.2019.02633.

Colomer-Lluch, M. *et al.* (2011) 'Bacteriophages carrying antibiotic resistance genes in fecal waste from cattle, pigs, and poultry', *Antimicrobial Agents and Chemotherapy*, 55(10), pp. 4908–4911. Available at: https://doi.org/10.1128/aac.00535-11.

Cookey, T. and Otokunefor, K. (2016) 'Poultry environment as a reservoir of antimicrobial resistant bacteria – A Nigerian story', *British Microbiology Research Journal*, 17(1), pp. 1–11. Available at: https://doi.org/10.9734/BMRJ/2016/28601.

Cui, M. *et al.* (2019) 'Prevalence and characterization of fluoroquinolone resistant salmonella isolated from an integrated broiler chicken supply chain', *Frontiers in Microbiology*, 10. Available at: https://doi.org/10.3389/fmicb.2019.01865.

'CVM-GFI--209-The-Judicious-Use-of-Medically-Important-Antimicrobial-Drugs-in-Food-Producing-Animals.pdf' (n.d.). https://www.fda.gov/media/79140/download.

Daehre, K. *et al.* (2018) 'ESBL-producing klebsiella pneumoniae in the broiler production chain and the first description of ST3128', *Frontiers in Microbiology*, 9, p. 2302. Available at: https://doi.org/10.3389/fmicb.2018.02302.

D'Angelantonio, D. *et al.* (2021) 'Bacteriophage therapy to reduce colonization of campylobacter jejuni in broiler chickens before slaughter', *Viruses*, 13(8), p. 1428. Available at: https://doi.org/10.3390/v13081428.

Das, Q. *et al.* (2020) 'Gut microbiota, blood metabolites, and spleen immunity in broiler chickens fed berry pomaces and phenolic-enriched extractives', *Frontiers in Veterinary Science*, 7, p. 150. Available at: https://doi.org/10.3389/fvets.2020.00150.

Das, Q. *et al.* (2021) 'Organic cranberry pomace and its ethanolic extractives as feed supplement in broiler: impacts on serum Ig titers, liver and bursal immunity', *Poultry Science*, 100(2), pp. 517–526. Available at: https://doi.org/10.1016/j.psj.2020.09.044.

Davies, Julian and Davies, Dorothy. (2010) 'Origins and evolution of antibiotic resistance', *Microbiology and Molecular Biology Reviews* 74(3), pp. 417–433.

De Mesquita Souza Saraiva, M. *et al.* (2022) 'Antimicrobial resistance in the globalized food chain: a one health perspective applied to the poultry industry', *Brazilian Journal of Microbiology*, 53(1), pp. 465–486. Available at: https://doi.org/10.1007/s42770-021-00635-8.

Dhanarani, T.S. *et al.* (2009) 'Study on acquisition of bacterial antibiotic resistance determinants in poultry litter', *Poultry Science*, 88(7), pp. 1381–1387. Available at: https://doi.org/10.3382/ps.2008-00327.

Diarra, M.S. *et al.* (2014) 'Antibiotic resistance and diversity of *Salmonella enterica* serovars associated with broiler chickens', *Journal of Food Protection*, 77(1), pp. 40–49. Available at: https://doi.org/10.4315/0362-028.JFP-13-251.

Diaz-Sanchez, S. *et al.* (2015) Available at: https://www.foodprotection.org/members/fpt-archive-articles/2015-11-antibiotic-use-in-poultry-a-driving-force-for-organic-poultry-production/ (Accessed: 21 March 2024).

Dong, S. *et al.* (2024) 'Improving quality of poultry and its meat products with probiotics, prebiotics, and phytoextracts', *Poultry Science*, 103(2), p. 103287. Available at: https://doi.org/10.1016/j.psj.2023.103287.

Donoghue, D.J. (2003) 'Antibiotic residues in poultry tissues and eggs: human health concerns?', *Poultry Science*, 82(4), pp. 618–621. Available at: https://doi.org/10.1093/ps/82.4.618.

Dutil, L. *et al.* (2010) 'Ceftiofur resistance in salmonella enterica serovar heidelberg from chicken meat and humans, Canada', *Emerging Infectious Diseases*, 16(1), pp. 48–54. Available at: https://doi.org/10.3201/eid1601.090729.

Farooq, M. *et al.* (2022) 'Antibiotic resistance genes occurrence in conventional and antibiotic-free poultry farming, Italy', *Animals*, 12(18), p. 2310. Available at: https://doi.org/10.3390/ani12182310.

Fatoba, Dorcas Oladayo, et al. (2022) 'Transmission of antibiotic-resistant Escherichia coli from chicken litter to agricultural soil', *Frontiers in Environmental Science*, 9, p. 751732.

Feeks, J. Available at: https://poultryhealthtoday.com/mobile/article/?id=7963 (Accessed: 21 March 2024).

Forsberg, K.J. *et al.* (2012) 'The shared antibiotic resistome of soil bacteria and human pathogens', *Science*, 337(6098), pp. 1107–1111. Available at: https://doi.org/10.1126/science.1220761.

Furtula, V. *et al.* (2010) 'Veterinary pharmaceuticals and antibiotic resistance of *Escherichia coli* isolates in poultry litter from commercial farms and controlled feeding trials', *Poultry Science*, 89(1), pp. 180–188. Available at: https://doi.org/10.3382/ps.2009-00198.

Furtula, V. *et al.* (2013) 'Antimicrobial resistance in enterococcus spp. Isolated from environmental samples in an area of intensive poultry production', *International Journal of Environmental Research and Public Health*, 10(3), pp. 1020–1036. Available at: https://doi.org/10.3390/ijerph10031020.

Gharbi, M. *et al.* (2018) 'Prevalence and antibiotic resistance patterns of *Campylobacter* spp. Isolated from broiler chickens in the North of Tunisia', *BioMed Research International*, 2018, p. e7943786. Available at: https://doi.org/10.1155/2018/7943786.

Ghosh, S. and LaPara, T.M. (2007) 'The effects of subtherapeutic antibiotic use in farm animals on the proliferation and persistence of antibiotic resistance among soil bacteria', *The ISME Journal*, 1(3), pp. 191–203. Available at: https://doi.org/10.1038/ismej.2007.31.

Goetting, V., Lee, K.A. and Telll, L.A. (2011) 'Pharmacokinetics of veterinary drugs in laying hens and residues in eggs: a review of the literature', *Journal of Veterinary Pharmacology and Therapeutics*, 34(6), pp. 521–556. Available at: https://doi.org/10.1111/j.1365-2885.2011.01287.x.

Gracia, M.I. *et al.* (2016) 'Efficacy of feed additives against Campylobacter in live broilers during the entire rearing period: part B', *Poultry Science*, 95(4), pp. 886–892. Available at: https://doi.org/10.3382/ps/pev346.

Gullberg, E. *et al.* (2011) 'Selection of resistant bacteria at very low antibiotic concentrations', *PLoS Pathogens*, 7(7), p. e1002158. Available at: https://doi.org/10.1371/journal.ppat.1002158.

Gyles, C.L. (2008) 'Antimicrobial resistance in selected bacteria from poultry', *Animal Health Research Reviews*, 9(2), pp. 149–158. Available at: https://doi.org/10.1017/S1466252308001552.

Haider, Z. *et al.* (2022) 'Isolation, toxinotyping and antimicrobial susceptibility testing of *Clostridium perfringens* isolated from Pakistan poultry', *Anaerobe*, 73, p. 102499. Available at: https://doi.org/10.1016/j.anaerobe.2021.102499.

Hassan, M.M. (2014) 'Antimicrobial resistance pattern against E. coli and Salmonella in layer poultry', *Research Journal for Veterinary Practitioners*, pp. 30–35. Available at: https://doi.org/10.14737/journal.rjvp/2014/2.2.30.35.

Hiki, M. *et al.* (2015) 'Decreased resistance to broad-spectrum cephalosporin in escherichia coli from healthy broilers at farms in Japan after voluntary withdrawal of ceftiofur', *Foodborne Pathogens and Disease*, 12(7), pp. 639–643. Available at: https://doi.org/10.1089/fpd.2015.1960.

Holmes, A.H. *et al.* (2016) 'Understanding the mechanisms and drivers of antimicrobial resistance', *The Lancet*, 387(10014), pp. 176–187. Available at: https://doi.org/10.1016/S0140-6736(15)00473-0.

Hu, N. *et al.* (2019) 'Pharmacokinetics of sanguinarine, chelerythrine, and their metabolites in broiler chickens following oral and intravenous administration', *Journal of Veterinary Pharmacology and Therapeutics*, 42(2), pp. 197–206. Available at: https://doi.org/10.1111/jvp.12729.

Huang, X. *et al.* (2013) 'Occurrence and distribution of veterinary antibiotics and tetracycline resistance genes in farmland soils around swine feedlots in Fujian Province, China', *Environmental Science and Pollution Research*, 20(12), pp. 9066–9074. Available at: https://doi.org/10.1007/s11356-013-1905-5.

Hussein, M., Ahmed, M.M. and Morshedy, A.M. (2016) 'Effect of cooking methods on some antibiotic residues in chicken meat', *Japanese Journal of Veterinary Research*, 64, pp. S225–S231.

Idris, U. *et al.* (2006) 'Dissemination of fluoroquinolone-resistant campylobacter spp. within an integrated commercial poultry production system', *Applied and Environmental Microbiology*, 72(5), pp. 3441–3447. Available at: https://doi.org/10.1128/AEM.72.5.3441-3447.2006.

Igbinosa, E.O. *et al.* (2022) 'Antimicrobial resistance and genetic characterisation of *Salmonella enterica* from retail poultry meats in Benin City, Nigeria', *LWT*, 169, p. 114049. Available at: https://doi.org/10.1016/j.lwt.2022.114049.

Iroha, 1 *et al.* (2016) 'Occurrence of multi-drug resistant gram negative bacteria from poultry and poultry products sold in Abakaliki metropolis, Nigeria'. Available at: https://www.semanticscholar.org/paper/Occurrence-of-Multi-drug-Resistant-Gram-Negative-in-Iroha-Aniokette/c3686280031c3aa1cdbc8ba95ca515015b98c710 (Accessed: 24 March 2024).

Islam, M.R. *et al.* (2019) 'Effects of wild blueberry (Vaccinium angustifolium) pomace feeding on gut microbiota and blood metabolites in free-range pastured broiler chickens', *Poultry Science*, 98(9), pp. 3739–3755. Available at: https://doi.org/10.3382/ps/pez062.

Jiang, L. *et al.* (2022) 'Characterization of the Lytic Phage Flora with a broad host range against multidrug-resistant escherichia coli and evaluation of its efficacy against E. coli biofilm formation', *Frontiers in Veterinary Science*, 9, p. 906973. Available at: https://doi.org/10.3389/fvets.2022.906973.

Johny, A.K., Hoagland, T. and Venkitanarayanan, K. (2010) 'Effect of subinhibitory concentrations of plant-derived molecules in increasing the sensitivity of multidrug-resistant Salmonella enterica serovar Typhimurium DT104 to antibiotics', *Foodborne Pathogens and Disease*, 7(10), pp. 1165–1170. Available at: https://doi.org/10.1089/fpd.2009.0527.

Just, N.A. *et al.* (2012) 'Potentially pathogenic bacteria and antimicrobial resistance in bioaerosols from cage-housed and floor-housed poultry operations', *The Annals of Occupational Hygiene*, 56(4), pp. 440–449. Available at: https://doi.org/10.1093/annhyg/mer105.

Kadykalo, S.V. *et al.* (2019) 'Resistance to antimicrobials used at hatchery level among clinical Escherichia coli isolates from Ontario broilers', *The Canadian Veterinary Journal*, 60(9), p. 945. Available at: https://www.ncbi.nlm.nih.gov/pmc/articles/PMC6697018/ (Accessed: 2 April 2024).

Kali, A. *et al.* (2013) 'Changing trends in resistance pattern of methicillin resistant staphylococcus aureus', *Journal of Clinical and Diagnostic Research : JCDR*, 7(9), pp. 1979–1982. Available at: https://doi.org/10.7860/JCDR/2013/6142.3377.

Kamboh, A.A. *et al.* (2018) 'Antimicrobial resistance in enterobacteriaceae isolated from liver of commercial broilers and backyard chickens', *Journal of Applied Poultry Research*, 27(4), pp. 627–634. Available at: https://doi.org/10.3382/japr/pfy045.

Karim, M.R. *et al.* (2023) 'Antimicrobial resistance profiles and co-existence of multiple antimicrobial resistance genes in mcr-harbouring colistin-resistant enterobacteriaceae isolates recovered from poultry and poultry meats in Malaysia', *Antibiotics*, 12(6), p. 1060. Available at: https://doi.org/10.3390/antibiotics12061060.

Karp, B.E. *et al.* (2017) 'National antimicrobial resistance monitoring system: two decades of advancing public health through integrated surveillance of antimicrobial resistance', *Foodborne Pathogens and Disease*, 14(10), pp. 545–557. Available at: https://doi.org/10.1089/fpd.2017.2283.

Kim, J.M. *et al.* (2010) 'Prevalence, antibiograms, and transferable tet(O) Plasmid of *Campylobacter jejuni* and *Campylobacter coli* isolated from raw chicken, pork, and human clinical cases in Korea', *Journal of Food Protection*, 73(8), pp. 1430–1437. Available at: https://doi.org/10.4315/0362-028X-73.8.1430.

Kollanoor Johny, A. *et al.* (2010) 'Antibacterial effect of trans-cinnamaldehyde, eugenol, carvacrol, and thymol on Salmonella Enteritidis and Campylobacter jejuni in chicken cecal contents in vitro', *Journal of Applied Poultry Research*, 19(3), pp. 237–244. Available at: https://doi.org/10.3382/japr.2010-00181.

Kollanoor-Johny, A. *et al.* (2012a) 'Caprylic acid reduces salmonella enteritidis populations in various segments of digestive tract and internal organs of 3- and 6-week-old broiler chickens, therapeutically', *Poultry Science*, 91(7), pp. 1686–1694. Available at: https://doi.org/10.3382/ps.2011-01716.

Kollanoor-Johny, Anup *et al.* (2012b) 'Reduction of Salmonella enterica Serovar Enteritidis Colonization in 20-day-old broiler chickens by the plant-derived compounds *trans*-cinnamaldehyde and Eugenol', *Applied and Environmental Microbiology*, 78(8), pp. 2981–2987. Available at: https://doi.org/10.1128/AEM.07643-11.

Kosznik-Kwaśnicka, K. *et al.* (2022) 'Biological aspects of phage therapy versus antibiotics against Salmonella enterica serovar Typhimurium infection of chickens', *Frontiers in Cellular and Infection Microbiology*, 12. Available at: https://doi.org/10.3389/fcimb.2022.941867.

Krauland, M.G. *et al.* (2009) 'Integron-mediated multidrug resistance in a global collection of nontyphoidal salmonella enterica isolates', *Emerging Infectious Diseases*, 15(3), pp. 388–396. Available at: https://doi.org/10.3201/eid1503.081131.

Kumar, D., Pornsukarom, S. and Thakur, S. (2019) 'Antibiotic usage in poultry production and antimicrobial-resistant salmonella in poultry', in K. Venkitanarayanan, S. Thakur, and S.C. Ricke (eds) *Food Safety in Poultry Meat Production*. Cham: Springer International Publishing (Food Microbiology and Food Safety), pp. 47–66. Available at: https://doi.org/10.1007/978-3-030-05011-5_3.

Kumar, Y. (2019) *Antimicrobial Resistance: A Global Threat*. BoD – Books on Demand.

Laube, H. *et al.* (2014) 'Transmission of ESBL/AmpC-producing Escherichia coli from broiler chicken farms to surrounding areas', *Veterinary Microbiology*, 172(3–4), pp. 519–527. Available at: https://doi.org/10.1016/j.vetmic.2014.06.008.

Laxminarayan, R. *et al.* (2013) 'Antibiotic resistance–the need for global solutions', *The Lancet Infectious Diseases*, 13(12), pp. 1057–1098. Available at: https://doi.org/10.1016/S1473-3099(13)70318-9.

Le Page, G. *et al.* (2017) 'Integrating human and environmental health in antibiotic risk assessment: a critical analysis of protection goals, species sensitivity and antimicrobial resistance', *Environment International*, 109, pp. 155–169. Available at: https://doi.org/10.1016/j.envint.2017.09.013.

Lemos, M.P.L. *et al.* (2020) 'The posthatch prophylactic use of ceftiofur affects the cecal microbiota similar to the dietary sanguinarine supplementation in broilers', *Poultry Science*, 99(11), pp. 6013–6021. Available at: https://doi.org/10.1016/j.psj.2020.06.078.

Levy, S.B., Fitzgerald, G.B. and Macone, A.B. (1976) 'Spread of antibiotic-resistant plasmids from chicken to chicken and from chicken to man', *Nature*, 260(5546), pp. 40–42. Available at: https://doi.org/10.1038/260040a0.

Li, Y. *et al.* (2022) 'A cocktail of three virulent phages controls multidrug-resistant salmonella enteritidis infection in poultry', *Frontiers in Microbiology*, 13. Available at: https://doi.org/10.3389/fmicb.2022.940525.

Ma, F. *et al.* (2021) 'Use of antimicrobials in food animals and impact of transmission of antimicrobial resistance on humans', *Biosafety and Health*, 3(1), pp. 32–38. Available at: https://doi.org/10.1016/j.bsheal.2020.09.004.

Mak, P.H.W. *et al.* (2022) 'Production systems and important antimicrobial resistant-pathogenic bacteria in poultry: a review', *Journal of Animal Science and Biotechnology*, 13(1), p. 148. Available at: https://doi.org/10.1186/s40104-022-00786-0.

Martel, A. *et al.* (2004) 'Susceptibility of clostridium perfringens strains from broiler chickens to antibiotics and anticoccidials', *Avian Pathology*, 33(1), pp. 3–7. Available at: https://doi.org/10.1080/0307945031000163291.

Miranda, J.M. *et al.* (2008) 'Comparison of antimicrobial resistance in *Escherichia coli*, *Staphylococcus aureus*, and *Listeria monocytogenes* strains Isolated from organic and conventional poultry meat', *Journal of Food Protection*, 71(12), pp. 2537–2542. Available at: https://doi.org/10.4315/0362-028X-71.12.2537.

Moreno, M.A. *et al.* (2019) 'Day-old chicks are a source of antimicrobial resistant bacteria for laying hen farms', *Veterinary Microbiology*, 230, pp. 221–227. Available at: https://doi.org/10.1016/j.vetmic.2019.02.007.

Morshdy, A.E.M.A. *et al.* (2013) 'Oxytetracycline residues in bovine carcasses slaughtered at Mansoura Abattoir, Egypt', *The Japanese Journal of Veterinary Research*, 61 Suppl, pp. S44–S47.

Mund, M.D. *et al.* (2017) 'Antimicrobial drug residues in poultry products and implications on public health: a review', *International Journal of Food Properties*, 20(7), pp. 1433–1446. Available at: https://doi.org/10.1080/10942912.2016.1212874.

Munita, J.M. and Arias, C.A. (2016) 'Mechanisms of antibiotic resistance', in *Virulence Mechanisms of Bacterial Pathogens*. John Wiley & Sons, Ltd, pp. 481–511. Available at: https://doi.org/10.1128/9781555819286.ch17.

Murshed, M., Abudabos, A.M. and Qaid, M.M. (2024) 'Effects of feeding eubiotics as antibiotic substitutes on growth performance, intestinal histomorphology and microbiology of broilers', *Italian Journal of Animal Science*, 23(1), pp. 65–75. Available at: https://doi.org/10.1080/1828051X.2023.2290192.

Musa, L. *et al.* (2021) 'Susceptibility of commensal E. coli isolated from conventional, antibiotic-free, and organic meat chickens on farms and at slaughter toward antimicrobials with public health relevance', *Antibiotics*, 10(11), p. 1321. Available at: https://doi.org/10.3390/antibiotics10111321.

Muyyarikkandy, M.S. and Amalaradjou, M. (2017) 'Lactobacillus bulgaricus, Lactobacillus rhamnosus and Lactobacillus paracasei Attenuate Salmonella Enteritidis, Salmonella Heidelberg and Salmonella Typhimurium Colonization and Virulence Gene Expression in vitro', *International Journal of Molecular Sciences*, 18(11), p. 2381. Available at: https://doi.org/10.3390/ijms18112381.

Mwansa, M. *et al.* (2023) 'Determination of antimicrobial resistance patterns of Escherichia coli isolates from farm workers in broiler poultry production and assessment of antibiotic resistance awareness levels among poultry farmers in Lusaka, Zambia', *Frontiers in Public Health*, 10. Available at: https://doi.org/10.3389/fpubh.2022.998860.

National Chicken Council. Available at: https://www.nationalchickencouncil.org/about-the-industry/statistics/per-capita-consumption-of-poultry-and-livestock-1965-to-estimated-2012-in-pounds/ (Accessed: 22 March 2024).

National Chicken Council | Chicken Industry Provides 1.5 Million Jobs and $417 Billion in Economic Impact: New Study, National Chicken Council. Available at: https://www.nationalchickencouncil.org/chicken-industry-provides-1-5-million-jobs-and-417-billion-in-economic-impact-new-study/ (Accessed: 1 April 2024).

Nelson, J.M. *et al.* (2007) 'Fluoroquinolone-resistant campylobacter species and the withdrawal of fluoroquinolones from use in poultry: a public health success story', *Clinical Infectious Diseases*, 44(7), pp. 977–980. Available at: https://doi.org/10.1086/512369.

Nesme, J. *et al.* (2014) 'Large-scale metagenomic-based study of antibiotic resistance in the environment', *Current Biology*, 24(10), pp. 1096–1100. Available at: https://doi.org/10.1016/j.cub.2014.03.036.

Nhung, N.T., Chansiripornchai, N. and Carrique-Mas, J.J. (2017) 'Antimicrobial resistance in bacterial poultry pathogens: a review', *Frontiers in Veterinary Science*, 4. Available at: https://doi.org/10.3389/fvets.2017.00126.

Nicolas, M. *et al.* (2023) 'In ovo administration of a phage cocktail partially prevents colibacillosis in chicks', *Poultry Science*, 102(11), p. 102967. Available at: https://doi.org/10.1016/j.psj.2023.102967.

Osman, K.M. *et al.* (2012) 'Prevalence of clostridium perfringens type A isolates in commercial broiler chickens and parent broiler breeder hens in Egypt', *Revue scientifique et technique (International Office of Epizootics)*, 31, pp. 931–41. Available at: https://doi.org/10.20506/rst.31.3.2169.

Osman, K.M. *et al.* (2018) 'Poultry hatcheries as potential reservoirs for antimicrobial-resistant Escherichia coli: a risk to public health and food safety', *Scientific Reports*, 8(1), p. 5859. Available at: https://doi.org/10.1038/s41598-018-23962-7.

Park, J.Y. *et al.* (2015) 'Characterization of *Clostridium perfringens* isolates obtained from 2010 to 2012 from chickens with necrotic enteritis in Korea', *Poultry Science*, 94(6), pp. 1158–1164. Available at: https://doi.org/10.3382/ps/pev037.

Patterson, J. and Burkholder, K. (2003) 'Application of prebiotics and probiotics in poultry production', *Poultry Science*, 82(4), pp. 627–631. Available at: https://doi.org/10.1093/ps/82.4.627.

Pelyuntha, W. *et al.* (2022) 'Broad lytic spectrum of novel Salmonella phages on ciprofloxacin-resistant Salmonella contaminated in the broiler production chain', *Veterinary World*, 15(8), pp. 2039–2045. Available at: https://doi.org/10.14202/vetworld.2022.2039-2045.

Peng, M. *et al.* (2020) 'Effectiveness of probiotics, prebiotics, and prebiotic-like components in common functional foods', *Comprehensive Reviews in Food Science and Food Safety*, 19(4), pp. 1908–1933. Available at: https://doi.org/10.1111/1541-4337.12565.

Pérez-Boto, D. *et al.* (2010) 'Drinking water as the source of Campylobacter coli infection in grandparent heavy breeders', *Avian Pathology*, 39(6), pp. 483–487. Available at: https://doi.org/10.1080/03079457.2010.518138.

Petersen, A. *et al.* (2006) 'Vertical transmission of a fluoroquinolone-resistant Escherichia coli within an integrated broiler operation', *Veterinary Microbiology*, 116(1–3), pp. 120–128. Available at: https://doi.org/10.1016/j.vetmic.2006.03.015.

Poudel, A. *et al.* (2024) 'Antimicrobial stewardship hindered by inadequate biosecurity and biosafety practices, and inappropriate antibiotics usage in poultry farms of Nepal–A pilot study', *PLoS One*, 19(3), p. e0296911. Available at: https://doi.org/10.1371/journal.pone.0296911.

Projahn, M. *et al.* (2018) 'Environmental adaptation and vertical dissemination of ESBL-/pAmpC-producing Escherichia coli in an integrated broiler production chain in the absence of an antibiotic treatment', *Microbial Biotechnology*, 11(6), pp. 1017–1026. Available at: https://doi.org/10.1111/1751-7915.13040.

Punchihewage-Don, A.J. *et al.* (2022) 'The outbreaks and prevalence of antimicrobial resistant Salmonella in poultry in the United States: an overview', *Heliyon*, 8(11). Available at: https://doi.org/10.1016/j.heliyon.2022.e11571.

Rafiq, K. *et al.* (2022) 'Antimicrobial resistance profile of common foodborne pathogens recovered from livestock and poultry in Bangladesh', *Antibiotics*, 11(11), p. 1551. Available at: https://doi.org/10.3390/antibiotics11111551.

Rau, R.B. *et al.* (2021) 'Antimicrobial resistance of Salmonella from poultry meat in Brazil: results of a nationwide survey', *Epidemiology & Infection*, 149, p. e228. Available at: https://doi.org/10.1017/S0950268821002156.

Rehman, M.A. *et al.* (2017) 'First detection of a fosfomycin resistance gene, fosA7, in Salmonella enterica Serovar Heidelberg isolated from broiler chickens', *Antimicrobial Agents and Chemotherapy*, 61(8). Available at: https://doi.org/10.1128/aac.00410-17.

Reig, M. and Toldrá, F. (2008) 'Veterinary drug residues in meat: concerns and rapid methods for detection', *Meat Science*, 78(1), pp. 60–67. Available at: https://doi.org/10.1016/j.meatsci.2007.07.029.

Ribeiro, J. *et al.* (2023) 'Antibiotic resistance among gastrointestinal bacteria in broilers: a review focused on Enterococcus spp. and Escherichia coli', *Animals*, 13(8), p. 1362. Available at: https://doi.org/10.3390/ani13081362.

Ricke, S.C. (2015) 'Potential of fructooligosaccharide prebiotics in alternative and nonconventional poultry production systems', *Poultry Science*, 94(6), pp. 1411–1418. Available at: https://doi.org/10.3382/ps/pev049.

Ricke, S.C. (2018) 'Impact of prebiotics on poultry production and food safety', *The Yale Journal of Biology and Medicine*, 91(2), pp. 151–159. Available at: https://www.ncbi.nlm.nih.gov/pmc/articles/PMC6020725/ (Accessed: 2 April 2024).

Roth, N. *et al.* (2019) 'The application of antibiotics in broiler production and the resulting antibiotic resistance in Escherichia coli: a global overview', *Poultry Science*, 98(4), pp. 1791–1804. Available at: https://doi.org/10.3382/ps/pey539.

Rothrock, M.J. *et al.* (2016) 'Antibiotic resistance patterns of major zoonotic pathogens from all-natural, antibiotic-free, pasture-raised broiler flocks in the Southeastern United States', *Journal of Environmental Quality*, 45(2), pp. 593–603. Available at: https://doi.org/10.2134/jeq2015.07.0366.

Sahin, O., Morishita, T.Y. and Zhang, Q. (2002) 'Campylobacter colonization in poultry: sources of infection and modes of transmission', *Animal Health Research Reviews*, 3(2), pp. 95–105. Available at: https://doi.org/10.1079/AHRR200244.

Saliu, E.-M., Vahjen, W. and Zentek, J. (2017) 'Types and prevalence of extended–spectrum beta–lactamase producing Enterobacteriaceae in poultry', *Animal Health Research Reviews*, 18(1), pp. 46–57. Available at: https://doi.org/10.1017/S1466252317000020.

Samanta, I. *et al.* (2014) 'Prevalence and antibiotic resistance profiles of*Salmonella* serotypes isolated from backyard poultry flocks in West Bengal, India', *Journal of Applied Poultry Research*, 23(3), pp. 536–545. Available at: https://doi.org/10.3382/japr.2013-00929.

Savin, M. *et al.* (2021) 'Antibiotic-resistant bacteria, antibiotic resistance genes, and antibiotic residues in wastewater from a poultry slaughterhouse after conventional and advanced treatments', *Scientific Reports*, 11(1), p. 16622. Available at: https://doi.org/10.1038/s41598-021-96169-y.

Scicchitano, D. *et al.* (2024) 'Routes of dispersion of antibiotic resistance genes from the poultry farm system', *Science of The Total Environment*, 912, p. 169086. Available at: https://doi.org/10.1016/j.scitotenv.2023.169086.

Silva, R.O.S. *et al.* (2009) 'Antimicrobial susceptibility of Clostridium perfringens strains isolated from broiler chickens', *Brazilian Journal of Microbiology*, 40, pp. 262–264. Available at: https://doi.org/10.1590/S1517-83822009000200010.

Singer, Randall S. (2023) 'Estimates of on-farm antimicrobial usage in egg production in the United States, 2016–2021', *Frontiers in Veterinary Science,* 10, p. 1135377.

Singer, R.S. *et al.* (2023) 'Antimicrobial usage in broiler chicken production in the United States, 2013–2021', *Frontiers in Veterinary Science*, 10. Available at: https://doi.org/10.3389/fvets.2023.1139908.

Son, I. *et al.* (2007) 'Antimicrobial resistance of *Arcobacter* and *Campylobacter* from broiler carcasses', *International Journal of Antimicrobial Agents*, 29(4), pp. 451–455. Available at: https://doi.org/10.1016/j.ijantimicag.2006.10.016.

Tawakol, M.M. *et al.* (2023) 'The potential role of migratory birds in the transmission of pathogenic Campylobacter species to broiler chickens in broiler poultry farms and live bird markets', *BMC Microbiology*, 23(1), p. 66. Available at: https://doi.org/10.1186/s12866-023-02794-0.

Upadhyaya, I. *et al.* (2013) 'Rapid inactivation of Salmonella Enteritidis on shell eggs by plant-derived antimicrobials', *Poultry Science*, 92(12), pp. 3228–3235. Available at: https://doi.org/10.3382/ps.2013-03126.

US Outbreaks of Zoonotic Diseases Spread between Animals & People | Healthy Pets, Healthy People | CDC (2023). Available at: https://www.cdc.gov/healthypets/outbreaks.html (Accessed: 1 April 2024).

van Boven, M. *et al.* (2003) 'Rapid selection of quinolone resistance in Campylobacter jejuni but not in Escherichia coli in individually housed broilers', *Journal of Antimicrobial Chemotherapy*, 52(4), pp. 719–723. Available at: https://doi.org/10.1093/jac/dkg402.

van Dijk, H.F.G. and Verbrugh, H.A. (2022) 'Resisting disinfectants', *Communications Medicine*, 2(1), pp. 1–5. Available at:https://doi.org/10.1038/s43856-021-00070-8

Van Immerseel, F. *et al.* (2009) 'Strategies to control *Salmonella* in the broiler production chain', *World's Poultry Science Journal*, 65(3), pp. 367–392. Available at: https://doi.org/10.1017/S0043933909000270.

Varga, C. *et al.* (2019) 'Antimicrobial resistance in fecal Escherichia coli and Salmonella enterica isolates: a two-year prospective study of small poultry flocks in Ontario, Canada', *BMC Veterinary Research*, 15(1), p. 464. Available at: https://doi.org/10.1186/s12917-019-2187-z.

Vela, J. *et al.* (2012) 'Characterization of *Staphylococcus xylosus* isolated from broiler chicken barn bioaerosol', *Poultry Science*, 91(12), pp. 3003–3012. Available at: https://doi.org/10.3382/ps.2012-02302.

Venkitanarayanan, K., Thakur, S. and Ricke, S.C. (2019) *Food Safety in Poultry Meat Production*. Cham: Springer (Food microbiology and food safety. Practical approaches).

Veterinary Feed Directive (VFD), *FDA* [Preprint]. Available at: https://www.fda.gov/animal-veterinary/development-approval-process/veterinary-feed-directive-vfd (Accessed: 21 March 2024).

Wagle, B.R. *et al.* (2017) 'Application of β-resorcylic acid as potential antimicrobial feed additive to reduce campylobacter colonization in broiler chickens', *Frontiers in Microbiology*, 8. Available at: https://doi.org/10.3389/fmicb.2017.00599.

Wales, A. and Davies, R. (2020) 'Review of hatchery transmission of bacteria with focus on Salmonella, chick pathogens and antimicrobial resistance', *World's Poultry Science Journal*, 76(3), pp. 517–536. Available at: https://doi.org/10.1080/00439339.2020.1789533.

Wallinga, D. *et al.* (2022) 'A review of the effectiveness of current US policies on antimicrobial use in meat and poultry production', *Current Environmental Health Reports*, 9(2), pp. 339–354. Available at: https://doi.org/10.1007/s40572-022-00351-x.

Walsh, C. (2010) 'The problem of antimicrobial resistance in the foodchain', *Reports* [Preprint]. Available at: https://arrow.tudublin.ie/schfsehrep/8.

Wang, L. *et al.* (2022) 'Isolation, characterization and whole genome analysis of an Avian Pathogenic Escherichia coli Phage vB_EcoS_GN06', *Veterinary Sciences*, 9(12), p. 675. Available at: https://doi.org/10.3390/vetsci9120675.

Wang, Y. *et al.* (2019) 'Antibiotic resistance gene reservoir in live poultry markets', *Journal of Infection*, 78(6), pp. 445–453. Available at: https://doi.org/10.1016/j.jinf.2019.03.012.

Waters, A.E. *et al.* (2011) 'Multidrug-resistant staphylococcus aureus in US meat and poultry', *Clinical Infectious Diseases*, 52(10), pp. 1227–1230. Available at: https://doi.org/10.1093/cid/cir181.

Wellington, E.M. *et al.* (2013) 'The role of the natural environment in the emergence of antibiotic resistance in Gram-negative bacteria', *The Lancet Infectious Diseases*, 13(2), pp. 155–165. Available at: https://doi.org/10.1016/S1473-3099(12)70317-1.

WHO (2024). *WHO List of Medically Important Antimicrobials: a Risk Management Tool for Mitigating Antimicrobial Resistance Due to Non-human Use*. Geneva, Switzerland: World Health Organization.

Xue, G.D. *et al.* (2017) 'Impact of a Macleaya cordata-derived alkaloid extract on necrotic enteritis in broilers', *Poultry Science*, 96(10), pp. 3581–3585. Available at: https://doi.org/10.3382/ps/pex164.

Yadav, S. and Jha, R. (2019) 'Strategies to modulate the intestinal microbiota and their effects on nutrient utilization, performance, and health of poultry', *Journal of Animal Science and Biotechnology*, 10(1), p. 2. Available at: https://doi.org/10.1186/s40104-018-0310-9.

Yang, J. *et al.* (2022) 'Metagenomic insights into chicken gut antibiotic resistomes and microbiomes', *Microbiology Spectrum*, 10(2), pp. e01907-21. Available at: https://doi.org/10.1128/spectrum.01907-21.

Yang, Y. *et al.* (2019) 'Review of antibiotic resistance, ecology, dissemination, and mitigation in U.S. broiler poultry systems', *Frontiers in Microbiology*, 10, p. 2639. Available at: https://doi.org/10.3389/fmicb.2019.02639.

Yin, X. *et al.* (2021) 'Antimicrobial resistance in non-typhoidal *Salmonella* from retail poultry meat by antibiotic usage-related production claims – United States, 2008–2017', *International Journal of Food Microbiology*, 342, p. 109044. Available at: https://doi.org/10.1016/j.ijfoodmicro.2021.109044.

Yossapol, M., Sugiyama, M. and Asai, T. (2017) 'The occurrence of CTX-M-25-producing Enterobacteriaceae in day-old broiler chicks in Japan', *The Journal of Veterinary Medical Science*, 79(10), pp. 1644–1647. Available at: https://doi.org/10.1292/jvms.17-0294.

Zhang, R. *et al.* (2017) 'Presence of VIM-positive pseudomonas species in chickens and their surrounding environment', *Antimicrobial Agents and Chemotherapy*, 61(7). Available at: https://doi.org/10.1128/aac.00167-17.

Chapter 7

Biotic alternatives to antibiotics for poultry gastrointestinal health and food safety applications: prebiotics, postbiotics, and associated compounds

S. C. Ricke, University of Wisconsin-Madison, USA; and A.V.S. Perumalla, Hill's Pet Nutrition, USA

1 Introduction

Foodborne pathogens that have been identified with poultry such as *Salmonella* and *Campylobacter* continue to be a major issue of public health concern (Dawoud et al., 2017; Yang et al., 2019a; Ricke, 2021; O'Bryan et al. 2022). Understanding their capabilities for becoming established in the gastrointestinal (GI) tracts of chickens and subsequent transmission in poultry flocks remains elusive (Dunkley et al., 2009; Horrocks et al., 2009; Ricke, 2017a,b; Sibanda et al., 2018; Gast et al., 2022; Obe et al., 2023). For example, there are increasing concerns about the risk of developing cross-resistance and multiple antibiotic resistance in pathogenic bacteria of the host animal. Historically, several antibiotics and chemotherapeutic agents have been used in prophylactic doses in poultry feed to improve poultry performance and

http://dx.doi.org/10.19103/AS.2024.0146.09

thereby achieve economic benefits but, at the same time, increasing the risk of resistance (Jones and Ricke, 2003; Wythe et al., 2023). As a result, there is potential for outbreaks of specific enteric diseases originating from food animal sources including poultry that would greatly compromise public health treatment efforts on a global scale (Chen et al., 2015; Aarestrup, 2015; Van Boeckel et al., 2017; Yang et al., 2019b; Ager et al., 2023; Wythe et al., 2023). These enteric diseases could lower productivity as well as increase bird mortality and associated contamination of poultry products for human consumption. In response to these concerns, current and proposed regulations have challenged the meat animal industry to seek a wide range of feed amendment alternatives to not only maintain live animal production proficiency but retain quality attributes of the resulting food product (Hajati and Rezai, 2010; Alloui et al., 2013; Ricke, 2015; Hume, 2011; Clavijo and Flórez, 2018; Dittoe et al., 2018; Ricke, 2018, 2021; Callaway and Ricke, 2023).

Among the attributes that are of concern is microbial contamination that can not only be a cause of diminished shelf life in poultry products but depending on the nature of the contaminant be a food safety risk. While several postharvest strategies are currently implemented to decontaminate and thus minimize the spoilage costs and economic losses caused by foodborne, diseases/recalls preharvest food safety options remain elusive (Ricke, 2021; Ricke et al., 2022; O'Bryan et al., 2022; Obe et al., 2023). Preharvest food safety control measures including vaccination, competitive exclusion cultures, and the use of natural antimicrobial feed additives such as plant extract compounds, bacteriophage, bacteriocins, and their combinations, just to name a few, have all either been proposed, experimentally tested, or currently implemented as means to limit either the colonization and/or carriage of foodborne pathogens such as *Salmonella* and *Campylobacter* associated with poultry (Joerger, 2003; Patterson and Burkholder, 2003; Revolledo et al., 2006; Vandeplas et al., 2010; Hume, 2011; O'Bryan et al., 2015; Aehle and Curtiss III, 2017; Micciche et al., 2019; Olson et al., 2022; Callaway and Ricke, 2023).

While there is a wide range of potential feed amendments that offer potential substitutes for antibiotics, a group of compounds collectively referred to as biotics have become extensively examined and commercially promoted in the animal and poultry industry. Members of the biotics family include probiotics, prebiotics, synbiotics, and postbiotics, and share the common functionality of delivering some form of health benefit to the host (Vinderola et al. 2023). In general, probiotics are those that are administered as live microorganisms (Hill et al., 2014). For poultry, a number of probiotics also known as competitive exclusion cultures or direct-fed microbials have been developed over the years some of which consist of complex consortia of microorganisms, while others rely only on a single organism such as *Bacillus*

(Stavric, 1992; Nisbet et al., 1996a, b; Nisbet, 2002; Ricke and Saengkerdsub, 2015; Deng et al., 2020; Perumalla et al., 2023).

Prebiotics was originally defined as 'a non-digestible food ingredient that beneficially affects the host by selectively stimulating the growth and/or activity of one or a limited number of gastrointestinal (GI) microflora' but that definition has been further refined as more has become understood about the interaction between the gut microbiome and dietary components (Gibson and Roberfroid, 1995; Wang, 2009; Bird et al., 2010; Hutkins et al., 2016; Gibson et al., 2017). More simply put, prebiotics can generally be considered a class of compounds that specifically provide substrates (energy) for the beneficial GI microbiome (Hajati and Rezai, 2010). Prebiotics are non-digestible ingredients; pass through the small intestine to the lower gut where they undergo fermentation by colonic and/or cecal bacteria to produce short-chain fatty acids (SCFA), primarily acetate, butyrate, and propionate that in turn can be absorbed by the host (Demigné et al., 1986; Józefiak et al., 2004; Wang, 2009; Bird et al., 2010). Conversely, postbiotics have been defined by Salminen et al. (2021) as essentially mixture(s) of nonliving organisms and their respective components that generate health benefits when fed. Synbiotics usually involve the combination of prebiotics and probiotics and have been defined by Swanson et al. (2020) as 'a mixture comprising live microorganisms and substrate(s) selectively utilized by host microorganisms that confers a health benefit on the host'. While probiotics have a relatively long history with poultry production, prebiotics, synbiotics, and postbiotics have also emerged as promising alternatives to growth-promoting antibiotics for administration to poultry (Roto et al., 2015; Teng and Kim, 2018; Ricke, 2015, 2021; Ricke et al., 2023). In recent years, the introduction of 16S rDNA microbiome sequencing, shotgun metagenomic sequencing, and metabolomics for microbial community and functional characterization has greatly expanded the understanding of the complex interaction between the poultry GI tract microbiota and the various biotics being fed as dietary amendments. In this review, the primary focus for application of biotics to poultry production will be an overview of prebiotics, synbiotics, and postbiotics along with related compounds. The current chapter represents an update of our previously published chapter on this topic as well as new developments that have occurred after our chapter was published (Ricke et al., 2017b).

2 Prebiotic sources and general mechanisms of action in the poultry host

Prebiotics used in poultry nutrition for preharvest safety includes non-digestible carbohydrates such as fructooligosaccharide (FOS) products (oligofructose, inulin-type fructans found in many vegetables), trans-galactooligosaccharides, glucooligosaccharides, glycooligosaccharides, lactulose, lactitol, maltooligosaccharides, soy-oligosaccharides, xylo-oligosaccharides, stachyose, raffinose, and yeasts cell walls (mannano-oligosachharides) and other fermentable carbohydrates which are not or only minimally digested in the small intestine (Monsan and Paul, 1995; Flickinger et al., 2003; Collins and Gibson, 1999; Ricke, 2015; Roto et al., 2015; Ricke, 2018; Teng, and Kim, 2018). In general, the most common oligosaccharides are inulin and its hydrolysates and oligofructans, present in chicory, onion, garlic, asparagus, artichoke, leek, bananas, tomatoes, and numerous other plants (Flickinger et al., 2003; Ricke, 2015). Such prebiotic oligosaccharides can be produced by either extracting from plant sources, microbial synthesis, or enzymatic hydrolysis of polysaccharides (Crittenden and Playne, 1996; Ricke, 2015; Ricke et al., 2023).

Potential beneficial effects of prebiotics include altering the GI microbiome, stimulating the immune system, preventing colon cancer, reducing pathogen invasion including pathogens such as *Salmonella* and *Escherichia coli,* and other health benefits (Cummings and Macfarlane, 2002; Bengmark, 2012; Roto et al., 2015; Micciche et al., 2018; Kim et al., 2019). In general, lactobacilli and bifidobacteria are the two common microbial groups specifically stimulated by prebiotic addition, but more recent microbiome research has identified additional poultry GI tract microorganisms that may also respond to prebiotic supplementation (Kaplan and Hutkins, 2000; Ricke, 2015, 2018; Ricke et al., 2020, 2023).

Prebiotics can act either by binding the pathogens, increasing the osmotic potential in the lumen of the intestine, or through metabolites that are produced by the intestinal microbiome which uses the prebiotic compounds as specific substrates for their own metabolism (Hajati and Rezaei, 2010; Ricke et al., 2020; Dittoe et al., 2022). Indigestible carbohydrates and proteins undergo hydrolysis and degradation to produce SCFA and branched chain fatty acids respectively, that generate ATP via fermentation pathways in beneficial bacteria harbored in the host (Schell et al., 2002; Ten Bruggencate et al., 2003; Apajalahti, 2005; Rossi et al., 2005; Ricke, 2015; Ricke et al., 2020; Dittoe et al., 2022). However, mannan oligosaccharides possess the ability to bind the mannose receptors and inhibit pathogen colonization of the gut (Spring et al., 2000; Patterson and Burkholder, 2003; Teng and Kim, 2018). Incorporating prebiotics as feed supplements facilitates antagonistic effects on enteric pathogens by competitive exclusion, promotion of enzyme reactions (Hajati and Rezai, 2010), immune stimulation (Monsan and Paul, 1995; Patterson and Burkholder, 2003; Teng and Kim, 2018; Swaggerty et al., 2019), and an increase in resistance to pathogen colonization (Bengmark, 2001; Patterson and Burkholder, 2003; Ricke, 2015; Micciche et al., 2018; Kim et al., 2019). Factors influencing the effect of prebiotics include the type of diet (amount of non-digestible carbohydrates), type and level of supplements used, animal parameters (species, age, sex, and stage of production) (Yang et al., 2009), and hygiene status of the farm (Verdonk et al., 2005; Yang et al., 2009). With the introduction of 16S rDNA microbiome sequencing approaches specific impacts of prebiotics and other factors on the GI tract microbiota of the bird have become much better understood (Feye et al., 2020a; Ricke et al., 2017b, 2020; Weinroth et al., 2022). Specific actions of oligosaccharides displaying prebiotic properties are discussed in more detail in the following sections.

3 Non-digestible carbohydrates

Non-digestible carbohydrates in their native form are non-usable by the host when they enter the small intestine; however, they can be hydrolyzed by colonic bacteria of the large intestine and cecal bacteria depending on the host and potentially in the upper GI tract such as the crop in birds (Józefiak et al., 2004; Ricke, 2015, 2018). The main categories of non-digestible oligosaccharides (NDOs) that are commonly used in prebiotic supplements include carbohydrates in which the monosaccharide units are fructose, galactose, glucose and/or xylose (Bird, 1999). Indigestible carbohydrates can act as prebiotics where they undergo saccharolytic fermentation in the hindgut and ceca to produce SCFA (Bird, 1999; Józefiak et al., 2004; Apajalahti, 2005; Bird et al., 2010; Hutkins et al., 2016; Ricke et al., 2020; 2023). These fatty acids exhibit beneficial effects including lowering the pH of the environment, improving mineral absorption, inhibiting acid-sensitive pathogens, and enhancing the proliferation of indigenous bacteria such as bifidobacteria (bifidogenic) (Patterson and Burkholder, 2003; Apajalahti, 2005; Bengmark, 2012; Roto et al., 2015; Dittoe et al., 2018).

Non-digestible carbohydrates include non-digestible oligosaccharides, non-starch polysaccharides, and resistant starch (Bird, 1999; Bird et al., 2010; Hutkins et al., 2016). In the beginning, NDOs were used as low-calorie, low-cariogenic, and sucrose substitutes as bulking agents in the food industry (Crittenden and Playne, 1996; Ricke, 2015). Commercially available prebiotics include: fructooligosaccahrides (FOS), lactulose, isomaltooligosaccharides (IMO), galactooligosaccharides (GOS), *trans*-galactooligosachharides (TOS), inulin and oligofructose (Collins and Gibson, 1999). Nevertheless, all these compounds except FOS did not necessarily meet the traditional criteria as per the prebiotic classification originally described by Gibson and Roberfroid (1995) and thus were not considered prebiotics because there was no specific mechanism of action of colonic fermentation of these complex carbohydrates. However, as the gut microbiota and their various metabolic interactions have become better understood the definition of prebiotics has evolved and thus become more inclusive (Bird, 1999; Patterson and Burkholder, 2003; Roberfoid,

2000, 2007; Bird et al., 2010; Bengmark, 2012; Hutkins et al., 2016; Gibson et al., 2017; Ricke, 2018).

There has been an increasing trend in poultry nutrition research to formulate feeds rich in NDOs so that they could be used as prebiotic supplements (Patterson and Burkholder, 2003; Hajati and Rezai, 2010; Alloui et al., 2013; Ricke, 2018). Since these relatively resistant oligosaccharides escape the small intestine and can reach the ceca somewhat intact, they could potentially support the growth of beneficial bacteria including bifidobacteria and lactobacilli as well as other fermentative bacteria and thus exhibit additional prebiotic properties as described in the previous section (Bird, 1999; Gomes and Malcata, 1999; Józefiak et al., 2004; Ricke et al., 2013; Lindberg, 2014). Furthermore, as nontraditional cereal grains, cereal coproducts, and protein crops have become a greater part of the primary nutritional sources for energy and protein in poultry, increases in the intake of fiber and nonstarch carbohydrates which may have detrimental impacts on bird performance could occur (Knudsen, 2014; Zhuang et al., 2017; Ricke, 2018; Bodie et al., 2019). However, perhaps with additional research on further processing and other extraction approaches, these feedstuffs may offer opportunities to recover components with prebiotic properties. For example, Herfel et al. (2013) demonstrated that stabilized rice bran improved weanling pig performance by exhibiting prebiotic characteristics and Kumar et al. (2012) reported a reduction of *S.* Typhimurium fecal shedding and an increase in colonization of indigenous lactobacilli in mice fed rice bran. Rubinelli et al. (2017) also reported a reduction in *S.* Typhimurium when incubated in a poultry cecal *in vitro* batch culture system for certain rice bran cultivars as well as shifts in microbiome compositional and metabolite profiles.

The opportunity to isolate such components with both *in vitro* screening incubation methods that simulate the GI tract as well as a more complete omics analyses of gut contents from poultry fed diets with these dietary ingredients should help to better refine the prebiotic characteristics and develop more targeted benefits (Bird, 1999; Rastall and Maitin, 2002; Donalson et al., 2007; 2008a; Dunkley et al., 2007a, b; Bengmark, 2012; Ricke et al., 2013; Lindberg, 2014; Rubinelli et al., 2017). More recent work with sorghum varieties suggests that direct comparisons of GI tract microbiome profiles with cereal seed grain genetics offer an opportunity to identify and align cereal grain genetic traits with optimized fermentation of potential candidate sources of prebiotics (Yang et al. 2022). Although this was applied to a human microbiome system, this approach could be used for nutritional dietary formulation to achieve better poultry and animal performances. Depending on the genetic resolution, this strategy could be further refined to select certain cereal grain cultivars which represent more predictable performance and gut health outcomes when included as dietary components.

4 Fructooligosaccharides

FOS are one of the more frequently used prebiotics (Flickinger et al., 2003; Ricke, 2015). Common sources of plants and vegetables include onions, Jerusalem artichokes, bamboo shoots, chicory roots, and bananas (Flickinger et al., 2003; Ricke, 2015). FOS are non-digestible and non-absorbable oligosaccharides consisting of short-chain polymers of 1–2 linked fructose units (short-chain fructooligosaccharide, SCFOs) that can be produced commercially either by hydrolysis of inulin or by enzymatic synthesis from sucrose or lactose (Patterson and Burkholder, 2003; Ricke, 2015). However, they are resistant to enzymatic degradation and absorption due to the presence of β-linkages and hence pass through the upper GI tract to reach the colon and ceca, where they undergo fermentation (Gibson and Roberfroid, 1995; Gibson, 1999; Flickinger et al., 2003; Xu et al., 2003; Józefiak et al., 2004). The endproducts of FOS fermentation include the SCFA, which possess the ability to inhibit enteric pathogens including *Salmonella*, *E. coli*, and *Clostridium perfringens* (Cummings et al., 2001; Cummings and Macfarlane, 2002; Flickinger et al., 2003; Patterson and Burkholder, 2003; Ricke 2003a; Dittoe et al., 2018, 2022). In addition, FOS can also act as a fermentable substrate and thus select for beneficial intestinal bacteria such as bifidobacteria and lactobacilli (Gibson and Wang, 1994; Ricke, 2015) which in turn reduce the pathogens by competitive exclusion (Gibson and Roberfroid, 1995; Flickinger et al., 2003; Xu et al., 2003; Ricke, 2015).

The use of FOS in poultry has been reviewed extensively elsewhere (Ricke, 2015) and will be discussed only briefly in the current review. FOS can be selectively fermented by most strains of bifidobacteria and support proliferation of bifidobacteria known to inhabit the poultry GI tract (Kaplan and Hutkins, 2000; Rossi et al., 2005; Fukuda et al., 2011; Ricke, 2015). Bifidobacteria are anaerobic, Gram-positive bacteria found in the GI tract of various warm-blooded animals. The presence of bifidobacteria has been demonstrated to have specific antibacterial impacts as they can suppress potential pathogens such as *E. coli* by producing antimicrobials or by lowering the pH through the rapid production of organic acids such as acetate (Fukuda et al., 2011).

Numerous studies have demonstrated the ability of FOS supplementation to reduce the colonization of *Salmonella* in the GI tract (Bailey et al., 1991; Fukata et al., 1999; Donalson et al., 2008b; Ricke, 2015). Dietary amendments to limit *Salmonella* colonization and infection in commercial birds has proven to be particularly critical for laying hens undergoing induced molts where historically the molting of commercial laying hens was initiated by complete removal of feed for a number of days to shut down the reproductive tract and cease egg production as a prelude to initiating an additional egg laying cycle (Ricke 2003b; 2017a, b). However, emptying of the intestinal tract creates a microenvironment favorable for *S.* Enteritidis colonization and subsequent systemic invasion into the reproductive tract (Durant et al., 1999; Ricke, 2003b; Ricke et al., 2013; Ricke 2017a, b). Supplying low energy high fiber-based diets such as alfalfa during the molt period proved to be effective in limiting *Salmonella* colonization and infection (Ricke et al., 2013). Donalson et al. (2008b) reported that laying hens fed alfalfa-based diets supplemented with FOS reduced the colonization of *S.* Enteriditis in ovaries and livers when compared to the hens subjected to feed withdrawal, although some inconsistencies among independent trials were observed. Clearly more needs to be understood about the impact of FOS on not only the gut microbiota but also the physiological status of the gut architecture along with parameters such as passage rate and how this impacts the residence time of dietary FOS (Ricke, 2015). For example, Hanning et al. (2012) reported increased villi length and intestinal crypt depth in pasture flock-raised Naked Neck birds fed FOS compared to control birds by the end of an 8-week growth cycle time period.

While there have been fewer studies reported, FOS in their various forms have also been used to limit the establishment of other pathogens in poultry as well. Short-chain FOS and mannanoligosaccharide supplementation (4 g/kg) of a dextrose-isolated soy protein diet decreased cecal *C. perfringens* population levels in some of the trials reported by Biggs et al. (2007). When feeding male broilers Xu et al. (2003) observed that adding 4.0 g/kg FOS enhanced the growth of *Bifidobacterium* and *Lactobacillus*, but inhibited *E. coli* populations in small intestinal and cecal contents. Yusrizal and Chen (2003) reported that supplementation of fructans in broiler diets resulted in an increase in lactobacilli counts in the gut and decreases in *Campylobacter* and *Salmonella* in some of the reported trials.

In summary, it would appear that the FOS prebiotic relationship between the selection of beneficial bacteria and creating an adverse gut microenvironment for *Salmonella* is becoming more mechanistically clear. This is much less the case for other pathogens partially because fewer studies have been done but also at least in some cases the relationship between the indigenous microbiome and the corresponding pathogen may differ from pathogens such as *Salmonella* and thus make the response to FOS less predictable. For example, Indikova et al. (2015) have suggested that *Campylobacter* may to some extent interact with the microbial consortia within mixed microbial populations. If these relationships consistently exist in the GI tract it will be critical to identify the individuals among these potentially supportive microbial communities and use this information to design prebiotics that more effectively enrich members antagonistic to *Campylobacter* or that limit those members which are most mutualistically supportive of *Campylobacter* (Kim et al., 2019).

5 Galactooligosaccharide and isomaltooligosaccharide

Galactooligosaccharides are generated from lactose by a glycosyl transfer of one or more D-galactosyl units onto a chain of galactose units (Crittenden and Playne, 1996; Mahoney, 1998; Boon et al., 2000; Tzortzis et al., 2005; Macfarlane et al., 2008). The composition of GOS varies depending on the degree of polymerization ranging from 2 to 8 monomeric units (Macfarlane et al., 2008). The impact of GOS on the host may depend on a number of factors but they are believed to preferentially stimulate the growth of organisms such as bifidobacteria and lactobacilli by providing a fermentable substrate and thus generating an inhibitory GI microenvironment for pathogens (Gomes and Malcata, 1999; Rastall and Maitin, 2002; Tzortzis et al., 2005; Macfarlane et al., 2008; Saminathan et al., 2011; Alloui et al., 2013; Ricke et al., 2023). Other inhibitory mechanisms may be in play as well. Shoaf et al. (2006) demonstrated that GOS could reduce adherence of enteropathogenic *E. coli* to epithelial cell tissue cultures leading them to suggest that GOS could act as a structural mimic of the pathogen binding sites on the GI epithelial surface. However, this may both be somewhat host specific as well as pathogen strain-specific as Naughton et al. (2001) did not observe decreases in *S.* Typhimurium or nonpathogenic *E. coli* association with pig intestinal cells when a commercial GOS product was included. Intestinal architecture could also be influenced as Hanning et al. (2012) reported that at 8 weeks pasture flock-raised Naked Neck broilers that were fed GOS exhibited the deepest intestinal crypt depth of the intestinal villi compared to birds fed either FOS, plum fibers or no feed additive. When comparing different prebiotics in pasture flock broilers, Park et al. (2017a) examined cecal contents with 16S rDNA microbiome analyses and identified *Eubacterium* and *Alcaligenaceae* genera as being harbored only in the GOS-fed birds.

Isomaltooligosaccharides are glucosyl saccharides linked by α 1-6 glucosidic bonds and occur in the form of compounds such as isomaltose, panose, and isomaltotriose (Kuriki et al., 1993). They possess a degree of polymerization anywhere from 2 to 6 and can be derived from a variety of sources including starch, maltose, dextran, and sucrose (Crittenden and Playne, 1996; Aslan and Tanriseven, 2007). IMO have been proposed as a potential prebiotic for poultry use (Chung and Day, 2004). Based on this premise, Chung and Day (2004) screened a branched α-IMO produced by *Leuconostoc mesenteroides* and demonstrated that unidentified poultry cecal isolates as well as *Bifodobacterium longum* could use the IMO but *E. coli* and *S.* Typhimurium could not. When cecal isolates and *S.* Typhimurium were grown in mixed culture with IMO the *Salmonella* populations were reduced (Chung and Day, 2004). In a follow-up *in vivo* study, broilers challenged with *S.* Typhimurium and fed either 1%, 2%, or 4% wt/wt IMO exhibited a 2 log CFU/ml reduction of cecal *Salmonella* only at the 1% IMO level versus the control (no IMO) while an increase in cecal bifidobacteria populations was similar for all levels of IMO (Thitaram et al., 2005). However, Zhang et al. (2003) reported that while IMO inclusion in broiler diets improved performance during the first 3 weeks, no detectable changes occurred in the levels of *Lactobacillus*, *E. coli,* and total aerobes from the crop and the ceca.

6 Yeast-derived components and mannan derivatives

Yeasts are eukaryotic and unicellular fungi that have been used for both preventive and therapeutic effects against diarrhea and other gastrointestinal disturbances (Auclair, 2000; Roto et al., 2015). *Saccharomyces cerevisae* cells have been extensively examined and commercialized as a potential probiotic source because of their previous applications in the food and beverage industry that made them readily available (Roto et al, 2015; Ricke et al., 2023). Live yeast cells gained early popularity as an animal feed additive with various health and animal performance benefits associated with administration either alone or in combination with other probiotic bacteria (Bradley et al., 1994; Auclair, 2000; Fonty and Chaucheyras-Durand, 2006; Roto et al., 2015; Vohra et al., 2016; Ricke et al., 2023). Most of the work has been focused on live yeasts as probiotics for ruminants (Roto et al., 2015; Vohra et al., 2016). Several benefits to rumen function have been noted including positive impacts on fiber degrading bacteria, stabilization of rumen pH, and enhancement of lactate utilizing ruminal bacteria (Jouany, 2001; Fonty and Chaucheyras-Durand, 2006; Vohra et al., 2016).

The presence of lactate utilizing ruminal bacteria can be critical for ruminants under certain dietary conditions. Lactate accumulation in the rumen occurs after sudden dietary shifts to highly fermentable carbohydrates that can be readily fermented by lactate producers such as *Streptococcus bovis* and other lactic acid bacteria (Herrera et al., 2009). If left unchecked, this can become a severe condition in ruminants leading to lactic acidosis, acidification of the rumen, and in some cases death of the animal (Herrera et al., 2009). When the lactate utilizing ruminal bacterium *Selenomonas ruminantium* (Ricke et al., 1996) was co-cultured with a commercial *S. cerevisae* filter sterilized filtrate, Nisbet and Martin (1991) observed increases in growth and uptake of radiolabeled lactate by the selenomonad. More recently, Ding et al. (2014) demonstrated that when live *S. cerevisae* were included in the diets of beef steers the percentages of lactate-producing bacteria were reduced while the percentage of *S. ruminantium* was increased.

In addition to serving as probiotics, yeast cell components can also be used as prebiotics. The principal components, glucans and mannans, present in the yeast cell wall along with the fermentation and metabolite products have been extensively examined for prebiotic properties (Roto et al., 2015; Ricke et al., 2023). The *S. cerevisiae* cell wall consists of an outer layer consisting mostly of glycosylated mannoproteins (90% carbohydrate) and an inner layer (50 to 60% of the cell wall dry weight) composed of β 1,3-glucan and chitin (Klis et al., 2002). Yeast mannans which are closely associated to the cell wall proteins are one of the primary soluble polysaccharides that can be extracted from the cell wall (Lee and Ballou, 1965; Stewart et al., 1968). Although mannan structures are yeast species-specific, they are generally characterized as macromolecular polymers in which the mannose backbone is linked via α-(1-6) bonds with side chain bearing short chains (average of 2–3 mannose units containing α-(1-2) and α-(1-3) linkages) attached to the backbone via α-(1-2) linkages (Jones and Ballou, 1968; Raschke et al., 1973).

MOS as an isolated component of the yeast outer cell wall was introduced as a feed additive in the early 1990s (Hooge, 2004). The exact mechanism of inhibition of pathogenic bacteria is still uncertain but several general modes of action have been proposed over the years and can be summarized as follows: (a) MOS may bind the bacteria with type-1 fimbriae thereby inhibiting them from binding to carbohydrate moieties of the intestinal lining (Hooge, 2004; Teng and Kim, 2018). Steer et al. (2000) have pointed out that this may be a key mechanism as most pathogens possess carbohydrate-specific binding proteins for attachment to host cells and any compound that interferes with that can (a) not only potentially prevent colonization, but dislodge already attached cells, (b) agglutination or clumping of pathogenic cells with type-1 fimbriae and subsequently expel them out of the solution thus reducing subclinical or lethal infection (Spring et al., 2000; Steer et al., 2000), (c) immune system modulation (Tizard et al., 1989; Hajati and Rezai, 2010; Alloui et al., 2013); (d) modulation of intestinal brush border and mucin enzyme expression (Hooge, 2004; Hajati and Rezai, 2010); and (e) mannan derivative induction of phenotypic and functional maturation in mouse dendritic cells (Sheng et al., 2006). The immune modulatory effects of MOS have been attributed to several aspects as summarized by Tizard et al (1989) including: (1) binding to mannose – binding proteins, (2) attenuating presentation of antigens (colonized pathogens) to activate macrophages, and (3) induction of interleukin-l release.

As a prebiotic, MOS has been examined for both bird health and performance as well as resistance to colonization by foodborne pathogens. Based on an overall statistical meta-analysis across broiler pen trials conducted globally from 1993 to 2003 Hooge (2004) concluded that dietary supplementation of MOS decreased mortalities, while achieving comparable body weights and feed conversion ratios (FCR) when compared to birds receiving antibiotic-supplemented diets. This retention of bird performance appears to hold true when MOS has been combined with immuno-based treatments such as vaccines or introduction of infective microorganisms. For example, Yang et al (2008b) reported improvements in FCR and/or body weight gains in birds fed MOS and challenged with pathogenic *E. coli* compared to the negative control. Likewise, when Nollet et al. (2007) fed MOS to birds vaccinated for *Eimeria* coccidia followed by a coccidian challenge they found that MOS improved the FCR of the vaccinated and *Eimeria*-challenged broilers from 15 to 42 days of age compared to the birds receiving the same coccidia/vaccination combination but no MOS while remaining the same as the FCR for nonvaccinated/nonchallenged birds.

Oyofo et al. (1989a, b) were able to demonstrate that D-mannose could block *in vitro* attachment on chicken small intestinal epithelial cells and reduce cecal colonization of *S.* Typhimurium in challenged birds. However, mannan oligosaccharides impact on cecal microbiota and the cecal microenvironment has been somewhat inconsistent. Spring et al. (2000) did see some reductions in *S.* Typhimurium and *S.* Dublin-challenged broiler chicks fed MOS but no effect on cecal concentrations of lactobacilli, enterococci, anaerobic bacteria, lactate, SCFA, or cecal pH. Populations of lactobacilli and coliforms were reduced in the ileum of birds fed 2 g/kg MOS but less impact was observed in the ceca (Yang et al., 2008a). Environmental management conditions may be a factor as well. For example, when birds were raised in the presence of new litter and fed 2 g/kg MOS, Yang et al. (2007) did not detect any differences in total anaerobes, *Clostridium perfringens* or lactic acid bacterial populations in the ileum, duodenum, or ceca. Using a *C. perfringens* challenge model Thanissery et al (2010) observed only a slight reduction of intestinal *C. perfringens* in broiler chicks fed a commercial yeast extract on days 1 and 7 postchallenge. Further work with yeast-based prebiotic sources remains complicated given the complexity of the cell wall structure and the fact that individual components may elicit different responses to the GI tract microbiota (Santovito et al., 2018; Ricke et al., 2023).

7 Postbiotics

As the understanding of yeast and yeast components has become more refined and beneficial effects associated with them as both a probiotic as well as a source of prebiotics identified, further applications have emerged. Given the multiple metabolic processes involving yeast fermentation of different grains for ethanol-based biofuels and fermented beverages, it was realized that the resultant fermentation metabolites could also provide beneficial attributes as feed supplements (Limayen and Ricke, 2012; Zhuang et al., 2017; Bodie et al., 2019; Perricone et al., 2022; Ricke et al., 2023). These fermentation products have been identified as postbiotics and defined by Salminen et al. (2021) as the 'preparation of inanimate microorganisms and/or their components that confers a health benefit on the host'. Sources of organisms for the generation of postbiotics vary and can be single or multiple organisms and do not have to be probiotic organisms (Salminen, 2021). However, the candidate organism(s) generating the postbiotic metabolites must be clearly identified as metabolites from undefined microbial mixtures that do not fit the definition (Salminen et al., 2021). In addition, they are defined as purposely inactivated organisms and may or may not include metabolites and/or cell components that provide a health benefit to the host but cannot be purified metabolites alone (Salminen et al., 2021).

Yeast fermentation cultures can be used for generating postbiotic products and are typically comprised of dead cells, yeast cell fragments, and various metabolites (Perricone et al., 2022; Ricke et al., 2023). *Saccharomyces cerevisiae* is the most common yeast used as a source for postbiotic products for feed additives in the animal and poultry industry (Perricone et al., 2022). Fermentation generally occurs via inoculation of the yeast culture onto a fermentable substrate such as a cereal grain, followed by ethanol production, and subsequently terminated by drying to lyse the yeast cells (Perricone et al., 2022; Ricke et al., 2023). This results in a product containing yeast cell wall components, biomass, and various metabolites that can be used as a feed supplement (Hancock et al., 2000; Buzzini and Vaughn-Martini, 2006; Perricone et al., 2022; Ricke et al., 2023). The types of metabolites vary and can include a myriad of enzymes, lipids, vitamins, organic acids, and polysaccharides among other compounds (Buzzini and Vaughn-Martini, 2006). Given the complexity and variability, it is not surprising that the mechanism(s) and the corresponding host responses might also be varied depending on the nature of the fermentation process and the resulting yeast postbiotic product (Reuben et al., 2021; Perricone et al., 2022; Ricke et al., 2023).

As commercial yeast-based postbiotics have become more available, several studies have been conducted to determine their impact on poultry performance as well as mitigation of foodborne pathogen establishment in the GI tract. In many cases, these products were referred to as yeast fermentates rather than the more recently defined postbiotic definition. A series of studies using *in vitro* poultry cecal studies have been used to test dosages and efficacy of yeast postbiotics. Rubinelli et al., (2016) demonstrated that a commercial yeast fermentate product identified as XPC (Diamond V Original XPC™, Cedar Rapids, IA) would reduce *S.* Typhimurium from an initial 6.81 $\log_{10}$ CFU/ml to 3.73 and 1.19 $\log_{10}$ CFU/ml after 24 and 48 hours incubation times with two-fold increases in acetate and butyrate after 24 h. However, a key finding was that the cecal mixed microbial population required 24 h preincubation in the presence of XPC prior to adding the *Salmonella* culture to exhibit a detectable impact on *Salmonella* levels. Follow-up studies with cecal inocula from different ages of broilers confirmed the reduction in *S.* Typhimurium by XPC as well as impacts on the cecal microbiome but indicated that the age of bird should be considered an influential factor as well (Park et al., 2017b). This XPC-mediated reduction has been seen with different *Salmonella* serovars such as Reading and Enteritidis as well as different sources of cecal inocula including turkeys and roosters (Feye et al., 2021; Costello et al., 2023). This *in vitro* cecal system has also been modified to accommodate *Campyobacter jejuni* growth under a microaerophilic atmosphere (Feye et al., 2020b). When *C. jejuni* was exposed to XPC-adapted cecal microbiota, reduction was observed in parallel with increased proportions of certain taxa groups such as *Lactobacillus* spp. and *Methanobrevibacter* spp. (Feye et al., 2020b).

Several *in vivo* studies have also been conducted with broilers and yeast products. Both *Salmonella* and *Campylobacter* reduction were observed in broilers receiving XPC but the impact on cecal microbiome profiles was minimal (Guyard-Nicodème et al., 2016; Park et al., 2017c; Roto et al., 2017; Nelson et al., 2020). The production parameters for the generation of yeast-based products may be a factor. Sun et al. (2020) demonstrated that the length of fermentation time allowed before harvesting yeast fermentates not only altered broiler cecal microbiome and metabolite profiles but improved broiler performance. This would indicate that yeast fermentation may vary over time and there are specific times during the fermentation cycle when a postbiotic product is collected that is optimal for poultry feed supplementation. It is also likely that the type of grain being fermented as well as the variety of yeast used in the fermentation process are factors as well. This may be an opportunity to apply cereal grain genetics to match the grain cultivar with the most appropriate fermentative organism.

8 Synbiotics

As discussed previously in this review, prebiotics and feed amendments with prebiotic properties have been used as feed additives in poultry to improve bird performance, health, and prevent colonization by foodborne pathogens. Likewise, probiotics containing various combinations of live organisms have been commercially developed to achieve similar benefits in poultry. Since many of the prebiotics serve as substrates for GI organisms that are beneficial to the host it is intuitive that combinations of selected prebiotics and some of these organisms might be considered as being even more effective. Consequently, the impact of prebiotics could be enhanced by combining them with selective probiotic cultures that would employ synergism in their action for better protective effects against foodborne pathogens. These combinations have been referred to as synbiotics and are defined as nutritional supplements which contain a mixture of prebiotics and probiotics that act synergistically and deliver beneficial effects to the host by improving the survival and establishment of live microbial dietary supplements in the GI tract (Gibson and Roberfroid,

1995; Schrezenmeir and de Vrese, 2001; Rastall and Maitin, 2002; Patterson and Burkholder, 2003; Rastall, 2004; Bengmark, 2012; Swanson et al., 2020). In short, synbiotics involve the provision of prebiotic and probiotic elements in the same product (Gibson and Roberfroid, 1995; Patterson and Burkholder, 2003). However, to achieve synergistic effects in the host, optimal administration of these biotic compounds will require the development of combinations of prebiotics and probiotics that maximize their combined efficacy (Rastall et al., 2005; Markowiak and Śliżewska, 2018; Fan et al., 2021; Ricke et al., 2023).

While conceptionally the idea of a synbiotic appears straightforward, mechanistically it can be more complicated as the inclusion of a prebiotic involves not only manipulation of the indigenous microbiome and host responses such as the immune system, but the type of prebiotic provided as a substrate as well as the specific bacteria serving as the target probiotic (Bomba et al., 2002; Bengmark, 2012). For example, *in vitro* mixed culture studies conducted by Fooks and Gibson (2002) demonstrated that oligosaccharides supported the growth of *Lactobacillus plantarum* and *Bifidobacterium bifidum* and inhibited *E. coli*, *Campylobacter jejuni* and *S.* Enteritidis but the extent of reduction was dependent on the type of oligosaccharide. Herrera et al. (2012) not only noted a similar carbohydrate specificity in *in vitro* co-cultures of *Streptococcus bovis* and *S.* Typhimurium but also observed that allowing time for the *Streptococcus bovis* to grow and ferment the carbohydrate prior to inoculating *S.* Typhimurium was critical to maximizing the reduction of the pathogen. Based on their *in vitro* screening study with poultry *Lactobacillus* isolates, Saminathan et al. (2011) concluded that not only were there differences in isolate responses to different prebiotic substrates but there were strain differences in the ability to grow on a particular prebiotic.

A novel way to resolve the dilemma of optimizing prebiotic/probiotic combinations is to preempt the vagueness associated with matching the right prebiotic with the right probiotic by simply employing the prebiotic to selectively isolate GI tract bacteria that are probiotic candidates possessing the desired properties of interest in the corresponding host animal. A classic example of this is using lactose as the carbon source to isolate a defined cecal population that would limit *Salmonella* colonization in poultry flocks. Lactose as a poultry feed candidate to target GI tract bacteria is attractive because utilization of it by the bird is generally low at all times nor can it be used by *S.* Typhimurium (Atkinson et al., 1957; Gutnick et al., 1969; Moran Jr., 1985; Waldroup et al., 1992; Kermanshahi and Rostami, 2006). Lactose either directly fed, added to the water or as a part of feed supplements such as whey was initially shown to inhibit *Salmonella* colonization in broilers and believed to be fermented by cecal bacteria to generate organic acids inhibitory to *Salmonella* (Bilgili and Moran Jr., 1990; Corrier et al., 1990a, b; DeLoach et al, 1990; Kermanshahi and Rostami, 2006). This relationship between the cecal microbiome and lactose was further substantiated when dietary lactose was combined with anaerobic cecal bacterial inocula and introduced to broiler chickens and turkey poults (Hinton et al., 1991; Ziprin et al., 1990; Corrier et al., 1991a, 1991b). The cecal inocula had been generated by incubating a mixed cecal culture for 24 h in a Viande Levure (VL) broth modified to exclude glucose and substitute with lactose (Corrier et al., 1991b). In order to stabilize the cecal inocula Nisbet et al. (1993a,b) took this a step further and used continuous flow (CF) cultures containing lactose VL broth in place of glucose to select for a cecal lactose utilizing a bacterial consortia that stabilized after reaching steady state following several growth vessel turnovers. When they inoculated chickens, significant *S.* Typhimurium reduction was achieved with the CF culture combined with lactose compared to control birds and these birds also yielded higher levels of cecal propionic acid (Nisbet et al., 1993b). As further refinements were introduced to avoid dietary dependence on relatively expensive lactose, it eventually became clear that developing and maintaining a metabolically stable cecal microbial consortia that could consistently generate propionic acid in the ceca of inoculated birds was a critical element for achieving potential commercial applications (Corrier et al., 1995; Nisbet et al., 1996a, b).

Other synbiotic combinations have also proven to be beneficial to the poultry host and/or inhibitory to pathogen colonization and some have since been commercialized. These studies are too numerous to list and discuss here. However, overall outcomes tend to be varied and thus it is difficult to compare individually or define mechanisms that are common across all studies. This is not necessarily surprising given the extra layers of complexity associated not just with the different chemical compositions of individual prebiotics but also considering the biological variability associated with the probiotic live cultures, and finally the immense and still poorly understood GI tract microbiome. Certainly, host factors such as the immune response to a shifting GI tract microbiome as the host animal matures are critical as well. Likewise, the interaction between the prebiotic and the probiotic when combined and fed to the animal introduces additional mechanistic complexities related to the stability of the prebiotic to the digestive tract and its relative fermentability by the GI microbiome. This interaction has potential commercial implications. One of the benefits of including prebiotics as part of the feed amendment is the potential ability to increase the efficacy of probiotic culture preparations by protecting probiotic bacteria from hostile host conditions such as the acidic gastric compartment as they travel to the upper intestinal tract; thus, ensuring arrival of viable probiotic cultures to the target areas of the lower GI tract (Rastall and Maitin, 2002). Obviously, the degree of protection will potentially influence the consistency of the synbiotic.

Finally, the individual pathogen strain responses to external environmental stresses and interventions employed in the food industry are not always consistent and strain variation must also be taken into account (Lianou and Koutsoumanis, 2013). This no doubt holds true for *Salmonella* responses to the presence of prebiotics and synbiotics as well. For example, *Salmonella* serovar and strain differences in susceptibility versus resistance to antagonistic metabolic products such as fermentation acids generated by probiotic cultures do occur and therefore should probably be considered (Ricke, 2003a; Dunkley et al., 2009; Dittoe et al., 2018). As interest in prebiotics and synbiotics grows, it will be important to screen the candidates against not only multiple *Salmonella* serovars but several representative strains for each individual serovar as well.

9 Conclusion and future trends

Biotic-based feed additives as a group represent variations in sources of biological matrices and include the broad categories of prebiotics, synbiotics, probiotics, and postbiotics. A common trait they all share is benefiting the host when they are added as a dietary supplement. The specific health benefit mechanism(s) associated with each type of biotic likely varies to some extent ranging from the introduction of live organisms (probiotics), substrates for probiotics and GI tract beneficial organisms (prebiotics), end products and cell components of some of these same organisms (postbiotics) as well as combinations of live organisms and prebiotic substrates (synbiotics). However, delineating the specific mechanism(s) for any of these biotic compounds remains elusive. Part of this is due to the variation in the composition and functionalities of different biotics. For example, prebiotics can vary considerably in chemical composition and yeast cells represent several different components each of which can impart independent prebiotic-like activities.

The other element that adds to the difficulty in understanding biotics is the complex nature of the vast array of microorganisms that inhabit the GI tract of animals including poultry. Not only do the GI tract organisms represent different substrate preferences and metabolic pathways but taxonomic composition can vary by GI tract location. In poultry, for example, the ceca are considered the primary fermentation site hence the extensive focus on the cecal microbial responses to biotic feed additives. However, the small intestine and other upper GT tract compartments such as the crop are likely to play a role albeit often overlooked role in the metabolism of biotics such as certain carbohydrate-based prebiotics (Ricke, 2018). Obviously, as more becomes known about the interactions and organisms involved in these different host microbiome GI tract compartments an understanding of the contribution that they may make to the overall metabolism of prebiotics and postbiotics can be achieved. Regardless of the type of biotic, the basic concept of biotic supplementation improving the growth performance, health status, and decreasing the foodborne pathogen load of the host animal remains the goal.

The age of the bird is also a factor. It is possible that early introduction of biotics to the young animal or bird that possesses a developing GI tract microbiota may have a greater impact versus an older bird with a more recalcitrant mature GI tract microbial population. For example, there is increasing interest in administering dietary additives to birds *in ovo* prior to hatching (Roto et al., 2016; Peebles, 2018; Siwek et al., 2018; Rubio, 2019; Das et al., 2021). Such an approach represents an opportunity to shape the GI tract microbiome and impact intestinal physiological development early in the bird's life cycle (Siwek et al., 2018; Rubio, 2019; Das et al., 2021). This approach may be one of the easier, more economical means to ensure consistent and uniform delivery of biotics on the massive scale required for the large numbers of birds being hatched for commercial flock placement. However, technical logistics would need to be worked out such as devising delivery mechanisms compatible for *in ovo* introduction. Delivery systems could vary depending on whether the biotic is in liquid form or is a solid feed supplement.

Optimizing combinations of prebiotics and probiotic cultures into synbiotics offers an opportunity to enhance the growth of beneficial bacteria such as bifidobacteria and lactobacilli spp. in a more controlled fashion that would ultimately produce more predictable synergistic effects in the host. However, such combinations, while conceptionally quite attractive, are fraught with complexities that require a better understanding not only of the GI microbiome but the host animal responses as well. Likewise, further research is needed to gain a better understanding of the mechanism(s) of action of postbiotics and prebiotics against foodborne pathogens such as *Salmonella* and *Campylobacter* in the GI tract and the fermentation responses of the indigenous GI tract microbiota to these externally introduced metabolites. It would not be surprising if several members of the GI tract population responded to prebiotics and postbiotics either directly as substrates or indirectly via cross-feeding with other members of the GI tract population. For example, given the carbohydrate polymers classified as prebiotics it is conceivable that certain GI tract microbiota directly hydrolyze the polymer whereas others only utilize some of the carbohydrate disaccharides and sugars released during hydrolysis.

Other factors need to be optimized as well to achieve more effectiveness for these types of products and to make them more commercially attractive. Certainly, further chemical refinement of prebiotic compounds to be more mechanistically selective and consistently predictable towards targeted GI tract microorganisms would be helpful. Designing prebiotics that are more resistant to degradation until they reach their respective target GI tract compartment would probably improve efficacy. This would in part support an increasing interest in more novel sources of NDO-based prebiotics. Likewise, aligning cereal grain genetics with optimized GI tract microbiome responses offers a more precise screening tool to evaluate potential sources of prebiotics and postbiotics. Regardless of the potential future research directions, prospects appear to be promising for biotics playing a major commercial role in poultry production systems.

10 Where to look for further information

Selected references:

- Callaway, T.R. and Ricke, S.C. (Eds.) (2023). *Direct Fed Microbials/Prebiotics for Animals: Science and Mechanisms of Action*, 2nd Edition. Springer Science, New York, NY, 348 pp.
- Gibson, G.R., et al. (2017), Expert consensus document: the international scientific association for probiotics and prebiotics (ISAPP) consensus statement on the definition and scope of prebiotics. *Nature Revs. Gastroenterol & Hepatol. 14*, 491–502. https://doi.org/10.1038/nrgastro.2017.75.
- Hill, C., et al. (2014), The International Scientific Association for Probiotics and Prebiotics consensus statement on the scope and appropriate use of the term probiotic *Nat. Rev. Gastroenterol. Hepatol. 11*, 506–514. https://doi.org/10.1038/nrgastro.2014.66.
- Hutkins, R.W., et al. (2016), Prebiotics: Why definitions matter. *Curr. Opin. Biotech. 37*, 1–7.
- Kim, S.A., et al. (2019), Potential for prebiotics as feed additives to limit foodborne *Campylobacter* establishment in the poultry gastrointestinal tract. *Front. Microbiol. 10*, 91. https://doi.org/10.3389/fmicb.2019.00091.
- Markowiak, P. and Śliżewska, K. (2018), The role of probiotics, prebiotics and synbiotics in animal nutrition. *Gut Pathog. 10*, 21. https://doi.org/10.1186/s13099-018-0250-0.
- Micciche, A.C., et al. (2018), A review of prebiotics against *Salmonella* in poultry: current and future potential for microbiome research application. *Frontiers Vet. Sci. 5*, 191. https://doi.org/10.3389/fvets.2018.00191.
- Perricone, V., et al. (2022), Yeast-derived products: the role of hydrolyzed yeast and yeast culture in poultry nutrition: a review. *Animals 12*, 1426. https://doi.org/10.3390/ani12111426.
- Reuben, R.C., et al. (2021), Prebiotics, probiotics and postbiotics for sustainable poultry production. *World's Poultry Sci. J. 77*, 825–882. https://doi.org/10.1080/00439339.2021.1960234.
- Ricke, S.C. (2018), Impact of prebiotics on poultry production and food safety. *Yale J. Biology and Med. 91*, 151–159.
- Salminen S., et al. (2021), The International Scientific Association of Probiotics and Prebiotics (ISAPP) consensus statement on the definition and scope of postbiotics. *Nature Revs. Gastroenterol. Hepatol. 18*, 649–667. https://doi.org/10.1038/ s41575-021-00440-6.

- Swanson, K.S., et al. (2020), The international scientific association for probiotics and prebiotics (ISAPP) consensus statement on the definition and scope of synbiotics. *Nature Revs. Gastroenterol. & Hepatol. 17*, 687–701. https://doi.org/10.1038/ s41575-020-0344-2.

11 References

Aarestrup, F.M. (2015), The livestock reservoir for antimicrobial resistance: a personal view on changing patterns of risks, effects of interventions and the way forward. *Phil. Trans. R Soc. B 370*, 20140085. https://doi.org/10.1098/rstb.2014.0085.

Aehle, S. and Curtiss III, R. (2017), Chapter 14. Current and future perspectives on development of Salmonella vaccine technologies. In: *Producing Safe Eggs-The Microbial Ecology of Salmonella.* Ricke, S.C. and Gast, R.K. (Eds.). Elsevier, Inc., San Diego, CA. pp. 281–299. https://doi.org/10.1016/B978-0-12-802582-6.00014-8.

Ager, E.O., et al. (2023), Global trends in antimicrobial resistance on organic and conventional farms. *Scientific Repts. 13*, 22608. https://doi.org/10.1038/s41598-023-47862-7.

Alloui, M.N., et al. (2013), The usefulness of prebiotics and probiotics in modern poultry nutrition: a review. *Ann. Anim. Sci., 13*, 17–32.

Apajalahti, J. (2005). Comparative gut microflora, metabolic challenges, and potential opportunities. *J. Appl. Poult. Res., 14*, 444–453.

Aslan, Y. and Tanriseven, A. (2007), Immobilization of *Penicillium lilacinum* dextranase to produce isomaltooligosaccharides from dextran. *Biochem. Eng. J. 34*, 8–12.

Atkinson, R.L., et al. (1957), Lactose in animal and human feeding. *J. Dairy Sci. 50*, 1114–1132.

Auclair, E. (2000), Yeast as an example of the mode of action of probiotics in monogastric and ruminant species. III Conference show feed manufacturing in the Mediterranean region improving food safety. From feed to food. *Reus, Espana. 54*, 45–53.

Bailey, J.S., et al. (1991), Effect of fructooligosaccharide on *Salmonella* colonization of the chicken intestine. *Poultry Sci. 70*, 2433–2438.

Bengmark S. (2001), Pre-, pro- and synbiotics. *Curr. Opin. Clin. Nutr. Metab. Care 4*, 571–579.

Bengmark, S. (2012), Integrative medicine and medicine-The role of pre-, pro- and synbiotics. *Clin. Translational Med. 1*(6), 1–13 (http://www.clintransmed.com/content/1/1/6).

Biggs, P., et al. (2007), The effects of several oligosaccharides on growth performance, nutrient digestibilities, and cecal microbial populations in young chicks. *Poultry Sci. 86*, 2327–2336.

Bilgili, S.F. and Moran, E.T. Jr. (1990), Influence of whey and probiotic-supplemented withdrawal feed on the retention of *Salmonella* intubated into market age broilers. *Poultry Sci. 69*, 1670–1674.

Bird, A.R. (1999), Prebiotics: a role for dietary fibre and resistant starch? *Asia Pacific J. Clin. Nutr. 8*(Suppl.), S32–S36.

Bird, A.R., et al. (2010), Resistant starch, large bowel ferementation and a broader perspective of prebiotics and probiotics. *Benef. Microbes 1*, 423–431.

Bodie, A.R., et al. (2019), Current trends of rice milling byproducts for agricultural applications and alternative food production systems. *Front. Sustain. Food Syst. 3*, 47. https://doi.org/10.3389/fsufs.2019.00047.

Bomba, A., et al. (2002), Improvement of the probiotic effect of micro-organisms by their combination with maltodextrins, fructooligosaccharides and polyunsaturated fatty acids. *Br. J. Nutr., 88*(Suppl., 1), S95–S99.

Boon, M.A., et al. (2000), Enzymatic synthesis of oligosaccharides: product removal during a kinetically controlled reaction. *Biotechnol. Bioeng. 70,* 411–420.

Bradley, G.L., et al. (1994), The effects of supplementing diets with *Saccharomyces cerevisiae* var. *boulardii* on male poult performance and ileal morphology. *Poultry Sci. 73,* 1766–1770.

Buzzini, P. and Vaughn-Martini, A. (2006), Yeast biodiversity and biotechnology. In: *Biodiversity and Ecophysiology of Yeasts,* Rosa C.A. and Péter, G. (Eds). Springer, Berlin, pp. 533–560.

Callaway, T.R. and Ricke, S.C. (Eds.). (2023), *Direct Fed Microbials/Prebiotics for Animals: Science and Mechanisms of Action*, 2nd Edition. Springer Science, New York, NY, 348 pp.

Chen, C.-Y., et al. (Eds.). (2015), *Antimicrobial resistance and food safety: Methods and techniques*. Academic Press/Elsevier Science, Amsterdam, The Netherlands.

Chung, C.-H. and Day, D.F. (2004), Efficacy of *Leuconostoc mesenteroides* (ATCC 13146) isomaltooligosaccharides as a poultry prebiotic. *Poultry Sci. 83,* 1302–1306.

Clavijo, V. and Flórez, M.J.V. (2018), The gastrointestinal microbiome and its association with the control of pathogens in broiler chicken production: A review. *Poultry Sci. 97,* 1006–1021.

Collins, M.D. and Gibson, G.R. (1999), Probiotics, prebiotics, and synbiotics: Approaches for modulating the microbial ecology of the gut. *Am. J. Clin. Nutr. 69 (suppl),* 1052S–1057S.

Corrier, D.E., et al. (1990a), Effect of dietary lactose on cecal pH, bacteriostatic volatile fatty acids, and *Salmonella typhimurium* colonization of broiler chicks. *Avian Dis. 34,* 668–676.

Corrier, D.E., et al. (1990b), Effect of dietary lactose on *Salmonella* colonization of market-age broiler chickens. *Avian Dis. 34,* 668–676.

Corrier, D.E., et al. (1991a), Effect of anaerobic cecal microflora and dietary lactose on colonization resistance of layer chicks to invasive *Salmonella enteritidis*. *Avian Dis. 35,* 337–343.

Corrier, D.E., et al. (1991b), Decreased *Salmonella* colonization in turkey poults inoculated with anaerobic cecal microflora and provided dietary lactose. *Poultry Sci. 70,* 1345–1350.

Corrier, D.E., et al. (1995), Control of *Salmonella typhimurium* colonization in broiler chicks with a continuous-flow characterized mixed culture of cecal bacteria. *Poultry Sci. 74,* 916–924.

Costello, M., et al. (2023), Comparison of yeast-derived commercial feed additives on *Salmonella* Enteritidis survival and microbiota populations in rooster cecal *in vitro* incubations. *PLoS ONE 18(12),* e0295657. https://doi.org/ 10.1371/journal.pone .0295657.

Crittenden, R.G. and Playne, M.J. (1996), Production, properties, and applications of food-grade oligosaccharides. *Trends Food Sci. Tech. 7,* 353–361.

Cummings, J.H. and Macfarlane, G.T. (2002), Gastrointestinal effects of prebiotics. *Br. J. Nutr. 87,* S145–S151.

Cummings, J.H., et al. (2001), Prebiotic digestion and fermentation. *Am. J. Clin. Nutr. 73,* S415–S420.

Das, R., Mishra, P., and Jha, R. (2021), *In ovo* feeding as a tool for improving performance and gut health of poultry: a review. *Front. Vet. Sci. 8*, 754246. https://doi.org/10.3389/fvets.2021.754246.

Dawoud, T.M., et al. (2017), Chapter 7. Overview of salmonellosis and foodborne Salmonella: Historical and current perspectives. In: *Producing Safe Eggs-The Microbial Ecology of Salmonella.* Ricke, S.C. and Gast, R.K. (Eds.). Elsevier, Inc., San Diego, CA. pp. 113–138. https://doi.org/10.1016/B978-0-12-802582-6.00007-0.

DeLoach, J.R., et al. (1990), Reduction of *Salmonella typhimurium* concentration in broiler chickens by milk or whey. *Avian Dis., 34,* 389–392.

Demigné, C., et al. (1986), Effects of absorption of large amounts of volatile fatty acids on rat liver metabolism. *J. Nutr. 116,* 77–86.

Deng, W., et al. (2020), Current perspectives and potential of probiotics to limit foodborne *Campylobacter* in poultry. *Front. Microbiol. 11*, 583429. https://doi.org/10.3389/fmicb.2020.583429.

Ding, G., et al. (2014), Effect of *Saccharomyces cerevisiae* on alfalfa nutrient degradation characteristics and rumen microbial populations of steers fed diets with different concentrate – to – forage ratios. *J. Anim. Sci. Biotechnol. 5,* 24. http://www.jasbsci.com/content/5/1/24.

Dittoe, D.K., et al. (2018), Organic acids and potential for modifying the avian gastrointestinal tract and reducing pathogens and disease. *Front.Vet. Sci. 5*, 216. https://doi.org/10.3389/fvets.2018.00216.

Dittoe, D.K., et al. (2022), Impact of the gastrointestinal microbiome and fermentation metabolites on broiler performance. *Poultry Sci. 101*,101786. https://doi.org/10.1016/j.psj.2022.101786.

Donalson, L.M., et al. (2007), *In vitro* anaerobic incubation of *Salmonella enterica* serotype Typhimurum and laying hen cecal bacteria in poultry feed substrates and a fructooligosacharide prebiotic. *Anaerobe 13,* 208–214.

Donalson, L.M., et al. (2008a). In vitro fermentation response of laying hen cecal bacteria to combinations of fructooligosaccharide (FOS) prebiotic with alfalfa or a layer ration. *Poultry Sci. 87,* 1263–1275.

Donalson, L.M., et al. (2008b), The influence of a fructooligosacharide prebiotic combined with alfalfa molt diets on the gastrointestinal tract fermentation, *Salmonella* Enteritidis infection and intestinal shedding in laying hens. *Poultry Sci. 87,* 1253–1262.

Dunkley, K.D., et al. (2007a), Comparison of in vitro fermentation and molecular microbial profiles of high-fiber feed substrates (HFFS) incubated with chicken cecal inocula. *Poultry Sci. 86,* 801–810.

Dunkley, K.D., et al. (2007b), Molting in *Salmonella* Enteritidis-challenged laying hens fed alfalfa crumbles.II. Fermentation and microbial ecology response. *Poultry Sci., 86,* 2101–2109.

Dunkley, K.D., et al. (2009), Foodborne *Salmonella* ecology in the avian gastrointestinal tract. *Anaerobe 15,* 26–35.

Durant, J.A., et al. (1999), Feed deprivation affects crop environment and modulates *Salmonella enteritidis* colonization and invasion of leghorn hens. *Appl. Environ Microbiol., 65,* 1919–1923.

Fan, R., et al. (2021), Purification of crude fructo-oligosaccharide preparations using probiotic bacteria for the selective fermentation of monosaccharide byproducts. *Front. Microbiol 11*, 620626. https://doi.org/10.3389/fmicb.2020.620626.

Feye, K.M., et al. (2020a), Influential factors on the composition of the conventionally raised broiler gastrointestinal microbiomes. *Poultry Sci. 99*, 653–659. https://doi.org/10.1016/j.psj.2019.12.013.

Feye, K.M., et al. (2020b), The preliminary development of an *in vitro* poultry cecal culture model to evaluate the effects of original XPC™ for the reduction of *Campylobacter jejuni* and its potential effects on the microbiota. *Front. Microbiol. 10*, 3062. https://doi.org/10.3389/fmicb.2019.03062

Feye, K.M., et al. (2021), Yeast fermentate-mediated reduction of *Salmonella* Reading and Typhimurium in an *in vitro* turkey cecal culture model. *Front. Microbiol. 12*, 645301. https://doi.org/10.3389/fmicb.2021.645301

Flickinger, E.A., et al. (2003), Nutritional responses to the presence of inulin and oligofructose in the diets of domesticated animals: A review. *Crit. Revs. Food Sci. Nutr. 43,* 19–60.

Fonty, G. and Chaucheyras-Durand, F. (2006), Effects and modes of action of live yeasts in the rumen. *Biologia, Bratislava, 61/6,* 741- 50.

Fooks, L.J. and Gibson, G.R. (2002), In vitro investigations of the effect of probiotics and prebiotics on selected human intestinal pathogens. *FEMS Microbiol. Ecol. 39,* 67–75.

Fukata, T., et al. (1999), Inhibitory effects of competitive exclusion and fructooligosaccharide, singly and in combination, on *Salmonella* colonization of chicks. *J. Food Prot. 62,* 229–233.

Fukuda, S., et al. (2011), Bifidobacteria can protect from enteropathogenic infection through production of acetate. *Nature 469,* 543–547.

Gast, R.K., et al. (2022) *Salmonella* in eggs and egg-laying chickens: Pathways to effective control. *Crit. Revs. In Microbiol.* https://doi.org/10.1080/1040841X.2022.2156772.

Gibson, G.R. and Wang, X. (1994), Inhibitory effects of bifidobacteria on other colonic bacteria. *J. Appl. Bacteriol., 77,* 412–420.

Gibson G.R. and Roberfroid, M.B. (1995), Dietary modulation of the human colonic microflora: Introducing the concept of prebiotics. *J. Nutr., 125,* 1401–1412.

Gibson, G.R. (1999). Dietary modulation of the human gut microflora using the prebiotics oligofructose and inulin. *J. Nutr., 129,* 1438–1441.

Gibson, G.R., et al. (2017), Expert consensus document: The International Scientific Association for Probiotics and Prebiotics (ISAPP) consensus statement on the definition and scope of prebiotics. *Nature Revs. Gastroenterol & Hepatol. 14*, 491–502. https://doi.org/10.1038/nrgastro.2017.75.

Gomes, A.M.P. and Malcata, F.X. (1999), *Bifidobacterium* spp. and *Lactobacillus acidophilus*: Biological, biochemical, technological and therapeutical properties relevant for use as probiotics. *Trends Food Sci. Technol. 10,* 139–157.

Gutnick, D., et al. (1969), Compounds which serve as the sole source of carbon or nitrogen for *Salmonella typhimurium* LT-2. *J. Bacteriol. 100,* 215–219.

Guyard-Nicodème, M., et al. (2016), Efficacy of feed additives against *Campylobacter* in live broilers during the entire rearing period. *Poultry Sci. 95*, 298–305. https://doi.org/10.3382/ps/pev303.

Hajati, H. and Rezaei, M. (2010), The application of prebiotics in poultry production. *Int. J. Poultry Sci. 9*, 298–304.

Hancock, R.D., et al. (2000), Biosynthesis of L-ascorbic acid (vitamin C) by *Saccharomyces cerevisiae*. *FEMS Microbiology Letts. 186*, 245–250. https://doi.org/10.1111/j.1574-6968.2000.tb09112.x.

Hanning, I., et al. (2012), Assessment of production performance in two breeds of broilers fed prebiotics as feed additives. *Poultry Sci. 91,* 3295–3299.

Herfel, T., et al. (2013), Stabilized rice bran improves weaning pig performance via a prebiotic mechanism. *J. Anim. Sci. 91,* 907–913.

Herrera, P., et al. (2009), Ecology and pathogenicity of gastrointestinal *Streptococcus bovis. Anaerobe 15,* 44–54.

Herrera, P., et al. (2012), Growth response of *Salmonella enterica* Typhimurium in co-culture with ruminal bacterium *Streptococcus bovis* is influenced by time of inoculation and carbohydrate substrate. *Food Res. Int. 45,* 1054–1057.

Hill, C., et al. (2014), The International Scientific Association for Probiotics and Prebiotics consensus statement on the scope and appropriate use of the term probiotic *Nat. Rev. Gastroenterol. Hepatol. 11,* 506–514. https://doi.org/10.1038/nrgastro.2014.66.

Hinton Jr., A., et al. (1991), Comparison of the efficacy of cultures of cecal anaerobes as inocula to reduce *Salmonella typhimurium* colonization in chicks with or without dietary lactose. *Poultry Sci. 70,* 67–73.

Hooge, D.M. (2004), Meta-analysis of broiler chicken pen trials evaluating dietary mannan oligosaccharide, 1993–2003. *Int. J. Poultry Sci. 3,* 163–174.

Horrocks, S.M., et al. (2009), Incidence and ecology of *Campylobacter* in animals. *Anaerobe 15,* 18–25.

Hume, M.E. (2011), Historic perspective: Prebiotics, probiotics, and other alternatives to antibiotics. *Poultry Sci. 90,* 2663–2669.

Hutkins, R.W., et al. (2016), Prebiotics: Why definitions matter. *Curr. Opin. Biotech. 37,* 1–7.

Indikova, I., et al. (2015), Survival with a helping hand: *Campylobacter* and microbiota. *Front. Microbiol. 6,* 1266, 1–6. https://doi.org/10.3389/fmicb.2015.01266

Joerger, R.D. (2003), Alternatives to antibiotics: Bacteriocins, antimicrobial peptides and bacteriophages. *Poultry Sci. 82,* 640–647.

Jones, F.T. and Ricke, S.C. (2003), Observations on the history of the development of antimicrobials and their use in poultry feeds. *Poultry Sci. 82,* 613–617.

Jones, G.H. and Ballou, C.E. (1968), Isolation of an α-mannosidase which hydrolyzes yeast mannan – Structure of the backbone of yeast mannan. *J. Biol. Chem. 243,* 2442–2446.

Jouany, J.P. (2001), A new look at yeast cultures as probiotics for ruminants. *Feed Mix 9,* 17–19.

Józefiak, D., et al. (2004), Carbohydrate fermentation in the ceca: a review. *Anim. Feed Sci. Technol. 113,* 1–15.

Kaplan, H. and Hutkins, R.W. (2000), Fermentation of fructooligosaccharides by lactic acid bacteria and bifidobacteria. *Appl. Environ. Microbiol. 66,* 2682–2684.

Kermanshahi, H. and Rostami, H. (2006), Influence of supplemental dried whey on broiler performance and cecal flora. *Int. J. Poultry Sci. 5,* 538–543.

Kim, S.A., et al. (2019), Potential for prebiotics as feed additives to limit foodborne *Campylobacter* establishment in the poultry gastrointestinal tract. *Front. Microbiol. 10,* 91. https://doi.org/10.3389/fmicb.2019.00091.

Klis, F.M., et al. (2002), Dynamics of cell wall structure in *Saccharomyces cerevisiae. FEMS Microbiol. Revs. 26,* 239–256.

Knudsen, K.E.B. (2014), Fiber and nonstarch polysaccharide content and variation in common crops used in broiler diets. *Poultry Sci. 93,* 2380–2393.

Kumar, A., et al. (2012), Dietary rice bran promotes resistance to *Salmonella enterica* serovar Typhimurium 4 colonization in mice. *BMC Microbiol. 12,* 71 (http://www.biomedcentral.com/1471-2180/12/71).

Kuriki, T., et al. (1993), A new way of producing isomalto-oligosaccharide syrup by using the transglycosylation reaction of neopullulanase. *Appl. Environ. Microbiol. 59,* 953–959.

Lee, Y.-C. and Ballou, C.E. (1965), Preparation of mannobiose, mannotriose, and a new mannotetraose from *Saccharomyces cerevisiae* mannan. *Biochemistry 4,* 257-264.

Lianou, A. and Koutsoumanis, K.P. (2013), Strain variability of the behavior of foodborne bacterial pathogens: A review. *Int. J. Food Microbiol. 167,* 310–321.

Limayen, A. and Ricke, S.C. (2012), Lignocellulosic biomass for bioethanol production: Current perspectives, potential issues and future prospects. *Prog. Energy Comb. Sci. 38,* 449–467. https://doi.org/10.1016/j.pecs.2012.03.002.

Lindberg, J.E. (2014), Fiber effects in nutrition and gut health in pigs. *J. Anim. Sci. Biotechnol. 5,* 15. http://www.jasbsci.com/content/5/1/15.

Macfarlane, G.T., et al. (2008), Bacterial metabolism and health-related effects of galacto-oligosaccharides and other prebiotics. *J. Appl. Microbiol. 104,* 305–344.

Mahoney, R.R. (1998), Galactosyl-oligosaccharide formation during lactose hydrolysis: A review. *Food Chem. 63,* 147–154.

Markowiak, P. and Śliżewska, K. (2018), The role of probiotics, prebiotics and synbiotics in animal nutrition. *Gut Pathog. 10,* 21. https://doi.org/10.1186/s13099-018-0250-0.

Micciche, A.C., et al. (2018), A review of prebiotics against *Salmonella* in poultry: Current and future potential for microbiome research application. *Frontiers Vet. Sci. 5,* 191. https://doi.org/10.3389/fvets.2018.00191.

Micciche, A., et al. (2019), Essential oils as an intervention strategy to reduce *Campylobacter* in poultry production: A review. *Front. Microbiol. 10,* 1058. https://doi.org/10.3389/fmicb.2019.01058.

Monsan, P. and Paul, F. (1995), Enzymatic synthesis of oligosaccharides. *FEMS Microbiol. Revs 16,* 187–192.

Moran Jr., E.T. (1985), Digestion and absorption of carbohydrates in fowl and events through perinatal development. *J. Nutr. 115,* 665–674.

Naughton, P.J., et al. (2001), Effects of nondigestible oligosaccharides on *Salmonella enterica* serovar Typhimurium and nonpathogenic *Escherichia coli* in the pig small intestine in vitro. *Appl. Environ. Microbiol. 67,* 3391–3395.

Nelson, J.R., et al. (2020), Effects of yeast fermentate supplementation on cecal microbiome, plasma biochemistry and ileal histomorphology in stressed broiler chickens. *Livestock Sci. 240,* 104149. https://doi.org/10.1016/j.livsci.2020.104149.

Nisbet, D.J. and Martin, S.A. (1991), Effect of a *Saccharomyces cerevisiae* culture on lactate utilization by the ruminal bacterium *Selenomonas ruminantium*. *J. Anim Sci. 69,* 4628–4633.

Nisbet, D.J., et al. (1993a), Effect of mixed cecal microflora maintained in continuous culture and of dietary lactose on *Salmonella typhimurium* colonization in broiler chicks, *Avian Dis., 37,* 528–535.

Nisbet, D.J., et al. (1993b), Effect of a defined continuous-flow derived bacterial culture and dietary lactose on *Salmonella typhimurium* colonization in broiler chickens *Avian Dis., 37,* 1017–1025.

Nisbet, D.J., et al. (1996a), Maintenance of the biological efficacy in chicks of a cecal competitive-exclusion culture against *Salmonella* by continuous-flow fermentation. *J. Food Prot., 59,* 1279–1283.

Nisbet, D.J., et al. (1996b), Cecal propionic acid as a biological indicator of the early establishment of a microbial ecosystem inhibitory to *Salmonella* in chicks. *Anaerobe, 2,* 345–350.

Nisbet, D. (2002), Defined competitive exclusion cultures in the prevention of enteropathogen colonisation in poultry and swine. *Antonie van Leeuwenhoek., 81,* 481–486.

Nollet, L., et al. (2007), Effect of dietary mannan oligosaccharide (Bio-Mos) on live performance of broiler chickens given an anticoccidial vaccine (Paracox) followed by a mild coccidial challenge. *J. Appl. Poultry Res. 16,* 397–403.

Obe, T., et al. (2023), Controlling *Salmonella*: Strategies for feed, the farm, and the processing plant. *Poultry Sci. 102*, 103086. https://doi.org/10.1016/j.psj.2023.103086.

O'Bryan, C.A., et al. (2015), Potential of plant essential oils and their components in animal agriculture – *in vitro* studies on antibacterial mode of action. *Front. Vet. Sci. 2,* 35. https://doi.org/10.3389/fvets.2015.00035.

O'Bryan, C.A., et al. (2022) Public health impact of *Salmonella* spp. on raw poultry: Current concepts and future prospects in the United States. *Food Control 132*, 108539. https://doi.org/10.1016/j.foodcont.2021.108539.

Olson, E.G., et al. (2022) Application of bacteriophages to limit *Campylobacter* in poultry production. *Front. Microbiol. 12*, 458721. https://doi.org/10.3389/fmicb.2021.458721.

Oyofo, B.A., et al. (1989a), Prevention of *Salmonella typhimurium* colonization of broilers with D-mannose. *Poultry Sci. 68,* 1357–1360.

Oyofo, B.A., et al. (1989b), Inhibition by mannose of *in vitro* colonization of chicken small intestine by *Salmonella typhimurium. Poultry Sci. 68,* 1351–1356.

Park, S.H., et al. (2017a), Pasture flock chicken cecal microbiome responses to prebiotics and plum fiber feed amendments. *Poultry Sci. 96*, 1820–1830. https://doi.org/10.3382/ps/pew441.

Park, S.H., et al. (2017b), Original XPC™ effect on *Salmonella* Typhimurium and cecal from three different ages of broiler chickens when incubated in an anaerobic *in vitro* culture system. *Front. Microbiol. 8*, 1070. https://doi.org/10.3389/fmicb.2017.01070.

Park, S.H., et al. (2017c), Effects of feeding Original XPC™ to broilers with a live coccidiosis-vaccine under industrial conditions: Part 2. Cecal microbiota analysis. *Poultry Sci. 96,* 2400–2411. https://doi.org/10.3382/ps/pex014.

Patterson, J.A. and Burkholder, K. (2003), Application of prebiotics and probiotics in poultry production. *Poultry Sci. 82,* 627–631.

Peebles, E.D. (2018), In ovo applications in poultry: A review. *Poultry Sci. 97*, 2322–2338. https://doi.org/10.3382/ps/pey081.

Perricone, V., et al. (2022), Yeast-derived products: The role of hydrolyzed yeast and yeast culture in poultry nutrition - A review. *Animals 12*, 1426. https://doi.org/10.3390/ani12111426.

Perumalla, A.V.S., et al. (2023), Chapter 7. *Probiotics in poultry preharvest food safety: Historical developments and current prospects. In: Direct Fed Microbials/Prebiotics for Animals: Science and Mechanisms of Action*, 2nd Edition, Callaway. T.R. and Ricke, S.C. (Eds.), Springer Science, New York, NY. pp. 127–166. https://doi.org/10.1007/978-3-031-40512-9_6.

Raschke, W.C., et al. (1973), Genetic control of yeast mannan structure. Isolation and characterization of mannan mutants. *J. Biol. Chem. 248,* 4660–4666.

Rastall, R. (2004), Bacteria in the gut: friends and foes and how to alter the balance. *J. Nutr. 134,* 2022S- 2026S.

Rastall, R.A. and Maitin, V. (2002), Prebiotics and synbiotics: towards the next generation. *Curr. Opin. Biotechnol. 13,* 490–496.

Rastall, R.A., et al. (2005), Modulation of the microbial ecology of the human colon by probiotics and synbiotics to enhance human health: An overview of enabling science and potential applications. *FEMS Microbiol Ecology 52,* 145–152.

Reuben, R.C., et al. (2021), Prebiotics, probiotics and postbiotics for sustainable poultry production. *World's Poultry Sci. J. 77,* 825–882. https://doi.org/10.1080/00439339.2021.1960234.

Revolledo, L., et al. (2006), Prospects in *Salmonella* control: Competitive exclusion, probiotics, and enhancement of avian intestinal immunity. *J. Appl. Poultry Res.15,* 341–351.

Ricke, S.C., et al. (1996), Ecology, metabolism, and genetics of ruminal selenomonads. *Crit. Revs. Microbiol. 22,* 27–65.

Ricke, S.C. (2003a), Perspectives on the use of organic acids and short chain fatty acids as antimicrobials. *Poultry Sci. 82,* 632–639.

Ricke, S.C. (2003b), The gastrointestinal tract ecology of *Salmonella* Enteritidis colonization in molting hens. *Poultry Sci. 82,* 1003–1007.

Ricke, S.C., et al. (2013), A review on development of novel strategies for controlling *Salmonella* Enteritidis colonization in laying hens: Fiber – based molt diets. *Poultry Sci. 92,* 502–525.

Ricke, S.C. (2015), Potential of fructooligosaccharide prebiotics in alternative and nonconventional poultry production systems. *Poultry Sci. 94,* 1411–1418.

Ricke, S.C. and Saengkerdsub, S. (2015), Chapter 19. *Bacillus probiotics and biologicals for improving animal and human health: Current applications and future prospects In: Beneficial Microbes in Fermented and Functional Foods*, Ravishankar Rai, V. and Jamuna Bai, A. (Eds.) CRC Press/Taylor & Francis Group, Boca Raton, FL pp. 341–360.

Ricke, S.C. (2017a), Insights and challenges of *Salmonella* infections in laying hens. *Curr. Opin. in Food Sci. 18,* 43–49. https://doi.org/10.1016/j.cofs.2017.10.012.

Ricke, S.C. (2017b), Chapter 16. *Gastrointestinal ecology of Salmonella* Enteritidis *in laying hens and intervention by prebiotic and non-digestible carbohydrate dietary supplementation. In: Producing Safe Eggs -The Microbial Ecology of Salmonella*. Ricke, S.C. and Gast, R.K. (Eds.). Elsevier, Inc., San Diego, CA. pp. 323–345. https://doi.org/10.1016/B978-0-12-802582-6.00016-1.

Ricke, S.C., et al. (2017a), Chap. 19. *Unravelling food production microbiomes: Concepts and future directions. In: Food and Feed Safety Systems and Analysis*. Ricke, S.C., Atungulu, G.G., Park, S.H., and Rainwater, C.E. (Eds.) Elsevier Inc., San Diego, CA pp. 347–374. https://doi.org/10.1016/B978-0-12-811835-1.00019-1.

Ricke, S.C., et al. (2017b) Chapter 5. Alternatives to antibiotics in preventing zoonoses and other pathogens in poultry: prebiotics and related compounds. In *Achieving Sustainable Production of Poultry Meat- Volume 1 Safety, Quality and Sustainability*. Ricke, S.C. (Ed.) Burleigh Dodd Publishing, Cambridge, UK. pp. 87–108. https://doi.org/10.19103/AS.2016.0010.06.

Ricke, S.C. (2018), Impact of prebiotics on poultry production and food safety. *Yale J. Biology and Med. 91,* 151–159.

Ricke, S.C. (2021), Strategies to improve poultry food safety, a landscape review. *Ann. Rev. of Anim. Biosciences 9*, 379 - 400. https://doi.org/10.1146/annurev-animal-061220- 023200.

Ricke, S.C., et al. (2020), Prebiotics and the poultry gastrointestinal tract microbiome. *Poultry Sci. 99*, 670–677. https://doi.org/10.1016/j.psj.2019.12.018.

Ricke, S.C., et al. (2022), Practical opportunities for microbiome analyses and bioinformatics in poultry processing. *Poultry Sci. 101*, 101787. https://doi.org/10.1016/j.psj.2022.

Ricke, S.C., et al. (2023), Chapter 6. *Prospects for prebiotics and postbiotics in poultry. In Direct Fed Microbials/Prebiotics for Animals: Science and Mechanisms of Action, 2nd Edition*, Callaway, T.R. and Ricke, S.C. (Eds.), Springer Science, New York, NY. pp. 103–124. https://doi.org/10.1007/978-3-031-40512-9_6.

Roberfroid, M. (2007), Prebiotics: the concept revisited. *J. Nutr. 137*, 830S–837S.

Roberfroid, M.B. (2000), Prebiotics and probiotics: are they functional foods? *Am J Clin Nutr. 71*, 1682–1687.

Rossi, M., et al. (2005), Fermentation of fructooligosaccharides and inulin by bifidobacteria: A comparative study of pure and fecal cultures. *Appl. Environ. Microbiol. 71*, 6150– 6158.

Roto, S.M., et al. (2015), An introduction to the avian gut microbiota and the effects of yeast-based prebiotic compounds as potential feed additives. *Front. Vet. Sci. 2*, 28. https://doi.org/10.3389/fvets.2015.00028.

Roto S.M., et al. (2016), Applications of *in ovo* technique for the optimal development of the gastrointestinal tract and the potential influence on the establishment of its microbiome in poultry. *Frontiers Vet. Sci. 3*, 63. https://doi.org/10.3389/fvets.2016.00063.

Roto, S.M., et al. (2017), Effects of feeding Original XPC™ to broilers with a live coccidiosis-vaccine under industry conditions: Part 1. Growth performance and *Salmonella* inhibition. *Poultry Sci. 96*, 1831–1837. https://doi.org/10.3382/ps/pew445.

Rubinelli, P., et al. (2016), Reduction of *Salmonella* Typhimurium by fermentation metabolites of Diamond V Original XPC in an *in vitro* anaerobic mixed chicken cecal culture. *Front. Vet. Sci. 3*, 83. https://doi.org/10.3389/fvets.2016.00083.

Rubinelli, P.M., et al. (2017), Differential effects of rice bran cultivars to limit *Salmonella* Typhimurium in chicken cecal in vitro incubations and impact on the cecal microbiome and metabolome. *PLoS ONE 12(9)*, e0185002. https://doi.org/10.1371/journal.pone.0185002.

Rubio, L.A. (2019), Possibilities of early life programming in broiler chickens via intestinal microbiota modulation. *Poultry Sci. 98*, 695–706. https://doi.org/10.3382/ps/pey416.

Salminen S., et al. (2021), The International Scientific Association of Probiotics and Prebiotics (ISAPP) consensus statement on the definition and scope of postbiotics. *Nature Revs. Gastroenterol. Hepatol. 18*, 649–667. https://doi.org/10.1038/s41575-021-00440-6.

Saminathan, M., et al. (2011), Effect of prebiotic oligosaccharides on growth of *Lactobacillus* strains used as a probiotic for chickens. *African J. Microbiol. Res.*, *5*(1), 57–64.

Santovito, E., et al. (2018), Eubiotics for food security at farm level: yeast cell wall products and their antimicrobial potential against pathogenic bacteria. *Foodborne Path. Dis. 15*, 531–537. doi.10.1089/fpd.2018.2430.

Schell, M.A., et al. (2002), The genome sequence of *Bifidobacterium longum* reflects its adaptation to the human gastrointestinal tract. *PNAS 99,* 14422–14427.

Schrezemmeir, J. and de Vrese, M. (2001), Probiotics, prebiotics, and synbiotics: approaching a definition. *Am. J. Clin. Nutr. 73,* 361S–364S.

Sheng, K.-C., et al. (2006), Mannan derivatives induce phenotypic and functional maturation of mouse dendritic cells. *Immunology 118,* 372–383.

Shoaf, K., et al. (2006), Prebiotic galactooligosaccharides reduce adherence of enteropathogenic *Escherichia coli* to tissue culture cells. *Infect. Immun. 74,* 6920–6928.

Sibanda, N., et al. (2018), A review of the effect of management practices on *Campylobacter* prevalence in poultry farms. *Front. Microbiol. 9,* 2002. https://doi.org/10.3389/fmicb.2018.02002.

Siwek, M., et al. (2018), Prebiotics and synbiotics – *in ovo* delivery for improved lifespan condition in chicken. *BMC Vet. Res. 14*, 402. https://doi.org/10.1186/s12917-018-1738-z.

Spring, P., et al. (2000), The effects of dietary mannanoligosaccharides on cecal parameters and the concentrations of enteric bacteria in the ceca of *Salmonella*-challenged broiler chicks. *Poultry Sci. 79,* 205–211.

Stavric, S. (1992), Defined cultures and prospects. *Int. J. Food Microbiol. 15,* 245–263.

Steer, T., et al. (2000), Perspectives on the role of the human gut microbiota and its modulation by pro and prebiotics. *Nutr. Res. Rev., 13,* 229–254.

Stewart, T.S., et al. (1968), Preparation of a mannopentaose, mannohexaose, and mannoheptaose from *Saccharomyces cerevisiae* mannan. *Biochem. 7,* 1843–1854.

Sun, Z., et al. (2020), Effects of yeast cultures with different fermentation times on the growth performance, caecal microbial community and metabolite profile of broilers. *J. Anim. Physiol. and Animal Nutr. 104*, 212–223. https://doi.org/10.1111/jpn.13241.

Swaggerty, C.L., et al. (2019), Modulation of the immune response to improve health and reduce foodborne pathogens in poultry. *Microorganisms 7*, 65. https://doi.org/10.3390/microorganisms7030065.

Swanson, K.S., et al. (2020) The International Scientific Association for Probiotics and Prebiotics (ISAPP) consensus statement on the definition and scope of synbiotics. *Nature Revs. Gastroenterol. & Hepatol. 17*, 687–701. https://doi.org/10.1038/s41575-020-0344-2.

Ten Bruggencate, S.J.M., et al. (2003), Dietary fructo-oligosaccharides dose-dependently increase translocation of *Salmonella* in rats. *J. Nutr. 133,* 2313–2318.

Teng, P.-Y. and Kim, W.K. (2018), Review: Roles of prebiotics in intestinal ecosystem of broilers. *Front. Vet. Sci. 5*, 245. https://doi.org/10.3389/fvets.2018.00245.

Thanissery, R., et al. (2010), Evaluation of the efficacy of NuPro yeast extract in reducing intestinal *Clostridium perfringens* levels in broiler chickens. *Poultry Sci. 89,* 2380–2388.

Thitaram, S.N., et al. (2005), Isomaltooligosaccharide increases cecal bifidobacteria population in young broiler chickens. *Poultry. Sci. 84,* 998–1003.

Tizard, I.R., et al. (1989), The biological activities of mannans and related complex carbohydrates. *Mol. Biother. 1,* 290–296.

Tzortzis, G., et al. (2005), Synthesis of prebiotic galactooligosaccharides using whole cells of a novel strain, *Bifidobacterium bifidum* NCIMB 41171. *Appl. Microbiol. Biotech. 68,* 412–416.

Van Boeckel, T.P., et al. (2017), Reducing antimicrobial use in food animals. *Science 357(6358)*, 1350–1352. https://doi.org/10.1126/science.aao1495.

Vandeplas, S., et al. (2010), *Salmonella* in chicken: Current and developing strategies to reduce contamination at farm level. *J. Food Prot. 73*, 774–785.

Verdonk, J.M.A.J., et al. (2005), Application of inulin-type fructans in animal feed and pet food. *Br. J. Nutr. 93*, S125–S138.

Vinderola, G., et al. (2023), Fermented foods: a perspective on their role in delivering biotics. *Front. Microbiol. 14*, 1196239. https://doi.org/10.3389/fmicb.2023.1196239.

Vohra, A., et al. (2016), Probiotic yeasts in livestock sector. *Anim. Feed Sci. Technol. 219*, 31–47. https://doi.org/10.1016/j.anifeedsci.2016.05.019 0377-8401.

Waldroup, A.L., et al. (1992), Effects of dietary lactose on incidence and levels of Salomonellae on carcasses of broiler chickens grown to market age. *Poultry Sci. 71*, 288–295.

Wang, Y. (2009), Prebiotics: Present and future science and technology. *Food Res. Int. 42*, 8–12.

Weinroth, M.D., et al. (2022), Considerations and best practices in animal science 16S rRNA gene sequencing microbiome studies. *J. Anim. Sci. 100*, 1–18 https://doi.org/10.1093/jas/skab346.

Wythe, L.A., et al. (2023), Chapter 4. Advancements in poultry nutrition and genetics, The role of antibiotic growth promoters, and the introduction of feed additive alternatives. In *Direct Fed Microbials/Prebiotics for Animals: Science and Mechanisms of Action*, 2nd Edition, Callaway, T.R. and Ricke, S.C. (Eds.), Springer Science, New York, NY. pp. 59–79. https://doi.org/10.1007/978-3-031-40512-9_4.

Xu, Z.R., et al. (2003), Effects of dietary fructooligosaccharide on digestive enzyme activities, intestinal microflora and morphology of male broilers. *Poultry Sci. 82*, 1030–1036.

Yang, Q., et al. (2022), Genetic analysis of seed traits in *Sorghum bicolor* that affect the human gut microbiome. *Nature Comm. 13*, 5641. https://doi.org/10.1038/s41467-022-33419-1.

Yang, Y., et al. (2019a), A historical review on antibiotic resistance of foodborne *Campylobacter*. *Front. Microbiol. 10*,1509. https://doi.org/10.3389/fmicb.2019.01509.

Yang, Y., et al. (2019b), Review of antibiotic resistance, ecology, dissemination, and mitigation in U.S. broiler poultry systems. *Front. Microbiol. 10*, 2639. https://doi.org/10.3389/fmicb.2019.02639.

Yang, Y., et al. (2007), Effects of mannanoligosaccharide on growth performance, the development of gut microflora, and gut function of broiler chickens raised on new litter. *J. Appl. Poultry Res. 16*, 280–288.

Yang, Y., et al. (2008a), Effects of mannanoligosaccharide in broiler chicken diets on growth performance, energy utilisation, nutrient digestibility and intestinal microflora. *Br. Poultry Sci. 49*, 186–194.

Yang, Y., et al. (2008b), Effects of mannanoligosaccharide and fructooligosaccharide on the response of broilers to pathogenic *Escherichia coli* challenge. *Br. Poultry Sci. 49*, 550–559.

Yang, Y., et al. (2009), Dietary modulation of gut microflora in broiler chickens: A review of the role of six kinds of alternatives to in-feed antibiotics. *World's Poultry Sci. J. 65*, 97–114.

Yusrizal, Y. and Chen, T.C. (2003), Effect of adding chicory fructans in feed on fecal and intestinal microflora and excreta volatile ammonia. *Int. J. Poultry Sci. 2*, 188–194.

Zhang, W.F., et al. (2003), Effects of isomalto-oligosaccharides on broiler performance and intestinal microflora. *Poultry Sci. 82*, 657–663.

Zhuang, X., et al. (2017), Chapter 10. Cereal grain fractions as potential sources of prebiotics: Current status, opportunities, and potential applications. In: *Food and Feed Safety Systems and Analysis*. Ricke, S.C., Atungulu, G.G., Park, S.H., and Rainwater, C.E. (Eds.) Elsevier Inc., San Diego, CA pp. 173–191. https://doi.org/10.1016/B978-0-12-811835-1.00010.

Ziprin, R.L., et al. (1990), Intracloacal *Salmonella typhimurium* infection of broiler chickens: Reduction of colonization with anaerobic organisms and dietary lactose. *Avian Dis. 34*, 749–753.

Chapter 8

Optimising microbiological and food safety in free-range and organic poultry meat production

Léni Corrand, Anibio Group Vétérinaire, France; Richard Jackson, St David's Veterinary Team, UK; and Paul McMullin, Poultry Health International, UK

1 Introduction

There has been considerable growth in free-range poultry meat production systems, mainly because of consumer demand for more animal-welfare-friendly production, though this demand for premium products is known to vary between markets and fluctuate according to economic cycles. These systems involve slower growth, longer production cycles, and less environmental control than for housed poultry. As a result of these factors, environmental sustainability may be less good than for housed poultry by some measures.[1] Our objective is to maintain good poultry health, predictable and uniform growth, and good control of pathogens of public health significance.

Free-range poultry meat production systems can vary considerably between farms and companies, more so than in indoor production. It is essential that any veterinarians providing care to birds in a free-range system, and scientists involved in research in this area, familiarise themselves with the given system to better understand potential disease and food safety issues. While many of

http://dx.doi.org/10.19103/AS.2024.0146.10

the principles developed in relation to conventionally housed poultry apply in similar ways, the relatively long growing cycles, and, consequently, the use of specialist strains developed for this purpose, bring with them differences above and beyond the access to range. This chapter aims to give an overview of free-range poultry production systems and the consequent potential meat product issues that can occur.

2 Free-range poultry meat production systems

Figure 1 illustrates in general terms free-range poultry meat production systems. There is considerable power of multiplication in poultry production systems, so the required breeding population (shown on the left) is commonly small in comparison with the commercial level birds; nevertheless, they contribute substantially to the health of the system. Breeding birds are typically fully housed to maintain their high health status, and are also vaccinated to reduce the risk of vertical transmission, and to provide useful levels of maternal antibody, in a similar way as colostrum for mammalian species. After a period of incubation of 21–33 days according to the species, the commercial birds are supplied to the growing farms.

Commercial-level day-olds may be moved directly to farms that grow birds to slaughter, or first to specialist brooding farms from which they are transferred to growing farms with range ('brood and move'). In either case, chicks will be started in houses equipped to provide supplemental heat and in which the 'pop-holes' can be fully sealed, and in which good biosecurity can be applied with respect to any older birds and the range (if present). They need to be in houses with the required access to range by the specified maximum age to start ranging (typically to give range access during the daytime for at least half of their life span) before finally being moved to specialist processing plants.

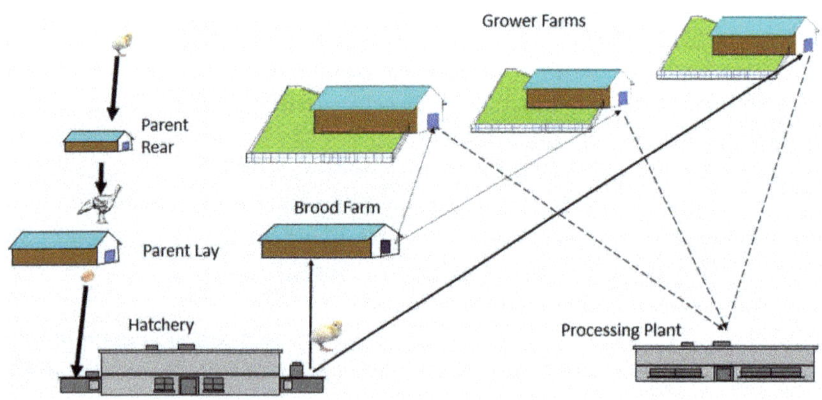

Figure 1 Flow chart of free-range poultry meat production systems.

The arrows in Fig. 1 indicate the normal flow of birds and some associated microbes through the production system. Cleaning and disinfection of vehicles and crates in a biosecure fashion are also required to avoid the 'reverse-flow' of pathogens. The long-term health status of the individual farm links in the production system depends primarily on the degree of 'multi-ageness' of each unit. The larger the age difference on a farm, the greater the disease challenge and the tendency to acquire successively more challenges.[2] This is because clinically recovered flocks commonly continue to harbour pathogens and become a source for the younger flocks coming through. While as veterinarians we normally advise against multi-age production on the same farm, producers may sometimes need this to satisfy the needs of their customer base. Many disease challenges/infections are of no public health significance, but some of these, such as Marek's disease, weaken the immune system and have the potential to increase the burden of pathogens which are. Brood farms should, as far as possible, operate on a single age, or as small an age range as possible.[2] Grow-on units should ideally get birds from only a single brood farm and apply robust separation of flocks of different ages, if not single-age.

Farm structure, equipment, and operating procedures will vary substantially according to the species and particular production rules applicable to them. In the European Union (EU), for instance, there are detailed rules laid down in relation to the poultry meat marketing regulation (Reg 543/2008[3]) to the different meat-producing species with distinct free-range production categories, as summarised in Table 1. The requirements for 'enhanced indoor/barn' production are shown for comparison with three standards of free-range poultry meat production, as well as the requirements for organic production, which were laid down in regulation 2092/91.[4] It may be noted that legal standards, required to be able to use particular terms in describing products, mainly focus, currently, on space allocation, both in the housed birds and in the various free-range classes. They are quite complex and have various adjustments as described in the extensive table footnotes. The legislation includes a range of other requirements, particularly for traditional free-range or organic classification, such that, for example, the total usable area of poultry houses for meat production on any single production unit must not exceed 1600 m^2.

In general, free-range flocks are expected to have access to a defined minimum area of range during daylight hours for at least half of their life. Guinea fowl are exceptional in that the range can be replaced with a 'perchery' and they can still be described as free-range. There are some other exemptions to range access, such as when veterinary authorities mandate flock housing for disease control purposes. Currently, this exemption is limited to a period of 12 weeks for any particular flock, but it is planned, in light of changing Avian Influenza epidemiology, to remove this upper time limit, in line with what was

Table 1 Key production standards in the EU according to species and class of poultry meat

	Extensive indoor ('Barn-reared')				Free range				Traditional free range and total freedom[a]					Organic				
	Birds/ m²	kg/ m²	Minimum slaughter age (days)	Range (m²/ bird)	Birds/ m²	kg/ m²	Minimum slaughter age (days)	Range (m²/ bird)	Birds/ m²	kg/ m²	Maximum birds/ house	Minimum slaughter age (days)	Range (m²/ bird)	Birds/ m²	kgs/ m²	Maximum birds/ house	Minimum slaughter age (days)	Range (m²/ bird)
Chickens	15	25	56	n/a	13	27.5	56	1	12[b]	25[c]	4800	81	2	10[i]	21[j]	4800	81[k]	4
Capons	15	25	140	n/a	7.5	27.5	140	2	6.25[d]	35	2500	150	4	10[i]	21[j]	2500	150[k]	4.5
Ducks, Peking, female	–	25	112	n/a	–	25	112	2	10	25	4000	49	2	10[i]	21[j]	4000	49[k]	4.5
Ducks, Peking, male	–	25	112	n/a	–	25	112	2	8	35	3200	49	2	10[i]	21[j]	3200	49[k]	4.5
Ducks, Muscovy, female	–	25	70	n/a	–	25	70	2	10	25	4000	70	2	10[i]	21[j]	4000	70[k]	4.5
Ducks, Muscovy, male	–	25	84	n/a	–	25	84	2	8	35	3200	84	2	10[i]	21[j]	3200	84[k]	4.5
Ducks, Mulard	–	25	65	n/a	–	25	65	2	8	35	3200	92	3	10[i]	21[j]	–	–	–
Guinea Fowl	–	25	82	n/a	–	25	82	1	13	25	5200	94	2	10[i]	21[j]	5200	94[k]	4
Turkeys	–	25	70	n/a	–	25	70	4	6.25[e]	35	2500	98[g]	6	10[i]	21[j]	2500	–	10
Geese	–	15	112	n/a	–	15	112	4	5[f]	30	2500	140[h]	10	10[i]	21[j]	2500	–	15
Goslings	–	15	60	n/a	–	15	60	4	5[f]	30	2500	60	10	10[i]	21[j]	2500	–	–

[a]For 'total freedom' must have continuous daytime access to open-air runs of unlimited area.
[b]20/m², if in mobiles ≤ 150 m², which remain open at night.
[c]40 kg/m² if in mobiles ≤ 150 m², which remain open at night.
[d]May stock at 12/m² to 91 days.
[e]Up to 7 weeks, 10/m².
[f]Up to 6 weeks, 10/m², but 3/m² during the last 3 weeks if kept in claustration (confined).
[g]For females for cutting up, 126 days; for males for cutting up, 140; for whole birds for roasting.
[h]Figure is for birds marketed whole for roasting, reduced to 95 days in geese for foie gras and magret (breast fillet) production.
[i]All fattening poultry, except can be 16/m² if in mobiles ≤ 150 m², which remain open at night.
[j]All fattening poultry, except can be 30 kg/m² if in mobiles ≤ 150 m², which remain open at night.
[k]Earlier age of slaughter may be permitted when using slow-growing breeds.

done for laying chickens. Organic flocks are also subject to more rigid control of the use of feed and medicines, as laid down in paras 4.11–4.18, 5.7, and 5.8 of the regulation.[4] The effect of this is that if allopathic medicinal products need to be used in an organic flock, the withdrawal period before sale of the product is twice that normally required, or if no withdrawal period is specified then 48 h is required. Furthermore, repeated treatments may invalidate organic classification.

Other countries may have detailed local production standards for one or more of the poultry species. Those for Brazil apply to chicken production.[5] In contrast, in the United States it is possible to label poultry from any species as free range by providing a description of the housing conditions to the U.S. Department of Agriculture for approval and ensuring that there is continuous, free access to the out-of-doors for over 51% of the animals' lives.[6] In addition to local legal requirements, which will often be different in other jurisdictions, poultry producers commonly have to meet specific assurance requirements under a scheme required by their customers, which will typically have more detailed requirements, with the Freedom Foods Scheme in the United Kingdom being one such example.[7] Scientists and veterinarians dealing with free-range poultry production need to have a thorough understanding of both legal requirements and additional assurance scheme requirements and how these interact with pathogen controls.

Figure 2 shows a cross-section of a typical fixed house for free-range poultry. It is similar to conventional poultry housing and, if used only in the brooding phase, may be identical to conventional housing. If intended to be used for day-old to death, then it will need 'pop-holes' of sufficient size, location, and number to satisfy the legislation, and standard if applicable, to allow birds to range during the daytime for the required proportion of their lives. For the periods of non-range access during brooding and overnight in later life, these buildings offer as good separation from wild birds as conventional housing, as

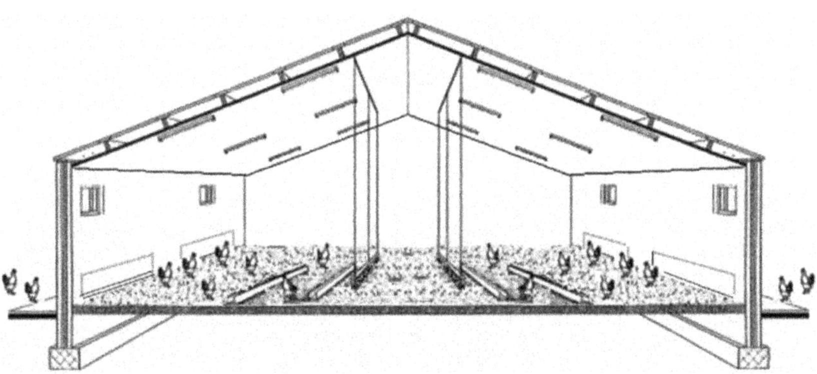

Figure 2 Typical fixed housing for free-range meat poultry.

long as the appropriate biosecurity systems are in place and they are properly maintained. Systems that use 'brood and move' commonly use mobile arks or houses for the post-brooding phase. These may be of wood or metal framework and may incorporate mechanisms to facilitate moving the building by dragging. In this situation, there will be no concrete floor. They may have as good separation of poultry from wild birds as a permanent building but are much more challenging with respect to the exclusion of wild mammals, which can result in losses from predation and be a source of significant infections.

Regardless of production system, we have made great progress in both identifying and controlling the major single-factor infectious diseases in commercial poultry, though new ones emerge or are newly recognised from time to time. Figure 3 is adapted from that in McMullin[8] to add potential transmission by aerosol and airborne fomites. This illustrates the many pathways whereby infections can arise in farmed poultry, including direct spread from infected livestock and wildlife and a range of indirect mechanisms. It is based on an approach used in the risk assessment of High Pathogenicity Avian Influenza (HPAI), generalised to include other sources/infections. In this model, it is obvious that certain pathways are primarily seen in free-range systems, whereas others may be found equally well in housed poultry production. We will return below to consider in general terms the role of biosecurity in controlling infections of both public and animal health importance in free-range poultry.

The two infections that have significant food safety implications that we need to discuss are Campylobacter and Salmonella. We will also briefly consider other infections such as Erysipelas and avian influenza, whose public health risk is primarily due to exposure of farm staff. Although nematode infections are not a public health risk, they do raise issues of sustainability of free-range

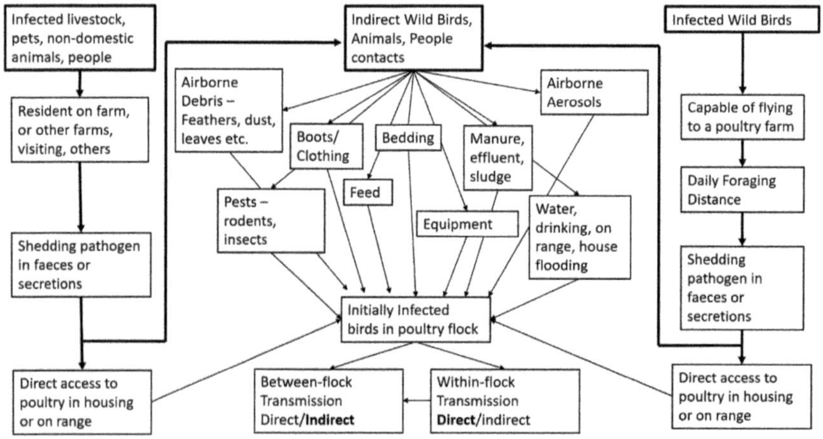

Figure 3 Potential infection routes.

production and so need to be discussed. Finally, we will consider the particular importance of biosecurity in free-range systems.

3 Campylobacter

The topic of Campylobacter is covered in Chapter 1. However, it is worth noting some of the key points of difference in free-range production vs indoor production. Many of the known risk factors for Campylobacter infection in poultry appear to be more commonly found in free-range production systems. These factors include:

- A greater age at kill is associated with a greater likelihood of a flock being Campylobacter positive (Note in the United Kingdom and European Union, free-range broilers must not be slaughtered before 56 days of age to comply with special marketing terms).[9]
- More challenging biosecurity setting. The range cannot be completely disinfected between flocks and relies mostly on UV light to destroy pathogens. Furthermore, rodent and insect control can be more challenging on free-range units.[9]
- Many free-range units have a greater number of smaller sheds. The greater the number of sheds, the greater the likelihood of Campylobacter infection).[9]
- Some free-range units can be multispecies, which increases the risk of being positive for Campylobacter.[9]
- Some free-range units can be multi-age with older birds acting as a potential reservoir for infection.[9]

Despite the risk factors described earlier suggesting that free-range poultry should have a higher prevalence of Campylobacter at slaughter, there is conflicting literature available. Figures produced in 2018 by the Food Standards Agency (FSA) in the United Kingdom[10] show free-range and organic poultry to be no more likely to be positive for Campylobacter than conventional broilers. However, Vandeplas et al.[11] indicated that free-range flocks were more likely than indoor flocks to be positive for Campylobacter. There are several possible reasons for a lower likelihood of Campylobacter being recovered from carcasses at slaughter for free-range broilers.

Although an older slaughter age is associated with Campylobacter-positive status, once birds reach 56–63 days of age, the prevalence and infectious burden of Campylobacter decrease as their immune system matures and starts to clear the infection. With free-ranging birds not processed until at least 56 days of age, many flocks have started to test negative again after previously being positive.[12]. In the United Kingdom at least, there are processing factors that reduce the number of positive carcasses. In large-scale abattoirs, which are not specialised in handling free-range birds, the free-range poultry are often processed first in the day (at the request of their customers) and to ensure catching occurs during the hours of darkness, as free-range birds, which are usually a slower-growing breed, can be flightier. Because of this, the abattoir is cleaner at the time of slaughter. In small-scale abattoirs, the processing line speed is often slower, leading to less contamination at evisceration. Indeed, some smaller abattoirs, particularly those processing turkeys, even use three-point hanging at evisceration to reduce contamination.[13] Emanowicz et al.[14] demonstrate that evisceration is one of the key points for contamination of carcasses with Campylobacter, and as such, manufacturers of three-point suspension systems use reduced contamination of carcasses as a selling point. A UK study showed there was little genetic similarity between the Campylobacter found in wild birds and that found in poultry and that ranging behaviour was not associated with an increased risk of being Campylobacter positive.[15] Thinning (partial depopulation) is a considerable risk factor for Campylobacter infection,[16] but since free-range poultry are marketed as higher welfare, many free-range producers do not thin or are not allowed by their customers to do so.

Considering the factors described earlier, free-range poultry meat production does not necessarily equate to an increased risk of Campylobacter positive status at kill; however, there is certainly more potential for risk factors to occur on a free-range farm compared to a conventional one. Each free-range farm should be assessed based on the risk factors that apply to it.

4 Salmonella

As is the case with Campylobacter, the topic of Salmonella is covered earlier in this book (Chapter 2). Given the biosecurity challenges faced in free-range systems, the presumption might be that free-range poultry have a higher incidence of Salmonella. Bailey and Cosby[17] found in the United States that free-range poultry did indeed have a higher incidence of Salmonella. However, in 2001, the UK FSA found no statistically significant difference between the incidence of Salmonella in chicken carcasses based on production system.[18] Alali et al.[19] in a small-scale comparison of Salmonella prevalence in organic and conventional birds in the United States found much lower prevalence of Salmonella in fecal samples but also in feed samples in the organic flocks. The prevalence of Salmonella was also compared between production systems in a study in Sweden[20] in which free-range broilers were no more likely to be Salmonella positive than indoor broilers. Many of the risk factors for Salmonella infection are similar to those for Campylobacter, such as free-range holdings being more likely to be multispecies, poorer rodent control, and access to the range, which allows for contact with wild birds.

It may be the case that, because free-range broilers live longer than indoor ones, exposed flocks have time to clear the infection pre-slaughter. However, more likely is the fact that free-range broilers have better intestinal health than indoor broilers, leading to the intestine being more resistant to Salmonella colonisation through the well-documented process of competitive exclusion.[21]

The control measures for Salmonella in free-range poultry are, in many respects, similar to those for indoor poultry. Good range management and rotation may contribute to a reduction in exposure. The use of temporary or mobile housing can increase risk unless there is diligent attention to the control of wild rodents. Wild birds may also be a source of infection, particularly in relation to *Salmonella typhimurium*. Farmed mammals are also a common hazard, in part because these are not routinely tested to the extent that poultry are, so infection is often only identified when clinical salmonellosis occurs or when a food-borne disease outbreak occurs.[22] The use of feed and water acidification, prebiotics, and competitive exclusion products all can have roles in controlling *Salmonella* infections. Free-range systems have some advantages, particularly when they are 'brood and move'. The brooding phase can be used to establish healthy gut flora in a low-challenge environment, and vaccination may be used. The relatively long lifespan and relatively high value of the product facilitates the use of vaccines, allowing time for immunity to develop and complying with the required withdrawal periods prior to slaughter. There is also growing evidence that providing some forms of fibre in the diet can significantly enhance the cecal flora, which produces fatty acids that help control *Salmonella* infection [23].

5 Other infections: erysipelas, avian influenza, and parasitic infections

5.1 Erysipelas

Erysipelas, caused by *Erysipelothrix rhusopathiae*, can cause significant disease in a wide range of farmed poultry, in particular turkeys[24] and guinea fowl, but also broiler chickens,[25] geese,[26] and other species. The organism is commonly associated with pig and sheep production and appears to survive very well in soil. The route of infection for birds, mammals, and people exposed to infection is via contamination of skin wounds or abrasions. The disease in humans is known as erysipeloid and is usually a local infection, typically producing intense inflammation and swelling in the area affected, usually fingers, due to exposure to affected carcasses, wool, or hides, it can also cause serious systemic infection. This condition has been well reviewed by Abd El-Gahny[27] from a 'One-Health' perspective. Erysipelas is not a new problem, but it is re-emerging with the re-establishment of free-range poultry production, most probably because of its long-term survival in soil and the difficulty of its control by disinfection. The human disease is easily treated if identified promptly, so it is important that veterinarians, farm staff, and slaughter plant workers are aware of this hazard. Its control tends to be based on vaccination for longer-lived birds, good management to avoid skin damage, and avoidance of contact with farmed mammals, especially pigs and sheep. Poultry ranges should not be used for spreading manures in general but particularly not from these species.

5.2 Avian influenza

H5N1 HPAI appears to have undergone a substantial change in epidemiology since 2019,[28,29] resulting in an unprecedented number of cases and enormous economic damage in poultry industries, particularly in Europe and North America. Significant wild bird mortality occurred in migratory species and in sea-bird colonies, with a clear increase of risk of introduction by wild bird contacts with poultry sites. Human cases of H5N1 have been rare recently, with only four reported to WHO in the period December 2022 to December 2023[30] but there is concern that more cases are being identified in terrestrial and marine mammals. There is little public health concern with respect to poultry products in the food chain, but the occurrence of any form of HPAI in farmed poultry has a major impact on international trade because of rules designed to limit the international spread of the disease. Economic losses stem both from actual lost production, mortality, and culling of affected flocks, and from many of the complex disease control measures applied to reduce the spread of the disease. Poultry cases have been occurring in both housed poultry, even in some relatively high-biosecurity breeding flocks, as well as free-range flocks, but there is a clearly increased risk to flocks where biosecurity procedures are poor or where poultry have access to range. This may be because of direct contact with sick wild birds or carcasses, or indirectly as a result of contact with infected faeces, feathers, litter, etc. In recognition of this, veterinary authorities have regularly mandated periods of housing free-range flocks when the risk of exposure is increased. Routine pre-move and pre-depletion testing of poultry in affected areas is helpful in limiting the spread of disease due to the movement of sub-clinically infected flocks. Given the increasing difficulty of HPAI control, we discuss later in detail the importance of biosecurity, even in free-range systems. The European Union has recently relaxed restrictions in relation to HPAI vaccination, but still within the context of 'stamping out' the disease in commercial poultry.[31] There are various constraints on the practical application of vaccination, mainly in relation to trade restrictions imposed by trading partners when it is permitted and achieving surveillance for lower prevalence infection in vaccinated birds at a sustainable cost.

5.3 Parasitic infections

Unlike some mammalian nematodes, there are no direct public health issues relating to poultry nematodes, though larval migration in intestinal tissues or mucosal damage by coccidia may facilitate the migration of zoonotic pathogens from the intestine to the tissues. The longer production cycle of free-range birds makes the adoption of vaccination for coccidiosis practical for those species in which vaccines are available (chickens, and in some places turkeys). There are no commercially available vaccines for helminth parasites.

There are some important nematode infections that are more challenging to control in free-range production, in particular *Ascaridia galli*, but sometimes *Heterakis* spp. and *Capillaria* sp. The longer the period with access to the range, the greater the potential for completion of the life cycle of the nematodes with the production of increasingly large numbers of eggs until crowding effects and immunity reduce helminth egg output. Occasionally other nematodes may cause clinical problems because of the greater contact with wild bird populations, for example, *Syngamus trachea*, particularly where the range includes woodland where corvids roost. Because the worm eggs produced are extremely resistant to degradation, and some species, in particular *Heterakis* spp., can use earthworms as transport hosts, pasture rotation is relatively ineffective in their control unless mobile houses can be moved onto land that has not previously had free-range poultry. So far, there are no commercially available nematode vaccines, so control relies mainly on treatment with anthelmintics at the appropriate time to block egg production. Some modern anthelmintics have efficacy against developing larvae and worm eggs, which allows a significant reduction in the need to re-treat. Anthelmintic resistance is rare in poultry. Because of the long-term persistence of eggs on the range, it is usually advised not to rely on faecal worm egg counting to decide when to treat. Unless sampling daily, by the time worm eggs are found in samples, there has been considerable contamination of the environment. Histomonosis is a parasitic infection of poultry that causes particularly high mortality in turkey flocks (up to 100%) and can also cause disease in chickens. Significant disease can also occur in farmed and wild game birds, and infection is maintained in a broad range of wild species. As the protozoa can be transmitted via *Heterakis gallinarum* larvae and eggs,[32] free ranging poultry are at much greater risk of infection compared to their housed counterparts, particularly where soil types and climate favour the persistence of *Heterakis* and earthworms. In the past, nitroimidazole products were successfully used to control such infections. However, across Europe and North America, such products have been banned, and there are no equally effective options currently available.

Ranges heavily contaminated by helminth eggs and associated parasites may eventually make their use for free-range poultry production unsustainable.

6 Other food safety concerns: veterinary therapeutic residues and toxins

6.1 Veterinary therapeutic residues

Most of these are subject to defined meat-withdrawal periods, which are more rigorous in birds produced to organic standards. Many free-range units rely on smaller sheds and therefore have much lower water consumption than indoor sheds. This means that in many free-range farms, header tanks are used to administer medication to broilers rather than proportioners. If the outlet pipes on header tanks are positioned to the side of the tank, medicinal residues are much more likely to remain in the drinker system compared to setups in which in-line proportioner dosers are used and are thoroughly cleaned after use. Another hazard to consider is the use of antiparasitic products intended and approved for other species, in poultry. These may be unsuitable for administration through water systems, resulting in persistent low-dose exposure and result in residues. Occasionally such products are not approved for food animals. Detection of such residues, which is increasingly likely with more sensitive testing methodology, can have severe consequences for the producer.

6.2 Toxins: dioxin, PCBs, and lead

There have been a few large-scale food residue incidents in different animal species in different countries. Usually, these have been associated with feed contamination with an industrial product. However, in recent years, there has been concern about the possibility of significant residues in animal products because of soil levels which meet normal regulatory requirements. Weber et al.[33] review this topic in detail by species and recommend that soil used for the production of free-range eggs should ideally contain less than 2 ng TEQ/kg dm both for PCDD/Fs and for dl–PCBs. They extend this recommendation to free-range poultry meat production.

Lead poisoning is rarely seen in poultry. It may sometimes arise in either free-range or housed poultry due to gross contamination of the bedding material. If shooting of game birds is practised on farmland, then lead shot must not be permitted, however.

7 The importance of biosecurity in free-range systems

The importance of biosecurity procedures for controlling infections in poultry is very well recognised and has been described in considerable detail by Vaillancourt et al.[34] though the focus is on conventionally housed poultry production. Vaillancourt et al.[35] does include 'Outdoor access' on a useful list of risk factors and suggest that this risk requires measures designed to separate healthy flocks from sources of infection, regional measures, such as coordinated movement of vehicles and personnel, and human measures, such as dealing with lack of compliance. On the other hand, Laing[36] took the view that 'it is impossible to prevent new infections being introduced by wild birds and animals, including insects and other invertebrates,' and that 'living in balance with the background infection on the unit is necessary for commercial success'. It is, of course, true that many poultry flocks do live in balance with a range of background infections and that this can contribute to their commercial success. However, this does *not* mean that biosecurity has no role in free-range production.

The primary reason to apply biosecurity procedures is to reduce the burden of new infections. We know that there are significant quantitative effects associated with parasitic and even infectious diseases[8] such that even where exposure cannot be totally prevented, some aspects of biosecurity can reduce the challenge to the point that the flock is not adversely affected. Many normal aspects of good management of poultry flocks, in particular those detailed by Sossidou et al.[37] as it relates to range/pasture management, have the effect of reducing disease challenges and improving resistance to diseases. Even though the prime focus of these authors is organic production systems, much of the guidance provided on range management applies equally well to non-organic free-range production systems. Areas discussed include maximising foraging behaviour, feeding, pasture rotation, stocking densities, feather pecking/aggression, mitigation of climate variation, predation control and housing. Of particular importance is advice with respect to avoidance of long-term water puddles on range, especially close to the pop-holes. These are attractive as a source of drinking and bathing water for both poultry and wild birds.

The second reason to apply biosecurity in free-range flocks at the entrance is that it helps contain any new infections within that flock, slowing down spread to other flocks on the same farm and to other farms under the same management. In the case of a serious notifiable disease, this will mean that a higher proportion of birds are covered by compensation. For other diseases, it allows more time to diagnose and implement treatment or prevention in other flocks.

The third, and perhaps most important, reason to implement biosecurity procedures is the reduction in disease spread in the local area from people and vehicles that have contact with different farms and other farms within the same production system/company supply base.

The dramatic change in HPAI epidemiology in many parts of the world in 2020–2023 has been characterised by long-distance movements of HPAI in migratory birds but also substantial outbreaks in indigenous species, particularly seabird colonies. An assessment of case flocks with more than 500 susceptible birds in Great Britain over a 6-month period at the height of the epidemic[28] found 55 conventional meat or egg production farms (45%), 39 free-range farms (32%), 18 breeding farms (15%), 11 mixed-species farms (9%), and three game bird premises (2%). This suggests the need for a broad reassessment of biosecurity procedures in the broadest sense, as well as careful remediation of failures in the application of existing procedures. Figure 3 provides a framework for listing the various possible infection pathways which may be relevant in each farm situation. Airborne spread of HPAI between farms is considered to be low risk except where distances are very short[38] and direct contact with infected wild birds, their feathers, and secretions is likely to be a significant route of spread where birds have access to range. Good practice to encourage widespread and even use of the available range mandates the provision of shelters, either artificial or natural on the range, and this is in some cases a legislative requirement, yet there appears to have been little research regarding the impact of these requirements on ranging poultry interaction with wild birds. Bestman et al.[39] found that tree cover on range reduced the number of highest risk wild bird species (geese, ducks, waders) as compared to not having trees and being in an area of open country. All farms in the study with no trees were surrounded by open country, so it was not possible to separate these effects. The prevalence of lower HPAI risk species (raptors and corvids) was unaffected by tree cover. Few birds of prey were identified, and corvids were found on all farms. Bestman and Bikker-Ouwijen[40] did find significant predation (losses, on average of 4%) of free-range laying flocks by northern goshawks (*Accipiter gentilis*) and common buzzards (*Buteo buteo*). These predators were beginning attack flights from perches on trees or fence posts, so it is possible that free-range birds are exposed to their faeces and regurgitated food residue pellets. They were, so far, unable to identify an effective control measure for such predation and contact with birds of prey.

Contact with wild and domesticated mammals also needs to be considered in detail, particularly with respect to the control of Salmonella infections. Having domestic animals on range is considered helpful in keeping grass height down to that used by poultry, and vegetation control to reduce rodent harbourages, as well as for attracting birds onto the range. Permission to co-graze may be required in accordance with local regulations, and assurance schemes may have more specific requirements. Because of the risk of periparturient increases in pathogen excretion, it is prudent to consider the age and reproductive status of any animals used in co-grazing, as well as their contact with markets and other farms.

8 Conclusion and future trends

Free-range and organic poultry meat production have developed from traditional poultry production methods for the supply of premium products. It is economically important in various countries, often concentrated in regions where the climate and soil types are most appropriate for the activity, often as a means of diversification on mixed farms. Breeding companies are now commonly supplying hybrid stock specifically targeted for use in the sector. While the general principles of poultry management, health maintenance, and zoonoses prevention apply, care is required to reduce the risk of certain infections. Research in relation to optimal nutrition, where the diet is supplemented with pasture and associated invertebrates, may be helpful in delivering both health benefits for the birds and support for marketing differentiation. The ecology of poultry ranges is currently receiving more attention. Avoidance of attraction, or even human discouragement of access, of high and intermediate HPAI risk species onto range is a key objective. Internal parasite control, and in particular *Histomonas* control, is a key issue in free-range poultry, so research on improved control measures applicable to the range of production systems would be welcome. HPAI has necessitated periods of mandated housing in recent years, so research is needed on how to minimise the need for this, and/or manage this process to maintain welfare and further reduce the risk of exposure to HPAI.

The need for cost-benefit assessment of the different components of biosecurity systems has been highlighted in light of the serious increases in losses as a result of HPAI. While the benefits of such systems extend to the prevention of a wide range of infections, the relative importance of different aspects is likely to vary according to the epidemiological characteristics of specific infections.

'Intestinal health', usually achieved by the rapid development of adult-like flora, particularly in the caeca, is well recognised to be important for poultry health and food pathogen exclusion. It is our experience that intestinal health tends to be more robust in free-range production systems, though there do not appear to have been comparisons between gut flora under different management systems. There is certainly potential for free-range birds to be exposed to a wider range of microbes, whether pathogens or beneficial, than housed poultry. Much progress has been made in reducing the need for antimicrobial use in commercial poultry production. This provides greater opportunity for the use of approaches based on competitive exclusion analogous to the use of colonic transplantation of flora used in human medicine. Given the premium nature of products from free-range systems, their greater risk of exposure to pathogens in the environment, and the variability of environments, there will be increasing interest in using such approaches.

A better understanding of the infection risks posed by wild bird contacts of different types is merited for the development of practical approaches to reducing risk for both housed and free-range poultry.

9 Where to look for further information

9.1 Poultry Health in General

Poultry health optimisation in general is covered in detail in a separate book.[41] Much applies equally well to free-range production.

9.2 Biosecurity

Publications by the University of Quebec – especially as referred to in Vaillancourt et al.[34][35]

The OIE pages dedicated to Avian Influenza[42] and its WAHIS reports relevant to your country and region.[43]

The OIE Terrestrial Animal Health Code sets standards for the improvement of animal health and welfare, and veterinary public health worldwide.[44] At the time of writing, a new chapter on biosecurity is being developed for Section 4 of the code.

9.3 Intestinal Health

Rychlik et al.[45] have reviewed and summarised what is known about the microbiota associated with chickens and confirm that many of the more beneficial components, though present in parent stock and on the hatching eggs, do not survive through egg sanitisation, storage, and incubation.

Van Immerseel[46] reviewed the topic of intestinal health from the perspective of inflammation, and the various mechanisms which may be used to minimise problems.

9.4 Vaccination

Use of vaccination is a keystone of our efforts to improve poultry health and food safety. Pharmaceutical companies provide technical information in the form of product brochures and documents supporting proper administration and monitoring of their efficacy. In the United Kingdom, the Responsible Use of Medicines Alliance (RUMA) has provided a general paper on the use of vaccines in poultry production[47] which remains relevant.

Vaccination has been widely used for both Low and High Pathogenicity Avian Influenza in countries where these diseases are endemic in poultry. It is much less commonly used in countries with significant export markets for poultry breeding stock and poultry products because importing countries commonly apply restrictions on imports. Recently, the European Union has laid down conditions to facilitate more use of HPAI vaccine,[31] and an HPAI control programme incorporating obligatory vaccination has been initiated in the duck sector in France.[48]

10 References

1. Leinonen, I. (2023). Life cycle assessment (LCA) of intensive poultry production systems, Chapter 17.
2. Heier, B.T. & Jarp, J. (2000). Risk factors for Marek's disease and mortality in White Leghorns in Norway. *Preventive Veterinary Medicine* 44(3–4), 153–165. https://doi.org/10.1016/S0167-5877(00)00106-9
3. Anon (2008). COMMISSION REGULATION (EC) No 543/2008 of 16 June 2008 laying down detailed rules for the application of Council Regulation (EC) No 1234/2007 as regards the marketing standards for poultrymeat. https://eur-lex.europa.eu/legal-content/EN/TXT/PDF/?uri=CELEX:02008R0543-20130701&qid=1697461753732
4. Anon (2007). Consolidated text: Council Regulation (EEC) No 2092/91 of 24 June 1991 on organic production of agricultural products and indications referring thereto on agricultural products and foodstuffs. https://eur-lex.europa.eu/legal-content/EN/TXT/?uri=CELEX%3A01991R2092-20080514
5. Anon (2015). NBR16389 DE 08/2015 Poultry Production - Production, slaughter, processing and identification of frango 'caipira, colonial ou capoeira'. https://www.normas.com.br/visualizar/abnt-nbr-nm/35138/nbr16389-avicultura-producao-abate-processamento-e-identificacao-do-frango-caipira-colonial-ou-capoeira
6. AskUSDA (2023). https://ask.usda.gov/s/article/Does-the-label-free-range-pertain-only-to-poultry-or-also-to-meats
7. Anon. (2017). Accessed 16/10/2023, RSPCA welfare standards for MEAT CHICKENS https://science.rspca.org.uk/documents/1494935/9042554/RSPCA+welfare+standards+for+meat+chickens+%288.48+MB%29.pdf/e7f9830d-aa9e-0908-aebd-2b8fbc6262ea?t=1557668435000
8. McMullin, P.F. (2022). Infectious diseases in free-range compared to conventional poultry production. *Avian Pathology*, 51(5), 424–434. https://doi.org/10.1080/03079457.2022.2086448
9. Newell, D.G., et al. (2011). Biosecurity-based interventions and strategies to reduce Campylobacter spp. on poultry farms. *Applied and Environmental Microbiology*, 77(24). https://doi.org/10.1128/AEM.01090-10
10. Jorgensen, F., et al. (2018). A microbiological survey of Campylobacter contamination in fresh whole UK-produced chilled chickens at retail sale. https://www.food.gov.uk/sites/default/files/media/document/campylobacter-in-chilled-chickens-year-3-2016-2017.pdf
11. Vandeplas, S., et al. (2010). Prevalence and sources of Campylobacter spp. contamination in free-range broiler production in the southern part of Belgium. *Base*, 14(2). https://popups.uliege.be/1780-4507/index.php?id=5259
12. EFSA Panel on Biological Hazards (BIOHAZ) (2011). Scientific Opinion on Campylobacter in broiler meat production: control options and performance objectives and/or targets at different stages of the food chain. *EFSA Journal*, 9(4): 2105. https://www.adiveter.com/ftp_public/A4150411.pdf
13. https://marel.com/en/products/tp-t-three-point-suspension-system

14. Emanowicz, M., et al. (2021). The impact of key processing stages and flock variables on the prevalence and levels of Campylobacter on broiler carcasses. *Food Microbiology*, 95, 103688. https://doi.org/10.1016/j.fm.2020.103688
15. Colles, F.M., et al. (2008). Campylobacter infection of broiler chickens in a free-range environment. *Environmental Microbiology*, 10(8), 2042–2050. https://www.ncbi.nlm.nih.gov/pmc/articles/PMC2702501
16. Hertogs, K., et al. (2021). The effect of partial depopulation on Campylobacter introduction in broiler houses. *Poultry Science*, 100(2), 1076–1082. https://www.ncbi.nlm.nih.gov/pmc/articles/PMC7858079
17. Bailey, J.S. & Cosby, D.E. (2005). Salmonella prevalence in free-range and certified organic chickens. *Journal of Food Protection*, 68(11), 2451–2453. https://pubmed.ncbi.nlm.nih.gov/16300088/
18. Food Standards Agency. (2003). UK-wide survey of salmonella and campylobacter contamination of fresh and frozen chicken on retail sale. https://www.food.gov.uk/sites/default/files/media/document/b180002finreport.pdf
19. Alali, W., et al. (2010). Prevalence and distribution of salmonella in organic and conventional broiler poultry farms. *Foodborne Pathogens and Disease*, 7(11). https://doi.org/10.1089/fpd.2010.0566
20. Wierup, M., et al. (2017). Occurrence of Salmonella spp.: a comparison between indoor and outdoor housing of broilers and laying hens. *Acta Veterinaria Scandinavica*, 59, Article number 13. https://actavetscand.biomedcentral.com/articles/10.1186/s13028-017-0281-4
21. Kempf, F., et al. (2020). Gut microbiota composition before infection determines the Salmonella super- and low-shedder phenotypes in chicken. *Microbial Biotechnology*, 13(5), 1611–1630. https://ami-journals.onlinelibrary.wiley.com/doi/full/10.1111/1751-7915.13621
22. Adamson, J., et al. (2023). An outbreak of Salmonella Typhimurium following an Eid al-Adha celebration barbecue in Wales (UK), July 2021, Medrxiv Preprint. https://www.medrxiv.org/content/10.1101/2023.01.03.22283848v1.full.pdf
23. Vermeulen, K., et al. (2018). Reduced-particle-size wheat bran is efficiently colonized by a lactic acid-producing community and reduces levels of enterobacteriaceae in the cecal microbiota of broilers. *Environmental Microbiology*, 84(21), e01343-18. https://doi.org/10.1128/AEM.01343-18
24. Jerstad, A.C. & Dickinson, E.M. (1956). Erysipelas of turkeys – Extension Bulletin 756, Oregon State College. https://ir.library.oregonstate.edu/downloads/6682x765n
25. Basnet, H.B. & Kwon, H.J. (2004). A case report on erysipelas in a broiler chicken flock. *Nepalese Veterinary Journal.* http://nva.org.np/downloadsfiles/NVJ-2015-Vol_32-1653893410.pdf#page=119
26. Bobrek, K. & Gaweł, A. (2015). Erysipelas outbreaks in flocks of Geese in Poland–biochemical and genetic analyses of the isolates. *Avian Diseases*, 59(3), 436–439. https://doi.org/10.1637/11050-030115-Case.1
27. Abd El-Ghany, W.A. (2023). A review article on avian Erysipelas infection: an occupational disease of one health importance. *Veterinary Integrative Sciences*, 21(2), 481–494. https://he02.tci-thaijo.org/index.php/vis/article/view/260765/178774
28. McMullin, P. (2023). *H5N1 Highly pathogenic avian influenza: towards the management of conflicting goals – Food security, biosecurity and biodiversity*. 22nd

Congress of the World Veterinary Poultry Association, Verona Italy 2–8 Sept 2023. Book of Abstracts S5-020.

29. Anon (2022). EFSA Avian influenza overview September – December 2022. https://efsa.onlinelibrary.wiley.com/doi/epdf/10.2903/j.efsa.2023.7786
30. Anon (2023). Investigation into the risk to human health of avian influenza (influenza A H5N1) in England: technical briefing 1 (updated 14/07/2023). https://www.gov.uk/government/publications/avian-influenza-influenza-a-h5n1-technical-briefings/investigation-into-the-risk-to-human-health-of-avian-influenza-influenza-a-h5n1-in-england-technical-briefing-1
31. EFSA (2023). Vaccination of poultry against highly pathogenic avian influenza – Available vaccines and vaccination strategies. https://www.efsa.europa.eu/en/news/vaccination-poultry-against-highly-pathogenic-avian-influenza-available-vaccines-and#
32. McDougald, L.R. (2005). Blackhead disease (histomoniasis) in poultry: a critical review. Avian Diseases, 49(4), 462–76. https://pubmed.ncbi.nlm.nih.gov/16404985/
33. Weber, R., et al. (2018). Reviewing the relevance of dioxin and PCB sources for food from animal origin and the need for their inventory, control and management. *Environmental Sciences Europe*, 30, 42. https://doi.org/10.1186/s12302-018-0166-9
34. Vaillancourt, J.-P., et al. (2022). *Improving biosecurity in poultry flocks) of Optimising poultry flock health*, S. de Wit, editor. Burleigh Dodds Science Publishing Limited, p. 414. ISBN-13: 9781786768872.
35. Vaillancourt, J.-P., et al. (2023). *Biosecurity: What works, what does not, what has been ignored and how to address it*. Book of Abstracts, 22nd Congress of the WVPA, Verona, Italy.
36. Laing, P.W. (1988). Diseases of free-range birds. *World's Poultry Science Journal*, 44(1), 72–74. https://doi.org/10.1079/WPS19880012
37. Sossidou, E.N., et al. (2015). Effects of pasture management on poultry welfare and meat quality in organic poultry production systems. *World's Poultry Science Journal*, 71(2), 375–384. https://www.researchgate.net/profile/Alessandro-Bosco-5/publication/279163942_Effects_of_pasture_management_on_poultry_welfare_and_meat_quality_in_organic_poultry_production_systems/links/559277a108ae15962d8e72dc/Effects-of-pasture-management-on-poultry-welfare-and-meat-quality-in-organic-poultry-production-systems.pdf
38. James, J., et al. (2023). Role of Airborne particles (viruses) the role of airborne particles in the epidemiology of clade 2.3.4.4b H5N1 high pathogenicity avian influenza virus in commercial poultry production units. *Viruses*, 15(4), 1002. https://www.mdpi.com/1999-4915/15/4/1002
39. Bestman, M., et al. (2018). *Less Avian Influenza risk birds in poultry free-range areas covered by trees*. 4th European Agroforestry Conference Agroforestry as Sustainable Land Use. https://www.repository.utl.pt/bitstream/10400.5/18598/1/EURAFIVConf_Bestman_M_et_all_page_447_451.pdf
40. Bestman, M. & Bikker-Ouwejan, J. (2020). Predation in organic and free-range egg production. *Animals*, 10(2), 177. https://doi.org/10.3390/ani10020177
41. De Wit, J. (ed.) (2022). *Optimising poultry flock health*, S. de Wit, editor. Burleigh Dodds Science Publishing Limited, ISBN-13: 9781786768872. https://shop.bdspublishing.com/store/bds/detail/workgroup/3-190-109526
42. OIE Avian Inluenza. https://www.woah.org/en/disease/avian-influenza/

43. OIE WAHIS: World Animal Health Information System. https://wahis.woah.org/#/home
44. OIE Terrestrial Animal Health Code. https://www.woah.org/en/what-we-do/standards/codes-and-manuals/terrestrial-code-online-access/
45. Rychlik, I., et al. (2023). Review article: microbiota of chickens and their environment in commercial production. *Avian Diseases*, 67(1), 1–9. https://doi.org/10.1637/aviandiseases-D-22-00048
46. Van Immerseel, F. (2023). *Microbial changes in the broiler intestine induced by inflammation, and how to restore these*. Book of Abstracts, 22nd Congress of the WVPA, Verona, Italy.
47. Anon (2006). Responsible use of vaccines and vaccination in poultry production. https://ruma.org.uk/wp-content/uploads/2022/08/Responsible-use-of-vaccnes-and-vaccination-in-poultry-production.pdf
48. Anon (2023). Avian Influenza - The Vaccination Plan in France (In French and English) https://agriculture.gouv.fr/tout-ce-quil-faut-savoir-sur-le-plan-daction-vaccination-iahp-en-france#:~:text=On%2022%20December%202022%2C%20the,the%20conditions%20have%20been%20met

Part 3

Postharvest control along the poultry supply chain

Chapter 9

Safety management and monitoring in poultry slaughter operations: a United States perspective

Dianna V. Bourassa, Auburn University, USA

1 Introduction

Poultry processing safety monitoring refers to the practice of regularly checking and evaluating the safety protocols and procedures in place during the processing of poultry products. This monitoring helps ensure that the processing facility complies with food safety regulations and industry best practices. The ultimate goal of safety management and monitoring is to prevent contamination and reduce risks of poultry products while also maintaining worker health and safety.

The safety management and monitoring process is conducted using a systematic approach with both facility-tailored and defined program plans, including Good Manufacturing Practices (GMPs), Sanitation Standard Operating Procedures (SSOPs), and Hazard Analysis and Critical Control Point (HACCP) plans. These plans typically involve conducting regular inspections and audits of the entire poultry processing facility, including verifying food safety protocols, checking the cleanliness and hygienic conditions of processing areas, equipment and utensils, as well as the workers' personal hygiene and

http://dx.doi.org/10.19103/AS.2024.0146.11

protective measures. Potential hazards are identified through collecting and analyzing samples of poultry products for microbial, chemical, or physical contaminants to allow corrective actions to be taken promptly. Inspections and audits are conducted by personnel trained in food safety regulations that can accurately assess compliance. Management should establish regular training and education schedules to ensure employee knowledge is current.

By conducting regular safety monitoring, poultry processing facilities can identify and address potential risks and hazards, maintain high hygiene standards, and ensure the safe production of poultry products for consumers.

2 The commercial poultry processing environment

The poultry processing environment poses potential physical, chemical, and biological risks (e.g. bone fragments, residues, and foodborne pathogens) both to employees within the processing plant and downstream to consumers who purchase the products produced in poultry processing facilities. The conversion of muscle to meat typically follows the same general flow (Fig. 1). After birds have been harvested from the farm, they are transported to the processing plant, where they are held before processing. The process of catching, loading, transporting, and unloading birds is known as live haul. The period when birds are kept in a dedicated area before they are processed is known as lairage. When it is time for processing, birds enter the processing plant itself, where they are then stunned before exsanguination. The head is subsequently removed, and birds are scalded with hot water to aid the removal of feathers during picking. Following picking, the feet are removed, and the carcasses are transferred to a new shackle line, opened, and eviscerated to remove the digestive tract and other organs. The carcass is then chilled to reduce the temperature of the meat from approximately 104°F (40°C) at the time of slaughter to below 40°F (4.4°C). After the meat has been cooled to below 40°F, the carcass can be packaged and sold as whole or deboned. Poultry parts (e.g. wings, leg quarters, breast

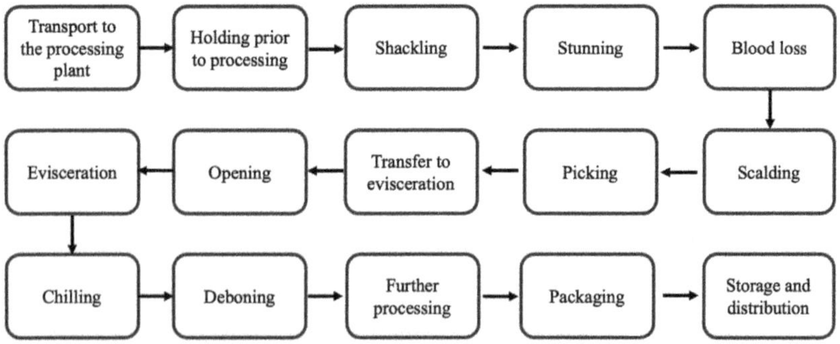

Figure 1 The general flow of a poultry processing facility.

Published by Burleigh Dodds Science Publishing Limited, 2025.

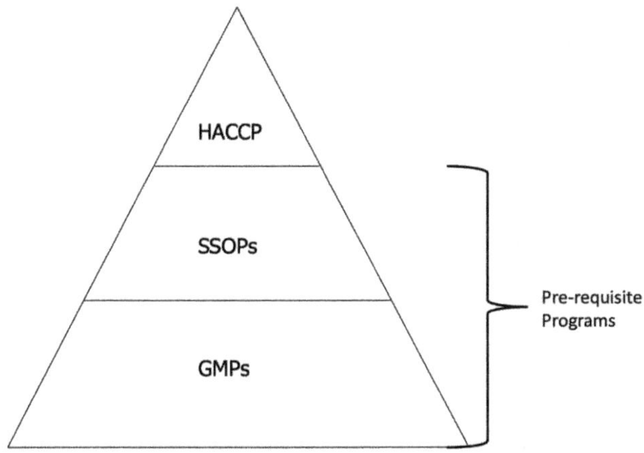

Figure 2 Good Manufacturing Practices (GMPs) and Sanitation Standard Operating Procedures (SSOPs) are the base Prerequisite Programs needed for an effective Hazard Analysis and Critical Control Points (HACCP) plan.

fillets, etc.) can be packaged and sold as Not-Ready-to-Eat or further processed through marination, batter/breading, and cooking processes.

This process, from live bird receiving through the final product, poses multiple risks to personnel and end consumers. Numerous systems which build upon each other have been put into place to mitigate these risks. Pre-requisite programs are the foundational building blocks of HACCP (Fig. 2). Without effective pre-requisite programs, many foodborne hazards in the processing environment would result in foodborne illnesses. Prerequisite programs include GMPs and SSOPs. Although both regulations contain similar language, the implementation is quite different.

GMPs underlie the prerequisite programs for most plants with a broad range of goals that produce an environment conducive to food processing. GMPs typically address extensive issues such as plant layout, pest control, and employee hygiene. SSOPs narrow this focus. SSOPs involve more specific concerns, such as the condition and cleanliness of particular food contact surfaces, actual sanitation procedures, and protection of food from contamination with fuel, pesticides, and other adulterants. USDA-FSIS has mandated the implementation of written SSOPs in HACCP plants.

3 Good manufacturing practices

GMPs are guidelines and principles established to ensure the quality, safety, and consistency of food products. GMPs are designed to be applied throughout the production process, from the arrival of the birds and other materials at the processing plant to final product distribution. They are essential for preventing contamination, maintaining product integrity, and protecting consumer health.

The core principles of GMPs typically include facility design and maintenance, hygiene and sanitation, personnel training, packaging and labeling, allergen control, waste management, traceability, and equipment calibration and maintenance. As well as ensuring product safety and quality for the protection of consumers, GMPs also seek to ensure employee safety.

3.1 Employee safety risks

Poultry processing involves various tasks, some of which can pose worker safety risks due to the nature of the work environment and the equipment used. There are multiple potential health risks for personnel working in a poultry processing environment. The conversion of live animals into food products has inherent challenges. When birds arrive at the processing plant, they are held in sheds to await a designated order for processing. While these areas can be partially temperature controlled with fans and misters or heaters, this can become a significantly hot or cold environment, depending on the time of year. Good environmental control is important for both bird welfare (which can have an impact on meat quality) and employee safety.

As birds enter the processing plant, they are either stunned using electrical waterbath stunning or controlled atmosphere stunning, which utilizes carbon dioxide. High levels of carbon dioxide during controlled atmosphere stunning pose an inhalation risk to employees. However, these risks are mitigated through carbon dioxide sensors which, if activated, will trigger both alarms and a ventilation system to flush the room with fresh air, removing the excess carbon dioxide.

Cuts and lacerations can occur for workers in poultry-processing plants who often use sharp tools and knives to cut and debone poultry. In addition to knives, other equipments like slicers, grinders, and cutters can pose risks if not used carefully. Improper handling of knives or machinery can lead to cuts and lacerations. Musculoskeletal injuries or ergonomic hazards can also occur due to repetitive motions, such as cutting, packaging, and lifting. The processing area is inherently wet due to the water needed to process poultry which could increase the risk of slips, trips, and falls. Poultry processing plants can also be noisy environments, potentially damaging hearing if workers are not adequately protected.

Workers may come into contact with cleaning agents, sanitizers, and other chemicals used for sanitation, which can cause skin irritation or respiratory issues if not handled properly. Dust and particles from poultry feathers and processing materials can lead to respiratory problems if workers are not provided with appropriate respiratory protection. Workers may be exposed to pathogens present in poultry, which can lead to foodborne illnesses or infections if proper hygiene and sanitation practices are not followed. Lifting and moving heavy

Published by Burleigh Dodds Science Publishing Limited, 2025.

loads of poultry or equipment without appropriate techniques or equipment can result in injuries.

To mitigate these risks and ensure worker safety, poultry-processing facilities implement comprehensive health and safety programs as part of their GMPs. This includes providing personal protective equipment (PPE) such as cut-resistant gloves, aprons, hearing protection, and respiratory equipment. Training workers in proper handling techniques and safety procedures is crucial as is maintaining a clean and well-organized work environment. Regular safety inspections, hazard assessments, and worker health monitoring are essential components of an effective safety program in poultry-processing plants.

3.2 Other aspects of good manufacturing practices

Good facility design and continued maintenance are essential to ensure that the processing facility is designed, constructed, and maintained to minimize the risk of contamination. This includes proper sanitation, pest control, and separation of different processing areas to prevent cross-contamination.

Hygiene and sanitation not only apply to food processing but also need to be addressed from a facility and employee perspective. Implementing strict hygiene practices for personnel, equipment, and workspaces will minimize the risk of microbial contamination. This involves regular handwashing, wearing appropriate PPE, and cleaning and sanitizing all surfaces and equipment.

Comprehensive training of processing plant workers on GMPs, hygiene practices, proper handling of poultry, and recognizing and reporting any issues or deviations from standard procedures is critical to operations. All employees, regardless of position, should be trained by a knowledgeable individual in basic food safety and personal hygiene upon hiring and then periodically thereafter. Understanding why GMPs are essential is just as important as knowing and following the plant's GMPs and policies. Teaching about why a policy or procedure is necessary is known to reduce indifference and encourage buy-in of a culture-promoting food and employee safety. Employees should be provided hard copies of plant GMPs and sign forms stating that they have read and understood those GMPs, and there should be periodic verification that GMPs are being followed.

A basic understanding of commonly encountered bacteria is also essential for plant workers. Food safety training on the role of bacteria in causing foodborne illness, how bacteria impact food spoilage, and the variety of ways that those bacteria can be spread helps reinforce why following GMPs is essential. For example, employees should understand that foodborne pathogens enter the processing plant with birds and are highly concentrated in fecal material. Understanding growth parameters such as what temperatures slow or encourage the growth of bacteria will help justify why the processing environment and product temperatures need to be maintained and monitored.

Personal hygiene practices are another critical component of poultry processing plant GMPs to protect food and employee safety. Hands and fingernails should be clean and maintained. Employees are often restricted from wearing nail polish or artificial nails or must keep their hands gloved if needed to avoid the potential for foreign materials inadvertently entering the product flow. Hairnets and beard nets are required to prevent foreign material contamination. Proper clothing and footwear are specified to protect both the employee and the product. In vehicle traffic areas, e.g. where birds are unloaded, employees are often required to wear high-visibility safety vests. In the slaughter area of the processing plant, employees wear protective smocks, aprons, face coverings, and bump caps to reduce contact with bird fecal material and blood. Employees typically wear designated smocks and bump caps in the different remaining downstream areas of the processing plant from picking through to product distribution (often using different colors to designate a particular processing area). Using non-slip, sometimes steel-toe, footwear is also an important GMP due to the wet environment inherent to poultry processing. Other than a plain wedding band, jewelry is restricted to avoid foreign materials and for employee safety. Other activities, including eating, chewing, and spitting, are prohibited on the production floor. Many processing facilities provide break rooms, cafeteria areas for eating, and outdoor areas designated for smoking or other tobacco use.

Cleaning and maintenance of the poultry processing facility are also included in GMPs. Some examples include ensuring that employee accommodations such as break areas and restrooms are appropriate for the number of employees. These areas also need to be regularly cleaned and maintained.

Production area facility design and maintenance can include maintaining surface integrity and durability of floors and walls, using materials and equipment with cleanable surfaces with non-porous materials with smooth welds, and using covers on light fixtures to avoid the possibility of broken bulbs. Facility design factors that can aid in cleaning and maintenance also include floors sloped toward the drains, adequate drainage to allow for the anticipated rate of processing and sanitation water flow, and spacing of equipment to allow for good airflow, ease of cleaning, and accessibility for maintenance.

Clean and organized storage areas for ingredients, dry consumables, and product storage are addressed in GMPs. Storage areas should be designed and arranged to reduce cross-contamination by separating raw and finished products. Access to chemical storage areas should be restricted to authorized personnel due to the inherent danger of concentrated chemicals. The potential chemical and physical hazards of items in storage areas should be considered for product and employee safety. Infrequently used or obsolete equipment should be kept in a separate storage area. An inventory and log sheet should be kept in these storage areas. Establishing proper procedures for receiving, storing, and handling raw materials and components will help prevent contamination and errors.

Pest management and prevention protocols are also commonly included in GMPs to minimize the risk of pests contaminating the processing area. Multiple different pests, including rodents, birds, and insects, are managed both inside and outside the facility. Processing and storage areas are cleaned and maintained to avoid attracting pests. Physical deterrents can be used to prevent wild birds from perching or nesting around the facility. Doors should remain closed to prevent pests from entering the facility. Rodent control programs including rotating bait-in-bait stations are also part of a thorough GMP program. Insect traps near the doorway are also beneficial to minimize flying insects. When a pest control program is designed and implemented, procedures such as proper storage and application of any rodenticides or insecticides are also important to avoid any potential adulteration of product or creation of insanitary conditions.

Once a GMP plan is in place, documentation of the execution of that plan is needed. This documentation includes comprehensive and accurate records of all quality control tests and any deviations or corrective actions taken. Conducting periodic self-inspections and audits will identify any non-compliance with GMPs and allow for the implementation of corrective actions.

By adhering to these poultry processing GMPs, facilities can produce safe, high-quality poultry products that meet regulatory requirements and consumer expectations. Regulatory authorities or third-party organizations may conduct inspections and audits to verify compliance with these GMPs and ensure the safety and quality of the final poultry products.

It is also important to note that all individuals, including company employees, contractors, regulatory agency personnel, and visitors passing through processing areas, must comply with all company policies. In many poultry-processing facilities, before a visitor can enter the facility, they are required to sign a form stating that they have read, understood, and will comply with the GMPs detailed within the form.

4 Sanitation standard operating procedures

Sanitation Standard Operating Procedures are known as SSOPs. In poultry processing, SSOPs are part of a facility's prerequisite program and encompass a set of detailed written procedures and guidelines that outline the specific steps and measures to be followed for maintaining a clean and sanitary environment in the processing facility. The primary purpose of SSOPs is to prevent contamination, control the dissemination of pathogens, and ensure the production of safe and high-quality poultry products.

Sanitation is a critical aspect of poultry processing because it directly affects the safety and quality of the final products. SSOPs cover a wide range of activities and procedures related to cleaning and sanitation in the facility. Sanitation activities are written in detail, describing the sequence of events an establishment employee must conduct to prevent direct product contamination and adulteration.

The details of sanitation procedures written in the SSOPs must include *what* is being cleaned, *who* is responsible for completing the task *when* the task is to be completed with both frequency and time, and *how* the cleaning task is to be accomplished. SSOPs are typically divided into two sections: pre-operational SSOPs and Operational SSOPs.

Pre-operational SSOPs are the steps to be taken before the start of each processing day to ensure that all equipment and work areas are adequately cleaned and sanitized. This also includes the cleaning and sanitation that occurs at the end of production. When developing SSOPs, the seven steps of sanitation need to be considered:

1 Dry cleanup;
2 Rinse;
3 Detergent application and scrubbing;
4 Final rinse;
5 Inspect and spot-clean;
6 Sanitize; and
7 Air-dry.

Cleaning is the removal of soil particles from surfaces by mechanical, manual, or chemical means. Dry cleanup begins with the removal of large particles of debris on equipment and flooring. In a poultry processing facility, this often consists of collecting and disposing of feathers, offal, and other materials on equipment or the floor. Collection and disposal of these materials is essential to minimize the organic material that goes down the drain and enters the wastewater treatment systems. Next, all surfaces from top to bottom, including floors, walls, equipment, and support structures, are rinsed to remove smaller particles. This rinse step is conducted with warm, but not hot, water to avoid protein adherence to surfaces. After rinsing, detergents are applied to all surfaces, and scrubbing with brushes removes organic matter and prevents biofilm formation. The final rinse using hot water is conducted to remove the remaining detergent and loosened organic matter. After cleaning, the facility is inspected by sanitation personnel, and if any areas are missed, they are

Published by Burleigh Dodds Science Publishing Limited, 2025.

spot-cleaned. With experience, sanitation personnel become acquainted with areas known to be difficult to clean and can give closer attention to these areas during the cleaning and sanitation process.

Sanitizing is the reduction of microorganisms to a safe level. It is important to note that the first five cleaning and sanitation steps must be completed before the sanitation step can be completed. If organic matter remains on surfaces, then sanitation will be ineffective.

Training and record keeping are also critical when conducting the seven steps of sanitation. Training requirements and records for employees involved in sanitation activities are part of the comprehensive record-keeping, which demonstrates compliance with SSOPs. It is important to note that poultry processing SSOPs should be specific to each processing facility, considering the unique layout, equipment, and processes involved. Regular monitoring and verification are essential to ensure SSOPs are implemented correctly and effectively to maintain a safe and sanitary processing environment. Regular internal and external audits may be conducted to assess compliance with SSOPs and ensure the highest sanitation standards in the poultry-processing facility.

5 Hazard analysis and critical control point systems: overview

HACCP systems for poultry provide a systematic approach to food safety that is specifically applied to producing and processing poultry products. The HACCP system is an internationally recognized model for effectively preventing and controlling food safety hazards. This preventive system is designed to identify, assess, and control potential hazards that could compromise the safety of the final food product. It focuses on identifying CCPs in the production process where hazards can be prevented, eliminated, or reduced to an acceptable level. In 1996, the United States Department of Agriculture, Food Safety and Inspection Service (USDA-FSIS) passed the FSIS Pathogen Reduction/ HACCP Regulation. By 1998, the HACCP system was mandatory in the meat and poultry industries. The HACCP Model for Poultry Slaughter guideline (FSIS-GD-2020-0013) provides a generic model for a poultry slaughter HACCP plan.

There are five preliminary tasks for developing a HACCP plan. The National Advisory Committee on the Microbiological Criteria of Food (NACMCF) has designated the following preliminary tasks for developing a HACCP plan:

1 Assemble a HACCP team;
2 Describe the food and its distribution;
3 Describe the intended use and the consumers of the food;

4 Construct a flow diagram that describes the process; and
5 Verify the flow diagram.

These preliminary tasks should be accomplished before considering the seven HACCP principles. The purpose of preliminary tasks is to build the foundation for which HACCP principles are applied. A HACCP team should comprise a team of individuals with the best expertise, technical knowledge, and diversity available to construct a food safety plan. The leader of an HACCP team should have in-depth knowledge and experience with food safety plans and good facilitation and decision-making skills. This knowledge base comes from HACCP certification from an accredited HACCP training program, on-the-job training, and experience implementing a HACCP plan within a poultry facility. HACCP training programs for both Introductory HACCP and Advanced HACCP are accredited through the International HACCP Alliance and are offered both in person and virtually through universities and private organizations. The rest of the team should be recruited to include persons with familiarity with the processing plant and the process for the product that the HACCP plan is being created for, with expertise encompassing a broad scope of the processing operations, including engineering, production, sanitation, management, and research and development and a variety in knowledge, skills, and experiences. Including persons encompassing broad perspectives will enhance communications and a sense of ownership in developing a HACCP plan. In particular, including management and production personnel is crucial for ensuring that the HACCP plan will be assimilated as best as possible since these persons will ultimately be held accountable for successful follow-through with the HACCP plan.

When describing the food product and its distribution, all the ingredients should be listed and a description provided on how the product is prepared. The list of ingredients will help identify which ingredients pose food safety hazards regarding allergens, susceptibility to microbiological growth, and any other considerations that might be regulated. Describing the product in terms of how the ingredients are handled and combined in cooking processes will help identify conditions that allow for microbial growth or contamination. Also, the product in terms of shelf life should be described.

The next step is to describe the intended use of the product. Defining the intended end user is essential because these considerations will impact the HACCP plan. There are nine processing categories recognized by the USDA-FSIS and are listed in 9 Code of Federal Regulations (CFR) 417.2:

1 Slaughter all species;
2 Raw product ground;
3 Raw product not ground;

4 Thermally processed commercially sterile;
5 Not heat treated shelf stable;
6 Heat treated -shelf stable;
7 Fully cooked not shelf stable;
8 Heat-treated but not fully cooked not shelf stable; and
9 Product with secondary inhibitors not shelf stable.

After describing the product, a flow diagram is constructed. The flow diagram will often be referred to in subsequent steps in the HACCP plan development process, such as hazard analysis. A flow diagram is a diagram of the steps of each process that illustrates the product flow in the processing facility. Each step is depicted as a block, and the order of steps is represented by connecting the blocks with arrows. The flow diagram should be simple. Complicated processes should be broken down into individual steps. Also, the flow diagram should include every step in the process, typically from receiving the ingredients to the point at which the product is taken by a third party, such as a distributor. After the flow diagram is constructed, it needs to be evaluated by the HACCP team to look for gaps.

6 Hazard analysis and critical control point principles and their implementation

The HACCP system consists of seven principles:

1 Hazard analysis;
2 Determine the CCPs;
3 Establish critical limits;
4 Establish a system to monitor control of the CCP;
5 Establish the corrective action to be taken when monitoring indicates that a particular CCP is not under control;
6 Establish procedures for verification to confirm that the HACCP system is working effectively; and
7 Establish documentation concerning all procedures and records appropriate to these principles and their application.

These are outlined in the following sub-sections.

6.1 Hazard analysis

The hazard analysis is conducted to identify all potential biological, chemical, and physical hazards that could occur during the production process of poultry products. When evaluating hazards, one should consider opportunities in which hazards could be introduced in typical food processes. Some of these hazards might include raw materials and ingredients, packaging materials, metal shavings from processing equipment, chemical contaminants from processing, such as refrigerants, lubrication, cleaning agents, etc., and growth of bacteria due to improper heating, chilling, or formulation. Biological hazards can encompass bacteria, viruses, parasites, and molds. Pathogenic bacteria are the most commonly identified biological hazard in poultry processing facilities. Chemical hazards can occur naturally, with widely used additives like antimicrobials or unintentionally added chemicals. Physical hazards can result in choking or injury. These hazards may come from materials naturally present in the food, such as bone fragments. Other materials may come from handling or processing equipment such as staples, twist ties, pop-up timers, broken glass shards, bolts, nuts, and saw blade pieces.

For each step in the flow chart process, any potential hazards that may occur or be introduced in the product or ingredients are identified as biological, chemical, or physical. Whether the potential hazard is likely to happen in the presence of properly applied prerequisite programs is subsequently evaluated. The decision of whether the hazard is reasonably likely to occur needs to be justified. For any hazard that is found to be reasonably likely to occur, control measures that may be used to control the hazard are identified. Prerequisite programs reduce the likelihood of the occurrence of a hazard, while the application of a control measure will prevent it, eliminate it, or reduce it to a safe level. Each step of this process is documented in the hazard analysis.

6.2 Critical control points

Once the potential hazards have been determined, Critical Control Points (CCPs) are defined. CCPs are specific points in the production process where controls can be applied to prevent, eliminate, or reduce hazards to an acceptable level. A CCP is a point, step, or procedure in a process at which it is essential that a control can be applied to prevent, eliminate a hazard, or reduce it to a safe level. A single CCP can control multiple hazards, such as the cooking step. Sometimes several CCPs may be needed, for instance, to prevent allergen adulteration. HACCP decisions are made in response to an identified hazard that is likely to occur in the absence of its control. This determination of which hazards will be controlled at a CCP is made using hazard analysis methods. CCPs are not added to the HACCP plan to control quality or other factors. An exception to this may be a regulatory-mandated CCP, such as the zero fecal tolerance inspection required by USDA-FSIS at slaughter. Sometimes, it may not be possible to entirely eliminate or prevent the occurrence of a hazard. In some processes, especially those dealing with poultry slaughter, reducing the occurrence of the hazard as far as possible or to a preset regulatory limit may be the only reasonable goal of the HACCP plan.

The most common CCP is a step that provides thermal destruction of pathogens. This is also one of the most effective in that it eliminates the hazard. Other CCPs may be similarly effective, such as adding acid to reduce the pH or incorporating chemical preservatives. A cooked product's holding and cooling time is also a common CCP. Usually, the CCP directly controls the process at that point, as in heating or pH adjustment. Still, sometimes it can be ancillary to the process, such as an allergen cleanup.

6.3 Critical limits

Critical limits are the maximum or minimum values that must be met at each CCP to ensure that the biological, chemical, or physical hazard is effectively controlled. It is the value placed on an attribute of a control measure at a CCP. At this chosen value, the control measure effectively prevents, eliminates, or reduces the hazard to a safe level. An accurate, critical limit is crucial to the effectiveness of a HACCP plan, as the remaining principles of HACCP are concerned with the ability of the processor to manage its CCPs and prerequisite programs. More specifically, the remaining parts of the HACCP plan (concerning that CCP) are only relevant if a critical limit is sufficient to control food safety hazards.

The critical limits in HACCP distinguish between safe and unsafe operating conditions at a CCP. They can indicate when a process is trending toward being out of control, especially when documented on a control chart during processing. This allows the operator to adjust the process before there is a loss of control. Because this is such a useful tool, plants often establish operating limits to let the operator know when actions should be taken to prevent this loss of control. Critical limits have become a regulatory mandate, and deviations from the critical limits require follow-up action and documentation. Critical limits are defined in the HACCP plan while operating limits are not included in the HACCP plan to avoid conflict and confusion.

Good critical limits should be easily and rapidly measurable, inexpensive to measure, use observation, or chemical or physical parameters that can be measured in real-time; they should not interfere with or contaminate the process; measurement procedures should be easily taught and understood, be accurate and reproducible, and be backed by verification data.

6.4 Monitoring

Monitoring a CCP consists of using a planned sequence of observations or measurements to establish whether a critical control point (CCP) is under control and to produce accurate records for future use and verification.

Monitoring is essential to ensure that the critical limits are consistently met. In addition, monitoring alerts the operator when measurements indicate that the process is trending out of control. The monitor must conduct observations or measurements according to a predetermined plan and at the specific frequency indicated by the plan. The monitor is responsible for establishing that the measured control parameters are within the critical limits and recording the information. If there is a deviation, another part of the monitor's responsibility is to begin the corrective action process. A deviation is a reading at a CCP that fails to fall within the critical limits. A written real-time record of the results of monitoring must be made.

A monitoring plan addresses who will monitor, what precisely the person is monitoring, how to conduct the monitoring, and the frequency the monitoring is to be conducted. Usually, the identity of the monitor will be specified according to the position, such as the operator, shift supervisor, or QC technician. Before being assigned responsibility, the monitor must understand the importance of the job and be trained in the correct procedures to be used. Training should include the calibration of monitoring equipment, if necessary. Also, the monitor must understand their responsibilities should a deviation occur.

6.5 Corrective actions

Defined corrective actions are to be taken when monitoring indicates that a CCP is not within the established critical limits. A critical limit is the point at which a parameter is no longer acceptable. These limits are often determined by regulatory requirements or data previously collected within the process. For example, if a cook step were defined as no less than 250°F for 30 min, and the cooking temperature was recorded at 245°F, then this would be considered a deviation from the critical limit. When a deviation occurs, corrective action must be taken to ensure that no product for which safety cannot be assured reaches the consumer. An obvious goal of corrective action is to stop producing products for which safety cannot be guaranteed. Corrective action should be taken according to a predetermined plan. If an appropriate corrective action plan is in place and authority is established, corrections can be applied expeditiously to restart production, even if plant management and quality control personnel are absent.

Corrective actions include several components. First, the processing and production of products under a deviation must be discontinued. The suspect product must be isolated and stored or disposed of to ensure it will not reach the consumer, and adjustments must be made to the process to bring it back into control. A determination of the disposition of the suspect product must be made. An accelerated monitoring and off-line testing frequency will often be instituted to ensure the adjustment or correction holds, and the corrective

Published by Burleigh Dodds Science Publishing Limited, 2025.

action is documented for later verification activities. In cases where a repeat deviation occurs, a follow-up investigation also needs to be conducted to determine the root cause of the deviation and to determine if modification of the hazard analysis or the HACCP plan is necessary.

6.6 Verification

Verification is a collection of activities that determine the HACCP plan's validity and whether the system is operating according to the plan. Verification processes occur during the development and implementation of the HACCP plan and the maintenance of the HACCP system. Verification activities encompass all aspects of the HACCP plan, beginning with the hazard analysis through the record-keeping activities. This includes reviewing records, conducting periodic audits, and validating the HACCP plan.

The key in the verification activities is the evaluation of records to ensure that the technical aspects of the plan and the assumptions behind it are valid. The principle of verification is divided into two major aspects. Validation determines whether the plan will be effective when applied. Verification determines whether the plan is correctly followed. Validation refers to those elements of verification which focus on using scientific and technical information to determine whether an adequately implanted HACCP plan will effectively control the identified food hazards.

The components of verification are applied at various times in the process. The initial validation of the plan is to determine that the plan is technically sound, that all the hazards have been identified, and that the HACCP plan will control these hazards when implemented. Pre-shipment signoffs are completed according to USDA-FSIS regulations. Daily calibrations and records are reviewed. Subsequent validations are conducted when there are unexplained system failures, changes in the product, process, or packaging, or if new potential hazards are identified. Scheduled periodic comprehensive verification of the HACCP system is completed and includes the results of findings of regulatory or third-party audits.

Validation focuses on collecting and evaluating scientific data to validate the appropriateness of the activities in the HACCP plan. Sometimes this involves standard, established, generic scientific information, and sometimes it comes from very specialized studies relating directly to the process in question. Often, it will need to be developed using carefully planned studies conducted in the plant. In all these situations, documentation of the relevant data is required for validation. These decision-making documents are part of the HACCP plan and should be maintained with it. Verification activities must be scheduled according to a plan. There will be verification activities for each CCP noted on the HACCP Plan Summary Form and comprehensive activities scheduled as an attachment to the form. All activities of the HACCP plan must be verified to ensure they are conducted according to plan, and all decisions must be validated with data showing that the plan, when implemented, will apply control measures that control the hazards.

6.7 Record keeping

Record keeping characterizes HACCP as much as any other principle. Records have a very practical aspect. Records archive decisions, reflect planning, standardize procedures, and document compliance. Written records standardize the procedures to be followed and provide a method for future review. Should a production problem be identified later, a review of accurate and complete records may limit the scope of products that must be recalled or destroyed. Record keeping allows the regulatory agency to change from the traditional 'snapshot in time' inspection to the HACCP audit during the entire time of operation. If litigation is forthcoming, records made in 'real-time' carry considerable weight as evidence. Unfortunately, records can document failure as well as compliance. Timely record reviews will allow immediate correction and follow-up.

Records document the application of the HACCP system. In addition to records that document the implementation of the SSOPs, these records are required by the various HACCP regulations. Processing and other pertinent information are entered into records at the time of observation using the actual values and observations obtained during monitoring. All required records, plans, and procedures are available for official review and copying at reasonable times. Confidentiality is maintained in accordance with regulations, but under some circumstances, records may be disclosed, so non-HACCP information about the company should be kept from required records.

7 The Food Safety Modernization Act

The Food Safety Modernization Act (FSMA) is a significant piece of legislation passed in the US to enhance the safety of the country's food supply. It was signed into law on January 4, 2011, and represents the most comprehensive reform of food safety regulations in several decades. The main objective of the FSMA is to shift the focus of food safety regulations from responding to contamination and foodborne illness outbreaks to preventing them from occurring in the first place.

The law addresses various aspects of the food supply chain. It places greater responsibility on food producers, processors, distributors, and importers to take preventive measures to ensure the safety of the food they handle. Many of the requirements under FSMA had already been addressed through the GMP and SSOP prerequisite programs and HACCP.

8 Conclusion

Using the systematic approaches of GMPs, SSOPs, and HACCP plans meets the need for the production of safe and wholesome poultry products. Through these safety management and monitoring systems muscle is converted to the poultry meat that feeds the world. Further information regarding HACCP plans can be found through the USDA-FSIS HACCP Guidance website or your state HACCP Coordinator.

9 References

United States Department of Agriculture-Food Safety and Inspection Service (USDA-FSIS). HACCP Model for Poultry Slaughter, FSIS GD-2020-0013.

United States Department of Agriculture-Food Safety and Inspection Service (USDA-FSIS). Pathogen Reduction; Hazard Analysis and Critical Control Point (HACCP) Systems. Docket No. 93-016F. July 25, 1996.

United States Department of Agriculture-Food Safety and Inspection Service (USDA-FSIS). Part 417 - Hazard Analysis and Critical Control Point (HACCP) Systems. 9 CFR 417.

United States Food and Drug Administration (FDA). Food Safety Modernization Act (FSMA). Public Law 111 – 353.

Published by Burleigh Dodds Science Publishing Limited, 2025.

Chapter 10

Official control of poultry processing operations: a European perspective

Janne Lundén, University of Helsinki, Finland; Kristiina Törmä, Finnish Food Authority, Finland; and Eija Kaukonen, University of Helsinki, Finland

1 Introduction

Official control is a cornerstone of food safety, fair trade, animal health, and welfare. In poultry processing, official control comprises of (a) meat inspection and (b) food safety inspections (Fig. 1). Meat inspection comprises detailed regulated procedures aiming to ensure that the poultry meat is fit for human consumption as well as to promote animal health and welfare (EU, 2017). The purpose of food safety inspections is to verify that the food business operators (FBOs) comply with food safety legislation (EU, 2017; Codex 2013). To be able to perform official control, the authorities must have the right to enter all production areas and inspect all operations and documents related to the production (EU, 2017). This is a precondition for conducting reliable and effective inspections and this right should be stipulated in the legislation.

Official veterinarians (OVs) are responsible for the official control, and depending on the official control organisation, OVs perform meat inspection themselves or mainly supervise assisting staff in meat inspection. The assisting staff can be official control personnel often called official auxiliaries (OAs) or the personnel of the slaughterhouse (Salines et al., 2018; Luukkanen et al., 2017). Whatever the organisational model is, the official control must possess a high

http://dx.doi.org/10.19103/AS.2024.0146.12

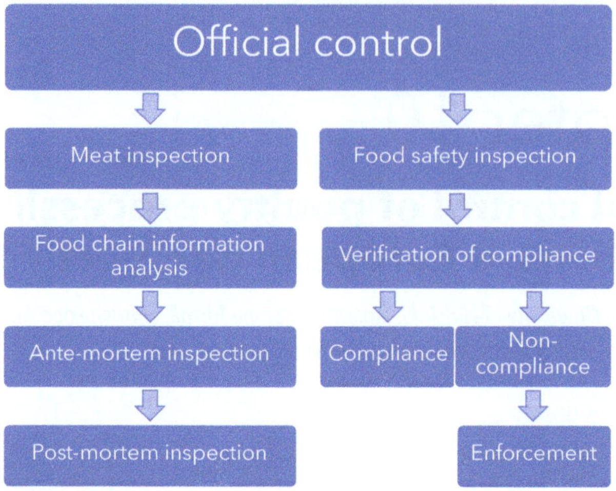

Figure 1 Elements of official control in poultry processing.

integrity. Incorruptibility and impartiality as prerequisites that must be priority for the organisation.

Official control should work efficiently, which is often emphasised in meat inspection. Work efficiency is important, but even more important is effectiveness which can be defined as the extent to which official control objectives and related outcomes are achieved (Codex, 2017). Poor effectiveness cannot be compensated with increasing efficiency highlighting the fundamental importance of effectiveness. The official control organisation should ensure the quality and effectiveness of their controls (FAO, 2021).

Poultry slaughtering and processing includes complex and challenging operations from a food safety point of view. The FBO is responsible for identifying food safety risks and managing them (FSMA, 2011; EC, 2002). The FBO is obligated to carry out self-checks to prevent noncompliance and correct any noncompliance occurring in the operations by their own initiative. The OV inspect the food business and make sure that food safety legislation is followed. To do this, the OV needs to have sufficient knowledge to carry out inspections and the corresponding tools for enforcement (EU, 2017; Lundén et al., 2007; Buntain, 2004).

Food safety inspections in poultry slaughterhouses is, in principle, similar to official control in other food establishments: the OV ensures that the operations are compliant with food safety legislation (EU, 2017). However, the complex production processes, the presence of pathogens in live animals and the usually continuous presence of official control in poultry slaughterhouses are a challenging combination for official control. Inspections should be carried out with such depth that the OV can get a comprehensive picture on how the FBO is managing the operations. Any noncompliance should be detected (EU,

2017; Evira, 2015), and the signs of food frauds should be recognised (EU, 2017; Obbink et al., 2014). To achieve these aims, the inspections must be of high quality. This chapter considers aspects important for performing high-quality meat inspection and food safety inspections in poultry processing.

2 Meat inspection

Meat inspection of poultry at poultry slaughterhouses includes the checking of food chain information and ante-mortem and post-mortem inspection. These official control procedures aim to ensure that the meat originating from the slaughtered poultry is not hazardous to health or unfit for human consumption (EU, 2017; EC, 2002). Official control also includes the protection of animal health and welfare and surveillance of animal by-products (EU, 2017) (Fig. 2).

2.1 Food chain information and ante-mortem inspection

Prior to the birds can being accepted to enter the slaughterhouse, the slaughterhouse operator checks and evaluates the food chain information of the poultry flock and makes the information available to the OV. The food chain information includes information on factors affecting meat safety, such as the health status of birds and possible medications during the rearing time. Food chain information can offer useful information for the risk-classification of poultry flocks (EFSA, 2012). However, the value of food chain information greatly depends on the information that the farmer is obligated to notify in the document. Currently, food chain information seems to be useful in acknowledging the *Salmonella*-status of the poultry flock, given that the information is included in the document. Also, the mortality rate of the flock provides important general knowledge of the flock health situation, but more detailed information, could be gathered with specific questions focusing on bird health. For example, reporting bird dirtiness and problems with bedding management such as wet litter related issues as part of the food chain information, would be useful when considering slaughter hygiene (Blomvall et al., 2023). The OV takes into consideration the food chain information when performing ante-mortem inspection and when making the decision of acceptance of the meat in the post-mortem inspection. Finally, the slaughterhouse operator is obligated to give feedback to the farmer about the number of birds condemned in ante-mortem or post-mortem inspection and the reasons for condemnations (FFA, 2020; Allain et al., 2018).

Ante-mortem inspection is performed by an OV as the inspection requires a profound understanding of animal health, welfare, and food safety. The purpose of ante-mortem inspection is to check whether there are signs in birds that would endanger their health or welfare. A further purpose is to ensure that the condition and health of the birds do not reveal any signs that would suggest

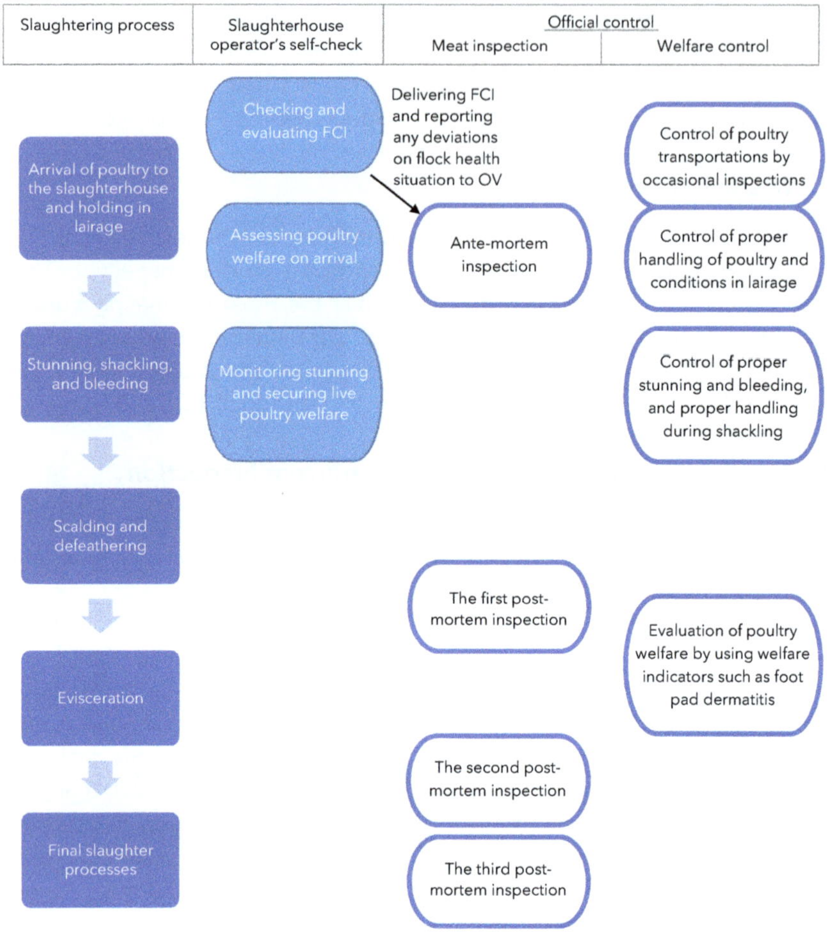

Figure 2 Organisation of meat inspection and welfare control in poultry slaughterhouses in the European Union. FCI = Food chain information.

the meat unfit for human consumption. The identification of the flock origin, assessing and securing the appropriate cleanliness of the birds and observing signs of contagious animal diseases are essential matters to be checked in ante-mortem inspection. Ante-mortem inspection is usually performed in the slaughterhouse lairage although the inspection can also be performed at the farm in which case the OV or OA verify bird welfare in the slaughterhouse (EU, 2017; 2019a). The welfare assessment covers signs pointing to compromised welfare on farm and during catching and transport. Suitable bird density in crates, also considering weather conditions, and the signs of thermal discomfort (panting or huddling) could serve as welfare indicators during transport, while wing fractures and fresh skin scratches could point to compromised welfare during catching (Jacobs et al., 2017). Since the birds are in crates when they arrive to the slaughterhouse, only a representative sample of birds can be inspected. Given that it is impossible to organise water supply for poultry while they are in crates, the ante-mortem inspection and slaughtering of birds must be organised so that the waiting time is not too long. During the waiting time, the conditions must be adequate for the birds and the checking of the lairage conditions is therefore important. Appropriate temperature, good ventilation, subdued lighting, and enough space are essential for poultry welfare during holding (Barker et al., 2004).

2.2 Post-mortem inspection

Post-mortem inspection is usually performed under the supervision of an OV. In the European Union (EU), it is possible to allow the poultry slaughterhouse staff to assist the OV in post-mortem inspection (EU, 2017). When working as meat inspectors, the slaughterhouse staff must fulfil the requirements of training and comply with the instructions of the OV. Moreover, the meat inspectors should be independent of the slaughterhouse, which should be clearly emphasised both to the slaughterhouse management and meat inspectors (Luukkanen et al., 2017). In the EU, the post-mortem inspection is performed in two or three stages during the slaughter process (Löhren, 2012). The Regulation (EU) 2019/627 requires that the post-mortem inspection is performed without delay after slaughter and all external surfaces of carcass and internal organs are inspected. If the post-mortem inspection is performed in three stages, the inspections are carried out (1) immediately after defeathering, (2) immediately after evisceration (carcass and visceral organs belonging to the carcass separately), and (3) final inspection of the eviscerated carcass and its body cavity before chilling (Fig. 2) (Löhren, 2012).

Different post-mortem inspection systems can be applied, such as the New Poultry Inspection System (NPIS) in slaughterhouses slaughtering young chickens and turkeys in the USA. In NPIS sorting, trimming, and reprocessing of poultry carcasses and their viscera are performed by the employees of the slaughterhouse before the inspection is performed by an official inspector (USDA, 2017). The United States Department of Agriculture, Food Safety and Inspection Service (FSIS) has only two inspectors for each evisceration line in the NPIS. One of the inspectors performs online carcass inspections before the chilling (USDA, 2017). The other inspector performs offline verification duties (USDA, 2017). However, in the USA, there are still other post-mortem inspection systems available where the FSIS food inspectors examine carcasses and internal organs after evisceration (Bilgili, 2010).

The poultry post-mortem inspection is visual although performing other examinations, such as palpation and incisions, are possible (EU, 2019b). However, in practice the high slaughter line speed, which can be up to 15 000 broilers per hour (Marel Poultry, 2023), and the position of inspection stations, may only allow visual inspection (Törmä et al., 2021). Laboratory tests can be carried out for diagnostic purposes or detecting the presence of zoonotic disease causing pathogens such as *Campylobacter* or *Salmonella*. Laboratory tests can also be used to detect chemical residues or contaminants in meat exceeding the permitted level, or to reveal the use of prohibited substances (EU, 2019b).

Effective post-mortem inspection requires sufficient inspection time per carcass and good practical arrangements in the post-mortem inspection stations. The faster the slaughter line the greater number of skilful meat inspectors is required to arrange proper inspection (EU, 2019b; Törmä et al., 2021). There must be appropriate facilities to wash hands (faucets with touchless function) and, when necessary, to sterilise knives on the inspection stations (EC, 2004b). Good working hygiene and avoiding cross-contamination between carcasses is essential for meat safety (EU, 2019b). Touch screen or push-button boards for recording carcass condemnations and alarm system with possibilities to stop the line in exceptional situations are needed particularly in broiler slaughterhouses where line speed is high (Törmä et al., 2021). A possibility to adjust height at inspection stations makes the inspection of carcass body cavities easier and improves the meat inspectors' working conditions. Proper lighting is particularly important. In the USA and Canada, there are requirements for lighting properties at inspection stations unlike in the EU legislation. Those requirements are for illuminance 2000 lx (or 200-footcandles meaning approximately 2153 lx) and for colour rendering index no less than 85 (USDA, 2019; CFIA, 2019). In addition to the illuminance and colour rendering index of the lighting, the colour temperature also affects how well colour differences on the carcasses and internal organs can be observed during post-mortem inspection (Collins and Worthey, 1984).

Meat inspection results are recorded, and the OV evaluates the results and take actions if the results indicate any disease or condition that may threat public or animal health or endangerment of the welfare of poultry (EU, 2017; EU, 2019b). The carcass and internal organ condemnations made in the meat inspection are always based on the legislation. According to the Regulation (EU) 2019/627, the meat derived from a bird with a generalised disease is unfit for human consumption. Also, if the carcass has pathological or organoleptic changes or there is soiling or faecal or other contamination, or the carcass is emaciated, the carcass including internal organs are condemned. The rejected carcasses and internal organs are collected, categorised, stored, treated for example acidified, and sent to a certified animal by-product facility. Rejected carcasses and internal organs are categorised according to the risks they could pose to human or animal health (EC, 2009). The role of the OV is to verify the proper handling of animal by-products (EU, 2017).

Meat inspection data is a valuable source for assessing poultry health and welfare issues (Huneau-Salaün et al., 2015; Stärk et al., 2014). However, the use of meat inspection data for those purposes is challenging because the condemnation reasons and the principles to collect meat inspection data differ between countries and slaughterhouses (Törmä et al., 2022). Also, the differences in the practical organisation of meat inspection prevent collecting uniform meat inspection data. Hence, the possibility to use condemnation rates as welfare indicators to a wider extent is limited. In a single slaughterhouse, however, the evaluation of the meat inspection data is important and useful. The number of condemned carcasses due to emaciation, ascites or leg problems can be used to evaluate on-farm welfare, while condemnations due to fresh factures, bruising and scratches are suitable indicators for assessing welfare during catching. If non-compliances are observed or the meat inspection results reveal the presence of disease or poor pre-harvest conditions threatening bird welfare and health, the OV informs the slaughterhouse operator, the farm, and the competent authorities responsible for the control of the farm (EU, 2019b).

Meat inspection should be risk-based and protect human health from hazards in the meat. However, important biological hazards including *Campylobacter* spp. and *Salmonella* spp., are impossible to detect by visual inspection in ante-mortem or post-mortem inspection of poultry (EFSA, 2012). The European Food Safety Authority (EFSA, 2012) has proposed a better use of food chain information and farm audits in risk-categorisation of the poultry flocks. Also, EFSA (2012) has proposed that the risk-categorisation of poultry slaughterhouses should be based on their capability of faecal contamination prevention or reduction. These are methods by which meat inspection and integrated meat safety securing systems can be developed towards a more risk-based approach.

While developing the risk-basis of meat inspection, it is also important to promote cost-effectiveness of the measures (Blagojevic et al., 2021). For that reason, automatic online inspection systems have been developed for poultry slaughterhouses that have high slaughter line speed. Spectral imaging -based machine vision technology has been developed to differentiate wholesome chickens from systemically diseased chickens (Yang et al., 2009), differentiate unhealthy broiler chicken viscera from healthy (Jørgensen et al., 2018), and for the detection of contamination on carcasses (Seo et al., 2019; Park et al., 2011). However, the use of those systems may be challenging in various environmental conditions in slaughterhouses (Chowdhury & Morey, 2020), and so far, in the EU, post-mortem inspection is not allowed to carry out by automatic inspection systems. However, the regulations permit slaughterhouse operators to carry out the detection and separation of birds that have abnormalities, contamination, or defects and only a representative sample of flock is inspected in post-mortem inspection (EU, 2019b). Slaughterhouse operators could use an automatic online inspection system for the detection and separation purposes if the system works to the satisfaction of the OV. On the other hand, export agreements may prevent the use of this sample-based post-mortem inspection in large slaughterhouses in the EU member states (Antunović et al., 2021).

3 Food safety inspections

3.1 Purpose of food safety inspections

Food safety inspections are carried out to make sure that poultry slaughterhouse operators comply with food safety legislation. This includes safeguarding the consumer, which is the most important task for the official control (Amery, 2023; EC, 2002). In addition to ensuring food safety, the official control must also ensure that consumers are not misled for example about the origin of animal species. The consumer has minimal possibilities to make sure that the food is safe and therefore rely on the integrity of the FBO and the independently operating official control. Although safeguarding consumer safety is the top priority, the OV has also an important role in facilitating food business. This is manifested especially in enabling international trade, which requires OV certifications for meat and products thereof to make sure that the products are wholesome and subject to official control (Kontio, 2014). In addition, the facilitation of business can include guidance to the FBO about legislative requirements. Guidance given by official control to FBOs can be particularly important for smaller businesses (Buckley, 2015; Yapp and Fairman, 2006) as they may have scarce knowledge of food safety requirements and risks.

An integral part of food safety inspections is not only to verify how the FBO is complying with food safety regulations but also any actions taken by the OV due to observed non-compliance. Food safety inspections should be carried out in an effective and consistent manner. In the following sections, we describe inspection prerequisites, inspection techniques, common non-compliance observed in poultry slaughterhouses and actions taken by the OV due to non-compliance.

3.2 Required knowledge and skills to perform food safety inspections

Performing high-quality food safety inspections requires wide knowledge of the poultry processing chain and official control (Fig. 3). The OV has a challenging task to make sure that food safety is maintained, but also to avoid unnecessary enforcement. The legislation is the basis for risk management and official control, but the legislation is not always detailed leaving much discretionary power to the OV. The goal of the legislation to produce safe food can be reached in different ways and the OV needs to be able to assess if the FBO's

Figure 3 Cornerstones of high-quality official inspections and enforcement of food safety legislation in poultry slaughterhouses.

approach is satisfactory (Evira, 2015). Therefore, the OV should have profound understanding of both poultry meat safety, risk management options and legislative requirements. Further, the OV need to have skills in using inspection techniques and applying control actions in different situations (EU, 2017; Nevas et al., 2013; Lundén et al., 2007; Buntain, 2004).

Control actions need to be exercised respecting good governance principles such impartiality, equality, fairness, consistency, and transparency (EU, 2017). Respecting these principles is important as they lay the ground for a credible and trustworthy official control. Trust in the official control is important as it may have a beneficial effect on interaction between the OV and the FBO. The importance of the interaction is emphasised by FBOs as they call for a supportive and collaborative approach by inspectors (Kettunen et al., 2017). Communication between official control and the FBO requires interactive skills from the OV. The OV should communicate the inspection results to the FBO intelligibly and in a motivational way to achieve a desired outcome at inspections (EU, 2017; Buckley, 2015; Nevas et al., 2013; Lundén et al., 2007).

3.3 The frequency and focus of food safety inspections

OVs are usually present in poultry slaughterhouses when slaughtering is ongoing to perform ante- and post-mortem inspections (EU, 2017). However, the frequency of food safety inspections can vary considerably. Several aspects must be considered when determining the frequency of food safety inspections. A mutual principle is that the frequency should be risk based (NFAS, 2023; EU, 2017; Evira, 2015; Codex, 2013; Gagnon et al., 2000) so that increased risks increase the inspection frequency. In practice, the frequency of food safety inspections is based on factors related to the size of the operations, the nature of the operations, and the history of compliance (EU, 2017; Evira, 2015). Prior compliance of the FBO is important as it reflects the food safety culture of the FBO. Food safety culture can be defined as long-term beliefs, behaviours and assumptions that are shared by the employees and impact the food safety performance (Sharman et al., 2020). It is important that the food safety culture is on a high level and the FBO should be able to provide evidence of an appropriate food safety culture (EU, 2017).

The frequency of inspections is important as an increase in inspection interval has been shown to increase serious non-compliances (Kosola et al., 2022). However, the effectiveness of the inspections cannot simply be measured by the number of performed inspections (Husu-Kallio, 2011). The correction of non-compliance, as shown in a restaurant setting, may not be improved by the increase of inspection frequency (Newbold et al., 2008). Even more important than the frequency is what the OV does during the inspection and how the OV reacts to possible non-compliance (Luukkanen and Lundén, 2016; Newbold et al., 2008).

Whatever the inspection frequency is, it should be such that the OV can obtain a comprehensive and transparent view of the operations. The OV should know for example the origin of animals and raw material as well as where food products are delivered (Fig. 4) but also focus on details to be able to verify compliance (Evira, 2015). The focus of the inspections should be on issues causing possible meat safety risks and misleading of business customers and consumers.

The implementation of Hazard Analysis and Critical Control Point (HACCP) is considered essential for managing the risks in slaughterhouses (Nørrung and Buncic, 2008) and is a high-priority control subject. Habib et al. (2012) observed a strong connection between operational hygiene scores (concerning for example, adequate space, sanitation, condition of surfaces, slaughter and working hygiene), given by official control and the *Campylobacter* prevalence in broiler slaughterhouses. Non-compliance with HACCP has also been observed to increase the risk for *Salmonella* contamination in poultry processing (Beczkiewicz and Kowalcyk, 2021). Such

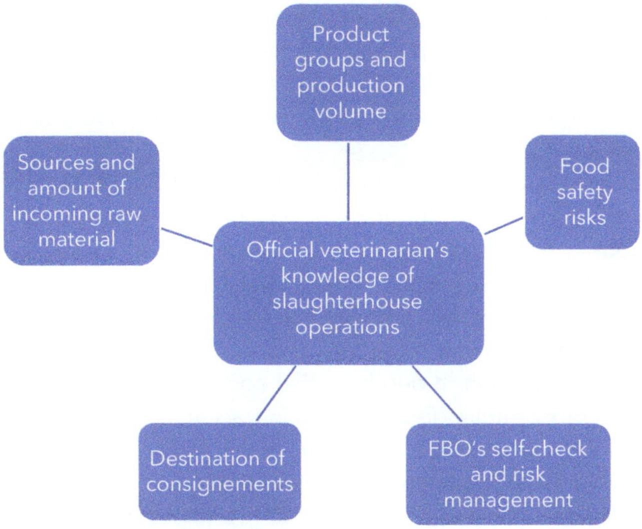

Figure 4 The official veterinarian should have a comprehensive view of the slaughterhouse operations. FBO = Food business operator.

relationships indicate that the focus of the inspections is relevant for food safety.

3.4 Inspection techniques

The inspection of the facilities and operations can be seen as a snapshot of the situation at that specific time (Gagnon et al., 2000), but inspections usually also are comprised elements that describe long-term activities of the food business. Such elements include the condition of surfaces, level of cleanliness, FBO's own-check documentation and the correction of earlier detected non-compliance. Inspections can be conducted using checklists or templates to aid the inspection process and to make sure that all relevant aspects are considered (Läikkö-Roto et al., 2015; Gagnon et al., 2000). Checklists or templates can enhance the consistency of inspections as they direct the inspector to inspect the same issues and to report the inspection results in a similar manner (Läikkö-Roto et al., 2015). Templates may also increase the effectiveness of the inspections by giving a certain mode of action and a structured way of carrying out the inspections and the subsequent reporting of inspections (Läikkö-Roto et al., 2015). Templates and given modes of action should, however, not lead to rigidity in the inspection process. The OV should be able to adapt to surprising situations in the slaughterhouse and refocus the inspection if necessary (Evira, 2015).

Slaughterhouse activities and meat trade in general involves large economic interests and therefore during inspections, the OV should also keep in mind the possibility of fraudulent activities. There is every possibility that fraudulent activities will take place in the meat business and slaughterhouses (Obbink et al., 2014; Premanandh, 2013). In 2021, the fourth most common request for cooperation in fraud suspicions in EU were related to poultry and poultry meat (EC, 2022).

Food frauds may have international dimensions (EC, 2022), but fraudulent acts always originate at the local level and therefore OVs inspecting slaughterhouses and other food premises are placed in key positions to prevent and detect food fraud (Premanandh, 2013). The routine inspections should be of such high quality that signs of fraudulent actions will be revealed at some point of time. This means that inspections need to be unexpected (EU, 2017), thorough and in-depth so that common fraudulent activities can be revealed. For example, discrepancies in traceability, mislabelling or other intentional violations (EC, 2022; Obbink et al., 2014; Premanandh, 2013) should be detected. Versatile inspection techniques are recommended to be used to verify compliance in a reliable way (Evira, 2015). Inspection techniques can be divided into two main categories depending on the source of the inspection data (Table 1). The OV can receive information from the FBO or by making own observations (EU, 2017). The OV cannot, however, rely solely on the information provided by the FBO even if the FBO has proven to be reliable and compliant.

Table 1 Inspection techniques

Source of inspection data	Inspection technique	Example
Food business operator	Interview/dialogInspection of self-check documentation	Discussion with manager/ other personnelChilling room temperature documentation
	Information on company internet pages or social media	Product spectrum
Inspector's own observations	Observation of facilities	Slaughter line
	Observation of operations Ability testing	Slaughter hygieneTraceability test
	Inspection of self-check plan	Temperature control plan
	Inspection of self-check documentation	Automatic temperature control logs
	Measurements	Temperature measurement of chiller
	Cross-inspection using data from two or more sources	Temperature self-check data and OV's own temperature measurement
	Sampling	Surface hygiene samples

EC, 2022; USDA, 2021; EU, 2017; Gagnon et al., 2000.

The inspection findings should convince the OV that the FBO is complying with requirements. If the OV is not satisfied with the collected information during the inspection and senses that something is amiss, further inspections should be performed using appropriate techniques. Following examples are given on the use of different inspection techniques.

3.4.1 Interviewing personnel

Information received from the FBO by talking to and interviewing personnel is valuable when gathering information about how the FBO handles the operations, recent updates in the operations and any hands-on problems influencing hygiene. It is important to interview staff in managerial position as well as staff at factory floor to attain a multifaceted view of operations. Non-specific answers and discrepancies in answers between workers should be elaborated by the FBO and documented proof displayed whenever possible. Both closed and open-ended questions are useful. The outcome of the interview should give the OV the impression that the issue is thoroughly inspected, and relevant information has emerged. If not, the OV should pursue the issue by further questions or other inspection techniques (Evira, 2015). Interviews, however, cannot be the only technique, because the OV must also collect information that is not provided by the FBO (EU, 2017).

3.4.2 Inspecting FBO's self-check documentation

The inspection of self-check documentation generated by the FBO is carried out to make sure that the FBO is executing the risk management plan (EU, 2017). The FBO documents the self-check procedures such as HACCP procedures by making entries of the monitored issue. This can include the documentation of temperature conditions, received raw material, traceability, and sampling results. Major differences can be found between the self-check results and the observations made by official control as noted in temperature control documentation at the retail level despite the assessments made by automatic self-check systems (Lundén et al., 2014). The self-check results may not reflect the actual situation due to many reasons like errors in the measurement device or in the use of it. The errors may be intentional or unintentional. The OV should consider the possible falsification of the self-check documents (EC, 2022; Gagnon et al., 2000) like entries marked for the next day or other falsification of the self-check outcome. Therefore, it is of utmost importance that the OV does not regard the self-check results as automatically correct, but also make own measurements (Lundén et al., 2014).

3.4.3 Review of internet presence

The exploring of the company's internet pages, or social media can reveal useful knowledge about the company operations. Internet pages can reveal issues related to products and marketing that have not been displayed at inspections. For example, the FBO may produce a product type that is not approved for the establishment but marketed on the Internet. It can therefore be beneficial to explore the internet pages occasionally.

3.4.4 Observation of facilities and operations

The observation of facilities and operations is important to gain a transparent picture of the premises and how food safety aspects are taken into consideration in the activities of the slaughterhouse. In particular, facility-related issues that may lead to the contamination of products such as inadequate size of the premises, poor possibility to work hygienically and poor condition of surfaces are observed during on-site inspections (EC SANTE, 2019, 2020, 2022). The principle should be that hygienic working is made as easy as possible. The space should be large enough and designed to enable hygienic working practices at the individual work posts (EC, 2004a). The routes of personnel, product and waste material should consider cross-contamination and surfaces should be in good condition. The most effective and reliable way to assess the management of cross-contamination is by following the activities in the premises for a sufficient amount of time. It can also be assessed by studying the layout and floor plan, but only by observing the operations reveal the actual practices on floor level. The OV should make sure that the inspection covers all facilities, processing machines and other equipment (EU, 2017; Evira, 2015).

In the slaughterhouses, the observation of slaughter hygiene is pertinent because meat can be contaminated with pathogenic bacteria such as *Salmonella* and *Campylobacter* due to poor hygiene practices (Nørrung and Buncic, 2008; WHO, 2006). Poor hygiene in slaughterhouses can also lead to serious outbreaks (Pennington, 2009) and should therefore be a frequent target for official control. Slaughter hygiene is inspected by observing the slaughter process and working hygiene. When observing worker hygiene, it is important to follow the work practices in such a way that a transparent picture is attained. Hand hygiene practices should be evaluated because they have an important role to play in the matter of food safety (Todd et al., 2010). Also, the use of knife sterilisers should be followed for some time to reveal the practices. Therefore, the inspection of operations requires more than just a walk-through of the slaughter line; it requires time that must be acknowledged beforehand when planning the resources of the official control. The inspection of cleanliness is best done after sanitation and before the initiation of the operations. This inspection should be comprehensive and more than just a 'look'. The sanitation of processing machines can be challenging due to complex design. Any accumulation of dirt should be recognised (Wirtanen and Salo, 2014) and eliminated by the FBO. In particular, conveyor belts and cutting devises have been shown to be difficult to clean (Wirtanen and Salo, 2014). This means that the OV needs to inspect processing machines not just on the outside but also on the inside and evaluate surfaces in direct and indirect contact with the meat, in particular.

3.4.5 Ability testing and traceability

Ability testing is an in-depth inspection method where the FBO is requested to display how a certain food safety procedure is managed. Traceability testing is a typical capability test where the FBO demonstrates how the traceability is conducted in practice. The FBO should have considerable interest in ensuring that the traceability is working in practice due to safety and economic reasons (Pozo and Schroeder, 2016), but the OV cannot rely on assumptions. Instead, the OV should inspect that the FBO can trace products in practice. This can be done by requesting the FBO to trace and display the list of consignees of a hand-picked lot. Mass-balance testing, where the FBO is asked to account for the amount of incoming raw material and produced products, is an advanced control technique that can be used in food fraud suspicions (Winkler et al., 2023). Fraud related to the origin of country, for example, may be revealed in mass balance testing when the amount of imported and domestic meat is compared to the amount of meat sold as foreign and domestic.

3.4.6 Inspection of the self-check plan

The inspection of the self-check plan gives the OV an understanding how the FBO has planned to manage the risks involved in production. The self-check plan includes written instructions on for example sanitation, pest control, sampling, traceability, and HACCP (EC, 2004a). The self-check plan should describe sufficiently possible risks in the production and how the risks are managed (EC, 2004a). Although the plan is important, the execution of the plan is even more important. A well-prepared plan does not mean that the activities are carried out adequately (Lundén et al., 2014). It is therefore important to inspect the actual activities described in the plan.

3.4.7 Cross-inspection

Acquiring data from two or more sources, that is, cross-inspection, can be a valuable technique in verifying the FBO's self-check results. Cross-inspection aims at acquiring data from another source besides the FBO. This is not always possible, but when possible and at situations when the OV is unsure about the reliability of the data generated by the FBO, cross-inspection is valuable. The simplest way of cross-inspection involves data generated by the OV such as temperature measurements and sampling which are then compared with the data produced by the FBO (Lundén et al., 2014). Data generated by other parties can be used when inspecting for example the disposal of non-edible by-products. Non-edible by-products must be disposed properly to eliminate any health threats and contagious animal disease agents (EC, 2009). Proper disposal incurs costs, however, and it may be tempting to dispose such material in an improper way. Cross-inspection can be used to make sure that the material is transported to the waste treatment plant by comparing the FBO's documentation of the amount of slaughtered animals and the amount of delivered non-edible by-products to the waste treatment plant. The OV can further request information from the waste management plant on how much non-edible by-products has been delivered from the slaughterhouse and compare that to the consignment documents sent from the slaughterhouse to the waste management plant. These inspection procedures will take time but will give a good idea about the correctness of the disposal of non-edible by-products.

3.4.8 Sampling and measurements

Sampling and measurements are useful in cases where visual observation of premises, activities and self-check need further assessment. Sampling by the OV can be used as verification of successful sanitation by taking surface samples for chemical or microbiological analysis indicating cleanliness (Wirtanen and Salo, 2014). Other samples taken by the OV can include microbiological surface samples of carcasses for assessment of the slaughter hygiene, samples to determine the amount of retained water in the carcasses and possible samples for determination of animal species in product preparations. A typical measurement that the OV should carry out is temperature measurements of the cold storage and the meat, as temperature can have a major impact on the growth of pathogens. Sampling and measurements done by the OV are independent from the slaughterhouse (EU, 2017) and describes the situation at the time of sampling and measuring.

The results of samples taken, and measurements performed by the FBO and OV are expected to align, but in cases of discrepancies it is important to understand the reasons for differences in self-check and official control results. Discrepancies have been observed for example in *Salmonella* samples from carcasses of various animal species and samples for *Campylobacter* from broiler carcases as shown in an EU-wide survey where *Campylobacter* samples were more frequently positive in official control samples (42.1% positive) than in self-check samples (27.2%) (EFSA and ECDC, 2022). The number of samples exceeding the limit of 1,000 CFU/g was also significantly higher in official samples (19.4%) than in self-check samples (7.3%). Discrepancies may be caused by several reasons such as differences in sampling technique, time, location, or reporting. Unintentional flaws in sampling, analysis and measurements or intentional misleading of results should also be considered.

3.5 Non-compliance and official control actions

3.5.1 Observed non-compliance in poultry slaughterhouses

Non-compliance related to slaughter and working hygiene and the condition of surfaces seem to be common in poultry slaughterhouses regardless of country and control organisation (EC SANTE, 2022, 2020, 2019; Pinillos and Jukes, 2008). Slaughter hygiene has been observed to be compromised by poor design of the slaughter line causing faecal contamination, poor possibilities for hand washing and poor working practices (EC SANTE, 2022, 2020, 2019). Faecal contamination is of major concern in poultry slaughterhouses and appears to be difficult to correct as it is a recurring problem (EC SANTE, 2022, 2019; Luukkanen and Lundén, 2016). Other observed non-compliance issues have been inadequate space, temperature abuse of carcasses, neglecting corrective actions described in the HACCP plan and general dirtiness of the premises (Luukkanen and Lundén, 2016; EC SANTE, 2022, 2019). Sometimes the non-compliances are related to food fraud. Most poultry and poultry meat product related food fraud suspicions concern mainly forged documentation, illegal trade of such goods or exceedance of water in the poultry meat (EC, 2022).

3.5.2 Actions taken by official control due to non-compliance

Official control needs to take appropriate actions upon noticing non-compliance at inspections (EU, 2017; Buntain, 2004). The actions taken depend on the nature of the observed non-compliance and the history of compliance of the FBO. The actions taken should be gradual and in proportion to the non-compliance: the actions should be efficient enough to make the FBO to correct the violation but not oversized (Fig. 5) (Kettunen et al., 2015; Lepistö and Hänninen, 2011). The actions taken should be swift when food safety is clearly threatened and milder when food safety is not directly challenged. Actions available for OVs are presented in Table 2. Actions may differ among countries, and also the terminology of the measures used can vary. The minimal action is to guide the FBO about the food safety requirements. Guiding may not be seen as a control action at all; however, the guidance of FBOs should be appreciated highly due to several reasons. In particular, small and medium-sized food businesses may experience challenges

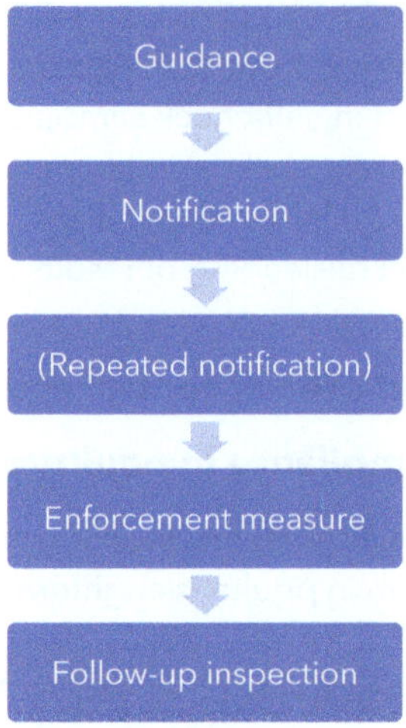

Figure 5 Stepwise use of control methods.

Table 2 Actions taken by official veterinarian upon a non-compliance

Action	Example
Guidance	Guiding concerning food safety requirements
Oral notification	Minor non-compliance for example concerning sanitation
Written notification (or warning)	Minor to major non-compliance
Oral or written notification with a time limit for correction of non-compliance	Any non-compliance requiring a time limit
Order	Ordering to improve sanitation
Prohibition	Prohibition of using or placing a product on the market
Seizure	Seizure of a food lot (taking the lot into the possession of the authorities)
Destruction of food stuffs	Destruction of a food lot unfit for human consumption
Suspension of operations or part of operations	Severe uncleanliness of premises, serious pest problem, unapproved operations
Economic sanctions	
Threat of a penalty payment	To enhance the effect of an order to correct surfaces in poor condition
Fine	Fining the FBO to increase compliance in the future
Withdrawal of approval	Major non-compliance that are not corrected by other measures
Notification to police	Serious and/or intentional non-compliance

USDA, 2021; EU, 2017; Luukkanen and Lundén, 2016; Kettunen et al., 2015.

in understanding the requirements of food safety legislation and therefore need guidance in the interpretation of the regulations (Kettunen et al., 2017; Buckley, 2015; Kotisalo et al., 2015; Nevas et al., 2013; Yapp and Fairman, 2006). Instead of immediately moving into enforcement measures, the first action taken should be guidance, unless stricter actions are warranted due to the nature of the non-compliance or previous compliance history (Kettunen et al., 2015; Buckley, 2015; Lindblad and Berking, 2013). An assistive approach with the first action being guidance is also appreciated by the FBOs (Buckley, 2015) and is considered efficient (Reske et al., 2007). In this way, the FBOs are given a possibility to correct the non-compliance without strict administrative procedures. Competent, high-quality guidance has also been shown to increase the safety of products by increasing the knowledge of the FBOs (Nevas et al., 2013).

Oral notification can be used while carrying out the inspection. Oral notification as the sole control action should be used only in very minimal non-compliance situations because it has not been observed to be as effective as other control actions (Luukkanen and Lundén, 2016). A much more effective control action is a written notification (also called a 'warning' in some countries) combined with a time limit for correction of the non-compliance (Luukkanen and Lundén, 2016; Läikkö-Roto et al., 2015). In slaughterhouses where OVs use more oral notifications without time limits, non-compliance tends to be more common (Luukkanen and Lundén, 2016). Therefore, it is important not only to use oral notification to communicate the observed non-compliance to the FBO at the inspection level, but also to document the non-compliance and to use time limits. The verification of correction of non-compliance and re-enforcement of uncorrected non-compliance is also important for reaching the desired impact on food safety (Luukkanen and Lundén, 2016; Läikkö-Roto et al., 2015) and to maintain the credibility of official control.

Enforcement measures such as orders, prohibitions, and economic sanctions should be used only when guiding and negotiation is not leading to correction of the non-compliance (Kettunen et al., 2015; Lindblad and Berking, 2013). The OV, however, must have the readiness and skills to use enforcement measures without delay when needed. The use of enforcement measures should also be gradual and proportional to the non-compliance (Kettunen et al., 2015; Lepistö and Hänninen, 2011). Enforcement measures have been used in non-compliance related to slaughter hygiene, working hygiene and poor condition of surfaces and are considered effective when used (Luukkanen and Lundén, 2016; Kettunen et al., 2015; Lindblad and Berking, 2013).

3.6 Documentation, analysis, and disclosure of the inspection results

Inspection results should always be documented accurately and without delay and conveyed to the FBO explicitly so that the FBO understands what has been inspected, what the non-compliance issues are and what is expected from the FBO (Yapp and Fairman, 2006). In the case of non-compliance, the OV should explain to the FBO the importance of correction of the non-compliance, as this has been observed to influence the motivation to improve FBOs' risk management (Läikkö-Roto et al., 2015).

The inspection can be paper based, that is the inspection findings are documented on paper, however, digital based documentation is preferred. Ideally, the inspection results are documented digitally at the inspection using for example a tablet that transfers the information to a database when Internet connection is available. Digital environment is recommended as it increases efficiency and decreases mistakes that can occur when transferring data from paper to digital form (Nabais et al., 2022). Digitalisation also facilitates analysis of inspection data. Inspections generate a large amount of data which can be used in developing official control, but efficient analyses require digitalisation (Labrique et al., 2013). The digitalisation of inspection data enables also trend analyses on non-compliances, official control actions and impact analyses of official control actions. Although, the use of digital based documentation can offer many advantages the applicability in the public sector is still underutilised (Meijer, 2015; Maiti and Awasthi, 2020).

Many countries disclose the results of the inspections (Anon., 2023, DVFA, 2023; FSA, 2022) to enhance the transparency of official control (EU, 2017). Some of the countries also publish poultry slaughterhouse inspection results on the internet (Anon., 2023; DVFA, 2023). The disclosure systems are suggested to have beneficial effects on compliance (DVFA, 2023; Kaskela et al., 2019; Waters et al., 2013) and decrease food bourn outbreaks (Kim et al., 2022) and should therefore be considered as an additional tool for official control.

4 Conclusion

The purpose of official control is to safeguard consumer health and to promote fair trade by verifying that food businesses are complying with food safety requirements. To achieve this in the poultry meat value chain, the official control performs ante- and post-mortem inspections as well as food safety inspections. Successful official control requires sufficient resources, access to FBO's premises and documents and effective inspection techniques. Official control should obtain a comprehensive and transparent picture of the operations and the management of the food safety risks.

The inspection techniques aim at collecting data from different sources. FBOs' interviews and inspection of the FBOs' self-check plan and documentation are used to assess the compliance of the slaughterhouse. Inspecting only data obtained from the FBO such as self-check documentation, however, may give an incorrect impression of the status of the food business. Therefore, the OV must also make own observations and measurements when applicable to get an objective picture of the premises and the operations. The inspection of slaughter and working hygiene, which often have been observed as insufficient in poultry slaughterhouses, should include the observation of the operations for a sufficient time to get a true picture of the hygiene practices. Other inspection techniques include capability testing such as FBOs' ability to trace products and cross-inspection, which utilises data from two or more sources. Also, sampling can be used to verify the adequacy of sanitation, animal species and water content of the meat.

Research has shown that soft control actions such as guiding the FBO have beneficial effects on compliance. Especially small- and medium-sized enterprises need more guidance in interpreting sometimes complex and ambiguous food safety legislation. Such guidance can be given at on-site inspections. In cases of non-compliance, the OV can use as a first control measure notifications combined with discussion about how and when the FBO intends to correct the non-compliance. Notifications should include a time limit for the correction of the non-compliance because it has been shown to facilitate compliance. Follow-up inspections should be done or conducted in some other way to verify that the non-compliance has been eliminated. When notifications are not leading to correction of the non-compliance or the non-compliance is causing a serious health threat, the OV should consider using more efficient enforcement measures such as orders, prohibitions, economic sanctions or suspension of operations. Negligence, intentional non-compliance and fraudulent activities can also be grounds for notification to the police. The inspection results are documented and required corrections should be well justified to the FBO to increase the motivation to correct the non-compliance. The outcomes of the inspection reports are published in some countries, and this can enhance compliance.

5 Future trends in research

Efficient and effective meat inspection remains an important research challenge in poultry industry. Efficiency is sought to decrease costs and effectiveness is pertinent for food safety and animal health and welfare. Poultry meat industry should continue to seek for solutions to prevent faecal contamination of carcasses and official control automated techniques that improve meat inspection. Risk-based meat inspection is advocated for, and this topic will remain in focus also in the future.

The actions of official control should lead to the correction of non-compliance. It appears, however, that official control should be more effective as non-compliances are observed frequently in slaughter hygiene. The use of control actions should be further elucidated, and new control actions developed that incentivize FBOs to comply.

The poultry meat business is vulnerable to fraudulent activities. Official control, including the OV on site, need to have efficient tools to recognise fraudulent activities. In particular, the recognition of early signals of fraud should be developed as well as increased use of intelligence.

The cooperation and often daily presence of the OV in the slaughterhouse is challenging. OVs need to cooperate with the FBO, but at the same time maintain a distance from the FBO to ensure objectivity at inspections. How the OV and the entire official control organisation handles this should be investigated. Furthermore, official control in slaughterhouses can be organised in different ways with a different amount of control responsibility shifted to the FBO (Luukkanen et al., 2015; Anon, 2014; Berzins et al., 2014). Research has shown that balancing between these responsibilities may lead to challenges and mistrust (Anon, 2014) and further investigation is needed to recognise well-functioning models.

6 Where to look for further information

Meat inspection is regulated in detail and country specific legislation can be found at government and competent authority internet sites. Basis for risk-based meat inspection is described in the document provided by the Food and Agriculture Organization (FAO, 2019) and Ferri et al. (2023) present challenges and solutions for developing risk-based meat inspection.

Scientific research concerning inspection techniques and their efficacy in slaughterhouses is sparse, but many studies focus on restaurant inspection. Results from those papers can in some cases be applied to the slaughterhouse context. However, this should be done with consideration and understanding of official control issues both in slaughterhouses and retail. The principles of inspecting are, in general, the same regardless of the type of food business, but the official control in slaughterhouses have many distinctive features such as the inspection frequency and presence on-site. Inspection manuals, guidelines, and checklists for carrying out inspections have been prepared by competent authorities in different countries and can be found on the internet. The textbook by Ninios et al. (2014) consider official control in slaughterhouses and offer deeper insight into the role and complexity of the control in slaughterhouses.

7 References

Allain, V., Salines, M., Le Bouquin, S. and Magras, C. (2018). Designing an innovative warning system to support risk-based meat inspection in poultry slaughterhouses. *Food. Control.*, 89, 177–186.

Amery, R.W. (2023). Federal meat and poultry inspection duties and requirements–part 1: history and current responsibilities. *J. Environ. Health*, 85, 22–25.

Anon. (2014). *Summary: risks in the meat supply chain.* Dutch Safety Boar. https://www.onderzoeksraad.nl/nl/media/inline/2018/7/10/38357a3466aerisico_s_vleesketen_en_web.pdf (accessed 26 May 2023).

Anon. (2023). *General information about Oiva.* Finnish Food Safety Authority. https://www.oivahymy.fi/en/for-consumers/general-information-about-oiva/ (accessed 26 May 2023)

Antunović, B., Blagojević, B., Johler, S., Guldimann, C., Vieira-Pinto, M., Vågsholm, I., Meemken, D., Alvseike, O., Georgiev, M. and Alban, L. (2021). Challenges and opportunities in the implementation of new meat inspection systems in Europe. *Trends Food Sci. Tech.*, 116, 460–467.

Barker, D., Lankhaar, J. and Stals, P. (2004). Primary processing of poultry. In G. C. Mead (Ed.). *Poultry meat processing and quality* (pp. 90–107). England: Woodhead Publishing Limited.

Beczkiewicz, A.T.E. and Kowalcyk, B.B. (2021). Risk factors for *Salmonella* contamination of whole chicken carcasses following changes in U.S. regulatory oversight. *J. Food Prot.*, 84, 1713–1721.

Berzins, A., Lundén, J. and Korkeala, H. (2014). Organization of official control. In T. Ninios, J. Lundén, H. Korkeala, and M. Fredriksson-Ahomaa (Eds.). *Meat inspection and control in the slaughterhouse.* Hoboken: Wiley Blackwell, pp. 556–561.

Bilgili, S., F. (2010). Poultry meat inspection and grading. In C. M. Owens, C. Z. Alvarado, and A. R. Sams (Eds.). *Poultry meat processing* (pp. 67–100). Boca Raton: Taylor & Francis Group.

Blagojevic, B., Nesbakken, T., Alvseike, O., Vågsholm, I., Antic, D., Johler, S., Houf, K., Meemken, D., Nastasijevic, I., Vieira Pinto, M., Antunovic, B., Georgiev, M. and Alban, L. (2021). Drivers, opportunities, and challenges of the European risk-based meat safety assurance system. *Food Control*, 124, 107870.

Blomvall, L., Kaukonen, E., Kurittu, P., Heikinheimo, A. and Fredriksson-Ahomaa, M. (2023). Food chain information and post-mortem findings in fattening Turkey flocks. *Food Control*, 150, 109739

Buckley, J. (2015). Food safety regulation and small processing: a case study of interactions between processors and inspectors. *Food Policy*, 51, 74–82.

Buntain, B. (2004). Emerging challenges in public health protection, food safety, and security: veterinary needs in the USDA' s food safety and inspection service. *J. Vet. Medic. Educ.*, 31, 334–340.

CFIA. (2019). *Guidance on Canadian Food Inspection Agency inspection stations for slaughter operation of food animals and meat products.* Canadian Food Inspection Agency. https://www.inspection.gc.ca/food-safety-for-industry/food-specific-requirements-and-guidance/meat-products-and-food-animals/cfia-stations-for-slaughter/eng/1550250803810/1550250952413?gf (accessed May 15 2023)

Chowdhury, E.U. and Morey, A. (2020). Application of optical technologies in the US poultry slaughter facilities for the detection of poultry carcase condemnation. *Br. Poult. Sci.*, 61(6), 646–652.

Codex 2013. Principles and guidelines for national food control systems. CAC/GL 82-2013. https://www.fao.org/fao-who-codexalimentarius/sh-proxy/es/?lnk=1&url=https%253A%252F%252Fworkspace.fao.org%252Fsites%252Fcodex%252FStandards%252FCXG%2B82-2013%252FCXG_082e.pdf (accessed 25 May 2023)

Codex. (2017). Principles and guidelines for monitoring the performance of national food control systems. CXG 91-2017. https://www.fao.org/fao-who-codexalimentarius/sh-proxy/en/?lnk=1&url=https%253A%252F%252Fworkspace.fao.org%252Fsites%252Fcodex%252FStandards%252FCXG%2B91-2017%252FCXG_091e.pdf (accessed 26 May 2023)

Collins, B. L. and Worthey, J. A. (1984). *The role of color in lighting for meat and poultry inspection*. Washington, DC: National Bureau of Standards, NBSIR 84-2829. https://nvlpubs.nist.gov/nistpubs/Legacy/IR/nbsir84-2829.pdf (accessed 26 May 2023)

DVFA. (2023). *About the Danish Smiley scheme*. The Danish Veterinary and Food Administration. https://www.findsmiley.dk/english/Pages/About.aspx (accessed 26 May 2023)

EC. (2002). Regulation (EC) No 178/2002 of the European Parliament and of the Council laying down the general principles and requirements of food law, establishing the European Food Safety Authority and laying down procedures in matters of food safety. http://eur-lex.europa.eu/LexUriServ/LexUriServ.do?uri=OJ:L:2002:031:0001:0024:en:PDF (accessed 26 May 2023).

EC. (2004a). Regulation (EC) No 852/2004 of the European Parliament and of the Council on the hygiene of foodstuffs. http://eur-lex.europa.eu/LexUriServ/LexUriServ.do?uri=OJ:L:2004:139:0001:0054:EN:PDF (accessed 25 May 2023)

EC. (2004b). Regulation (EC) No 853/2004 of the European Parliament and of the Council of 29 April 2004 laying down specific hygiene rules for food of animal origin. http://data.europa.eu/eli/reg/2004/853/2023-02-15 (accessed 25 May 2023)

EC. (2009). Regulation (EC) No 1069/2009 of the European parliament and of the council laying down health rules as regards animal by-products and derived products not intended for human consumption and repealing Regulation (EC) No 1774/2002 (Animal by-products Regulation). http://eur-lex.europa.eu/LexUriServ/LexUriServ.do?uri=OJ:L:2009:300:0001:0033:EN:PDF (accessed 25 May 2023)

EC. (2022). 2021 Annual report. Alert and Cooperation Network. European Commission. https://food.ec.europa.eu/system/files/2022-07/acn_annual-report_2021-final.pdf (accessed 26 May 2023)

EC SANTE. (2019). Final report of an audit carried out in the Czech Republic from 11 to 21 June 2019 in order to assess the implementation of the actions taken in response to the previous audits on official controls over production of poultry meat and products derived therefrom and the Salmonella national control programmes in particular poultry populations, European Commission, Directorate General for Health and Food Safety. https://ec.europa.eu/food/audits-analysis/audit-report/download/14666 (accessed 25 May 2023)

EC SANTE. (2020). Final report of an audit carried out in Chile from 9 to 19 March 2020 in order to evaluate the control system governing the production poultry meat products intended for export to the European Union, European Commission,

Directorate General for Health and Food Safety. https://ec.europa.eu/food/audits-analysis/audit-report/download/14982 (accessed 25 May 2023)

EC SANTE. (2022). Final report of an audit of Poland carried out from 11 to 22 October 2021 in order to follow-up the audit reports DG(SANTE)/2019-6843 on bovine meat, including traceability, and DG(SANTE)/2019-6671 on poultry meat and products derived therefrom, European Commission, Directorate General for Health and Food Safety. https://ec.europa.eu/food/audits-analysis/audit-report/download/15788 (accessed 25 May 2023)

EFSA. (2012). Scientific opinion on the public health hazards to be covered by inspection of meat from poultry. *EFSA J.* 10(6), 2741. https://efsa.onlinelibrary.wiley.com/doi/abs/10.2903/j.efsa.2012.2741 (accessed 26 May 2023)

EFSA and ECDC. (2022). The European Union One Health 2021 Zoonoses Report. European Food Safety Authority and European Centre for Disease Prevention and Control. *EFSA J.*, 20, 273 pp. https://doi.org/10.2903/j.efsa.2022.7666 (accessed 26 May 2023)

EU. (2017). Regulation (EU) 2017/625 of the European Parliament and of the Council of 15 March 2017 on official controls and other official activities performed to ensure the application of food and feed law, rules on animal health and welfare, plant health and plant protection products. https://eur-lex.europa.eu/legal-content/EN/TXT/?uri=CELEX%3A02017R0625-20191214&qid=1617129031298 (accessed 20 March 2023)

EU. (2019a). Commission delegated regulation (EU) 2019/624 of 8 February 2019 concerning specific rules for the performance of official controls on the production of meat and for production and relaying areas of live bivalve molluscs in accordance with Regulation (EU) 2017/625 of the European Parliament and of the Council. https://eur-lex.europa.eu/legal-content/EN/TXT/?uri=CELEX%3A32019R0624&qid=1617129432685 (accessed 26 March 2023)

EU. (2019b). Commission Implementing Regulation (EU) 2019/627 of 15 March 2019 laying down uniform practical arrangements for the performance of official controls on products of animal origin intended for human consumption in accordance with Regulation (EU) 2017/625 of the European Parliament and of the Council and amending Commission Regulation (EC) No 2074/2005 as regards official controls. https://eur-lex.europa.eu/legal-content/EN/TXT/?uri=CELEX%3A02019R0627-20210101&qid=1617129503276 (accessed 20 March 2023)

Evira. (2015). Instructions of the risk-based food safety control in food establishments, Evira Instructions 16044/1. https://www.ruokavirasto.fi/globalassets/tietoa-meista/asiointi/oppaat-ja-lomakkeet/yritykset/elintarvikeala/elintarvikehuoneistot/eviran_ohje_16044_1.pdf (accessed 26 May 2023)

FAO. (2019). *Technical guidance principles of risk-based meat inspection and their application*. Rome. https://www.fao.org/documents/card/en/c/ca5465en (accessed 26 May 2023)

FAO. (2021). Measuring food safety – Indicators to achieve Sustainable Development Goals (SDGs). Food safety technical toolkit for Asia and the Pacific No. 9. Bangkok. Thailand. https://www.fao.org/documents/card/en/c/cb4111en/ (accessed 26 May 2023)

Ferri, M., Blagojevic, B., Maurer, P., Hengl, B., Guldimann, C., Mojsova, S., Sakaridis, I., Antunovic, B., Gomes-Neves, E., Zdolec, N., Vieira-Pinto, M. and S. Johler. (2023).

Risk based meat safety assurance system – An introduction to key concepts for future training of official veterinarians. *Food Control*, 146, 109552.

FFA. (2020). Elintarvikeketjutietojen toimittaminen ja valvonta (Delivering and control of food chain informations). Finnish Food Authority instructions 4929/04.02.00.01/2020/6. https://www.ruokavirasto.fi/globalassets/tietoa-meista/asiointi/oppaat-ja-lomakkeet/yritykset/elintarvikeala/alkutuotanto/elintarvikeketjutietojen-toimittaminen-ja-valvonta.pdf (accessed 30 May 2023)

FSA. (2022). *Food hygiene rating schemes*. UK: Food Standards Agency. http://www.food.gov.uk/business-industry/hygieneratings (accessed 26 May 2023)

FSMA. (2011). Food safety modernization act. https://www.gpo.gov/fdsys/pkg/PLAW-111publ353/pdf/PLAW-111publ353.pdf (accessed 25 May 2023)

Gagnon, B., McEachern, V. and Bray, S. (2000). The role of the Canadian government agency in assessing HACCP. *Food Control*, 11, 359–364.

Habib, I., Berkvens, D., De Zutter, L., Dierick, K., Van Huffel, X., Spreybroeck, N., Geeraerd, A. and Uyttendaele, M. (2012). Campylobacter contamination in broiler carcasses and correlation with slaughterhouses operational hygiene inspection. *Food Microbiol.*, 29, 105–112.

Huneau-Salaün, A., Stärk, K. D. C., Mateus, A., Lupo, C., Lindberg, A. and Le Bouquin-Leneveu, S. (2015). Contribution of meat inspection to the surveillance of poultry health and welfare in the European Union. *Epidemiol. Infect.*, 143, 2459–2472.

Husu-Kallio, J. (2011). Audits and guides for good practice. *Slov. Vet. Res.*, 48(Suppl. 13), 77.

Jacobs, L., Delezie E., Duchateau L., Goethals K. and Tuyttens F.A.M. (2017). Impact of the separate pre-slaughter stages on broiler chicken welfare. *Poult. Sci.*, 96, 266–273.

Jørgensen A., Fagertun J. and Moeslund T. B. (2018). Classify broiler viscera using an iterative approach on Noisy Labeled training data. In G. Bebis et al. (Eds.). *Advances in visual computing*. ISVC 2018. Lecture Notes in Computer Science, vol 11241. Cham: Springer.

Kaskela, J., Vainio, A., Ollila, S. and Lundén, J. (2019). Food business operators' opinions on disclosed food safety inspections and occurrence of disagreements with inspector grading. *Food Control*, 105, 248–255.

Kettunen, K., Lundén, J., Läikkö, T. and Nevas, M. (2017). Towards more consistent and effective food control: learning from the views of food business operators. *Int. J. Environ. Health Res.*, 27, 215–229.

Kettunen, K., Nevas, M. and Lundén, J. (2015). Effectiveness of enforcement measures in local food control in Finland. *Food Control*, 56, 41–46.

Kim, T. N., Wildey, L., Gleason, B., Bleser, J., Firestone, M. J., Bare, G., Bliss, J., Dewey-Mattia, D., Stueven, H., Brown, L., Dyjack, D. and Hedberg, C. W. (2022). Foodborne outbreak rates associated with restaurant inspection grading and posting at the point of service: evaluation using national foodborne outbreak surveillance data. *J. Food Prot.*, 85, 1000–1007.

Kontio. (2014). International trade. In T. Ninios, J. Lundén, H. Korkeala, and M. Fredriksson-Ahomaa (Eds.). *Meat Inspection and Control in the Slaughterhouse*. Hoboken: Wiley Blackwell, pp. 643–650.

Kosola, M., Kiviniemi, K. and Lundén, J. (2022). Factors affecting effectiveness of food control inspections in food production establishments in Finland. *Sci. Rep.*, 12, 4230.

Kotisalo, N., Luukkanen, J., Fredriksson-Ahomaa, M. and Lundén, J. (2015). Effects of centralizing meat inspection and food safety inspections in Finnish small-scale slaughterhouses. *Food Policy*, 55, 15–21.

Labrique, A., Vasudevan, L., Kochi, E., Fabricant, R. and Mehl, G. (2013). mHealth innovations as health system strengthening tools: 12 common applications and a visual framework. *Glob. Health Sci. Practice.*, 1, 160–171.

Lepistö, O. and Hänninen, M. L. (2011). Effects of legal aspects on the use of compulsory procedures in environmental health and food control. *J. Environ. Health Res.*, 11, 127–134.

Lindblad, M. and Berking, C. (2013). A meat control system achieving significant reduction of visible faecal and ingesta contamination of cattle, lamb and swine carcasses at Swedish slaughterhouses. *Food Control*, 30, 101–105.

Lundén, J., Björkroth, J. and Korkeala, H. (2007). Meat inspection education in Finnish veterinary curriculum. *J. Vet. Medic. Educ.*, 34, 205–211.

Lundén, J., Vanhanen, V., Kotilainen, K. and Hemminki, K. (2014). Retail food stores' internet-based own-check databank records and health officers' on-site inspection results for cleanliness and food holding temperatures reveal inconsistencies. *Food Control*, 35, 79–84.

Luukkanen, J., Fredriksson-Ahomaa, M., Nevas, M. and Lundén, J. (2017). Prerequisites for high-quality official control in Finnish slaughterhouses. *Food Control*, 79, 50–56.

Luukkanen, J., Kotisalo, N., Fredriksson-Ahomaa, M. and Lundén, J. (2015). Distribution and importance of meat inspection tasks in Finnish high-capacity slaughterhouses. *Food Control*, 57, 246–251.

Luukkanen, J. and Lundén, J. (2016). Compliance in slaughterhouses and control measures applied by official veterinarians. *Food Control*, 68, 133–138.

Läikkö -Roto, T., Mäkelä, S., Lundén, J., Heikkilä, J. and Nevas, M. (2015). Consistency in inspection processes of food control officials and efficacy of official controls in restaurants in Finland. *Food Control*, 57, 341–350.

Löhren, U. (2012). *Overview on current practices of poultry slaughtering and poultry meat inspection*. Supporting Publications 2012:EN-298, 58 p. https://efsa.onlinelibrary.wiley.com/doi/pdf/10.2903/sp.efsa.2012.EN-298https://efsa.onlinelibrary.wiley.com/doi/pdf/10.2903/sp.efsa.2012.EN-298 (accessed 26 May 2023)

Maiti, D. and Awasthi, A. (2020). ICT exposure and the level of wellbeing and progress: a cross country analysis. Soc. Indic. Res., 147, 311–343.

Marel Poultry (2023). https://marel.com/en/poultry (accessed 7 May 2023)

Meijer, A. (2015). E-governance innovation: barriers and strategies. *Govern. Inform. Quart.*, 32, 198–206.

Nabais, P., Carmona, P., Monteiro, S., Melo de Vasconcelos, F., Tomé, S., Ravasco, F., Brazão, R., Dias, M. da G., Sokolic, D. and Basic, S. (2022). SSD2 and FoodEx2 compliant real-time registration and classification of food sampling data – Improving Data Quality for Risk Assessment (IDRisk). *EFSA Support. Publ.*, 19(10). https://doi.org/10.2903/sp.efsa.2022.en-7633 (accessed 26 May 2023)

Nevas, M., Kalenius, S. and Lundén, J. (2013). Significance of official food control in food safety; food business operators' perceptions. *Food Control*, 31, 59–64.

Newbold, K. B., McKeary, M., Hart, R. and Hall, R. (2008). Restaurant inspection frequency and food safety compliance. *J. Environ. Health.*, 71, 56–61.

NFAS. (2023). *Risk classification of establishments and operations from 2024 onwards*. Sweden: National Food Agency. https://kontrollwiki.livsmedelsverket.se/artikel/753

/riskklassning-av-anlaggningar-och-verksamheter-fran-och-med-2024 (accessed 26 May 2023).

Ninios, T., Lundén, J., Korkeala, H. and Fredriksson-Ahomaa, M. (2014). *Meat Inspection and Control in the Slaughterhouse*. Hoboken: Wiley Blackwell, UK, p. 698.

Nørrung, B. and Buncic, S. (2008). Microbial safety of meat in the European Union. *Meat Sci.*, 78, 14–24.

Obbink, N., Frissen, J. M. and Post, S. B. (2014). Food frauds. In T. Ninios, J. Lundén, H. Korkeala, and M. Fredriksson-Ahomaa (Eds.). *Meat inspection and control in the slaughterhouse*. Hoboken: Wiley Blackwell, UK, pp. 628–638.

Park, B., Yoon, S.-C., Windham, W. R., Lawrence, K. C., Kim, M.S. and Chao, K. (2011). Line-scan hyperspectral imaging for real-time in-line poultry fecal detection. *Sens. Instrument. Food Qual. Safety*, 5, 25–32.

Pennington, H. (2009). The public inquiry into the September 2005 outbreak of E. coli O157 in South Wales. http://www.reading.ac.uk/foodlaw/pdf/uk-09005-ecoli-report-summary.pdf (accessed 23 May 2023).

Pinillos, R. G. and Jukes, D. J. (2008). Hygiene assessment system (HAS) scores – an analysis of the available data from English slaughterhouses. *Food Control*, 19, 806–816.

Pozo, V. F. and Schroeder, T. C. (2016). Evaluating the costs of meat and poultry recalls to food firms using stock returns. *Food Policy*, 59, 66–77.

Premanandh, J. (2013). Horse meat scandal – a wake-up call for regulatory authorities. *Food Control*, 34, 568–569.

Reske, K., Jenkins, T., Fernandez, C., VanAmber, D. and Hedberg, C. (2007). Beneficial effects on implementing an announced restaurant inspection program. *J. Environ. Health*, 69, 27–34.

Salines, M., Allain, V., Magras, C. and Le Bouquin, S. (2018). Rethinking inspection in slaughterhouses: opportunities and challenges arising from a shared risk management system in poultry slaughterhouses. *Food Control*, 93, 48–55.

Seo, Y., Lee, H., Mo, C., Kim, M. S., Baek, I., Lee, J. and Cho, B.-K. (2019). Multispectral fluorescence imaging technique for on-line inspection of fecal residues on poultry carcasses. *Sensors*, 19, 3483.

Sharman, N., Wallace, A. C. and Jespersen, L. (2020). Terminology and the understanding of culture, climate, and behavioural change – Impact of organisational and human factors on food safety management. *Trends Food Sci. Tech.*, 96, 13–20.

Stärk, K. D. C., Alonso, S., Dadios, N., Dupuy, C., Ellerbroek, L., Georgiev, M., et al. (2014). Strengths and weaknesses of meat inspection as a contribution to animal health and welfare surveillance. *Food Control*, 39, 154–162.

Todd, E. C. D., Michaels, B. S., Greig, J. D., Smith, D., Holah, J. and Bartleson, C. A. (2010). Outbreaks where food workers have been implicated in the spread of foodborne disease. Part 7. Barriers to reduce contamination of food by workers. *J. Food Prot.*, 73, 1552–1565.

Törmä, K., Kaukonen, E., Lundén, J., Fredriksson-Ahomaa, M. and Laukkanen-Ninios, R. (2022). A comparative analysis of meat inspection data as an information source of the health and welfare of broiler chickens based on Finnish data. *Food Control*, 138, 109017.

Törmä, K., Lundén, J., Kaukonen, E., Fredriksson-Ahomaa, M. and Laukkanen-Ninios, R. (2021). Prerequisites of inspection conditions for uniform post-mortem inspection in broiler chicken slaughterhouses in Finland. *Food Control*, 130, 108384.

USDA. (2017). *New poultry inspection system: Post-mortem inspection and verification of ready-to-cook requirement. FSIS directive 6500.1*. Washington, DC: Food Safety and Inspection Service, U.S. Department of Agriculture. https://www.fsis.usda.gov/policy/fsis-directives/6500.1 (accessed 14 April 2023)

USDA. (2019). *USDA, Poultry products inspection regulations, title 9, part 381, section 36*. Washington, DC: Code of Federal Regulations. Food Safety and Inspection Service, U.S. Department of Agriculture. https://www.ecfr.gov/cgi-bin/text-idx?SID=dd06aefeecea6cc2dd956ed7d365fbd8&mc=true&node=se9.2.381_136&rgn=div8 (accessed 8 May 2023)

USDA. (2021). Inspection methods refresher course student handout. https://www.fsis.usda.gov/sites/default/files/media_file/2021-11/IMR-Student-Handout.pdf (accessed 26 May 2023)

Waters, A. B., Van Derslice, J., Porucznik, C. A., Kim, J., De Legge, R. and Durrant, L. (2013). Impact of internet posting of restaurant inspection scores on critical violations. *J. Environ. Health*, 75, 8–12.

WHO. (2006). *Five keys to safer food manual*. World Health Organization Department of Food Safety, Zoonoses and Foodborne Diseases. https://apps.who.int/iris/rest/bitstreams/51699/retrieve (accessed 26 May 2023).

Winkler, B., Maquet, A., Reeves-Way, E., Siegener, E., Cassidy, T., Valinhas De Oliveira, T., Verluyten, J., Jelic, M. and Muznik, A. (2023). *Fighting fraudulent and deceptive practices in the agri-food chain*. EUR 31436 EN, Luxembourg: Publications Office of the European Union. https://publications.jrc.ec.europa.eu/repository/handle/JRC131525 (accessed 26 May 2023)

Wirtanen, G. and Salo, S. (2014). Use of meat inspection data. In T. Ninios, J. Lundén, H. Korkeala, and M. Fredriksson-Ahomaa (Eds.). *Meat. inspection and control in the slaughterhouse* Hoboken: Wiley Blackwell, UK, pp. 453–471.

Yang, C. C., Chao, K. and Kim, M. S. (2009). Machine vision system for online inspection of freshly slaughtered chickens. *Sens. Instrument. Food Qual. Safety*, 3, 70–80.

Yapp, C. and Fairman, R. (2006). Factors affecting food safety compliance within small and medium sized enterprises: implications for regulatory and enforcement strategies. *Food Control*, 17, 42–51.

Chapter 11

Use of antimicrobials in poultry processing plants: the case of the United States

Dana K. Dittoe, University of Wyoming, USA; Elena G. Olson and Steven C. Ricke, University of Wisconsin-Madison, USA

1 Introduction

Over the last 25 years, substantial efforts have been made to reduce the contamination of raw poultry and its products. The Hazard Analysis Critical Control Point (HACCP) system and the US Food Safety Modernization Act (FSMA) of 2011 established essential regulations and guidelines designed to lower the occurrence of foodborne illnesses (Hulebak, 2002; FDA, 2011). Despite the positive impact of HACCP and FSMA on foodborne illness rates, more is still needed. Foodborne pathogens such as *Salmonella* and *Campylobacter* continue to be linked to poultry products despite current control efforts (Chai et al., 2017; Ricke, 2021; O'Bryan et al., 2022; Olson et al., 2022; Obe

http://dx.doi.org/10.19103/AS.2024.0146.13

et al., 2023). Part of the challenge lies in limiting contamination in broiler flocks before they enter the processing plant (Obe et al., 2023). Consequently, further mitigation strategies are necessary to effectively remove foodborne pathogens during poultry processing, necessitating the development of more effective interventions.

General methods to decontaminate poultry products in US processing facilities involve antimicrobial washes and sprays applied at various stages of primary and secondary processing (McKee, 2012). Chlorine and peracetic acid (PAA) are antimicrobials for use in chillers and post-chiller treatments employing techniques such as spray cabinets and part dips (McKee, 2012). Other antimicrobials, such as organic acids, phosphates, chlorine derivatives, and hydrogen peroxide solutions, have been suggested for use in the poultry and meat industries (Dincer and Baysal, 2004; Loretz et al., 2010). While some have been extensively studied under laboratory conditions, only a few are routinely employed in commercial poultry processing settings. Despite existing interventions, the risk of cross-contamination remains as processing progresses to later stages, with limited hurdles in place to mitigate this risk. The current review discusses recent US poultry processing procedures and the need for additional antimicrobials to further reduce pathogens and microbial load throughout all stages of poultry processing. The chapter discusses:

- Foodborne pathogens affecting poultry processing.
- Current operational conditions in US commercial processing of poultry.
- Intervention strategies employed by poultry processors.

The chapter will also explore the effects of current and novel acidifiers used or potentially available for poultry processing. Finally, it will propose future considerations for further developing antimicrobials in poultry processing.

2 Foodborne pathogens in poultry processing

Poultry products are an integral part of the US economy and diets, making their safety critical. *Salmonella*, *Campylobacter*, and *Escherichia coli* are the main pathogens associated with poultry production, resulting in gastroenteritis and severe gastrointestinal and systemic diseases in compromised individuals (World Health Organization, 2015). *Campylobacter* is the most frequent cause of diarrhea, while *Salmonella enterica* is most associated with foodborne-related deaths (World Health Organization, 2015). In 2015, *Salmonella*, *Campylobacter*, and *E. coli* incidents represented nearly a third of all US foodborne infections (CDC, 2017). From the late 1990s to today, over 20% of foodborne illnesses and outbreaks have been linked to contaminated poultry products (Painter et al., 2013; Chai et al., 2017). Whilst poor food handling and inadequate cooking by consumers are partly responsible, it is clearly essential for poultry processors to ensure safe poultry products at the point of purchase (Chai et al., 2017).

2.1 Salmonella

Salmonella is a Gram-negative, facultative aerobic bacterium that thrives at 37°C and can cause various illnesses, such as enteric and typhoidal fevers, gastroenteritis, and septicemia, depending on the serovar (Holt et al., 2000). Numerous foodborne illnesses are linked to the contamination of animal products (Barrow et al., 2012; Cosby et al., 2015; Crump et al., 2015). Globally, non-typhoidal *Salmonella* is responsible for millions of foodborne illness cases and over 150 000 deaths annually (Majowicz et al., 2010). In the United States alone there are over a million annual cases of non-typhoidal salmonellosis that result in hundreds of deaths (Scallan et al., 2011). The infectious dose for humans can range from 10^6 to 10^8 CFU, but it can be lower depending on the host's immune status, with the elderly and children being more vulnerable (Chen et al., 2013). In 2017, *Salmonella* was the leading cause of foodborne disease in the US, with Enteritidis, Typhimurium, Newport, and Javiana being the most common serotypes (CDC, 2017). The incidence of *Salmonella*, particularly Enteritidis, is closely associated with poultry and eggs (CDC, 2014a). The high prevalence of *Salmonella* in poultry is partly due to the bacterium's ability to colonize poultry without causing noticeable symptoms to birds themselves (subclinical infection) and to spread through horizontal and vertical transmission (Barrow et al., 2012; Cosby et al., 2015).

2.2 Campylobacter

Campylobacter is a Gram-negative, microaerophilic bacterium responsible for campylobacteriosis, an infectious gastrointestinal disease that can lead to acute paralysis (Holt et al., 2000). The most common strains associated with foodborne illness are *C. jejuni* and *C. coli* (CDC, 2017). Both strains thrive in a microaerophilic environment at approximately 42°C, similar to the body temperature of birds (41°C to 42°C) (CDC, 2014b; Holt et al., 2000). Like *Salmonella*, *Campylobacter* can be found in humans and animals' oral cavity, gastrointestinal tract, and reproductive organs (Holt et al., 2000). Previously, *Campylobacter* was considered a commensal organism in the gastrointestinal tract of poultry (Humphrey et al., 2014). However, more some findings have suggested that *Campylobacter* may cause pathogenesis (Van Deun et al., 2008; Lamb-Rosteski et al., 2008; Awad et al., 2014, 2015, 2016). Annually, campylobacteriosis affects over a million people and results in more than 70 deaths (CDC, 2006; CDC, 2014b).

2.3 Escherichia coli

Escherichia coli is a facultative-anaerobic, Gram-negative bacterium that inhabits the lower part of the gastrointestinal tract in animals (Holt et al., 2000). Certain strains of *E. coli* can produce enterotoxins and colonize the host, potentially causing gastroenteritis and systemic infections that may lead to hospitalizations and fatalities (Holt et al., 2000; Heiman et al., 2015). Over 60% of reported outbreaks in recent years have been transmitted through food (Heiman et al., 2015). While Enteropathogenic *E. coli* (ETEC) is more commonly found in poultry and processed poultry products than Shiga-toxin-producing *E. coli* (STEC), both types remain significant concerns for the poultry industry (Alonso et al., 2012).

2.4 Other bacterial contaminants

Not all bacterial contaminants on poultry carcasses are foodborne pathogens (Marmion et al., 2021). Numerous genera have been identified as part of the microbial consortia on poultry carcasses (Marmion et al., 2021). Traditionally, identification relied on culturing through nonselective plating and enumeration (Feye et al., 2020a). However, 16S rRNA gene amplicon sequencing has allowed for a comprehensive characterization of the poultry processing microbiome (Feye et al., 2020a; Marmion et al., 2021; Ricke et al., 2017, 2022).

Few studies have captured a microbial map across all the stages in poultry processing, particularly the overall microbial load of carcasses and microbiota changes during the initial processing stages (processing stages are discussed in more detail in the next section). Handley et al. (2018) sampled several re-hang, pre-chill, and post-chill slaughterhouses to create a bio-map of the first processing stage. They found that as the carcasses moved from re-hang to pre-chill and then to post-chill, there was a reduction in the total aerobic load and *Enterobacteriaceae* at each stage (Handley et al., 2018). Additionally, the microbiota composition of rinsates (fluids taken from carcasses immersed in buffered solutions for microbial analysis) shifted as the birds exited the post-chiller (Handley et al., 2018). Although the microbial load in the rinsates was almost undetectable by the post-chiller stage, it is essential to remember that processing is not complete until carcasses are packaged (Handley et al., 2018).

3 Poultry processing operations in the United States

In the United States, poultry is the top meat choice per person (NCC, 2020). The US is the world's largest poultry producer, raising billions of birds, and the world's highest per capita chicken consumer (NCC, 2020). The US broiler sector exports over 10% of its output, making it a leading exporter. The poultry industry in the US includes large, federally inspected companies that are vertically integrated, managing production from start to finish (NCC, 2020). These poultry firms produce different broiler sizes based on the intended use of the carcass, such as for further processing or whole use. The three principal broiler sizes are Cornish hens, mid-weight, and large broilers (Wages et al., 2019). While these birds may come from the same genetic line, their sizes are mainly determined by age. All broilers in the US go through a standardized processing method, turning poultry into meat and various processed products.

3.1 Live haul and live hang

Before reaching the processing plant, a crew of employees catches the birds, often using commercial mechanical chicken-catching systems (Ramasamy et al., 2004). These systems are believed to be more cost-effective and welfare-friendly, reducing leg bruising by nearly ten percent on average compared to manual catching (Lacy and Czarick, 1998). After being caught, poultry are transported in crates via tractor-trailers to the processing facility. Traditionally, these crates are open to the environment but, in winter, external coverings are added around the trailer's perimeter. Temperature control is crucial for transport lasting more than an hour to reduce death upon arrival (DOA) (Mitchell and Kettlewell, 1998; Kettlewell et al., 2000). The summer months pose the most concern as methods to keep the birds at proper temperatures are limited (Ritz et al., 2005). Many plants have loading areas where large box fans, sometimes with pressurized spray nozzles, surround the trailer upon arrival at the processing plants. These modified areas hold the birds until they are moved to the live hang area at the back of the facility.

Using modules instead of coop systems in US commercial processing reduces the need for manual labor during off-loading at the processing facility (Shackelford et al., 1981). Once broilers arrive, the module is typically released over a sizeable circular conveyor belt. Workers manually place the birds' hocks into overhead shackles in the dark. Blue light, rather than white light, is used to reduce stress and prevent pale, soft, and exudative (PSE) breast meat, as birds are photosensitive (Barbosa et al., 2013).

A significant concern during live haul and live hang is the stress on the birds and its impact on yield and hygiene. Numerous efforts aim to reduce stressors on the farm and during transport. While methods have been adjusted and diets supplemented with beneficial organisms or substrates, there is still room for improvement (Mitchell and Kettlewell, 1998; Ghareeb et al., 2008). Due to the stress during transportation and live hang, enteric foodborne pathogens can become problematic (Whyte et al., 2001). Pathogens such as *Salmonella* and *Campylobacter* detected in poultry feces can serve as potentially significant causes of foodborne illness in humans if carried through to the final poultry retail meat product (CDC, 2020). Cleaning transportation equipment and reducing stress in poultry during transport and live hang is therefore vital to minimizing pathogen transmission further down the processing chain (Ramesh et al., 2002; Berrang and Northcutt, 2005).

3.2 Slaughter

After the live hang, broilers are suspended upside down in shackles and transported via a conveyor belt system. Their heads are then submerged in a saline solution with an electrical current, causing unconsciousness and cardiac arrest (Savenije et al., 2000; Lambooij and Hindle, 2018). Typically, this electrical stunning induces epileptiform brain trauma because of brain injury (Savenije et al., 2000). Following electrical stunning, birds are euthanized through exsanguination and allowed to bleed out briefly before entering the scalding tank (Savenije et al., 2000). Scalding tanks feature a counterflow current, a triple-pass system (three separate tanks), and complete carcass immersion. Scalding loosens the feather follicles from the bird's skin, preparing the bird for feather removal by a picker. Scalding conditions vary by processing facility but generally involve either a soft scald or hard scald, depending on the duration and temperature of the process (McKee et al., 2008). Due to the high abundance of organic matter on the feathers from feces and the surrounding environment, bacterial levels can be high on both the carcasses and feathers (Barnes, 1975; Wilkerson et al., 1961; Lillard, 1989, 1990; Kotula and Pandya, 1995). Various additives have been considered to reduce bacterial loads. One method for lowering bacteria during or after scalding is using pre-scald brushes to reduce organic matter on carcass surfaces (Pacholewicz et al., 2016). Pacholewicz et al. (2016) demonstrated a significant reduction of *E. coli* and *Enterobacteriaceae*, although these reductions were not biologically substantial.

After the scalder, carcasses go through the picker, where defeathering occurs. The picker consists of rubber finger-like projections that vibrate to remove the feathers. The rubber fingers are commonly associated with bacterial cross-contamination and biofilm formation (Lillard, 1986; Arnold and Silvers, 2000; Hinton et al., 2004; Arnold, 2007; Arnold and Yates, 2009). Since the scalder loosens the feather follicles and can damage the intact skin, bacteria may reside beneath the skin's surface and proliferate (Thomas and McMeekin, 1980; Kim et al., 1993; Allen et al., 2003a, 2003b). Arnold (2007) found that parts of the picker can become rough and serve as a reservoir for biofilm formation, making it challenging to remove and eliminate bacterial populations and reduce cross-contamination. Antimicrobial intervention must occur before picking to improve sanitation procedures and pathogen control (Arnold, 2007).

Evidence indicates that scalding and defeathering can significantly contribute to carcass contamination (Dickens and Whittemore, 1997; Berrang et al., 2001; Allen et al., 2003a, 2003b; Bunic and Sofos, 2012). For example, feather follicles can serve as a reservoir for microbial contaminants during later processing stages, leading to cross-contamination (Zhang et al., 2020). Wages et al. (2019) found that the size of the bird at the processing plant influences the microbiota composition of poultry rinsates at post-scald and post-pick and affects the total aerobic bacteria and Enterobacteriaceae of rinsates. Significantly lower diversity and richness were observed within and between the carcass rinsates sampled at post-pick than post-scald (Wages et al., 2019). Post-scald and post-pick also affected the microbiota's overall composition, with fewer Proteobacteria and Firmicutes in the rinsates collected at the post-scald location (Wages et al., 2019). There remains a need for the ongoing exploration of microbial interventions during and after slaughter.

3.3 Primary processing: evisceration, chilling and cut-up

After the picker, the hocks are removed and re-hung on the evisceration floor, separated by a wall from the kill floor. The first step in evisceration is typically the removal of the preen gland. The bird's vent is removed ('buttonhole'), and the abdominal cavity is opened and widened. The eviscerate is subsequently extracted. In the case of larger poultry processing plants, the US Department of Agriculture Food Safety Inspection Service (USDA-FSIS) now has a network of online and offline inspection program personnel able to inspect eviscerate matched on a shackle with the carcass. If the eviscerated carcass passes FSIS inspection, it proceeds to have the remaining organs removed.

At this stage, it is crucial to prevent the digesta contained within the GIT from contaminating the carcass or equipment. Since poultry digesta often harbors pathogens harmful to humans, evisceration is a highly regulated process. *Salmonella* and *Campylobacter* primarily reside within the GIT of broilers and can contaminate the carcass through feces and GIT content (Hue et al., 2011; Pacholewicz et al., 2016). The abundance of *Campylobacter* in the ceca was observed to be proportional to the abundance of *Campylobacter* on the carcass's skin (Hue et al., 2011).

Based on this association between pathogens in GIT and on poultry carcasses, inside-out bird washers (IOBW) are commonly used to reduce fecal contamination of the carcass (USDA-FSIS, 1998; Rasekh et al., 2005). According to FSIS standards, carcasses must not be visibly contaminated by fecal matter before entering the pre-chiller. Although these IOBW reduce visible fecal matter, their use alone cannot eliminate pathogens. Northcutt et al. (2003) showed that using IOBW did not sufficiently remove coliforms and *E. coli* from the carcass and that antimicrobials must be added to achieve a significant effect.

Immersion chilling is the most commonly employed chilling method in the US. Immersion chilling is accomplished using three chiller tanks: pre-chiller, main chiller, and post-chiller. The chillers are designed to reduce the temperature of carcasses to 4.4°C within hours (USDA-FSIS, 2014a, 2014b). Following the IOBW stage, carcasses are loaded off their shackles and into the pre-chiller. The primary purpose of the pre-chiller is to reduce the organic load on carcasses by lowering their temperature before they enter the main chiller (Barbut, 2002; James et al., 2006). Maintaining the pre-chiller at low temperatures improves water quality and reduces the microbial burden on carcasses (Blevins et al., 2020). The main chiller consists of a large vat of chilled water, transporting the birds in a corkscrew motion. The chiller water has been reported to harbor various bacteria (Rothrock et al., 2016).

Historically, chlorine has been used in chillers to reduce the microbial load but, due to poultry's high organic load, free chlorine is quickly consumed (Tsai et al., 1992). More recently, peroxyacetic acid (PAA) has been used in the chiller system to reduce the microbial burden and has been noted to handle the organic load better than chlorine (Bauermeister et al., 2008a). Ultimately, carcasses leave the main chiller and enter the post-chiller for a shorter duration, during which the PAA is applied at higher concentrations to reduce the microbial load (Nagel et al., 2013). The reductions achieved at the post-chiller stage are crucial, as this may be the final opportunity to reduce pathogens using antimicrobial solutions before packaging.

3.4 Secondary processing: deboning and packaging

After carcasses exit the pre-chiller, they are transferred onto a ramp, where employees re-hang them on shackles. The carcasses may be packaged whole as roasters or divided into parts. In tray-pack facilities, the carcass is split in half below the breast but above the thighs on the back. The front half of the carcass is primarily deboned by hand, although some facilities use automated machinery, depending on its effectiveness at deboning without significant yield loss.

The deboning of the front half begins with the frame being placed on a deboning cone and a cut made at the shoulder joint. Next, cuts are made down either side of the breast against the ribcage. The breast is then pulled down, and the wings are cut off. The breast is pulled off the frame, and a "butterfly" cut is made by crisscrossing down the sides of the keel bone. Finally, the tenders are removed. This process is conducted while the cones are moving, involving a team of employees, each performing a part of the process to optimize yield and ensure efficiency. The lower half of the carcass is generally cut into parts by automated machinery and placed into a tray pack, deboned, or further processed. Thighs are deboned mainly by hand with machinery assistance. Due to the potential risk of pathogen introduction during this processing stage, the industry must implement additional mitigation strategies. One possible solution is using short-duration antimicrobial part dips, which can help mitigate pathogens and increase the water-holding capacity of the poultry meat parts (USDA-FSIS, 2015a).

4 Antimicrobials used in processing: general issues

Since pathogens are commonly present in the GIT of poultry, contaminated animals often enter the processing plant, making processing controls more crucial than ever. Antimicrobial washes and sprays are the most used methods to eliminate or mitigate contamination in poultry and poultry products within processing facilities (McKee, 2012). These antimicrobial treatments are applied throughout the initial processing stages to ensure a low bacterial load when entering secondary processing (McKee, 2012).

Various commonly used antimicrobials are utilized in the chiller, post-chiller, spray cabinets, and part dips during poultry processing, as illustrated in Figs 1 and 2. Additional antimicrobials have been investigated for poultry and food industrial applications, including organic acids, phosphates, chlorine, chlorine derivatives, ozone, and hydrogen peroxide solutions (Mead and Thomas, 1973; Khadre et al., 2001; Ricke, 2003; Dincer and Baysal, 2004; Loretz et al., 2010; McKee, 2012). Several mechanisms may contribute to the effectiveness of antimicrobials, such as functional metabolic inhibition, nucleic acid damage, membrane permeability alteration, and protein denaturation (Stearns et al., 2022). Access to multiple antimicrobials with different mechanisms is vital, as some Gram-negative species, such as *Salmonella*, can develop tolerance to organic acids (Shue and Freese, 1973; Ricke, 2003).

Acidifiers are promising antimicrobials for use in poultry processing. Acidifiers used in the poultry industry consist of organic and inorganic acids.

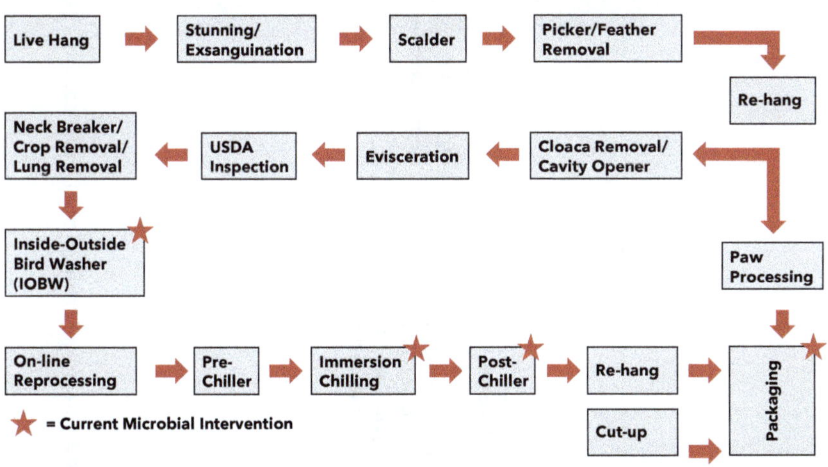

Figure 1 Typical flow of a commercial processing facility and conventional antimicrobial intervention location.

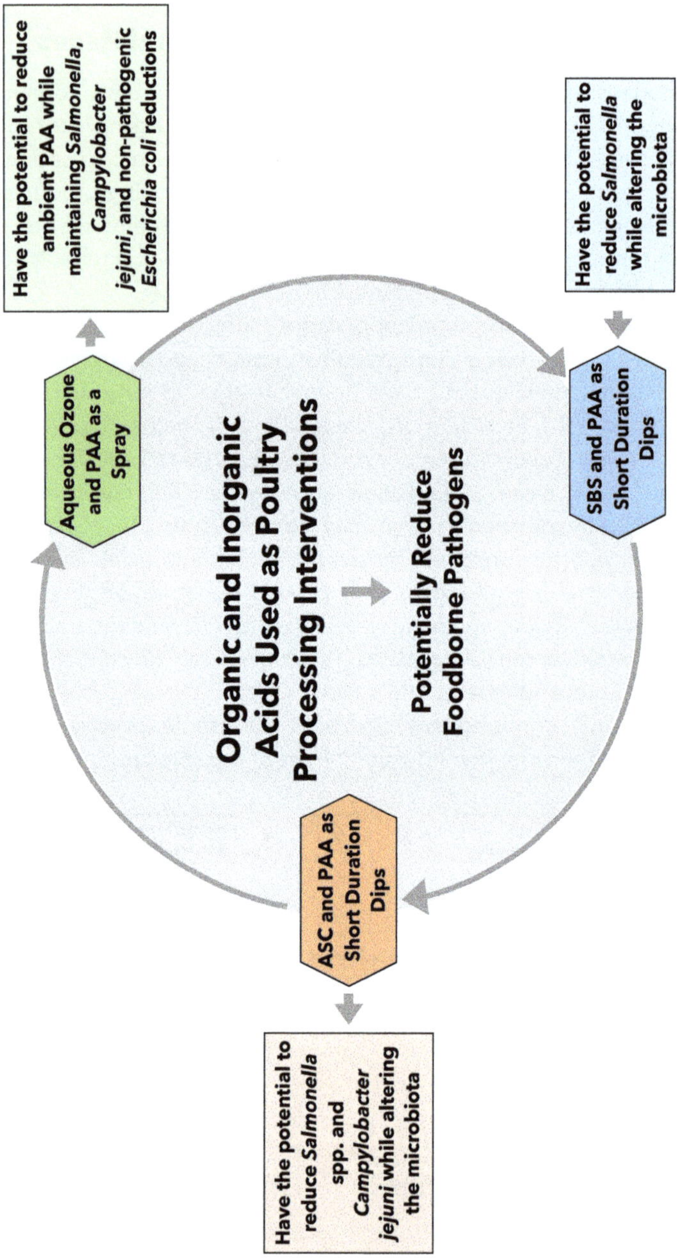

Figure 2 Strategies for antimicrobial applications.

However, due to the buffering effect associated with poultry skin, the efficacy of these antimicrobials can vary (Tan et al., 2014a, 2014b). Furthermore, bacteria residing on the surfaces of feather follicles (Zhang et al., 2020) make it challenging for antimicrobials to disinfect poultry carcasses adequately. Nonetheless, acidifiers alone or in combination with other antimicrobials, such as in short-duration rinses, washes, and dips, have achieved detectable success, with a 1 to 2-log reduction in pathogens being possible (Bauermeister et al., 2008b). Kim et al. (2017), using 16S rRNA gene sequencing, demonstrated that acidifiers could alter the microbiota of carcasses during the initial processing. However, there is limited research on the effect of acidifiers on the microbiota of skin-on poultry parts during secondary processing. Given the potential for recontamination during secondary processing, these applications are becoming more common in processing facilities. The following sections will discuss some of the more common antimicrobials currently used in poultry processing as well as recently proposed compounds (Fig. 2).

5 Individual antimicrobials: peroxyacetic acid (PAA)

Peroxyacetic acid (PAA), peracetic or peroxy acid, has been a germicide for over a century (Stearns et al., 2022). It is comprised of peroxyacetic acid, octanoic acid, acetic acid, hydrogen peroxide, peroxyoctanoic acid, and 1-hydroxyethylidene-1,1-diphosphonic acid (Cano et al., 2021). As a potent oxidizing and bacteriocidal agent, PAA oxidizes cell membranes and their components (Oyarzabal, 2005). This colorless acid is one of the more widely used antimicrobials for decontaminating poultry carcasses at 200 to 2000 ppm (McKee, 2012; Stearns et al., 2022). The US FDA has approved PAA for use up to a maximum concentration of 2000 ppm peroxy acid.

PAA is often employed in poultry chiller systems to reduce microbial load and is noted for better efficacy in the presence of organic loads than chlorine (Bauermeister et al., 2008a; Cano et al., 2021). Carcasses in the main chiller, where they remain for over an hour, are treated with smaller amounts of PAA of approximately 200 ppm (Bauermeister et al., 2008a). In the post-chiller, where carcasses are maintained for a shorter duration, PAA is applied at higher concentrations, 600 to 1000 ppm, to further reduce the microbial load (Nagel et al., 2013). Nagel et al. (2013) demonstrated that PAA in a post-chiller reduced *Salmonella* Typhimurium and *Campylobacter* by 2 Log_{10} CFU/mL, compared to chlorine, which reduced both pathogens to less than one log. These reductions are critical, as the post-chiller is often the last opportunity to reduce pathogens with antimicrobial solutions before packaging.

Over the past decade, PAA has replaced chlorine as the primary sanitizer in processing facilities due to chlorine's limited effectiveness in the presence of high organic loads. Research has shown that 30 ppm of chlorine reduced

Salmonella and *Campylobacter* by 43% and 13%, respectively, while 85 ppm of PAA reduced these same pathogens by 92% and 43% (Bauermeister et al., 2008b). PAA also reduced populations of *Staphylococcus*, *Listeria*, and generic *E. coli* by several log CFU in various food matrices (Brinez et al., 2006). Using 200 ppm in the chiller has reduced *Salmonella* and *Campylobacter* by approximately 1 Log_{10} CFU/mL more than 30 ppm chlorine (Bauermeister et al., 2008a).

Despite the effectiveness of PAA in reducing key pathogens on poultry carcasses during chilling and post-chilling, few studies have investigated its use during secondary processing. Nagel et al. (2013) found that post-chill application of 400 and 1000 ppm PAA for 20 seconds reduced the abundance of *Campylobacter jejuni* and *S.* Typhimurium on inoculated chicken breasts by 2 logs CFU/mL. However, applying 220 ppm PAA as a 15-s dip during second processing did not significantly reduce *Enterobacteriaceae* and coliforms on chicken legs compared to control (Del Río et al., 2007). The lack of significant effect during the second processing could be due to the low microbial load, as the study's authors did not inoculate the poultry parts with pathogens. Kumar et al. (2020) investigated PAA's effectiveness on split, skinless, boneless chicken breasts inoculated with *Salmonella* and *Campylobacter*. They found that various PAA concentrations ranging from 100 to 1000 ppm, used as a spray or a short-duration dip, reduced both populations. Specifically, a 30-s dip in 1000 ppm PAA reduced *Salmonella* and *Campylobacter* by approximately 2 logs compared to the control. Pathogen reduction was less significant when the skin's complex matrix is involved (Tan et al., 2014a, 2014b). Zhang et al. (2018) demonstrated that immersing various poultry parts in 700 or 1000 ppm PAA for 23 s reduced *S.* Typhimurium and *Campylobacter jejuni* by only one log, regardless of the part.

Despite the success of PAA in reducing pathogens during processing, concerns remain about its use in the poultry industry. PAA is highly corrosive and unstable. Although its decay kinetics leave no long-term residues on meat products, it poses potential hazards to human health, being an irritant to the upper respiratory tract, eyes, and skin (Merka and Urban, 1978; Fraser and Thorbinson, 1986; Janssen, 1989a, 1989b; Janssen and Van Doorn, 1994; Walsh et al., 2018). While good personal protective equipment (PPE) can prevent direct contact with the eyes and skin, protecting the upper respiratory tract from PAA vapors remains challenging (American Thoracic Society, 1996). As of 2020, there was no Occupational Safety and Health Administration (OSHA) limit on acute or long-term exposure to PAA for employees, although other regulatory agencies have set exposure limits. PAA has a 15-min Short-Term Exposure Limit (STEL) threshold of 0.04 ppm and an 8-h exposure limit for PAA vapor at 0.17, 0.51, and 1.3 ppm, whereas 0.55 ppm for PAA vapor has been suggested as a limit indicating immediate danger to life or health

(ACGIH, 2016; NAS, 2010; NIOSH, 2015). Recent efforts to combine PAA with agents such as aqueous ozone have shown promise in reducing ambient PAA while retaining its antimicrobial properties against pathogens such as *S.* Typhimurium and *C. jejuni* (Dittoe et al., 2019a).

The primary concerns about PAA are its effects on machinery and employees. There is growing evidence that PAA may select specific bacterial populations, contributing to spoilage and reducing poultry shelf life (Feye et al., 2020a). Dittoe et al. (2024) concluded that shelf life could be extended depending on the antimicrobial or antimicrobial combination. Brown et al. (2024) reported that, while PAA was generally effective as a short-duration antimicrobial spray, its efficacy against foodborne pathogens depended on the type of meat product and pathogen. Additionally, there is some indication that different commercial sources of PAA may qualitatively affect pathogen reduction in broiler carcasses (Feye et al., 2019). In summary, while PAA remains one of the most effective sanitizers for poultry processing, there is a need to develop complementary sanitizers that provide synergistic antimicrobial effects on pathogens without negatively impacting shelf life.

6 Individual antimicrobials: sodium bisulfate salt (SBS)

Similar to organic acids, inorganic acids used as short-duration dips can impact pH and microbial load and potentially induce resistance among pathogens to their antimicrobial effects (Micciche et al., 2019; Feye et al., 2020b; Dittoe et al., 2019b, 2024). For instance, *Salmonella* can tolerate a low environmental pH (Foster and Hall, 1991). However, for this resistance to be effective, *Salmonella* must initiate an acid tolerance response and produce acid-shock proteins (Foster and Hall, 1991; Foster, 1993; Bearson et al., 1997; Ricke, 2003). Although sodium bisulfate (SBS) can lower extracellular pH, *Salmonella* can adjust its cytoplasmic pH to maintain neutrality (Knueven, 1999; Hill et al., 1995). However, this response is energetically expensive and can result in cell death when the pH change is extreme (Hill et al., 1995). Therefore, SBS holds promise as an antimicrobial alternative to other organic and inorganic acids.

Originally, SBS was used in the poultry industry as a Poultry Litter Treatment (PLT®), a commercial litter amendment that acidifies the litter and reduces ammonia volatilization (Payne et al., 2002). SBS has proven effective in mitigating *Salmonella* while reducing ammonia volatilization in litter (Payne et al., 2002). When used as a dietary amendment in the feed, SBS reduced *Salmonella* in the feces and the litter (Ruiz-Feria et al., 2011). The Environmental Protection Agency has recognized SBS as a safe choice of antimicrobial and processing aid during poultry processing (United States EPA, 2018).

Although there is limited research on SBS as a processing aid, it holds potential as a productive alternative or addition to current practices. Dittoe et al. (2019b) utilized a range of 1 to 3% SBS alone or in combination with 200 ppm PAA as 15-s short antimicrobial dips on chicken drumsticks. All treatments reduced inoculated *Salmonella* Enteritidis. Specifically, 3% SBS resulted in the most significant *Salmonella* reduction.

More recent studies using microbiome characterization have focused on shelf life impact of SBS. Dittoe et al. (2024) applied either PAA or SBS alone or in combination as a 15-s dip on chicken wings. They monitored spoilage indicators, aerobic mesophiles (aerobic plate counts, APC), lactic acid bacteria plate counts, and 16S rRNA gene amplicon sequencing over 21 days of storage at 4°C. Based on APC counts, spoilage occurred by day 21; however, wings treated with SBS and SBS + PAA demonstrated a 7-day extended shelf-life compared to those treated with PAA alone. Microbiome analyses indicated an interaction between treatment and time on microbial diversity and composition, with those treated with SBS showing lower richness and evenness compared to controls (no treatment and tap water alone). On the 14th day, wings treated with SBS and SBS + PAA exhibited a lower relative abundance of typical spoilage populations while having a greater abundance of *Bacillus* spp. By day 21, *Bacillus* populations decreased below 10% in wings treated with SBS and SBS + PAA. Dittoe et al. (2024) concluded that SBS extended the shelf-life of chicken wings by seven days compared to untreated wings. On day 14, SBS was more effective in reducing culture-dependent (LAB plate data) and culture-independent (microbiome data) populations than the industry standard PAA. Still, by day 21, these effects had diminished. This study combined culture-based plate data with non-culture microbiome data to provide a comprehensive microbial ecological examination of spoilage progression in raw products.

7 Individual antimicrobials: acidified sodium chlorite (ASC) and organic acid blends

7.1 Acidified sodium chlorite (ASC)

As noted, the most used USDA-approved antimicrobial in poultry processing has been peroxyacetic acid (PAA) (Cano et al., 2021). However, there is growing interest in using acidifiers or weak acids to enhance the efficacy of currently approved antimicrobials. Over the past two decades, acidified sodium chlorite (ASC) has effectively reduced pathogens when used as a dip, spray, or post-chill intervention on chicken carcasses (Kemp et al., 2000, 2001, 2002).

Kemp et al. (2000) treated carcasses with 500, 850, and 1200 ppm of acidified sodium chlorite (ASC) as a spray or a 5-s dip, lowering pH with phosphoric or citric acid. The study found that the abundance of total aerobic bacteria, *E. coli*, and coliforms was most reduced using citric acid to lower pH. Specifically, reductions were 82.9 to 90.7% for total aerobic bacteria, 99.4

to 99.6% for *E. coli*, and 86.1 to 98.5% for total coliforms (Kemp et al., 2000). However, applying ASC as a dip was more effective at reducing the bacterial prevalence on carcasses than when utilized as a spray (Kemp et al., 2000).

The use of ASC in an IOBW showed a reduction in *Salmonella* and *Campylobacter*-contaminated carcasses compared to offline reprocessing (Kemp et al., 2001). *Salmonella* and *Campylobacter* presence was reduced by over 20% on carcasses treated with ASC (Kemp et al., 2001). Zhang et al. (2018) examined multiple acidifiers as 23-s rinses on *Salmonella* and *Campylobacter*-inoculated chicken things, drumsticks, wings, and breasts. The abundance of *Salmonella* and *Campylobacter* was reduced by less than one log by ASC, and the effect was not significantly different from tap water or 30 ppm chlorine.

7.2 Organic acid blends

Although less frequently employed in manufacturing facilities, organic acids and their blends represent viable alternatives to PAA. When applied as entire carcass sprays, organic acids such as lactic acid and acetic acid (1–2%) have successfully diminished bacterial presence on carcasses. However, caution is advised as higher concentrations may cause bleaching of the carcass's skin (Sohaib et al., 2016; Milillo and Ricke, 2010). Notably, at 37°C, the minimum inhibitory concentration (MIC) for sodium citrate and sodium lactate to inhibit *Salmonella* Typhimurium was found to be 1.25% and 2.5%, respectively (Milillo and Ricke, 2010). Citric, malic, and tartaric acids at 150.0 mM concentrations also reduced *E. coli* O157, *Listeria monocytogenes*, and *S.* Typhimurium by over 5, 2, and 4 Log10 CFU/g on chicken breast portions (Over et al., 2009). Chicxide, a commercially available antimicrobial comprising lactic and citric acids, was employed to treat artificially contaminated broiler carcasses (Laury et al., 2009). Application of Chicxide for 5–20 s differed in *Salmonella* reduction when applied as a spray or dip, irrespective of duration (Laury et al., 2009).

Ramirez-Hernandez et al. (2018) explored lactic acid, lactic acid–acetic acid blends, and PAA as antimicrobial sprays for poultry parts to decrease artificially introduced *Salmonella* from chicken thighs. These antimicrobials' effectiveness was evaluated on skin-on and skinless thighs at three temperatures, and the impact of PAA and lactic acid on *Salmonella* abundance on skinless chicken breasts was assessed (Ramirez-Hernandez et al., 2018). The study by Ramirez-Hernandez et al. (2018) indicated that lactic acid and buffered lactic acid were the most effective in reducing *Salmonella* on thighs regardless of temperature, and both lactic acid and PAA were effective in lowering Salmonella levels on skinless chicken breasts.

8 Factors affecting antimicrobial efficacy and multiple hurdle approaches to optimizing their use

Antimicrobials that decrease the pH of their surroundings show potential. However, Gram-negative bacteria such as *Salmonella* can develop resistance to organic acids due to lipopolysaccharide (LPS) in their cell walls (Shue and Freeze, 1973). *Salmonella* and related bacteria can also develop tolerance responses to stressful conditions including acidic environments (Ricke, 2003). An additional challenge with organic acids is that foodborne pathogens such as *Campylobacter jejuni* can utilize certain organic acids as substrates for their energy metabolism and biosynthetic activities (Kassem et al., 2013; Hofreuter, 2014). Studies have shown that *C. jejuni* can use organic acids as energy and growth sources (Line et al., 2010). Therefore, it is essential to consider microbial dynamics and metabolism when using a sole antimicrobial in processing, such as organic acids.

Though less common, inorganic acids can induce resistance in pathogens such as *Salmonella* to low environmental pH (Foster and Hall, 1991). *Salmonella* can lower their cytoplasmic pH to maintain a neutral state when experiencing a mild decrease in extracellular pH, but this response is metabolically demanding and can result in cell death (Hill et al., 1995). Hence, to prevent resistance development against common processing antimicrobials, novel antimicrobials and combinations need to be evaluated and developed. These antimicrobials should also be implemented at multiple points within the processing chain to reduce pathogen prevalence in raw poultry.

Cross-contamination from equipment to skin-on and skinless parts during secondary processing is also a concern. Pathogens like *Salmonella* can form biofilms on processing equipment, leading to rapid contamination of poultry meat (Carrasco et al., 2012). Despite constant cleaning and sanitization efforts, biofilms remain a significant issue in processing environments (Carrasco et al., 2012). Shi et al. (2019a) examined biofilm-forming genes, *bscA* and *csgD*, in multiple *Salmonella* strains isolated from US poultry processing facilities and found high biofilm-forming capabilities across all serovars examined. Extensive research has been conducted on biofilm formation among *Salmonella* strains (Obe et al., 2018; Dhakal et al., 2019; Nannapaneni et al., 2019; Shi et al., 2019a, 2019b). Further mitigation strategies are crucial, given the risk of pathogen reintroduction during secondary processing. One potential solution is short-duration antimicrobial dips lasting 10–30 s, which could reduce pathogens and increase the water-holding capacity of the parts (USDA-FSIS, 2015a).

Over the past 20 years, various chemical and physical interventions have been implemented during poultry processing to reduce microbial load before second processing (Leistner, 2000; Ricke et al., 2005). Critical control points where microbial burden can increase necessitate multiple interventions throughout evisceration and the first steps of poultry processing (Stopforth et al., 2007; Zweifel and Stephan, 2012; Sofos et al., 2013). The "multi-hurdle" approaches can significantly reduce the risk of recontamination/cross-contamination during harvest (Schuler and Badenhop, 1972; Mead et al., 1994; Stopforth et al., 2007; Carrasco et al., 2012). Multi-hurdle interventions are employed in the scalder, IOBW, and chiller systems, which have been extensively studied (Schuler and Badenhop, 1972; Mead, 2004; Stopforth et al., 2007; McKee et al., 2008; Zweifel and Stephan, 2012; Nagel et al., 2013).

9 Conclusion and future trends

While there is growing interest in using acidifiers to reduce common foodborne pathogens, limited research exists on their impact on poultry microbiota, especially when poultry parts are inoculated with pathogens. Future research should focus on understanding the effects of these antimicrobials on the microbiota of poultry carcasses and parts under high microbial or pathogen loads. Understanding microbial dynamics will help determine effective multi-hurdle approaches to maintaining a stable and healthy microbiota under various conditions.

Establishing a link between the host microbiota's composition and the resulting meat's microbiota is necessary to better prevent pathogens from entering processing facilities. Bio-mapping or tracking microbiota changes from hatchery through grow-out and processing will help identify critical control points and indicator organisms. Introducing microbiome 16S rRNA gene sequencing to monitor changes in the overall carcass microbial consortium through processing stages can provide an in-depth characterization of microbial taxa present and predict shelf life. Microbial assessment may be crucial for assessing antimicrobial choices that balance effective pathogen reduction without compromising shelf life.

10 Acknowledgements

A part of this chapter was modified from a Ph. D thesis submitted by Dana Dittoe to the University of Arkanus, Fayetteville, AR.

11 Where to look for further information

- Cano, C., Meneses, V. and Chaves, B. D. 2021. Application of peroxyacetic acid for decontamination of raw poultry products and comparison to other commonly used chemical antimicrobial interventions: a review. *J. Food Prot.* 84: 1772–1783. https://doi.org/10.43115/JFP-21-107.

- Carrasco, E., Morales-Rueda, A. and García-Gimeno, R. M. 2012. Cross-contamination and recontamination by *Salmonella* in foods: a review. *Food Res. Int.* 45: 545–556. https://doi.org/10.1016/j.foodres.2011.11.004.
- Mani-López, E., García, H. S. and López-Malo, A. 2012. Organic acids as antimicrobials to control *Salmonella* in meat and poultry products. *Food Res. Int.* 45: 713–721. https://doi.org/10.1016/j.foodres.2011.040043.
- Marmion, M., Ferone, M. T., Whyte, P. and Scannell, A. G. M. 2021. The changing microbiome of poultry meat; from farm to fridge. *Food Microbiol.* 99: 103823. https://doi.org/10.1016/j.fm.2021.103823.
- Owens, C. M., Alvarado, C. and Sams, A. R. (Eds.). 2010. *Poultry Processing*, 2nd ed. Taylor & Francis, London, UK, 448 pp.
- Stearns, R., Freshour, A. and Shen, C. 2022. Literature review for applying peroxyacetic acid and/or hydrogen peroxide to control foodborne pathogens on food products. *J. Agric. Food Res.* 10: 100442. https://doi.org/10.1016/j.jafr.2022.100442.
- Thames, H. T. and Sukumaran, A. T. 2022. A review of *Salmonella* and *Campylobacter* in broiler meat: emerging challenges and food safety measures. *Foods*. 9: 776. https://doi.org/10.3390/foods9060776.

12 References

Allen, V. M., Hinton, M. H., Tinker, D. B., Gibson, C., Mead, G. C. and Wathes, C. M. 2003a. Microbial cross-contamination by airborne dispersion and contagion during defeathering of poultry. *Br. Poultry Sci.* 44: 567–576. https://doi.org/10.1080/00071660310001616183.

Allen, V. M., Tinker, D. B., Hinton, M. H. and Wathes, C. M. 2003b. Dispersal of micro-organisms in commercial defeathering systems. *Br. Poultry Sci.* 44: 53–59. https://doi.org/10.1080/0007166031000085436.

Alonso, Z. M., Lucchesi, P. M. A., Rodríguez, E. M., Parma, A. E. and Padola, N. L. 2012. Enteropathogenic (EPEC) and Shiga toxigenic *Escherichia coli* (STEC) in broiler chickens and derived products at different retail stores. *Food Control* 23(2): 351–355. https://doi.org/10.1016/j.foodcont.2011.07.030.

American Conference of Governmental Industrial Hygienists, ACGIH®. 2016. *Annual TLVs® (Threshold Limit 4 Values) and BEIs® (Biological Exposure Indices) booklet*. Cincinnati, OH: ACGIH® Signature Publications.

American Thoracic Society. 1996. Respiratory protection guidelines. American Thoracic Society: Medical Section of the American Lung Association. Accessed 6 November 2020. https://www.thoracic.org/statements/resources/eoh/resp1-13.pdf.

Arnold J.W. and S. Silvers. 2000. Comparison of poultry processing equipment surfaces for susceptibility to bacterial attachment and biofilm formation. *Poultry Sci.* 79(8): 1215–1221. https://doi.org/10.1093/ps/78.8.1215.

Arnold, J. W. 2007. Bacterial contamination on rubber picker fingers before, during, and after processing. *Poultry Sci.* 86: 2671–2675. https://doi.org/10.3382/ps.2007-00187.

Arnold, J. W. and Yates, I. E. 2009. Interventions for control of *Salmonella*: clearance of microbial growth from rubber picker fingers. *Poultry Sci*. 88: 1292–1298. https://doi.org/10.3382/ps.208-00391.

Awad, W. A., Dublecz, F., Hess, C., Dublecz, K., Khayal, B., Aschenbach, J. R. and Hess, M. 2016. *Campylobacter jejuni* colonization promotes the translocation of *Escherichia coli* to extra-intestinal organs and disturbs the short-chain fatty acids profiles in the chicken gut. *Poultry Sci*. 95(10): 2259–2265 https://doi.org/10.3382/ps/pew151.

Awad, W. A., Smorodchenko, A., Hess, C. et al. 2015. Increased intracellular calcium level and impaired nutrient absorption are important pathogenicity traits in the chicken intestinal epithelium during *Campylobacter jejuni* colonization. *Appl. Microbiol. Biotechnol*. 99: 6431–6441 https://doi.org/10.1007/s00253-015-6543-z.

Awad, W. A., Aschenbach, J. R., Ghareeb, K., Khayal, B., Hess, C. and Hess, M. 2014. *Campylobacter jejuni* influences the expression of nutrient transporter genes in the intestine of chickens. *Vet. Microbiol*. 172(1–2): 195–201. https://doi.org/10.1016/j.vetmic.2014.04.001.

Barbosa, C. F., de Carvalho, R. H., Rossa, A., Soares, A. L., Coró, F. A. G., Shimokomaki, M. and Ida, E. I. 2013. Commercial preslaughter blue light ambience for controlling broiler stress and meat qualities. *Braz. Arch. Biol. Technol*. 56(5): 817–821. https://doi.org/10.1590/S1516-89132013000500013.

Barbut, S. 2002. *Poultry Products Processing: An Industry Guide* (first ed.), CRC Press, Boca Raton, FL, USA.

Barnes, E. M. 1975. The microbiological problems of sampling poultry carcass. Pages 23(1–8) in: *The Quality of Poultry Meat* (2nd European Symposium on Poultry Meat Quality). Oosterbeck, The Netherlands.

Barrow, P. A., Jones, M. A., Smith, A. L. and Wigley, P. 2012. The long view: *Salmonella*–the last forty years. *Avian Pathol*. 41(5): 413–420. https://doi.org/10.1080/03079457.2012.718071.

Bauermeister, L. J., Bowers, J. W. J., Townsend, J. C., McKee, S. R. 2008a. The microbial and quality properties of poultry carcasses treated with peracetic acid as an antimicrobial treatment. *Poultry Sci*. 87(11): 2390–2398. https://doi.org/10.3382/ps.2008-00087.

Bauermeister, L. J., Bowers, J. W. J., Townsend, J. C., McKee, S. R. 2008b. Validating the efficacy of peracetic acid mixture as an antimicrobial in poultry chillers. *J. Food Prot*. 71(6): 1119–1122. https://doi.org/10.4315/0362-028X-71.6.1119.

Bearson, S., Bearson, B. and Foster, J. 1997. Acid stress responses in enterobacteria. *FEMS Microbiol. Lett*. 147: 173–180. https://doi.org/10.1111/j.1574-6968.1997.tb10238.x.

Berrang, M. E. and Northcutt, J. K. 2005. Use of water spray and extended drying time to lower bacterial numbers on soiled flooring from broiler transport coops. *Poultry Sci*. 84(11): 1797–1801. https://doi.org/10.1093/ps/84.11.1797.

Berrang, M. E., Buhr, R. J., Cason, J. A. and Dickens, J. A. 2001. Broiler carcass contamination with *Campylobacter* from feces during defeathering. *J. Food Prot*. 64(12): 2063–2066. https://doi.org/10.4315/0362-028x-64.12.2063.

Blevins, R. E., Dittoe, D. K., Feye, K. M., Bench, L., Bench, B. J. and Ricke, S. C. 2020. *Campylobacter* and *Salmonella* levels and chemical composition of commercial whole bird carcass pre-chiller rinses. *J. Agric. Food Res*. 2: 100041. https://doi.org/10.1016/j.jafr.2020.100041.

Brinez, W. J., Roig-Sagues, A. X., Hernandez Herrero, M. M., Lopez-Pedemonte, T. and Guamis, B. 2006. Bactericidal efficacy of peracetic acid in combination with hydrogen peroxide against pathogenic and non pathogenic strains of *Staphylococcus* spp.,

Listeria spp. and *Escherichia coli*. *Food Control* 17: 516–521. https://doi.org/10.1016/j.foodcont.2005.02.014.

Brown, J. A., Dittoe, D. K., Austin, C. B., Mikkelson, K. B., Hughes, B. and Ricke, S.C. 2024. Utilization of microbiota analyses to assess the antimicrobial efficacy of peracetic acid on meat products. *Meat Muscle Biol.* 8(1): 16818. https://doi.org/10.22175/mmb.16818.

Buncic, S. and Sofos, J. 2012. Interventions to control *Salmonella* contamination during poultry, cattle and pig slaughter. *Food Res. Int.* 45(2): 641–655. https://doi.org/10.1016/j.foodres.2011.10.018.

Cano, C., Meneses, V. and Chaves, B. D. 2021. Application of peroxyacetic acid for decontamination of raw poultry products and comparison to other commonly used chemical antimicrobial interventions: a review. *J. Food Prot.* 84: 1772–1783. https://doi.org/10.43115/JFP-21-107.

Carrasco, E., A. Morales-Rueda, and R. M. García-Gimeno. 2012. Cross-contamination and recontamination by *Salmonella* in foods: A review. *Food Res. Int.* 45: 545–556. doi:10.1016/j.foodres.2011.11.004.

Center for Disease Control, CDC. 2006. Preliminary FoodNet data on the incidence of infection with pathogens transmitted commonly through food-10 states, United States, 2005. *Morb. Mort. Wkly. Rep.* 55: 393–395.

Center for Disease Control, CDC. 2014a. Incidence and trends of infection with pathogens transmitted commonly through food – foodborne diseases active surveillance network, 10 U.S. Sites, 2006–2013. *Morb. and Mort. Wkly. Rep.* 63(15): 328–332.

Center for Disease Control, CDC. 2014b. *Campylobacter*. Center for Disease Control and Prevention. www.cdc.gov/foodsafety/diseases.

Center for Disease Control, CDC. 2017. *Foodborne Disease Active Surveillance Network (Foodnet): Foodnet 2015 Surveillance Report (Final Data).* U.S. Department of Health and Human Services, CDC, Atlanta, Georgia.

Center for Disease Control, CDC. 2020. Chicken and Food Poisoning. Accessed 30 October 2020. https://www.cdc.gov/foodsafety/chicken.html.

Chai, S. J., Cole, D., Nisler, A. and Mahon, B. E. 2017. Poultry: the most common food in outbreaks with known pathogens, United States, 1998–2012. *Epidemiol. Infect.* 145(2): 316–325. https://doi.org/10.1017/S0950268816002375.

Chen, H. M., Wang, Y., Su, L. H. and Chiu, C. H. 2013. Nontyphoid *Salmonella* infection: microbiology, clinical features, and antimicrobial therapy. *Pediatr. Neonatol.* 54(3): 147–152. https://doi.org/10.1016/j.pedneo.2013.01.010.

Cosby, D. E., Cox, N. A., Harrison, M. A., Wilson, J. L., Buhr, R. J. and Fedorka-Cray, P. J. 2015. *Salmonella* and antimicrobial resistance in broilers: a review. *J. Appl. Poultry Res.* 24(3): 408–426. https://doi.org/10.3382/japr/pfv038.

Crump, J. A., Sjölund-Karlsson, M., Gordon, M. A. and Parry, C. M. 2015. Epidemiology, clinical presentation, laboratory diagnosis, antimicrobial resistance, and antimicrobial management of invasive *Salmonella* infections. *Clin. Microbiol. Rev.* 28(4): 901–937. https://doi.org/10.1128/CMR.00002-15.

Del Río, E., Panizo-Morán, M., Prieto, M., Alonso-Calleja, C. and Capita, R. 2007. Effect of various chemical decontamination treatments on natural microflora and sensory characteristics of poultry. *Int. J. Food. Microbiol.* 115(3): 268–280. https://doi.org/10.1016/j.ijfoodmicro.2006.10.048.

Dhakal, J., Sharma, C. S., Nannapaneni, R., Mcdaniel, C. D., Kim, T. and Kiess, A. 2019. Effect of chlorine-induced sublethal oxidative stress on the biofilm-forming ability of *Salmonella* at different temperatures, nutrient conditions, and substrates. *J. Food Prot.* 82(1): 78–92. https://doi.org/10.4315/0362-028X.JFP-18-119.

Dickens, J. A. and Whittemore, A. D. 1997. Effects of acetic acid and hydrogen peroxide application during defeathering on the microbiological quality of broiler carcasses prior to evisceration. *Poultry Sci.* 76(4): 657–660. https://doi.org/10.1093/ps/76.4.657

Dincer, A. H. and Baysal, T. 2004. Decontamination techniques of pathogen bacteria in meat and poultry. *Crit. Rev. Microbiol.* 30: 197–204. https://doi.org/10.1080/10408410490468803.

Dittoe, D. K., Feye, K. M., Ovall, C., Thompson, H. A. and Ricke, S. C. 2024. Exploiting the microbiota of organic and inorganic acid-treated raw poultry products to improve shelf-life. *Front. Microbiol.* 15: 1348159. https://doi.org/10.3389/fmicb.2024.1348159

Dittoe, D. K., Feye, K. M., Peyton, B., Worlie, D., Draper, M. J. and Ricke, S. C. 2019a. The addition of ViriditecTM to peracetic acid as an antimicrobial spray increases air quality while maintaining *Salmonella* Typhimurium, non-pathogenic *Escherichia coli*, and *Campylobacter jejuni* reduction on whole carcasses. *Front. Microbiol.* 9: 3180. https://doi.org/3389/fmmicb.2018.03180.

Dittoe, D. K., Atchley, J. A., Feye, K. M., Lee, J., Knueven, C. J. and Ricke, S. C. 2019b The efficacy of sodium bisulfate salt (SBS) alone and combined with peracetic acid (PAA) as an antimicrobial on whole chicken drumsticks artificially inoculated with *Salmonella* Enteritidis. *Front. Vet. Sci.* 6: 6. https://doi.org/10.3389/fvets.2019.00006.

Feye, K. M., Thompson, D. R., Rothrock, Jr., M. J., Kogut, M. H., and Ricke, S. C. 2020a. Poultry processing and the application of microbiome mapping. *Poultry Sci.* 99: 678–688. https://doi.org/10.1016/j.psj.2019.12.019.

Feye, K. M., Miccichi, A. C., Rubinelli, P. M., Knueven, C. J., Thompson, D. R., Kogut, M. H. and Ricke, S. C. 2020b. The effect of acid sanitizers on the microbiome of re-use water. *Front. Sust. Food. Syst.* 4: 85. https://doi.org/10.3389/fsufs.2020.00085.

Feye, K. M., Dittoe, D. K., Shi, Z., Woitte, J., Owens, C. M., Kogut, M. H. and Ricke, S. C. 2019. The reduction of pathogen load on Ross 708 broilers when using different sources of commercial peracetic acid sanitizers in a pilot processing plant. *Microorganisms* 7: 503. https://doi.org/10.3390/microorganisms7110503.

Food and Drug Administration (FDA). 2011. *Public Law 111–353 in – 111th Congress: FDA Food Safety Modernization Act*. FDA, Washington, DC.

Foster, J. W. 1993. The acid tolerance response of *Salmonella typhimurium* involves the synthesis of key acid shock proteins. *J. Bacteriol.* 75: 1981–1987. https://doi.org/10.1128/jb.175.7.1981-1987.1993.

Foster, J. W. and Hall, H. K. 1991. Inducible pH homeostasis and the acid tolerance response of *Salmonella typhimurium*. *J. Bacteriol.* 173: 5129–5135. https://doi.org/10.1128/jb.173.16.5129-5135.1991.

Fraser, J. A. L. and Thorbinson, A. 1986. *Fogging Trials with Tenneco Organics Limited (30th June, 1986) at Collards Farm*. Solvay Interox, Warrington, UK.

Ghareeb, K., Awad, W. A., Nitsch, S., Abdel-Raheem, S. and Böhm, J. 2008. Effects of transportation on stress and fear responses of growing broilers supplemented with prebiotic or probiotic. *Int. J. Poultry Sci.* 7(7): 676–685. https://doi.org/10.3923/ijps.2008.678.685.

Handley, J. A., Park, S. H., Kim, S. A. and Ricke, S. C. 2018. Microbiome profiles of commercial broilers through evisceration and immersion chilling during poultry slaughter and the identification of potential indicator microorganisms. *Front. Microbiol.* 9: 345. https://doi.org/10.3389/fmicb.2018.00345.

Heiman, K. E., Mody, R. K., Johnson, S. D., Griffin, P. M. and Gould, L. H. 2015. *Escherichia coli* O157 outbreaks in the United States, 2003–2012. *Emerg. Infect. Dis.* 21(8): 1293–1301. https://doi.org/10.3201/eid2108.141364.

Hill, C., O'Driscoll, B. and Booth, I. 1995. Acid adaptation and food poisoning microorganisms. *Int. J. Food Microbiol.* 28: 245–254.

Hinton A., Cason, J. A. and Ingram, K. D. 2004. Tracking spoilage bacteria in commercial poultry processing and refrigerated storage of poultry carcasses. *Int. J. Food Microbiol.* 91(2): 155–165. https://doi.org/10.1016/S0168-1605(03)00377-5.

Hofreuter, D. K. 2014. Defining the metabolic requirements for the growth and colonization capacity of *Campylobacter jejuni*. *Front Cell Infect. Microbiol.* 4: 137. https://doi.org/10.3389/fcimb.2014.00137.

Holt, J. G., Krieg, N. R., Sneath, P. H. A., Staley, J. T. and Williams, S. T. 2000. *Bergey's Manual® of Determinative Bacteriology: Ninth Edition*. Lippincott Williams and Wilkins, Philadelphia, PA.

Hue O., Allain, V., Laisney, M. J., Le Bouquin, S., Lalande, F., Petetin, I., Rouxel, S., Quesne, S., Gloaguen, P. Y., Picherot, M., Santolini, J., Bougeard, S., Salvat, G. and Chemaly, M. 2011. *Campylobacter* contamination of broiler caeca and carcasses at the slaughterhouse and correlation with *Salmonella* contamination. *Food Microbiol.* 28(5): 862–868. https://doi.org/10.1016/j.fm.2010.11.003.

Humphrey, S., Chaloner, G., Kemmett, K., Davidson, N., Williams, N., Kipar, A., Humphrey, T. and Wigley, P. 2014. *Campylobacter jejuni* is not merely a commensal in commercial broiler chickens and affects bird welfare. *mBio* 5(4): e01364-14. https://doi.org/10.1128/mBio.01364-14.

Hulebak, K. L. and Schlosser, W. 2002. Hazzard analysis and critical control point (HAACP) history and conceptual overview. *Risk Anal.* 22: 547–552. https://doi.org/10.1111/0272-4332.00038.

James, C., Vincent, C., de Andrade Lima, T. I. and James, S. J. 2006. The primary chilling of poultry carcasses–a review. *Int. J. Refrig.* 29(6): 847–862. https://doi.org/10.1016/j.ijrefrig.2005.08.003.

Janssen, P. J. M. 1989a. *Acute Inhalation Toxicity Studies of Proxitane 1507 in Male Rats (I)*. Report No. S. 8906, Int. Doc. No. 56645/25/89. Duphar B.V., Weesp, The Netherlands, and Solvay, Brussels, Belgium.

Janssen, P. J. M. 1989b. *Acute Inhalation Toxicity Studies of Proxitane 1507 in Male Rats (II)*. Report No. S. 8908, Int. Doc. No. 56645/34/89. Duphar B.V., Weesp, The Netherlands, and Solvay, Brussels, Belgium.

Janssen, P. J. M. and van Doorn, W. M. 1994. *Acute Inhalation Toxicity Study with Proxitane AHC in Male and Female Rats*. Report No. S. 9408, Int. Doc. No. 56345/48/94. Duphar B.V., Weesp, The Netherlands, and Solvay, Brussels, Belgium.

Kassem, I. I., Chandrashekhar, K. and Rajashekara, G. 2013. Of energy and survival incognito: a relationship between viable but non-culturable cells formation and inorganic polyphosphate and formate metabolism in *Campylobacter jejuni*. *Front Microbiol.* 4: 183. https://doi.org/10.3389/fmicb.2013.00183.

Kemp, G. K., Aldrich, M. L. and Waldroup, A. L. 2000. Acidified sodium chlorite antimicrobial treatment of broiler carcasses. *J. Food Prot.* 63(8): 1087–1092. https://doi.org/10.4315/0362-028x-63.8.1087.

Kemp, G. K., Aldrich, M. L., Guerra, M. L. and Schneider, K. R. 2001. Continuous online processing of fecal- and ingesta-contaminated poultry carcasses using an acidified sodium chlorite antimicrobial intervention. *J. Food Prot.* 64(6): 807–812. https://doi.org/10.4315/0362-028x-64.6.807

Kemp, G. K. and Schneider, K. R. 2002. Reduction of *Campylobacter* contamination on broiler carcasses using acidified sodium chlorite. *Dairy Food Environ. Sanit.* 22: 599–606.

Kettlewell, P. J., Hoxey, R. P. and Mitchell, M. A. 2000. Heat produced by broiler chickens in a commercial transport vehicle. *J. Agric. Engng. Res.* 75: 315–326.

Khadre, M. A., Yousef, A. E. and Kim, J.-G. 2001. Microbiological aspects of ozone applications in food: a review. *J. Food Sci.* 66(9): 1242–1252. https://doi.org/10.1111/j.1365-2621.2001.tb15196.x.

Kim J. W., Knabel, S. J. and Doores, S. 1993. Penetration of *Salmonella* Typhimurium into turkey skin. *J. Food Prot.* 56: 292–296. https://doi.org/10.4315/0362-028X-56.4.292.

Kim, S. A., Park, S. H., Lee, S. I., Owens, C. M. and Ricke, S. C. 2017. Assessment of chicken carcass microbiome responses during processing in the presence of commercial antimicrobials using a next generation sequencing approach. *Sci. Rep.* 7: 43354. https://doi.org/10.1038/srep43354.

Knueven, C. J. 1999. *Sodium Bisulfate as Acidulant in Foods. U.S. Patent No 5,958,491*. Patent US, and Trademark Office, Washington, DC.

Kotula, K. L. and Pandya, Y. 1995. Bacterial contamination of broiler chickens before scalding. *J. Food Prot.* 58(12): 1326–1329. https://doi.org/10.4315/0362-028X-58.12.1326.

Kumar, S., Singh, M., Cosby, D. E., Cox, N. A. and Thippareddi, H. 2020. Efficacy of peroxy acetic acid in reducing *Salmonella* and *Campylobacter* spp. populations on chicken breast fillets. *Poultry Sci.* 99(5): 2655–2661. https://doi.org/10.1016/j.psj.2019.12.045.

Lacy, M. P. and Czarick, M. 1998. Mechanical harvesting of broilers. *Poultry Sci.* 77: 1794–1797.

Lamb-Rosteski, J. M., Kalischuk, L. D., Inglis, G. D. and Buret, A. G. 2008. Epidermal growth factor inhibits *Campylobacter jejuni*-induced claudin-4 disruption, loss of epithelial barrier function, and *Escherichia coli* translocation. *Infect. Immun.* 76(8): 3390–3398; https://doi.org/10.1128/IAI.01698-07.

Lambooij, B. and Hindle, V. 2018. 4 - Electrical stunning of poultry. In *Food Science, Technology, and Nutrition, Advances of Poultry Welfare*. J. A. Mench, ed. Woodhead Publishing, Cambridge, U.K. https://doi.org/10.1016/B978-0-08-100915-4.00004-X.

Laury, A. M., Alvarado, M. V., Nace, G., Alvarado, C. Z., Brooks, J. C. and Echeverry, A. 2009. Validation of a lactic acid- and citric acid-based antimicrobial product for the reduction of *Escherichia coli* O157:H7 and *Salmonella* on beef tips and whole chicken carcasses. *J. Food Prot.* 72: 2208–2211. https://doi.org/10.4315/0362-028X-72.10.2208.

Leistner, L. 2000. Basic aspects of food preservation by hurdle technology. *Int. J. Food Microbiol.* 55: 181–186.

Lillard, H. S. 1986. Distribution of "attached" *Salmonella* Typhimurium cells between poultry skin and a surface film following water immersion. *J. Food Prot.* 49(6): 449–454. https://doi.org/10.4315/0362-028X-49.6.449.

Lillard, H. S. 1989. Factors affecting the persistence of *Salmonella* during the processing of poultry. *J. Food Prot.* 52(11): 829–832. https://doi.org/10.4315/0362-028X-52.11.829.

Lillard, H. S. 1990. Effect on broiler carcasses and water of treating chiller water with chlorine or chlorine dioxide. *Poultry Sci.* 59(8): 1761–1766. https://doi.org/10.3382/ps.0591761.

Line J. E., Hiett, K., Guard-Bouldin, J. and Seal, B. S. 2010. Differential carbon source utilization by *Campylobacter jejuni* 11168 in response to growth temperature variation. *J. Microbiol. Methods*. 80: 198–202. https://doi.org/10.1016/j.mimet.2009.12.011.

Loretz, M., Stephan, R. and Zweifel, C. 2010. Antimicrobial activity of decontamination treatments for poultry carcasses: a literature survey. *Food Control* 21: 791–804. https://doi.org/10.1016/j.foodcont.2009.11.007.

Majowicz, S. E., Musto, J., Scallan, E., Angulo, F. J., Kirk, M., O'Brien, S. J., et al. 2010. The global burden of nontyphoidal *Salmonella* gastroenteritis. *Clin Infect Dis*. 50(6): 882–889. https://doi.org/10.1086/650733.

Marmion, M., Ferone, M. T., Whyte, P. and Scannell, A. G. M. 2021. The changing microbiome of poultry meat; from farm to fridge. *Food Microbiol.* 99: 103823. https://doi.org/10.1016/j.fm.2021.103823.

McKee S. 2012. *Salmonella* control in poultry processing. In *Proceedings of the American Meat Science Association 65th Annual Reciprocal Meat Conference*. North Dakota State University Fargo, Fargo, SD.

McKee, S. R., Townsend, J. C. and Bilgili, S. F. 2008. Use of a scald additive to reduce levels of *Salmonella* Typhimurium during poultry processing. *Poultry Sci.* 87(8): 1672–1677. https://doi.org/10.3382/ps.2008.-00061.

Mead, G. C. 2004. Microbial hazards in production and processing. Pages 232–257 in Poultry Meat Processing and Quality. G. C. Mead, ed. Woodhead Publishing, Cambridge, UK.

Mead, G. C., Hudson, W. R. and Hinton, M. H. 1994. Use of a marker organism in poultry processing to identify sites of cross contamination and evaluate possible control measures. *Br. Poultry Sci.* 35: 345–354. https://doi.org/10.1080/00071669408417699.

Mead, G. C. and Thomas, N. L. 1973. Factors affecting the use of chlorine in the spin chilling of eviscerated poultry. *Br. Poultry Sci.* 1: 99–117. https://doi.org/10.1080/00071667308416000.

Merka, V. and Urban, R. 1978. Study of inhalation toxicity of performic, peracetic and perpropionic acid in mice. *J. Hyg. Epidemiol. Microbiol. Immunol.* 20(1): 54–60.

Micciche, A. C., Feye, K. M., Rubinelli, P. M., Lee, J., Knueven, C. J. and Ricke, S. C. 2019. Comparison of acid sanitizers on *Salmonella* Typhimurium inoculated commercial poultry processing reuse water. *Front. Sust. Food. Syst.* 2: 90. https://doi.org/10.3389/fsufs.2018.00090.

Milillo, S. R. and Ricke, S. C. 2010. Synergistic reduction of *Salmonella* in a model raw chicken media using a combined thermal and acidified organic acid salt intervention treatment. *J. Food Sci.* 75(2): M121–M125. https://doi.org/10.1111/j.1750-3841.2009.01510.x.

Mitchell, M. A. and Kettlewell, P. J. 1998. Physiological stress and welfare of broiler chickens in transit: solutions not problems! *Poultry Sci.* 77(12): 1803–1814. https://doi.org/10.1093/ps/77.12.1803.

Nagel, G. M., Bauermeister, L. J., Bratcher, C. L., Singh, M. and McKee, S. R. 2013. *Salmonella* and *Campylobacter* reduction and quality characteristics of poultry carcasses treated with various antimicrobials in a post-chill immersion tank. *Int. J. Food Microbiol.* 165(3): 281–289. https://doi.org/10.1016/j.ijfoodmicro.2013.05.016.

Nannapaneni, R., Bansal, M., Kode, D., Chang, K. S., Sharma, C. S., McDaniel, C. D. and Kiess, A. S. 2019. Rugose morphotype in *Salmonella* Typhimurium and *Salmonella* Heidelberg induced by sequential exposure to subinhibitory sodium hypochlorite aids in biofilm tolerance to lethal sodium hypochlorite on polystyrene and stainless steel surfaces. *Front. Microbiol.* 10: 2704. https://doi.org/10.3389/fmicb.2019.02704.

National Academies of Science, NAS. 2010. Chapter 7: peracetic acid-acute exposure guideline levels. In: acute exposure guideline levels for selected airborne chemicals: volume 8. http://www.epa.gov/opptintr/aegl/pubs/peracetic_acid_final_volume8_2010.pdf

National Chicken Council, NCC. 2020. Broiler Industry Key Facts 2019. Accessed 6 November 2020. https://www.nationalchickencouncil.org/about-the-industry/statistics/broiler-chicken-industry-key-facts/.

NIOSH (National Institute for Occupational Safety and Health). (2015). Immediately dangerous to life or health (IDLH) value profile for peracetic acid. External review draft. https://www.cdc.gov/niosh/docket/review/docket156a/pdfs/g1-013-peracetic-acid-cas-79-21-0.pdf.

Northcutt, J. K., Berrang, M. E., Smith, D. P. and Jones, D. R. 2003. Effect of commercial bird washers on broiler carcass microbiological characteristics. *J. Appl. Poultry Res.* 12(4): 435–438. https://doi.org/10.1093/japr/12.4.435.

Obe, T., Nannapaneni, R., Sharma, C. S. and Kiess, A. S. 2018. Homologous stress adaptation, antibiotic resistance, and biofilm forming ability of *Salmonella enterica* serovar Heidelberg ATCC8326 on different food-contact surfaces following exposure to sublethal chlorine concentrations. *Poultry Sci.* 97(3): 951–961. https://doi.org/10.3382/ps/pex346.

Obe, T., Boltz, T., Kogut, M., Ricke S.C., Brooks L.A., Macklin K., and Peterson, A. 2023. Controlling *Salmonella*: Strategies for feed, the farm, and the processing plant. *Poultry Sci.* 102:103086. doi.org/10.1016/j.psj.2023.103086.

O'Bryan, C. A., Ricke, S. C. and Marcy, J. A. 2022. Public health impact of *Salmonella* spp. on raw poultry: current concepts and future prospects in the United States. *Food Control* 132: 108539. https://doi.org/10.1016/j.foodcont.2021.108539.

Olson, E. G., Micciche, A. C., Rothrock Jr., M. J., Yang, Y. and Ricke, S. C. 2022. Application of bacteriophages to limit *Campylobacter* in poultry production. *Front. Microbiol.* 12: 458721. https://doi.org/10.3389/fmicb.2021.458721.

Over, K. F., Hettiarachchy, N., Johnson, M. G. and Davis, B. 2009. Effect of organic acids and plant extracts on *Escherichia coli* O157:H7, *Listeria monocytogenes,* and *Salmonella* Typhimurium in broth culture model and chicken meat systems. *J. Food Sci.* 74(9): M515–M21. https://doi.org/10.1111/j.1750-3841.2009.01375.x.

Owens, C. M., Alvarado, C. and Sams, A. R. (Eds.). 2010. *Poultry Processing*, 2nd ed. Taylor & Francis, London, UK, 448 pp.

Oyarzabal, O. A. 2005. Reduction of *Campylobacter* spp. by commercial antimicrobials applied during the processing of broiler chickens: a review from the United States perspective. *J. Food Prot.* 68(8): 1752–1760. https://doi.org/10.4315/0362-028X-68.8.1752.

Pacholewicz, E., Swart, A., Wagenaar, J. A., Lipman, L. J. and Havelaar, A. H. 2016. Explanatory variables associated with *Campylobacter* and *Escherichia coli* concentrations on broiler chicken carcasses during processing in two slaughterhouses. *J. Food Prot.* 79: 2038–2047. https://doi.org/10.4315/0362-028X.JFP-16-064.

Painter J. A., Hoekstra, R. M., Ayers, T., Tauxe, R. V., Braden, C. R., Angulo, F. J. and Griffin, P. M. 2013. Attribution of foodborne illnesses, hospitalizations, and deaths to food commodities by using outbreak data, United States, 1998–2008. *Emerg. Infect. Dis.* 19(3): 407–415. https://doi.org/10.3201/eid1903.111866.

Payne, J. B., Kroger, E. C. and Watkins, S. E. 2002. Evaluation of litter treatments on *Salmonella* recovery from poultry litter. *J. Appl. Poult. Res.* 11: 239–243. https://doi.org/10.1093/japr/11.3.239.

Rasekh, J., Thaler, A. M., Engeljohn, D. L. and Pihkala, N. H. 2005. Food safety and inspection service policy for control of poultry contaminated by digestive tract contents: a review. *J. Appl. Poultry Res.* 14: 603–611.

Ramasamy, S., Benson, E. R. and Van Wicklen, G. L. 2004. Efficiency of a commercial mechanical chicken catching system. *J. Appl. Poultry Res.* 13(1): 19–28. https://doi.org/10.1093/japr/13.1.19

Ramesh, N., Joseph, S. W., Carr, L. E., Douglass, L. W. and Wheaton, F. W. 2002. Evaluation of chemical disinfectants for the elimination of *Salmonella* biofilms from poultry transport containers. *Poultry Sci.* 81: 904–910. https://doi.org/10.1093/ps/81.6.904.

Ramirez-Hernandez, A., Brashears, M. M. and Sanchez-Plata, M. X. 2018. Efficacy of lactic acid, lactic acid–acetic acid blends, and peracetic acid to reduce *Salmonella* on chicken parts under simulated commercial processing conditions. *J. Food Prot.* 81(1): 17–24. https://doi.org/10.4315/0362-028X.JFP-17-087.

Ricke, S. C. 2021. Strategies to improve poultry food safety, a landscape review. *Annu. Rev. Anim. Biosci.* 9: 379–400. https://doi.org/10.1146/annurev-animal-061220-023200.

Ricke, S. C. 2003. Perspectives on the use of organic acids and short chain fatty acids as antimicrobials. *Poultry Sci.* 82: 632–639. https://doi.org/10.1093/ps/82.4.632.

Ricke, S. C., Dittoe, D. K., Brown, J. A. and Thompson, D. R. 2022. Practical opportunities for microbiome analyses and bioinformatics in poultry processing. *Poultry Sci.* 101: 101787. https://doi.org/10.1016/j.psj.2022.

Ricke, S. C., Hacker, J., Yearkey, K., Shi, Z., Park, S. H. and Rainwater, C. 2017. Chap. 19. Unravelling food production microbiomes: concepts and future directions. In: S. C. Ricke, G. G. Atungulu, S. H. Park, and C. E. Rainwater (Eds.) *Food and Feed Safety Systems and Analysis*. Elsevier Inc., San Diego, CA, pp. 347–374. https://doi.org/10.1016/B978-0-12-811835-1.00019-1.

Ricke, S. C., Kundinger, M. M., Miller, D. R., and Keeton, J. T. 2005. Alternatives to antibiotics: chemical and physical antimicrobial interventions and foodborne pathogen response. *Poultry Sci.* 84(4): 667–675. https://doi.org/10.1093/ps/84.4.667.

Ritz, C. W., Webster, A. B. and Czarick, M. 2005. Evaluation of hot weather thermal environment and incidence of mortality associated with broiler live haul. *J. Appl. Poultry Res.* 14(3): 594–602. https://doi.org/10.1093/japr/14.3.594.

Rothrock, M. J., Locatelli, A., Glenn, T. C., Thomas, J. C., Caudill, A. C. et al. 2016. Assessing the microbiomes of scalder and chiller tank waters throughout a typical commercial poultry processing day. *Poultry Sci.* 95(10): 2372–2382. https://doi.org/10.3382/ps/pew234.

Ruiz-Feria, C. A., Larrison, E., Davis, M., Farnell, M., Carey, J., Grimes, J. L. et al. 2011. Supplementation of feed grade sodium bisulfate in broiler diets improves feed efficiency. *Poultry Sci.* 10: 670–676. https://doi.org/10.3923/ijps.2011.670.676.

Savenije, B., Lambooij, E., Pieterse, C. and Korf, J. 2000. Electrical stunning and exsanguination decrease the extracellular volume in the broiler brain ad studied with brain impedance recordings. *Poultry Sci.* 79: 1062–1066. https://doi.org/10.1093/ps/79.7.1062.

Scallan E., Hoekstra, R. M., Angulo, F. J., Tauxe, R. V., Widdowson, M. A., Roy, S. L., Jones, J. L. and Griffin, P. M. 2011. Foodborne illness acquired in the United States- major pathogens. *Emerg. Infect. Dis.* 17(1): 7–15. https://doi.org/10.3201/eid1701.P11101.

Schuler, G. A. and Badenhop, A. F. 1972. Microbiology survey of equipment in selected poultry processing plants. *Poultry Sci.* 51: 830–835.

Shackelford, A. D., Holladay, J. H. and Whitehead, W. F. 1981. Coopless transport system speeds live handling. *Broiler Ind.* 44(6): 32–34.

Shi, Z., Dittoe, D. K., Feye, K. M., Kogut, M. and Ricke, S. C. 2019a. Short communication: preliminary differences identified in genes responsible for biofilm formation in poultry isolates of *Salmonella enterica* Heidelberg, Enteritidis, and Kentucky. Microorganisms 7(7): 196. https://doi.org/10.3390/microorganisms7070196.

Shi, Z., Dittoe, D. K. and Ricke, S. C. 2019b. Non-molecular characterization of pellicle formation by poultry *Salmonella* Kentucky strains and other poultry-associated *Salmonella* serovars in Luria Bertani broth. *J. Environ. Sci. Health B.* 54(12): 972–978. https://doi.org/10.1080/03601234.2019.1661210.

Shue, C. W. and Freese, E. 1973. Lipopolysaccharide layer protection of Gram-negative bacteria against inhibition by long-chain fatty acids. *J. Bacteriol.* 115: 869–875.

Sofos J. N., Flick, G., Nychas, G. J., O'Bryan, C. A., Ricke, S. C. and Crandall, P. G. 2013. Meat, poultry, and seafood. Pages 111–167. In *Food Microbiology: Fundamentals and Frontiers*, 4th ed, eds M. P. Doyle, R. L. Buchanan. ASM Press, Washington DC, US.

Sohaib, M., Anjum, F. M., Arshad, M. S. and Rahman, U. U. 2016. Postharvest intervention technologies for safety enhancement of meat and meat based products; a critical review. *J. Food Sci. Technol.* 53(1): 19–30. https://doi.org/10.1007/s13197-015-1985-y.

Stearns, R., Freshour, A. and Shen, C. 2022. Literature review for applying peroxyacetic acid and/or hydrogen peroxide to control foodborne pathogens on food products. *J. Agric. Food Res.* 10: 100442. https://doi.org/10.1016/j.jafr.2022.100442.

Stopforth, J. D., O'Connor, R., Lopes, M., Kottapalli, B., Hill, W. E. and Samadpour, M. 2007. Validation of individual and multiple-sequential interventions for reduction of microbial populations during processing of poultry carcasses and parts. *J. Food Prot.* 70: 1393–1401. https://doi.org/10.4315/0362-028X-70.6.1393.

Tan, S. M., Lee, S. M. and Dykes, G. A. 2014a. Buffering effect of chicken skin and meat protects *Salmonella enterica* strains against hydrochloric acid but not organic acid treatment. *Food Control* 42: 329–334. https://doi.org/10.1016/j.foodcont.2014.02.031.

Tan, S. M., Lee, S. M. and Dykes, G. A. 2014b. Fat contributes to the buffering capacity of chicken skin and meat but enhances the vulnerability of attached *Salmonella* cells to acetic acid treatment. *Food Res. Int.* 66: 417–423. https://doi.org/10.1016/j.foodres.2014.10.007.

Thomas C. J. and McMeekin, T. A. 1980. Contamination of broiler carcass skin during commercial processing procedures: an electron microscopic study. *Appl. Environ. Microbiol.* 40(1): 133–144.

Tsai, L., Schade, J. E. and Molyneux, B. T. 1992. Chlorination of poultry chiller water: chlorine demand and disinfection efficiency. *Poultry Sci.* 71: 188–196. https://doi.org/10.3382/ps.0710188.

United States Environmental Protection Agency (EPA). 2018. Safer Chemical Ingredients List. https://www.epa.gov/saferchoice/safer-ingredients (Accessed 25 September 2018).

USDA-FSIS. 1998. FSIS clarifies and strengthens enforcement of zero tolerance standard for visible fecal contamination of poultry. www.fsis.usda.gov/OA/background/zerofcl.htm.

USDA-FSIS. 2015. Consumer protection standards: raw products. Title 9. Code of Federal Regulations 441.1.

USDA-FSIS. 2014a. Poultry products inspection regulations. Title 9. Code of Federal Regulations 381.66.

USDA-FSIS. 2014b. Discontinuation of the qualitative (30 mL) *Campylobacter* analysis for young chickens. *Federal Register.* 79: 35.

Van Deun, K., Pasmans, F., Ducatelle, R., Flahou, B., Vissenberg, K., Martel, A., et al. 2008. Colonization strategy of *Campylobacter jejuni* results in persistent infection of the chicken gut. *Vet. Microbiol.* 130(3–4): 285–297. https://doi.org/10.1016/j.vetmic.2007.11.027.

Wages, J. A., Feye, F. M., Park, S. H., Kim, S. A. and Ricke, S. C. 2019. Comparison of 16S rDNA next sequencing of microbiome communities from post-scalder and post-picker stages in three different commercial poultry plants processing three classes of broilers. *Front. Microbiol.* 10: 972. https://doi.org/10.3389/fmicb.2019.00972.

Walsh, R. J., White, B., Hunker, L., Leishman, O., Hilgren, J. and Klein, D. 2018. Peracetic acid and hydrogen peroxide post-dip decay kinetics on red meat and poultry. *Food Prot. Trends.* 38: 96–103.

Whyte, P., Collins, J. D., McGill, K., Monahan, C. and O'Mahony, H. 2001. Th effect of transportation stress on excretion rates of campylobacters in market-age broilers. *Poultry Sci.* 80(6): 817–820. https://doi.org/10.1093/ps/80.6.817.

Wilkerson, W. B., Ayres, J. C. and Kraft, A. A. 1961. Occurrence of enterococci and coliform organisms on fresh and stored poultry. *Food Technol.* 15: 286–292.

World Health Organization. 2015. *WHO estimates of the global burden of foodborne diseases.* WHO/FOS/15.02.

Zhang, L., Garner, L. J., Mckee, S. R. and Bilgili, S. F. 2018. Effectiveness of several antimicrobials used in a postchill decontamination tank against *Salmonella* and *Campylobacter* on broiler carcass parts. *J. Food Prot.* 81(7): 1134–1141. https://doi.org/10.4315/0362-028X.JFP-17-507.

Zhang, X., Peng, Z., Li, P., Mao, Y., Shen, R., Tao, R., Diao, X., Liu, L., Zhao, Y. and Luo, X. 2020. Complex internal microstructure of feather follicles on chicken skin promotes the bacterial cross-contamination of carcasses during the slaughtering process. *Front. Microbiol.* 11: 571913. https://doi.org/10.3389/fmicb.2020.571913.

Zweifel, C. and Stephan, R. 2012. Microbial decontamination of poultry carcasses. Pages 60–95 in *Microbial Decontamination in the Food Industry*. A. Demirci and Ngadi, M. O. (Eds.) Woodhead Publishing, Cambridge, UK.

Chapter 12

Advances in microbiological assessment of the shelf life and spoilage of ready-to-eat (RTE) poultry meat products

Steven C. Ricke, Elena G. Olson and Lindsey A. Wythe, University of Wisconsin-Madison, USA; and Dana K. Dittoe, University of Wyoming, USA

1 Introduction

Ready-to-eat (RTE) meat and poultry products are defined by the USDA Food Safety and Inspection Service (FSIS, 2020) as being entirely cooked before packaging with no further cooking required by the consumer. Products in this category include hot dogs, luncheon meats, sausage, and other deli-type products (USDA, 2023). Unfortunately, these products are potentially vulnerable to contamination with foodborne pathogens after the cooking step and before packaging. These pathogens, including *Salmonella*, *Listeria monocytogenes*, and *Staphylococcus aureus*, have been implicated in several foodborne illness outbreaks around the world (Ward et al., 1997; Sim et al., 2002; Mead et al., 2006; Marler, 2022). One significant outbreak in Canada in 2008, traced back to RTE

http://dx.doi.org/10.19103/AS.2024.0146.15

meats contaminated with *L. monocytogenes*, resulted in 57 reported cases and 22 deaths (Canadian Food Inspection Agency, 2009). These RTE meat products are also potentially perishable and subject to eventual spoilage, primarily from microbiological causes (Bodie et al., 2024). When RTE meat products are heated to a temperature of 65°C to 75°C, most vegetative cells are killed, and it is recontamination after heat treatment that determines safety and shelf-life (Borch et al., 1996). During storage, environmental factors such as temperature, gaseous atmosphere, pH, and NaCl will select certain bacteria and affect their growth rate and activity (Gill and Molin, 1991; Blickstad and Molin, 1983).

Most antimicrobials that are applied to RTE meats tend to be broad-spectrum in their activities (Abdelhamid and El-Dougdoug, 2020). Historically, it has been challenging to characterize their effect on different microorganisms with conventional microbial culture-based approaches due to the limitations and selectivity of most media used for cultivation and enumeration. While some microorganisms have been identified as potential perpetrators and indicators of spoilage, these associations are generally relatively imprecise. A more thorough investigation of the microbial ecology of RTE meats is needed to properly assess the impacts of externally applied antimicrobials. In the past few years, developments in next-generation sequencing techniques (using the variable region of the 16S rRNA gene to generate molecular-based microbiome taxonomic identification of resident microbial populations on foods) have emerged as a viable means to assess overall microbial ecology responses in food products (Ricke et al., 2017). This chapter will discuss microorganisms affecting the safety and shelf life of RTE poultry products and their microbial spoilage ecology. Where research relating to poultry products is limited, non-poultry RTE product studies will be discussed. Finally, the use of microbiome 16S rRNA gene amplicon sequencing as a potential tool for gaining a better understanding of microbial ecology in RTE meats will be discussed along with specific applications.

2 Foodborne pathogens in ready-to-eat poultry meat products

RTE foods are, as the name implies, available for direct consumption with or without minimal further preparation (Wang et al., 2020; Temgire et al., 2021; USDA, 2023). However, these products have been linked with several foodborne pathogens such as *L. monocytogenes*, *Salmonella* spp., *Staphylococcus aureus*, *Bacillus cereus*, *Shigella*, *Escherichia coli*, and others, particularly in developing countries (Wang et al., 2020; Mengistu et al., 2022).

2.1 *Listeria monocytogenes*

Due to its ability to survive in environments associated with RTE meat processing and storage, *Listeria monocytogenes* is one of the primary foodborne pathogens of concern associated with poultry RTE products (Zhang et al., 2021; Forauer et al., 2021). *Listeria monocytogenes* can survive in cold environments and, therefore, may be a problem for foods held during refrigeration (Saldivar et al., 2017). *Listeria* can also tolerate high salt conditions, grow under anaerobic conditions, and form biofilms (Roberts et al., 2020; Forauer et al., 2021; Zhang et al., 2021). Some of these environmental stresses can accentuate the problem. For example, *Listeria monocytogenes* adhesion to abiotic surfaces has been shown to increase when the organism is experiencing cold stress (Lee et al., 2017).

Contamination of RTE poultry meats can occur from a variety of sources. *Listeria* species have been identified in poultry flocks (Rothrock et al., 2017). Non-conventional poultry production systems such as pasture flock operations and organic-based husbandry can present additional challenges, in part due to greater exposure to environmental sources of contamination, including *Listeria* species (Milillo et al., 2012; Jamshidi and Zeinali, 2019; Rothrock et al., 2017, 2019). Raw chicken products can become contaminated by *L. monocytogenes* during processing and remain viable under cold-temperature storage of poultry meat products (Schäfer et al., 2017; Oliveira et al., 2018; Rothrock et al., 2019). A higher bacterial load in raw meat increases the risk of pathogens surviving further processing.

Listeria has also been detected in RTE and cooked poultry products and can also become cross-contaminated after cooking by contact with contaminated surfaces, equipment, and personnel (Mena et al., 2004; Murphy et al., 2005; Osaili et al., 2011; Goh et al., 2012; Keeratipibul, and Techaruwichit. 2012; Chin et al., 2018). Comprehensive efforts are in place for monitoring *L. monocytogenes* RTE meats on a global scale and include a combination of advanced detection methods, increased surveillance, and are accompanied by regulatory enforcement (Jamshidi and Zeinali, 2019; Zhang et al., 2021).

Interventions for *L. monocytogenes* in poultry and other meat RTE products have been the subject of numerous studies over a long period of time. While various synthetic chemicals have been incorporated into RTE meat products to serve as antimicrobials, changing consumer attitudes have encouraged the development of what are considered more 'natural' antimicrobials (Zhang et al., 2021; Bodie et al., 2023). These include bacteriocins (generated by lactic acid and other bacteria) which are peptides or small proteins that possess antimicrobial activities (Gray et al., 2018; Bodie et al., 2023; Benítez-Chao et al., 2024). An example is nisin produced by *Lactobacillus lactis* which is commonly employed in the meat industry (And et al., 2003; De Arauz et al., 2009; Bodie et al., 2023). Bacteriophages (viruses infecting bacteria) and their hydrolytic enzymes have also been explored extensively for their specific ability to target *Listeria* (Gray et al., 2018). Some of these strategies have centered on attacking *L. monocytogenes* in biofilms. Biofilm intervention approaches have included antimicrobial botanical compounds, bacteriophages, competitive exclusion/inhibitory microbial cultures, and enzyme-based strategies focused on removal of biofilms (Gray et al., 2018; Bouarab et al., 2019; Zhang et al., 2021).

2.2 Salmonella

Foodborne *Salmonella* remains a significant public health concern, and poultry meat is one of the primary food sources of this pathogen (Chai et al., 2017; Thames and Sukumaran, 2020; O'Bryan et al., 2022; Obe et al., 2023). Most of the focus has been on *Salmonella* contamination of raw poultry meat, and considerable effort has been directed toward detection and interventions during processing (O'Bryan et al., 2022). *Salmonella* can still subsequently occur in raw poultry sold in retail markets (Thames and Sukumaran, 2020). However, *Salmonella* has also been detected in poultry RTE products. For example, poultry RTE samples collected from supermarkets in China were contaminated with *Salmonella,* attributed to poor hygienic practices by retail sales staff (Wang et al., 2019). Akbar and Anal (2015) isolated *Salmonella* from poultry RTE meat products and detected resistance to antibiotics such as ampicillin, chloramphenicol, tetracycline, and nalidixic acid. This is a particular cause for concern if *Salmonella* associated with RTE poultry products possesses antibiotic resistance genes that lead to difficulties in the clinical treatment of salmonellosis and can become even more of a concern if the *Salmonella* isolates exhibit multiple drug resistance capabilities.

Salmonella can overcome some environmental stress conditions associated with RTE meats and their subsequent storage. *Salmonella* are able to withstand thermal shock via a series of sigma factors and heat shock proteins that enable the microorganism to adapt to heat exposure (Dawoud et al., 2017). This ability allows *Salmonella* to survive a wide range of temperatures and can lead to cross-protection against other stressors (Dawoud et al., 2017). Heat resistance can also vary among *Salmonella* serovars and strains within a serovar (Humphrey et al., 1990; Doyle and Mazzotta, 2000; Lianou and Koutsoumanis, 2013). This variance can complicate applying thermal processing as a control measure and is further complicated when frozen raw chicken products such as breaded chicken nuggets and strips are presumed to be fully cooked and inadequately heated (Jarvis et al., 2016). Since *S. enterica* serotype Seftenberg 775 W is considerably more thermally tolerant than most of the other *Salmonella* serovars, it has been used as a candidate to represent the upper threshold for testing the lethality of heat treatment against *Salmonella* (Lianou and Koutsoumanis, 2013).

Foodborne *Salmonella* present on RTE poultry meat can also be exposed to sudden lowered temperatures as RTE meat products are stored in cold storage. *Salmonella* possesses a series of genes that allow it to express cold shock proteins in the presence of cold temperature shock (Ricke et al., 2018a). Adaptation to cold temperatures involves a combination of intracellular accumulation of low molecular weight solutes, increasing the fluidity of the cell membrane by desaturating fatty acids and optimizing RNA processing in concert with changes in DNA supercoiling and continued transcription of cold-induced genes to cold-induced proteins (Theiringer et al., 1998; Shivaji and Prakash, 2010; Barria et al., 2013; Ricke et al., 2018a).

When examined in frozen poultry RTE products, *Salmonella* can withstand freezing temperatures for at least several weeks. Dominguez and Schaffner (2009) inoculated fully cooked chicken nuggets with *S.* Typhimurium and *S.* Kentucky, followed by storage in a –20°C freezer (freezing rate 10.5°C per hour) over 16 weeks with periodic enumeration on minimal, nonselective and selective media. Viable *Salmonella* populations remained relatively the same throughout the 16-week storage period, when counts were estimated on minimal and nonselective media, but were decreased on selective media. This finding led the authors to conclude that *Salmonella* cells were partially damaged during cold storage, with selective media failing to recover these cells and thus underestimating viable cells. Survival of viable *Salmonella* during long-term cold storage presents a potential problem if not correctly handled.

Akbar and Anal (2015) isolated a *Salmonella* strain from poultry RTE meats. They determined that, depending on the setting, microwave heating was ineffective for eliminating these *Salmonella* isolates, and that the shape and weight of the RTE poultry products influenced survival. In addition, they observed that *Salmonella* from these products could survive when incubated in simulated gastric juices held at pH levels of 3, 4, and 7. This finding would suggest that *Salmonella* originating from these environments can tolerate multiple environmental stresses.

3 Ecology of microbial spoilage in RTE poultry meat products

Pathogens such as *L. monocytogenes* and *Salmonella* have received considerable scrutiny in terms of the safety of RTE meat products. However, meats serve as an ecological niche for other microorganisms that are not necessarily foodborne pathogens. While their presence may not lead to human diseases, they can still decrease meat quality and increase perishability. In general, shelf life and subsequent spoilage of meat products are determined by factors such as overall microbial population levels, the types of microorganisms present in the meat, and their respective metabolic activities that, collectively, contribute to a spoiled state that discourages human consumption (Odeyemi et al., 2020).

Not surprisingly, fresh meat harbors a wide range of microorganisms, particularly those belonging to genera such as *Acinetobacter*, *Pseudomonas*, *Brochothrix*, *Flavobacterium*, *Psychrobacter*, *Moraxella*, *Staphylococcus*, *Micrococcus*, lactic acid bacteria, and *Enterobacteriaceae* along with numerous minor bacterial genera together with yeasts and molds (Doulgeragi et al., 2012; Sofos et al., 2013; Odeyemi et al., 2020). However, once RTE processing is introduced, the microbial ecology of the meat matrix will harbor a much more selective community of organisms.

Cold temperatures select for a dominant psychrotrophic microbial population on chilled meats (Dainty and Mackey, 1992). Specific microbial composition varies under cold temperatures during storage, depending on water, activity, meat pH, or use of modified gas atmospheric packaging (Dainty and Mackey, 1992). Likewise, the production processes associated with different RTE products, such as drying or cooking, can exert different external selective pressures, leading to different microbial populations on the RTE product (Bodie et al., 2024).

Patsias et al. (2006a) compared a chilled precooked chicken product stored in air versus modified atmospheres of various carbon dioxide and nitrogen mixtures. Plate enumeration was used to track total viable counts of lactic acid bacteria, *Brochothrix thermosphacta*, pseudomonads, yeasts, molds, and *Enterobacteriaceae* over 20 days. Total plate counts reached 7 log colony forming units (CFU) per gram by day 12 in air samples and day 16 in modified 30% carbon dioxide/70% nitrogen atmospheres. In contrast, samples at higher atmospheric carbon dioxide levels did not reach these levels at 20 days of storage. These findings matched the results of sensory tests for taste and odor – lactic acid bacteria dominated in both air-held samples and all gas mixtures. The growth of *Brochothrix thermosphacta* was restricted at higher carbon dioxide levels, and the development of the other organisms was only supported in the air package refrigerated precooked chicken, except for *Enterobacteriaceae*, which did not grow under any of the packaging conditions.

In a second study, Patsias et al. (2006b) compared refrigerated precooked chicken packaged with air or 30% carbon dioxide/70% nitrogen to form biogenic amines that could be used as potential chemical indicators for the onset of spoilage. The results supported the previous observation that lactic acid bacteria dominated under modified atmospheric conditions and delayed putrescine and tyramine formation from reaching spoilage limits to 12 days versus 8 days for air-packaged chicken. In conclusion, these studies involving enumerating selected bacteria under refrigerated storage demonstrate that modified packaging can influence the microbial composition of RTE products, with consequences for extending shelf life and delaying the onset of spoilage.

More recent studies have used proteomics to elucidate individual spoilage microbial responses to RTE environmental conditions. Kolbeck et al. (2020) used an *in vitro* meat simulation model with different modified atmospheric mixtures of nitrogen, oxygen, and carbon dioxide to compare the proteomic responses of *Brochothrix thermosphacta*, and the lactic acid bacteria *Carnobacterium divergens*, *C. maltaromaticum*, *Leuconostoc gelidum gelidum*, and *L. gelidum gasicomitatum* during incubation at 25°C for 48 h. The authors observed that the two *Leuconostoc* strains tolerated modified atmosphere conditions. Other bacteria elicited adaptive responses to carbon dioxide, including enhanced oxidative stress reduction and shifts in pyruvate metabolism, oxygen consumption, intracellular pH homeostasis, osmotic balance maintenance, and alteration in fatty acid composition.

Kolbeck et al. (2021) also used a proteomic approach to demonstrate that meat spoilage *Pseudomonas* species could adapt to anaerobic conditions via arginine fermentation and could also use nitrite as an alternative electron acceptor for anaerobic respiration. Similarly, Yi and Xie (2021) used comparative proteomics to identify metabolic responses of *Shewanella putrefaciens* related to spoilage under refrigerated conditions. These proteomic studies demonstrate meat spoilage organisms' metabolic versatility and ability to adapt and function under modified atmospheric conditions. However, this only represents a partial meat spoilage microbial response and does not consider the entire microbial community. More recently developed comprehensive molecular methods offer the opportunity to retrieve a complete microbial consortium response and identify the impact of several factors on the shelf life and storage stability of RTE poultry products.

4 Molecular analyses of microbial populations in RTE poultry meat products

The use of molecular techniques such as polymerase chain assays, biosensors, whole genome sequencing, and other rapid approaches to identify and quantify foodborne pathogens in food and meat products has been extensively reviewed elsewhere and will not be discussed here (Kuchta et al., 2014; Chen and Park, 2016; Ricke et al., 2018b; Leonardo el., 2021; Quintela et al., 2022). Molecular techniques for identifying non-pathogenic organisms in food and meat products (by using 16S rRNA gene sequencing of their respective microbiomes) are now providing a means for achieving a collective snapshot of entire microbial communities and identifying individual members within them (Ricke et al., 2017). Sequencing technologies are being increasingly applied in food and animal sectors such as poultry processing (Ricke et al., 2017, 2022; Weinroth et al., 2022).

At the pre-harvest stage of poultry production, the microbiota of the poultry gastrointestinal tract (GIT) provides insight into growth performance, animal health, and even future food safety risks (Dittoe et al., 2022). This is also important for poultry meat processing as the presence of specific foodborne pathogens in the GIT can potentially be linked to the processing plant. GIT contents can become contaminated during the first stage of processing, posing a contamination risk throughout the rest of the operation (Ricke et al., 2022). There is also a high risk of contamination from spoilage microorganisms during slaughter. This is a concern as spoilage occurs in fresh poultry meat relatively rapidly (Odeyemi et al., 2020). During subsequent processing stages, contamination from pathogens and spoilage microorganisms from the processing environment pose a further threat to shelf life, and post-cooking contamination from psychrotrophic bacteria can significantly shorten the shelf life of meat under refrigerated temperatures and anaerobic conditions (Dainty and Mackey, 1992; ICMSF, 2005; Odeyemi et al., 2020).

Antimicrobials are typically used to reduce pathogens and total microbial load. However, there is a possibility of certain antimicrobials, such as acidifying agents, potentially selecting acid-tolerant species, pathogenic or otherwise (Quintavalla and Vicini, 2002; Ricke, 2003; Theron and Lues, 2007; Hauser et al., 2016; Bodie et al., 2024). Understanding the microbiota of the carcass and processing environment at a molecular level can help the industry better characterize relationships between indigenous and contaminant microorganisms and their effects on final products.

Standard microbiological techniques utilized within processing plants and by third-party laboratories can determine the presence of specific pathogens and estimate shelf-life through plate counts before processing into RTE products. However, conventional techniques cannot describe entire microbial populations, nor can they fully detail potential shelf-life outcomes. Molecular techniques such as microbiome sequencing can help the poultry industry understand the shifts occurring within indigenous microbial populations in response to antimicrobials and better define spoilage outcomes over the entire poultry meat supply chain. The following sections provide a general discussion of the descriptive terminology, metrics, and bioinformatic analyses involved in characterizing microbial communities using 16S rRNA gene sequencing and potential applications for RTE poultry meats.

5 Microbiome analyses using sequencing techniques: principles and applications

Of the molecular methods for characterizing microbial ecology, 16S rRNA gene amplicon sequencing has started to be used more routinely in the food industry due to increasing affordability, ongoing streamlining of bioinformatic pipelines, and development of practical applications (Ricke et al., 2017). A number of reviews have described the methodologies currently used when performing taxonomical analysis within poultry processing, but there is a current need to fully outline the proper steps required specifically for RTE poultry meat products (Feye and Ricke, 2019; Sabater et al., 2021; Ricke et al., 2022, Weinroth et al., 2022).

Appropriate protocols are being developed to sample poultry products during processing in a way that accurately captures the product's surface and internal microbiota, using sample material such as homogenates, rinsates, or sampling sponges, along with proper lysing steps in response to the sample type being sequenced (Feye and Ricke, 2019; Feye et al., 2020; Ricke et al., 2022). This makes it possible to generate an extensive data set to reveal patterns in microbial taxa during RTE meat production, including taxa that can potentially withstand cook temperatures or the use of antimicrobials. Knowing the taxa already present within a cooked, packaged product can, in turn, help the industry more readily define shelf-life and spoilage outcomes, particularly since the more prominent species sequenced can then be further characterized. In turn, the metabolic pathways and use of the various components of the product matrix by specific microorganisms, or the entire community together, can be described.

Once microbial population dynamics are understood, other techniques can deduce gene expression, metabolism, and other primary changes occurring throughout spoilage in response to the experimental or actual treatments (Dutta et al., 2022). Evaluating these changes provides details that traditional culture plate-based enumeration cannot, including genetic changes in virulence and survival mechanisms, along with other metabolic and genetic-related potential shelf-life behaviors. Understanding the effects of up- and down-regulation of genes and changes in metabolism and subsequent metabolite production can be used to develop multi-hurdle antimicrobial treatments that specifically address these changes.

5.1 Diversity and taxonomical analysis

When analyzing microbiota data, alpha and beta diversity are determined through various metrics and typically reported with a minimum of two metrics each (Table 1). Alpha diversity is used to describe the number of species within a sample. This number can be characterized as richness, evenness, or both. Richness is a descriptor of the number of unique species or amplicon sequencing variants (ASVs; previously operational taxonomical units, OTUs), while evenness compares the number of members of a unique species within an ecosystem (Weinroth et al., 2022). The alpha diversity of one sample can subsequently be compared to another, and this diversity does not change based on the sample being compared against. In contrast, beta diversity evaluates the evenness and richness of a sample by describing how dissimilar it is to another sample, following a distance matrix between the samples. As such,

Table 1 Diversity metrics used to describe alpha and beta diversity.

Metric		Description
Alpha		
	Observed ASV/OTU features	Counts the number of unique ASVs (richness)
	Pielou's evenness	Describes how balanced the distribution of species is within a sample (evenness)
	Shannon's index	Describes the diversity by calculating the evenness and richness together
	Faith's phylogenetic diversity	Measures the richness of a sample by weighting the diversity along a phylogenetic tree
Beta		
	Jaccard's distance	Richness is evaluated by calculating the number of shared species in a sample
	Bray–Curtis distance	Evenness is evaluated between samples by describing the common species between samples
	Unweighted UniFrac	Considers ASV presence and absence and calculates the fraction of the phylogenetic tree branch lengths unique to each community
	Weighted UniFrac	Considers ASV abundances to weight branch lengths with differences in abundance

two diverse samples may not be different, but one of these diverse samples, compared to a low-diversity sample, may differ drastically. Both alpha and beta diversity can incorporate phylogeny, which helps account for evolutionary differences along a phylogenetic tree. Phylogenetic diversity helps parse out functional differences in diversity.

Statistical analyses can be performed on treatment groups for alpha and beta diversity. Several bioinformatics platforms exist that streamline these analyses, including R, Mothur, and Qiime2 (R Core Team, 2013; Schloss et al., 2009; Bolyen et al., 2019). All pipelines have nuanced differences, and it is up to the researcher to decide which pipeline best fits their comprehension level and still sufficiently answers research questions. A non-parametric analysis of variance (ANOVA) technique can be performed for multivariate main effects on all alpha diversity metrics, with pairwise comparisons typically being performed with Kruskall–Wallis analyses to adequately correct the number of comparisons performed (Kruskall and Wallis, 1952). Alternatively, beta diversity is generally analyzed using the permutational multivariate analysis of variance (PERMANOVA) for singular main effects or the ADONIS test for multivariate main effects (Anderson, 2017). These two methods are used to control permutations within a distance matrix.

Taxonomical analysis goes further than diversity analysis by describing a sample's microbial constituents. Researchers employ machine learning algorithms for microbial ecology using 16S rRNA gene amplicon sequencing to classify amplicon sequence variants (ASVs) against databases, as in QIIME2 (Weinroth et al., 2022). ASVs can be accurately classified to the family or genus level, and differential abundance testing can subsequently be employed. Differential abundance means all abundances of all taxa are described as relative to other taxa present in the sample, such that all abundances sum to 100% (Mandal et al., 2015). Analysis of the composition of microbiomes (ANCOM) is a more recently described approach that calculates abundances based on log ratios and limits the false discovery rate (Mandal et al., 2015). This method can be used to compare differences in relative abundance between samples, making it a valuable tool for exploring taxonomical changes occurring due to various treatments and interactions of treatments.

6 Using microbiome mapping to understand spoilage in RTE poultry meat products

Environmental microbiome or bio-mapping is one method that aims to describe the microbiota throughout the processing plant, particularly during harvesting and raw product handling (Feye et al., 2020; Ricke et al., 2022). These environmental communities can be compared to the microbiota inhabiting the carcass and the resulting meat products (De Filippis et al., 2013; Hultman et al., 2015). To date, limited work has examined RTE meat spoilage using molecular methods, and the majority has been performed on raw products.

De Filippis et al. (2013) explored the microbiome of beefsteaks from half carcasses through spoiled products and compared the communities to various processing and slaughter facilities. Based on two experiments, the group reported storage time and carcass parts affected the composition of the microbial communities. Both experiments revealed complex communities that differed in both experiments and that became less diverse and ultimately dominated by specific genera by the time of spoilage. Similarly, environmental swabs in both experiments were similar in composition and diversity to beefsteaks at time zero, but microbiota diverged over spoilage time. The results led the group to suggest that the processing environment can provide a contamination source for spoilage microorganisms. Storage parameters after processing are more likely to select microbial species that out-compete other organisms initially associated with raw meat.

Hultman et al. (2015) analyzed populations throughout the processing of vacuum-packaged RTE sausages. They compared populations found in raw meat, final, and end-of-shelf-life products. The group found limited differences in populations on surfaces between areas of the plant. The group reported phylogenetic clustering differences and varying relative abundances between plant areas. However, similar taxonomic groups were identified throughout the plant. Raw meat cut selections, fat, and rind had similar populations to those found throughout the processing plant, with varying abundances between the unit and the materials. However, the communities within the sausage emulsion, as opposed to the raw materials themselves, exhibited similar community profiles to the final cooked product, with the order *Lactobacillales* having the highest relative abundance.

Overall population diversity within the meat appeared to decrease as the materials moved from raw to cooked products, with the relative abundance of *Lactobacillales* increasing throughout processing. The similarities between the microbial population of the emulsified product and the final product are surprising, as the sausages were cooked, which is considered a lethality step. This finding highlights the importance of molecular analysis. Various processing parameters, as opposed to any single step, can select specific microorganisms to dominate in products, and this may not be fully shown by standard microbiological techniques.

Zwiritz et al. (2020) used high-throughput full-length 16S rRNA gene sequencing to track the origin and path of spoilage organisms through a pork processing facility. Among the key findings was that the transfer of microorganisms from knives and aprons was generally low; however, a large portion of the microorganisms on the meat at the last station (going into the truck) was from the workers' gloves used in eviscerating carcasses. They also determined that air transfer of microorganisms was minimal, and most contamination came from surfaces. They also determined that several microorganisms occupied specific niches in the processing environment, although some had multiple points of origin. For instance, *Moraxella* most commonly originated from the polishing tunnel whips, gloves, and railings, but the origins of *Pseudomonas* species remained unknown.

Microbiome analysis can also be used to examine antimicrobials and their effects on the indigenous microbiota of raw and cooked products. Most microbiota analyses in response to the use of meat processing antimicrobials have involved raw products, and products inoculated with a pathogen. Bodie et al. (2022) analyzed RTE organic, uncured beef frankfurters inoculated with *Listeria monocytogenes*. They were treated with various levels of sodium bisulfate, sodium lactate, or combined levels of both. The frankfurters were dissolved in 100 ml of nBPW before DNA was extracted. The authors identified significant differences in diversity and relative abundances of taxa within the frankfurter homogenates. Based on Kruskal–Wallis analyses, pairwise differences revealed that frankfurters washed with tap water had a higher microbial diversity than both the no-treatment control and those treated with 1.56% sodium lactate.

ANOSIM analysis was also used to determine pairwise differences in beta diversity. Frankfurters treated as the control were found to be significantly different from treated frankfurters. When homogenates were analyzed by Bodie et al. (2022) using ANCOM, they detected the presence of various spoilage genera including *Bacillus*, *Leuconostoc*, *Pseudomonas*, as well as the inoculated *Listeria*. In this analysis, *Listeria* was the only significantly different genera in relative abundance across treatments, consistent with its use to inoculate the frankfurters. Nonetheless, numerical differences showed that frankfurters treated with sodium bisulfate or sodium lactate resulted in numerical differences in spoilage microorganisms. In particular, 0.39% or 0.78% sodium bisulfate had the lowest relative abundance of *Lactobacillus* but the highest numerical relative abundance of *Pseudomonas*. In contrast, products treated with sodium lactate had numerically greater abundances of *Lactobacillus*.

Microbial tolerance to various environmental conditions, such as acidity, is essential for antimicrobial food treatments, including RTE products. While an antimicrobial may successfully inhibit specific pathogens, any overall effect on quality and spoilage depends on overall effects on the microbiological community as a whole. Plate counts with media such as tryptic soy agar that are only selective for aerotolerant species in shelf-life studies provide an overarching picture of product growth responses over time. However, 16S rDNA sequencing can identify the individual members of the microbial community potentially responsible for spoilage. Therefore, harnessing 16S rRNA gene sequencing on products to evaluate antimicrobials to meet growing consumer demands will be necessary for the meat industry to understand the overall effects of microbial communities on products and maximize shelf-life potential.

7 The use of transcriptomics to analyze microbial reactions to different environmental conditions

Along with the potential for 16S rRNA gene sequencing to describe population changes occurring within products undergoing spoilage from a taxonomical perspective, other analyses can explore functional microbial changes (Dutta et al., 2022). Current 16S rRNA gene amplicon sequencing cannot differentiate between live and dead cell DNA. There is therefore a need to apply other molecular techniques such as transcriptomics. For transcriptomic-based studies, the RNA transcripts generated under specific circumstances, such as in response to antimicrobial treatments, modifying the atmosphere within the packaging, or cold storage, could provide in-depth data on microbial ecological outcomes upon exposure to these external factors.

Hultman et al. (2020) performed a longitudinal study on raw meat products, exploring the overall population and metatranscriptomic changes occurring during spoilage. The products were separated through sensory analysis into acceptable product (AP), early spoilage (ES), and late spoilage (LS). The group reported that the active microbial communities were mainly within the families *Leuconostocacea*, *Streptococcaceae*, and unclassified bacilli, with abundant variations as the products spoiled over time. The transcriptomic changes were reported by following three previously identified spoilage organisms or those with a high relative abundance of genes in the products: *Lactococcus gelidum* subsp. *gasicomitatum*, *L. piscium*, and *Dellaglioa algida*. The group reported these species as exhibiting a significantly greater relative abundance of cold-shock genes in AP than in the remainder of the experiment.

The study showed a notable prevalence of active gene expression for carbohydrate metabolism, suggesting that available glucose concentrations did not limit spoilage microorganisms in experimental conditions. Once the products entered ES, an increased relative abundance of oxidative stress genes continued into the LS stage. While these stress genes were reported to increase numerically only slightly over time, these genes were more actively expressed overall after day 6. Amino acid metabolism also increased over time, with a 7-fold increase reported from ES to LS. In addition, the group also reported that most reads mapped to *L. gelidum* subspecies *gasicomitatum* were involved in a metabolic pathway previously reported to produce diacetyl, a metabolite implicated in meat spoilage. Finally, the group reported metabolic increases in the role of yeasts in product spoilage, specifically from the genus *Debaryomycetacceae*. Overall, the reported findings of this study suggested that high-oxygen MAP supported the growth and activity of psychrotrophic lactic acid bacteria (LAB) from the *Leuconostoc* and *Lactococcus* genera while inhibiting the growth of other LAB that are less aerotolerant. These findings provide an essential metabolic understanding of the microbial spoilage in the products, and employing the same methods in future studies will be vital for better predictions for all products. Furthermore, these changes can provide insight into novel ways to inhibit product bacterial growth.

8 Conclusion

Poultry RTE products are a significant market segment which has been affected by consumer trends, such as the interest in 'clean label' products containing only natural ingredients for purposes such as improved quality and shelf-life (Santiesteban-López et al., 2022; Bodie et al., 2023, 2024). However, this trend poses a significant challenge as it often involves removing traditional ingredients crucial for food safety, such as lactate and diacetate. Despite this shift, consumer-friendly ingredients must continue to ensure food safety and comply with regulatory guidelines. While past research has primarily focused on pathogens, there is a growing need to understand the microbial ecology of meat product shelf life and spoilage.

Traditional isolation and identification methods make identifying and characterizing the organisms responsible for meat spoilage challenging. Therefore, more research is needed to determine the effects of clean-label antimicrobials on spoilage microbial communities in meat using advanced microbiome characterization techniques. Several studies have utilized molecular techniques to explore meat spoilage in meat and poultry RTE products. For instance, 16S rRNA gene amplicon sequencing has emerged as a valuable tool to examine microbial communities in RTE poultry meats. This technique makes it possible to compare the diversity of microbial populations across different treatments or environmental conditions, such as modified atmospheric packaging, and provides taxonomic identification of individual community members active in spoilage processes.

As more microbiome studies on poultry RTE meats are conducted, there will be opportunities to establish standard protocols and baseline data sets for comparison across independent studies. This will enhance our understanding of the microbial ecology of RTE products, including spoilage mechanisms. Integrating other omics approaches, such as proteomics and metabolomics, will also help understand metabolic interactions within these microbial communities. Recent proteomics studies e.g. have characterized primary spoilage organisms (Kolbeck et al., 2020, 2021; Yi and Xie, 2021), while metabolomics has been used to compare meat products from different sources (van Vliet et al., 2021). However, a more comprehensive strategy combining metagenomic sequencing, transcriptomics, proteomics, and metabolomics in RTE poultry meat and spoilage studies, is necessary for a fully integrated understanding. Implementing such an omics strategy will further improve our understanding of spoilage mechanisms and enable more precise modeling and prediction of the shelf life of poultry RTE products. Ultimately, this can optimize interventions to extend the shelf life of these meat products.

9 Where to look for further information

- An introduction to poultry and meat processing can be found within C. M. Owens, C. Alvarado, and A. R. Sams, Poultry Processing and F. Toldrá, Handbook of Meat Processing (See below for full details). As well, additional details regarding the current topic of RTE poultry products can be found on page 448 of C. M. Owens, C. Alvarado, and A. R. Sams, Poultry Processing.
    - Owens, C. M., Alvarado, C. and Sams, A. R. (Eds.). 2010. *Poultry Processing*, Second edition. Taylor & Francis, London, UK
    - Toldrá, F. (Ed.). 2010. Handbook of Meat Processing, Second edition. Wiley-Blackwell, Ames, Iowa, US
- For more information regarding common foodborne pathogens present on poultry products and the current challenges they pose to the industry and consumers please refer to Thames and Sukumaran (2020).

- Thames, H. T. and Sukumaran, A. T. 2020. A review of *Salmonella* and *Campylobacter* in broiler meat: emerging challenges and food safety measures. *Foods*. 9: 776. https://doi.org/10.3390/foods9060776. For more information regarding *Listeria monocytogenes* in ready-to-eat poultry products and in the retail deli environment, please refer to the following secondary journal articles:
  - Zhang, X., S. Wang, S., X. Chen, X. and C. Qu, C. 2021. Review controlling Listeria monocytogenes in ready-to-eat meat and poultry products: An overview of outbreaks, current legislations, challenges, and future prospects. Trends in Food Sci. & Technol. 116: 24–-35. doi.org/https://doi.org/10.1016/j.tifs.2021.07.014.
  - Forauer, E., Wu, S. T. and Etter, A. J. 2020. *Listeria monocytogenes* in the retail deli environment: a review. *Food Control*. 119: 107443. https://doi.org/10.1016/j.foodcont.2020.107443.
- There are several relevant secondary journal articles that articulate the need to understand the microbial ecology of foods to optomally mitigate food spoilage. As well, these articles express the notion of using next generation sequencing to further extrapolate the microbial ecology of these food products, a deeper understanding of common microbial ecology analyses, and what "best practices" to employ.
  - Ricke, S. C., Hacker, J., Yearkey, K., Shi, Z., Park, S. H. and Rainwater, C. 2017. Chapter 19. Unravelling food production microbiomes: Concepts and future directions. In: S. C. Ricke, G. G. Atungulu, S. H. Park, and C. E. Rainwater (Eds.), *Food and Feed Saf Systems and Analysis. Elsevier Inc.* San Diego, CA. pp. 347–374. https://doi.org/10.1016/B978-0-12-811835-1.00019-1.
  - Weinroth, M. D., Belk, A. D., Dean, C., Noyes, N., Dittoe, D. K., et al. 2022. Considerations and best practices in animal science 16S ribosomal RNA gene sequencing microbiome studies. *J. Anim. Sci*. 100(2). https://doi.org/10.1093/jas/skab346.
  - Odeyemi, O. O., Alegbeleye, O. O., Strateva, M. and Stratev, D. 2020. Understanding spoilage microbial community and spoilage mechanisms in foods of animal origin. Compr. Revs. Food Sci. Food Saf. 19: 311–331. https://doi.org/10.1111/1541-4337.12526.

10 References

Abdelhamid, A. G. and El-Dougdoug, N. K. 2020. Controlling foodborne pathogens with natural antimicrobials by biological control and antivirulence strategies. *Heliyon*. 6: e05020. https://doi.org/10.1016/j.heliyon.2020.e05020.

Akbar, A. and Anal, A. K. 2015. Isolation of *Salmonella* from ready-to-eat poultry meat and evaluation of its survival at low temperature, microwaving and simulated gastric juices. *J. Food Sci. Technol.* 52(2): 3051–3057. https://doi.org/10.1007/s13197-014-1354-2.

And, H. C. and Hoover, D. G. 2003. Bacteriocins and their food applications. *Comp. Rev. Food Sci. Food Saf.* 2: 82–100.

Anderson, M. J. 2017. Permutational Multivariate Analysis of Variance (PERMANOVA). In: N. Balakrishnan, T. Colton, B. Everitt, W. Piegorsch, F. Ruggeri, and J. L. Teugels (Eds.), *Wiley StatsRef: Statistics Reference Online*. John Wiley & Sons, Ltd., Hoboken, NJ. pp. 1–15. https://doi.org/10.1002/9781118445112.stat07841.

Barria, C., Malecki, M. and Arraiano, C. M. 2013. Bacterial adaptation to cold. *Microbiol.* 159: 2437–2443.

Benítez-Chao, D. F., León-Buitimea, A., Lerma-Escalera, J. A. and Morones-Ramírez, J. R. 2021. Bacteriocins: an overview of antimicrobial, toxicity, and biosafety assessment by in vivo models. *Front. Microbiol.* 12: 630695. https://doi.org/10.3389/fmicb.2021.630695.

Blickstad, E. and Molin. G. 1983. The microbial flora of smoked pork loin and frankfurter sausage stored in different gas atmospheres at 4°C. *J. Appl. Bacteriol.* 54: 45–56.

Bodie, A. R., Wythe, L. A., Dittoe, D. K., Rothrock, Jr. J. M., O'Bryan, C. A. and Ricke, S. C. 2024. Alternative additives for organic and natural ready-to-eat meats to control spoilage and maintain shelf life: current perspectives in the United States. *Foods.* 13: 464. https://doi.org/10.3390/foods13030464

Bodie, A. R., O'Bryan, C. A., Olson, E. G. and Ricke, S. C. 2023. Natural antimicrobials for *Listeria monocytogenes* in ready-to-eat meats: current challenges and future prospects. *Microorganisms.* 11: 1301. https://doi.org/10.3390/microorganisms11051301.

Bodie, A. R., Dittoe, D. K., Feye, K. M., Knueven, C. J., Ovall, C. and Ricke, S. C. 2022. Comparison of ready-to-eat "organic" antimicrobials, sodium bisulfate, and sodium lactate, on *Listeria monocytogenes* and the indigenous microbiome of organic uncured beef frankfurters stored under refrigeration for three weeks. *PLoS One.* 17(1): e0262167. https://doi.org/10.1371/journal.pone.0262167.

Borch, E., Kant-Muermans, M. L. and Blixt, Y. 1996. Bacterial spoilage of meat and cured meat products. *Int. J. Food Microbiol.* 33: 103–120.

Bolyen, E., Rideout, J. R., Dillon, M. R., Bokulich, N. A., Abnet, C. C., et al. 2019. Reproducible, interactive, scalable and extensible microbiome data science using QIIME2. *Nat. Biotechnol.* 37: 852–857. https://doi.org/10.1038/s41587-019-0209-9.

Bouarab, C. L., Degraeve P., Ferhout, H., Bouajila, J. and Plant, O. N. 2019. Antimicrobial polyphenols as potential natural food preservatives. *J. Sci. Food Agric.* 99: 1457–1474. https://doi.org/10.1002/jsfa.9357.

Callahan, B., McMurdie, P. and Holmes, S. 2017. Exact sequence variants should replace operational taxonomic units in marker-gene data analysis. *ISME J.* 11: 2639–2643. https://doi.org/10.1038/ismej.2017.119.

Canadian Food Inspection Agency. 2009. Report of the independent investigator into the 2008 listeriosis outbreak. Available at: http:// www.cmc-cvc.com/sites/default/files/files/Lister ialndependentInvestigatorReport_July212009.pdf. Accessed 27 April 2017.

Chai, S. J., Cole, D., Nisler, A. and Mahon, B. E. 2017. Poultry: The most common food in outbreaks with known pathogens, United States, 1998–2012. *Epidemiol. Infect.* 145: 316–325. https://doi.org/10.1017/S0950268816002375.

Chen, J. and Park, B. 2016. Recent advancements in nanobioassays and nanobiosensors for foodborne pathogenic bacteria detection. *J. Food Prot.* 79: 1055–1069. https://doi.org/10.4315/0362-028X.JFP-15-516.

Chin, P. S., Ang, G. Y., Yu, C. Y., Tan, E. L., Tee, K. K., Yin, W. F., Chan, K. G., Yuan, G. and Tan, A. 2018. Prevalence, antimicrobial resistance, and genetic diversity of *Listeria* spp. isolated from raw chicken meat and chicken-related products in Malaysia. *J. Food Prot.* 81: 284–289. https://doi.org/10.4315/0362-028X.JFP-17-186.

Dawoud, T. M., Davis, M. L., Park, S. H., Kim, S. A., Kwon, Y. M., Jarvis, N., O'Bryan, C., Shi, Z., Crandall, P. G. and Ricke, S. C. 2017. *Salmonella* thermal resistance – Molecular responses. *Front. Vet. Sci.* 4: 93. https://doi.org/10.33889/fvets.2017.00093.

De Filippis, F., La Storia, A., Villani, F. and Ercolini, D. 2013. Exploring the sources of bacterial spoilers in beefsteaks by culture-independent high-throughput sequencing. *PLoS One*. 8(7): e70222. https://doi.org/10.1371/journal.pone.0070222.

De Aaruz, L. J., Jozala, A. F., Mazzola, P. G. and Penna, T. C. V. 2009. Nisin biotechnological production and application: a review. *Trends Food Sci. Technol.* 20: 146–154.

Dittoe, D. K., Olson, E. G. and Ricke, S. C. 2022. Impact of the gastrointestinal microbiome and fermentation metabolites on broiler performance. *Poult. Sci.* 101: 101786. https://doi.org/10.1016/j.psj.2022.101786.

Dominguez, S. A. and Schaffner, D. W. 2009. Survival of *Salmonella* in processed chicken products during frozen storage. *J. Food Prot.* 72: 2088–2092.

Doulgeraki, A. I., Ercolini, D., Villani, F. and Nychas, G. J. E. 2012. Spoilage microbiota associated to the storage of raw meat in different conditions. *Int. J. Food Microbiol.* 157: 130–141. https://doi.org/10.1016/j.ijfoodmicro.2012.05.020.

Doyle, M. E. and Mazzotta, A. S. 2000. Review of studies on the thermal resistance of *Salmonellae*. *J. Food Prot.* 63: 779–795.

Dutta, B., Lahiri, D., Nag, M., Abukhader, R., Sarkar, T., Pati, S., Upadhye, V., Pandit, S., Amin, M. F. M., Al Tawaha, A. R. M. S., Kumar, M. and Ray, R. R. 2022. Multi-omics approach in amelioration of food products. *Front. Microbiol.* 13: 955683. https://doi.org/10.3389/fmicb.2022.955683.

Feye, K. M. and Ricke, S. C. 2019. Chapter 18. Establishment of a standardized 16S rDNA library preparation to enable analysis of microbiome in poultry processing using Illumina MiSeq platform. In: A. Bridier (Ed.), *Foodborne Bacterial Pathogens: Methods and Protocols. Methods in Molecular Biology.* Springer Science, New York, NY, pp. 213–227. https://doi.org/10.1007/978-1-4939-9000-9_18.

Feye, K. M., Thompson, D. R., Rothrock, M. J., Kogut, M. H. and Ricke, S C. 2020. Poultry processing and the application of microbiome mapping. *Poult. Sci.* 99: 678–688. https://doi.org/10.1016/j.psj.2019.12.019.

Forauer, E., Wu, S. T. and Etter, A. J. 2021. *Listeria monocytogenes* in the retail deli environment: a review. *Food Control.* 119: 107443. https://doi.org/10.1016/j.foodcont.2020.107443.

FSIS. 2020. Microbiological testing program for RTE meat and poultry and pasteurized egg products. Available at: https://www.fsis.usda.gov/science-data/data-sets-visualizations/microbiology/microbiological-testing-program-rte-meat-and-3 (accessed on 6 June 2023).

Gill, C. O. and Mohn, G. 1991. Modified atmospheres and vacuum packaging. In: N. J. Russell and G. W. Gould (Eds.), *Food Preservatives.* Blackie, Glasgow and London, pp. 172–199.

Goh, S. G., Kuan, C. H., Loo, Y. Y., Chang, W. S., Lye, Y. L., Soopna, P., Tang, J. Y. H., Nakaguchi, Y., Nishbuchi, M., Afsah, L. and Son, R. 2012. *Listeria* in retailed raw chicken meat in Malaysia. *Poult. Sci.* 91: 2686–2690. https://doi.org/10.3382/ps.2012-02349.

Gray, J. A., Chandry, P. S., Kaur, M., Kocharunchitt, C., Bowman, J. P. and Fox, E. M. 2018. Novel biocontrol methods for *Listeria monocytogenes* biofilms in food production facilities. *Front. Microbiol.* 9: 605. https://doi.org/10.3389/fmicb.2018.00605.

Hauser, C., Thielmann, J. and Muranyl, P. 2016. Organic acids: usage and potential in antimicrobial packaging. In: J. Barroes-Velazquez (Ed.), *Antimicrobial Food Packaging*. Academic Press, Cambridge, MA, pp. 563–580.

Hultman, J., Johansson, P. and Björkroth, J. 2020. Longitudinal metatranscriptomic analysis of a meat spoilage microbiome detects abundant continued fermentation and environmental stress responses during shelf life and beyond. *Appl. Environ. Microbiol.* 86: e01575-20. https://doi.org/10.1128/AEM.01575-20.

Hultman, J., Rahkila, R., Ali, J., Rousu, J. and Björkroth, K. J. 2015. Meat processing plant microbiome and contamination patterns of cold-tolerant bacteria causing food safety and spoilage risks in the manufacture of vacuum-packaged cooked sausages. *Appl. Environ. Microbiol.* 81(20): 7088–7097. https://doi.org/10.1128/AEM.02228-15.

Humphrey, T. J., Chapman, P. A., Rowe, B. and Gilbert, R. J. 1990. A comparative study of the heat resistance of salmonellas in homogenized whole egg, egg yolk or albumen. *Epidemiol. Infect.* 104: 237–241.

ICMSF. 2005. *Microorganisms in Foods 6 Microbial Ecology of Food Commodities*. 2nd edition. Kluwer Academic & Plenum Publishers, New York. ISBN: 0-306-48675-X.

Jamshidi, A. and Zeinali, T. 2019. Significance and characteristics of *Listeria monocytogenes* in poultry products. *Int. J. Food Sci.* 2019: 7835253. https://doi.org/10.1155/2019/7835253.

Jarvis, N. A., O'Bryan, C. A., Dawoud, T. M., Park, S. H., Kwon, Y. M., Crandall, P. G. and Ricke, S. C. 2016. An overview of *Salmonella* thermal destruction during food processing and preparation. *Food Control.* 68: 280–290. https://doi.org/10.1016/j.foodcont.2016.04.006.

Keeratipibul, S. and Techaruwichit, P. 2012. Tracking sources of *Listeria* contamination in a cooked chicken meat factory by PCR-RAPD-based DNA fingerprinting. *Food Control.* 27: 64–72. https://doi.org/10.1016/j.foodcont.2012.02.026.

Kolbeck, S., Abele, M., Hilgarth, M. and Vogel, R. F. 2021. Comparative proteomics reveals the anaerobic lifestyle of meat-spoiling *Pseudomonas* species. *Front. Microbiol.* 12: 664061. https://doi.org/10.3389/fmicb.2021.664061.

Kolbeck, S., Ludwig, C., Meng, C., Hilgarth, M. and Vogel, R. F. 2020. Comparative proteomics of meat spoilage bacteria predicts drivers for their coexistence on modified atmosphere packaged meat. *Front. Microbiol.* 11: 209. https://doi.org/10.3389/fmicb.2020.00209.

Kruskal, W. H. and Wallis, W. A. 1952. Use of ranks in one criterion variance analysis. *J. Am. Stat. Assoc.* 47: 583–621. https://doi.org/10.1080/01621459.1952.10483441.

Kuchta, T., Knutsson, R., Fiore, A., Kudirkiene, E., Höhl, A., Horvatek Tomic, D., Gotcheva, V., Pöpping, B., Scaramagli, S., To Kim, A., Wagner, M. and De Medici, D. 2014. A decade with nucleic acid-based microbiological methods in safety control of foods. *Lett. Microbiol.* 59: 263–271. https://doi.org/10.1111/lam.12283.

Lee, B.-H., Hébraud, M. and Bernardi, T. 2017. Increased adhesion of *Listeria monocytogenes* strains to abiotic surfaces under cold stress. *Front. Microbiol.* 8: 2221. https://doi.org/10.3389/fmicb.2017.02221.

Leonardo, S., Toldrà, S. and Campàs, M. 2021. Biosensors based on isothermal DNA amplification for bacterial detection in food safety and environmental monitoring. *Sensors* 21: 602. https://doi.org/10.3390/s21020602.

Lianou, A. and Koutsoumanis, K. P. 2013. Strain variability of the behavior of foodborne bacterial pathogens: a review. *Int. J. Food Microbiol.* 167: 310–321. https://doi.org/10.1016/j.ijfoodmicro.2013.09.016.

Mandal, S., Treuren, W. V., White, R. A., Eggesbø, M., Knight, R. and Peddada, S. D. 2015. Analysis of composition of microbiomes: a novel method for studying microbial composition. *Microb. Ecol. Health Dis.* 26: 27663. https://doi.org/10.3402/mehd.v26.27663.

Marler, B. 2022. Publisher's Platform: The world's largest and deadliest *Listeria* outbreak is turning 4 in March. Available at: https://www.foodsafetynews.com/2022/01/publishers-platform-the-worlds-largest-and-deadliest-listeria-outbreak-is-turning-4-in-march/. Accessed 14 March 2023.

Mead, P. S., Dunne, E. F., Graves, L., Wiedmann, M., Patrick, M., Hunter, S., Salehi, E., Mostashari, F., Craig, A., Mshar P., Bannerman, T., Sauders, B. D., Hayes, P., Dewitt, W., Sparling, P., Griffin, P., Morse, D., Slutsker, L. and Swaminathan, B. 2006. Listeria outbreak working group. Nationwide outbreak of listeriosis due to contaminated meat. *Epidemiol. Infect.* 134: 744–751. https://doi.org/10.1017/S0950268805005376.

Mena, C., Almeida, G., Careiro, L., Teixeira, P., Hogg, T. and Gibbs, P. A. 2004. Incidence of *Listeria monocytogenes* in different food products commercialized in Portugal. *Food Microbiol.* 21: 213–216.

Mengistu, D. A., Belami, D. D., Tefera, A. A. and Asefa, Y. A. 2022. Bacteriological quality and public health risk of ready-to-eat foods in developing countries: systematic review and meta analysis. *Microbiol. Insights* 15: 1–11. https://doi.org/1177/11786361221113916.

Milillo, S. R., Stout, J. C., Hanning, I. B., Clement, A., Fortes, E. D., den Bakker, H. C., Wiedemann, M. and Ricke, S. C. 2012. *Listeria monocytogenes* and hemolytic *Listeria innocua* in poultry. *Poult. Sci.* 91: 2158–2163. https://doi.org/ 10.3382/ps.2012-02292.

Murphy, R. Y., Hanson, R. E., Duncan, L. K., Feze, N. and Lyon, B. G. 2005. Considerations for post-lethality treatments to reduce *Listeria monocytogenes* from fully cooked bologna using ambient and pressurized steam. *Int. J. Food Microbiol.* 22: 359–365.

Obe, T., Boltz, T., Kogut, M., Ricke, S. C., Brooks, L. A., Macklin, K. and Peterson, A. 2023. Controlling *Salmonella*: strategies for feed, the farm, and the processing plant. *Poult. Sci.* 102: 103086. https://doi.org/10.1016/j.psj.2023.103086.

O'Bryan, C. A., Ricke, S. C. and Marcy, J. A. 2022. Public health impact of *Salmonella* spp. on raw poultry: current concepts and future prospects in the United States. *Food Control.* 132: 108539. https://doi.org/10.1016/j.foodcont.2021.108539.

Odeyemi, O. O., Alegbeleye, O. O. Strateva, M. and Stratev, D. 2020. Understanding spoilage microbial community and spoilage mechanisms in foods of animal origin. *Compr. Revs. Food Sci. Food Saf.* 19: 311–331. https://doi.org/10.1111/1541-4337.12526.

Oliveira, T. S., Varjao, L. M., da Silva, L. N. N., de Castro Lisboa Pereira, R., Hofer, E., Vallim, D. C. and de Castro Almeida, R. C. 2018. *Listeria monocytogenes* at chicken slaughterhouse: occurrence, genetic relationship among isolates and evaluation of antimicrobial susceptibility. *Food Control* 88: 131–138. https://doi.org/10.1016/j.foodcont.2018.01.015.

Patsias, A., Chouliara, I., Badeka, A., Savvaidis, J. N. and Kontominas, M. G. 2006a. Shelf-life of a precooked chicken product stored in air and under modified atmospheres: microbiological, chemical, sensory attributes. *Food Microbiol.* 23: 423–429. https://doi.org/10.106/j.fm.2005.08.004.

Patsias, A., Chouliara, I., Paleologos, E. K., Savvaidis, I. and Kontominas, M. G. 2006b. Relation of biogenic amines to microbial and sensory changes of precooked chicken meat stored aerobically and under modified atmosphere packaging at 4°C. *Eur. Food Res.* 223: 683–689. https://doi.org/10.1007/s00217-006-0253-9.

Osaili, T. M., Alaboudi, A. R. and Nesiar. E. A. 2011. Prevalence of *Listeria* spp. and antibiotic susceptibility of *Listeria monocytogenes* isolated from raw chicken and ready-to-eat chicken products in Jordan. *Food Control.* 22: 586–590. https://doi.org/10.1016/j.foodcont.2010.10.008.

Quintavalla, S. and L. Vicini, L. 2002. Antimicrobial packaging in meat industry. *Meat Sci.* 62: 373–380.

Quintela, I. A., Vasse, T., Lin, C. S. and Wu, V. C. H. 2022. Advances, applications, and limitations of portable and rapid detection technologies for routinely encountered foodborne pathogens. *Front. Microbiol.* 13: 1054782. https://doi.org/10.3389/fmicb.2022.1054782.

R Core Team. 2013. *R: A Language and Environment for Statistical Computing*. R Foundation for Statistical Computing, Vienna, Austria. http://www.R-product.org/.

Ricke, S. C. 2003. Perspectives on the use of organic acids and short chain fatty acids as antimicrobials. *Poult. Sci.* 82: 632–639.

Ricke, S. C., Dittoe, D. K., Brown, J. A. and Thompson, D. R. 2022. Practical opportunities for microbiome analyses and bioinformatics in poultry processing. *Poult. Sci.* 101: 101787. https://doi.org/10.1016/j.psj.2022.

Ricke, S. C., Dawoud, T. M., Kim, S. A., Park, S. H. and Kwon, Y. M. 2018a. Chapter 1. *Salmonella* cold stress response: mechanisms and occurrence in foods. In: G. M. Gadd and S. Sariaslani (Eds.), *Advances in Applied Microbiology,* 104: 1–38. https://doi.org/10.1016/bs.aambs.2018.03.001.

Ricke, S. C., Kim, S. A. and Park, S. H. 2018b. Molecular-based identification and detection of *Salmonella* in food production systems: current perspectives. *J. Appl. Microbiol.* 125: 313–327. https://doi.org/10.1111/jam.13888.

Ricke, S. C., Hacker, J., Yearkey, K., Shi, Z., Park, S. H. and Rainwater, C. 2017. Chapter 19. Unravelling food production microbiomes: Concepts and future directions. In: S. C. Ricke, G. G. Atungulu, S. H. Park, and C. E. Rainwater (Eds.), *Food and Feed Safety Systems and Analysis*. Elsevier Inc., San Diego, CA. pp. 347–374. https://doi.org/10.1016/B978-0-12-811835-1.00019-1.

Rothrock Jr., M. J., Micciche, A. C., Bodie, A. R. and Ricke, S. C. 2019. *Listeria* occurrence and potential control strategies in alternative and conventional poultry processing and retail. *Front. Sustain. Food Syst.* 3: 33. https://doi.org/10.3389/fsufs.2019.00033.

Rothrock Jr., M. J., Davis, M. L., Locatelli, A., Bodie, A., McIntosh, T. G. and Ricke, S. C. 2017. *Listeria* occurrence in poultry flocks: detection and potential implications. *Front. Vet. Sci.* 4: 125. https://doi.org/10.3389/fvets.2017.00125.

Roberts, B. N., Chakravarty, D., Gardner III, J. C., Ricke, S. C. and Donaldson, J. R. 2020. *Listeria monocytogenes* response to anaerobic environments. *Pathogens* 9: 210. https://doi.org/10.3390/pathogens9030210.

Sabater, C., Cobo-Diaz, J. F., Álvarez-Ordóñe, A., Ruas-Madiedo, P., Ruiz, L. and Margolles, A. 2021. Novel methods of microbiome analysis. *Int. Microbiol.* 24: 593–605. https://doi.org/10.1007/s10123-021-00215-8.

Saldivar, J., Davis, M. L., Johnson, M. J. and Ricke, S. C. 2017. Chapter 13. *Listeria monocytogenes* and growth at low temperatures: Mechanisms and implications for foodborne disease. In: S. C. Ricke, G. G. Atungulu, S. H. Park, and C. E. Rainwater (Eds.), *Food and Feed Safety Systems and Analysis*. Elsevier Inc., San Diego, CA. pp. 227–248. https://doi.org/10.1016/B978-0-12-811835-1.00013-0.

Santiesteban-López, N. A., Gomez-Salazar, J. A., Santos, E. M., Campagnol, P. C. B., Teixeira, A., Lorenzo, J. M., Sosa-Morales, M. E. and Dominguez, R. 2022. Natural antimicrobials: a clean label strategy to improve the shelf life and safety of reformulated meat products. *Foods*. 11: 2613. https://doi.org/10.3390/foods11172613.

Schäfer, D. F., Steffens, J., Barbosa, J., Zeni, J., Paroul, N., Valduga, E., Junges, A., Backes, G. T. and Cansian, R. L. 2017. Monitoring of contamination sources of *Listeria monocytogenes* in a poultry slaughterhouse. *LWT - Food Science and Technol.* 86: 393–398. https://doi.org/10.1016/j.lwt.2017.08.024

Schloss, P. D., Westcott, S. L., Ryabin, T., Hall, J. R., Hartmann, M., et al. 2009. Introducing mothur: open-source, platform-independent, community-supported software for describing and comparing microbial communities. *Appl. Environ. Microbiol.* 75: 7537–7541. https://doi.org/10.1128/AEM.01541-09.

Shivaji, S. and Prakash, J. S. S. 2010. How do bacteria sense and respond to low temperature? *Arch. Microbiol.* 192: 85–95.

Sim, J., Hood, D., Finnie, L., Wilson, M., Graham, C., Brett, M. and Hudson, J. A. 2002. Series of incidents of *Listeria monocytogenes* noninvasive febrile gastroenteritis involving ready-to-eat meats. *Lett. Appl. Microbiol.* 35: 409–413.

Sofos, J. N., Flick, G., Nychas, G. -J., O'Bryan, C. A., Ricke, S. C. and Crandall, P. G. 2013. Chapter 6. Meat, poultry, and seafood. In: M. P. Doyle, and R. L. Buchanan (Eds.), *Food Microbiology-Fundamentals and Frontiers. American Society for Microbiology*, 4th edition, Washington, DC, pp. 111–167. https://doi.org/10.1128/9781555818463.ch6.

Temgire, S., Borah, A. Kumthekar, S. and Idate, A. 2021. Recent trends in ready to eat/cook food products: a review. *J. Pharm. Innov.* 10: 211–217.

Thames, H. T. and Sukumaran, A. T. 2020. A review of *Salmonella* and *Campylobacter* in broiler meat: emerging challenges and food safety measures. *Foods* 9: 776. https://doi.org/10.3390/foods9060776.

Theron, M. M. and Lues, J. F. 2007. Organic acids and meat preservation: a review. *Food Rev. Int.* 23: 141–158.

Thieringer, H. A., Jones, P. G. and Inouye, M. 1998. Cold shock and adaptation. *BioEssays*. 20: 49–57.

USDA. 2023. What are ready-to-eat meat and poultry products? Available at: https://ask.usda.gov/s/article/What-are-ready--to--eat-RTE-meat-and-poultry-products. Accessed on 14 December 2023.

van Vliet, S., Bain, J. R., Muehlbauer, M. J., Provenza, F. D., Kronberg, S. L., Pieper, C. F. and Huffman K. M. 2021. A metabolomics comparison of plant-based meat and grass-fed meat indicates large nutritional differences despite comparable nutrition facts panels. *Sci. Rep.* 11: 13828. https://doi.org/10.1038/s41598-021-93100-3.

Wang, S. K., Fu, L. M., Chen, G. W., Xiao, H. M., Pan, D., Shi, R. F. and Yang, L. G. 2020. Multistate survey of bacterial contamination in ready-to-eat meat products throughout the cooking and selling processes in urban market, Nanjing, China. *Food Sci, Nutr.* 8: 2427–2435. https://doi.org/10.1002/fsn3.1532.

Ward, K., Hammond, R., Katz, D. and Hallman, D. 1997. Outbreak of Staphylococcal food poisoning associated with precooked ham- Florida, 1997. *Morb. Mortal. Wkly. Rep.* 46: 1189–1191.

Weinroth, M. D., Belk, A. D., Dean, C., Noyes, N., Dittoe, D. K., et al. 2022. Considerations and best practices in animal science 16S ribosomal RNA gene sequencing microbiome studies. *J. Anim. Sci.* 100(2). https://doi.org/10.1093/jas/skab346.

Yi, Z. and Xie, J. 2021. Comparative proteomics reveals the spoilage-related factors of *Shewanella putrefaciens* under refrigerated condition. *Front. Microbiol.* 12: 740482. https://doi.org/10.3389/fmicb.2021.740482.

Zhang, X., Wang, S., Chen, X. and Qu, C. 2021. Review controlling *Listeria monocytogenes* in ready-to-eat meat and poultry products: an overview of outbreaks, current legislations, challenges, and future prospects. *Trends Food Sci. Technol.* 116: 24–35. https://doi.org/10.1016/j.tifs.2021.07.014.

Zwirzitz, B., Wetzels, S. U., Dixon, E. D., Stessl, B., Zaiser, A., Rabanser, I., Thalguter, S., Pinior, B., Roch, F.-F., Strachan, C., Zanghellini, J., Dzieciol, M., Wagner, M. and Selberherr, E. 2020. The sources and transmission routes of microbial populations throughout a meat processing facility. *NPJ Biofilms Microbiomes.* 6: 26. https://doi.org/10.1038/s41522-020-0136-z.

Part 4

Sustainability

Chapter 13

Life cycle assessment (LCA) of poultry meat production

Ilkka Leinonen, Kirsi Usva and Sanna Hietala, Natural Resources Institute Finland (Luke), Finland

1 Introduction

Livestock farming has been recognized as producing significant environmental impacts. These include a significant role in global warming through greenhouse gas (GHG) emissions, substantial resource utilization, including land use change, nutrient leakage, water use and impacts on biodiversity (Steinfeld et al. 2006, FAO 2006, 2020, Gerber et al. 2013). Emissions and impacts from agriculture arise from diverse direct and indirect sources and can be challenging to address effectively. For instance, ruminant production generates substantial direct methane emissions (via enteric fermentation) and necessitates vast areas of grazing land which then contribute to global warming via deforestation.

Non-ruminant systems, like poultry and pig production, also have their own environmental issues. In general, poultry production has been suggested to be more environmentally sustainable than many other livestock systems, according to assessments based on several environmental indicators (Williams et al. 2006; de Vries and de Boer 2010). Nonetheless, challenges concerning environmental sustainability also persist within non-ruminant systems (Eriksson

http://dx.doi.org/10.19103/AS.2024.0146.16

et al. 2005, van der Werf et al. 2005; Pelletier 2008). These systems rely heavily on external feed production, especially protein sources like soya, the production of which is limited to specific regions and has its own environmental impacts. Some regions producing soya and other feeds are linked to land use changes, such as deforestation, which strongly contribute to global greenhouse gas emissions (Opio et al. 2013) and biodiversity loss (Kyttä et al. 2023). Additionally, poultry systems, in particular, exhibit distinctive characteristics requiring attention in terms of their environmental impacts, including nitrogen emissions, most notably the emissions of ammonia to the air, nitrous oxide emissions contributing to global warming, and nitrate leaching leading to eutrophication (Usva et al. 2023).

A fundamental prerequisite for enhancing the environmental sustainability of poultry production is the availability of systematic tools capable of comprehensively quantifying the range of environmental impacts originating from the production chain. Such tools should also be capable of pinpointing potential areas for improvement. Currently, the consensus leans toward the use of a method known as life cycle assessment (LCA) for quantifying the environmental impacts in agricultural production systems (MacLeod et al. 2013, FAO 2016). In this context, recent studies on the environmental sustainability of poultry production are reviewed here, focusing on those applying the LCA framework to demonstrate the contributions of different subsystems of poultry production (feed production, emissions from housing, breeding activities, etc.) to overall environmental impacts. These studies also suggest measures that could make poultry production more ecologically sustainable.

2 Background and principles of the life cycle assessment framework

The overall environmental sustainability of agricultural production is a result of a combination of direct and indirect environmental impacts. Direct emissions include so-called Scope 1 and Scope 2 greenhouse gas (GHG) emissions defined by the Greenhous Gas Protocol, the most widely accepted accounting standards for GHG emissions (https://ghgprotocol.org/). These might include reducing energy consumption on farms or reducing other emissions directly released from the farms to the environment, such as ammonia and methane from animal housing or fields (MacLeod 2013, Leinonen et al. 2014, FAO 2016).

However, a substantial proportion of the overall environmental impacts related to livestock production are actually produced by indirect activities (including so-called Scope 3 GHGs). For instance, producing feed for animals requires cultivation of feed crops. This process generates emissions for example through energy consumption of field operations, the use of fertilizers, and energy consumption in production of those fertilizers. Additional emissions

Published by Burleigh Dodds Science Publishing Limited, 2025.

are generated by processing and blending feed ingredients, and transporting them, sometimes across considerable distances (Hietala et al. 2023, Usva et al. 2023).

Moreover, the environmental impacts arising from both direct and indirect sources can lead to complex interactions. Enhancing the environmental sustainability of one sub-system of production might inadvertently result in detrimental indirect consequences on other links within the production chain. Furthermore, certain actions in livestock production can simultaneously yield environmental benefits and drawbacks. For instance, altering the composition of livestock feed simultaneously impacts nutrient emissions from manure and the environmental impacts connected to feed production, but can also have impact on animal performance and production efficiency (Huuskonen et al. 2023). Furthermore, employing livestock manure as fertilizer could mitigate the environmental impacts linked to manufacturing fertilizers, yet it could simultaneously elevate nutrient leaching and emissions into the atmosphere.

Navigating such intricate interactions and trade-offs demands a systematic, quantitative, and comprehensive approach to assess environmental impacts within agricultural systems. In recent times, system-based LCA models have been designed specifically for agricultural production (Williams et al. 2010, Leinonen et al. 2012a, 2012b, 2016, Hietala et al. 2023, Usva et al. 2023). Following the general principles of LCA modelling (BSI 2006), these methods encompass all environmental burdens across the production cycle, from raw material extraction to final products. However, a distinctive aspect of this systems-based methodology is its fusion of structural models of agricultural industries with crop and livestock production models, which sometimes can be based on rather complex, process-based approaches. This fusion is designed to effectively manage interactions between different subsystems. Within the systems models, quantities of inputs and outputs are intricately connected to other subsystems. For instance, producing a certain quantity of broiler meat requires a defined number of living animals which, in turn, necessitates a specific number of parent birds and a certain amount of feed and other inputs to production. All these interactions need to be quantified through model calculations.

As an outcome, an LCA model typically assesses emissions into the environment, together with the consumption of resources, and quantifies them in relation to a specifically selected "functional unit." The determination of the functional unit is dependent on the intended use of the assessment, and it is especially critical for achieving comparability between different products or services being analyzed. For example, if the aim is to compare one broiler meat production system to another broiler system, the functional unit can be simply '1 kg of broiler meat delivered to consumer'. However, especially in food production systems, mass- or volume-based functional units rarely provide sufficient comparability between food products, the actual functioning of which can strongly differ from each other. For example, it is not meaningful to compare 1 kg broiler meat to 1 kg of milk or 1 kg of bread, as the nutritional composition and intended use of these quantities differ strongly from each other.

In food products, therefore, it would be useful to specify the function of different items, for example based on their nutritional properties. As an example, the concentration and quality of protein has been used for formulation of so called "nutritional functional units" for food products (e.g. McAuliffe et al. 2023). However, challenges arise in creating an indicator covering comprehensively the nutritional values of a product and therefore these kinds of LCA studies are still rare (McLaren et al. 2021).

3 LCA environmental impact categories

When all the resources use and material flows (including emissions to the environment) related to the production system have been quantified, they can be aggregated into environmentally functional impact categories (e.g., Guinée et al. 2002, Leinonen et al. 2012a, Hietala et al. 2021, 2023, Usva et al. 2023). Several impact assessment methods exist to define the mathematical formulas used to produce the result of the impact category indicator from the input data, e.g. ReCiPe (Huijbregts et al. 2016) and IMPACT World+ (Bulle et al. 2019). There are two types of indicators: midpoint indicators measure the potential impact on certain environmental problem (e.g. climate change) and endpoint indicators measure the overall impact on certain area of protection summing up several midpoint indicators.

Probably the most widely used environmental impact category is the global warming potential (GWP), or "carbon footprint". This indicator measures the climate change impact in terms of greenhouse gas emissions released into the atmosphere, and technically should also quantify the amount of greenhouses gases possibly removed from the atmosphere. Key sources of these emissions in poultry production include:

- carbon dioxide (CO_2) from fossil fuels and land use change (LUC); and
- nitrous oxide (N_2O), and methane (CH_4) from feed crop production and animal farming.

In LCA models, the partial losses of ecosystem carbon storage resulting from land use conversion must be included in CO_2 emissions (so called Land Use and Land Use Change, or LULUC emissions), thereby affecting the global warming potential of the system (IPCC 2006, BSI 2011).

Eutrophication potential is a metric used to evaluate the excessive supply of nutrients, often termed as unnatural fertilization, which occurs when these nutrients find their way into water systems through leaching, run-off, or atmospheric deposition. The main contributors to this potential are the leaching of nitrate (NO^{3-}) and phosphate (PO_4^{3-}) into water bodies, as well as the emission of ammonia (NH_3) into the air. The eutrophication potential is sometimes divided to marine eutrophication, which is driven by nitrogen, and freshwater eutrophication, which is driven by phosphorus.

Acidification potential predominantly serves as an indicator of the potential decline in soil pH. The primary sources of this potential include ammonia emissions and sulfur dioxide (SO_2) stemming from the combustion of fossil fuels. Furthermore, in poultry production ammonia (NH_3) is also a significant source of acidification.

In addition to assessing emissions into the environment, LCA can gauge the utilization of various resources linked to the production of the specific functional unit under examination. These resources encompass elements like primary energy consumption, water usage, and land occupation. These factors are quantified in terms of the cumulative utilization of resources, including diverse energy sources (which include also the energy necessary for their extraction and distribution), such as gas, oil, coal, nuclear, and renewable energy.

Similarly, resource consumption can also be measured in terms of fresh water withdrawn by humans. Water scarcity impact is used to assess the contribution of water consumption in the product chain to the local water scarcity. The indicator (Boulay et al. 2018) refers to the share of the consumed water to the local freshwater resource after the necessary water needs of ecosystems and humans are fulfilled.

Land occupation describes the area of the land required to produce a unit of the product. In the case of poultry production, this mainly consists of the arable land for producing crops for feed. Closely related to land occupation is also the impact on biodiversity. In addition to the area of occupied land, this indicator can also take into account the land use changes and the location, type, and management of the occupied land (Kyttä et al. 2023).

4 Assessing the environmental impacts of different stages of the poultry production chain

The type and the magnitude of environmental impacts of poultry meat products differ between the life cycle stages. These separate stages include for example production of animal feed and other inputs to poultry system, animal farming, slaughtering and manufacturing the products, logistics, and consumer's actions. Systems-based LCA models for poultry production chains have been applied by several authors, to understand the magnitude of environmental impacts of the overall production systems and specifically the impacts of these different stages

(Costantini et al. 2021, Martinelli et al. 2020, Skunca et al. 2018, Gonzáles-García et al. 2014, Cesari et al. 2017, Prudêncio da Silva et al. 2014, Leinonen et al. 2012a, b). These studies have also given insight into the possibilities of improving the environmental sustainability in the entire poultry production chain.

By far the biggest direct impacts within the production cycle of poultry meat are related to primary production, i.e. production of feed and animal farming. Probably the most comprehensive, national scale study in this area was produced recently by Usva et al. (2023), who published an LCA study on Finnish broiler meat with detailed production data provided by the poultry industry and individual farms, covering almost 90% of broiler chicken production in Finland. The research focused on the animal farming and feed production stages of the Finnish broiler production chain and studied climate and water scarcity impacts.

Even though the direct impacts of slaughtering and meat processing, mainly arising from energy and water consumption, are relatively small (e.g. Costantini et al. 2021), the efficiency of the material flow in terms of utilized portion of the bird's live weight determines the number of birds needed to produce the required amount of meat and, therefore, also the quantities of inputs and emissions related to production. The yield of meat is affected both by the slaughtering process and by the genetics of the bird and by farming practices like the slaughtering age.

In addition to the various stages of production of poultry meat, the use (consumption) stage also has both direct and indirect impacts on the sustainability of poultry systems. Whether poultry meat is consumed in the restaurant or at home affects impacts from activities such as cooking, cleaning, and the logistics related to purchasing. However, in terms of environmental impacts, the relative share of these actions is low compared to the production of meat. Furthermore, consumer actions hardly ever focus solely on any specific food item such as poultry meat since, in most cases, meals consist of several food items. Any environmental assessment of human food consumption should therefore focus on the meal or on diet, rather than looking at single food ingredients alone. In general, although the use stage produces relatively low direct emissions, aggregate consumer choices have a major global environmental impact (e.g. for higher meat consumption) as well as the proportion of total food produced food that is actually consumed relative to the proportion wasted (given that the retail and consumption stages in agrifood supply chains can be responsible for up to 30% of food produced being wasted). It should be kept in mind therefore that the demand and preferences of consumers have a significant impact on overall food production chains, including the production of poultry meat.

4.1 The environmental impacts of feed production

When considering primary production, the findings of e.g. Usva et al. (2023) and Leinonen et al. (2012a, 2012b) found that, in all broiler production

Published by Burleigh Dodds Science Publishing Limited, 2025.

systems, feed production is the primary component responsible for several environmental impact categories, including global warming, eutrophication, acidification and water scarcity potentials. Feed production encompasses activities such as cultivating feed crops and processing and transporting ingredients and feed. In intensive broiler production systems in Finland, the contribution of feed production without land use change accounted for over 70% of the climate emissions of primary production, and this figure was nearly 80% when land use change was included (Usva et al. 2023). This is consistent with other recent studies. According to previous research, the significant role of feed in the climate change impact of broiler meat has been demonstrated by e.g. Martinelli et al. (2020), Skunca et al. (2018), Cesari et al. (2017), Kalhor et al. (2016), Prudêncio da Silva et al. (2014), González-García et al. (2014), Leinonen et al. (2012a), and Pelletier (2008).

The GHG emissions arising from poultry feed production originate mainly from the use of fossil energy in fertilizer manufacturing, consumption of fossil fuels in field operations in crop cultivation (e.g. from operating equipment such as tractors), direct (mainly nitrogen) emissions from fertilizer application and, to a smaller extent, transport and processing of feed (Leinonen et al. 2012a, 2012b, Hietala et al. 2023, Usva et al. 2023). In addition to direct and indirect N_2O emissions resulting from the cultivation of feed crops (significantly contributing to the global warming potential), nutrient leaching also affects the eutrophication potential (Leinonen et al. 2012a, 2012b; Hietala et al. 2022a).

Land use change (LUC) and related GHG emissions are strongly associated with feed production, and especially with the production of soya and palm oil in South America and South Asia (Leinonen et al. 2012a, 2012b, 2013). The magnitude of the climate change impact due to land use change depends on the diet of poultry and origin of the feeds, and can be significant. For example, LUC emissions account for 23% of all climate change impacts of primary production of Finnish broiler meat (Usva et al. 2023). Within the LCA framework, LUC emissions are generally considered an important indicator to include when assessing the climate change impacts of livestock production, and they are also included in commonly applied LCA guidelines such as the European Commission Product Environmental Footprint (PEF) framework (European Commission 2021). In addition to CO_2 emissions arising from the land use and management changes, it would be also possible to take into account the carbon sequestration (negative CO_2 emissions) obtained, for example as a result of improved farming practices. However, due to uncertainties associated with the method and the lack of data on the assessment of such LUC effects, carbon uptake is often recommended to be reported separately (e.g. European Commission 2021).

The loss of biodiversity associated with diets consumed in Finland was studied by Kyttä et al. (2023). Their main conclusion was that the majority of the biodiversity impacts arose from land-use change. Poultry has a relatively high biodiversity impact due to the consumption of soy feed and, for this reason, broiler meat was found to be one of the most significant drivers of biodiversity loss associated with food consumption in Finland (Kyttä et al. 2023).

4.2 The environmental impacts of poultry farming

In addition to the emissions related to feed production, other significant environmental impacts arise from animal farming and its inputs, including production of chicks, manure management at farms, and the consumption of other inputs such as energy and water on farms. These other sources account for about 20% of the climate and water scarcity impacts represented by each slaughtered broiler, when looking at broiler meat production chain from cradle to slaughterhouse in Finland (Usva 2023).

Direct emissions resulting from housing and manure management play a role in contributing to both eutrophication potential and acidification potential in broiler production systems (Leinonen et al. 2012a, 2012b). The primary factor behind this impact is the release of ammonia, which contributes to both eutrophication and acidification potentials. Additionally, nitrate leaching following land application affects eutrophication potential. Furthermore, housing and manure management also contribute to the global warming potential, primarily due to direct and indirect nitrous oxide emissions. However, the share of manure management in broiler production is less than 10% of the overall global warming potential (Usva et al. 2023). According to recent studies, this is smaller share than, for example, in beef or pork production (13–15% for beef and 17% for pork, Hietala et al. 2021, Hietala, Submitted) due to overall efficiency of intensive broiler production and relatively low amount of excreted nitrogen compared to the amount of meat produced. Improved feed efficiency as a result of breeding has reduced the amount of nitrogen excreted by the birds and, over decades, this has reduced the global warming, eutrophication and acidification caused by poultry production. This aspect is further discussed below.

Despite the substantial emissions, poultry manure offers some environmental benefits. When used in agricultural crop production, it can replace a portion of manufactured fertilizers, thereby partially offsetting the greenhouse gas emissions resulting from poultry production (Leinonen et al. 2012a, 2012b).

Farm energy consumption is often considered a key target in reducing the overall environmental impact of poultry meat production. Historically, it made a significant contribution to the GHG emissions related to poultry production. (Leinonen et al. 2012a). However, over the last decades, the energy efficiency of poultry farming has increased considerably. In addition, the electricity used (especially for ventilation, lighting and feeding in broiler farms) is becoming progressively "greener" in some countries. As a result, the contribution of direct farm energy use is currently very small in many countries compared to other components of the environmental impacts of poultry production. As an example, in Finland energy use at farm level was less than 4% of overall emissions, mainly because of the overall shift to renewable energy sources in Finland (Usva et al. 2023). In the case of eutrophication and acidification potential or water scarcity, the contribution of farm energy is even lower.

Water is consumed at poultry farms for both drinking and washing. This direct water consumption is the most visible part of the total water consumption, but it contributes only to about 2% to the total water scarcity impact of poultry production (Usva et al. 2023), with the major contribution to the water scarcity arising from feed production. The magnitude of the water scarcity impact depends strongly on the region where the water is withdrawn and what the water availability is in that location. This means that there is huge variation in the magnitude of the water scarcity impact in poultry production systems all over the globe due to local conditions both at poultry farms and the sites of feed crop cultivation. The volume of irrigation water used for feed crop production can be high. However, in the Finnish broiler production system, e.g. most locally produced feeds originate from the rain-fed cultivation systems (Usva et al. 2023). As an example, even though the share of imported, irrigated maize is less than 2% in an average diet for broilers in Finland, the contribution of maize to the total water scarcity impact of an average industrial feed mixture is about 15% (Usva et al. 2023). This example illustrates the significance of the cultivation system in terms of the water scarcity impact.

4.3 Comparing intensive and extensive systems

The intensity of the production has also a major role in environmental sustainability. Leinonen et al. (2012a, 2012b) conducted a comparative analysis of different broiler production systems, which included intensive indoor production, free-range, and organic production. Their findings revealed that the variations in environmental impacts were largely associated with the resource efficiency of each system. The study observed a consistent trend: less intensive poultry systems tended to have higher environmental impacts than more intensive systems, though there was significant variation between farms within each system (Leinonen et al. 2012a, 2012b). These differences were largely associated with the differences in the bird genotypes, their growth rate and associated feed efficiency. This is further discussed in Section 5.

The lower efficiency of organic systems, both in the production of organic feed crops and in animal production itself (feed efficiency), is a significant factor affecting the differences between the organic and conventional poultry systems. For example, it was found that the energy use in organic arable farming is 10–30% and in organic vegetable farming 40–50% higher than in their respective conventional counterparts. Therefore, the GHG emissions in organic arable and vegetable farming were estimated to be 0–15% and 35–40% higher, respectively (Bos et al. 2014).

However, it is essential to acknowledge that the relatively low productivity and high emissions of organic and other extensive systems can be partially offset by the lower input of resources required for these systems, including for example fertilizer inputs to organic feed crop production. While studies like Leinonen et al. (2012a, 2012b) have reported higher acidification potential and eutrophication potential in organic poultry systems compared to other systems, contradictory results have also been observed. Some studies indicate that extensive forms of livestock production can reduce the use of fossil fuels, fertilizers, and other inputs (Haas et al. 2001, Basset-Mens et al. 2009, Boggia et al. 2010) or have lower emissions from housing (Dekker et al. 2011). Additionally, organic production's requirement for certified feed sources allows for the avoidance of crops from "unsustainable" origins, thus potentially partially reducing GHG emissions related to direct land use change (LUC) and, consequently, the global warming potential.

5 Options to improve the environmental sustainability of poultry production

Given the importance of its range of environmental impacts, this section focuses first on improvements relating to the role of feed in poultry production. It then looks at ways of addressing non feed-related environmental impacts.

5.1 Addressing feed-related environmental impacts

As discussed in the above section, the direct influence of feed consumption per unit of produced poultry meat dominates overall environmental impacts across different poultry systems (Leinonen et al. 2012a, 2012b, 2016, Williams et al. 2006, Usva et al. 2023), and therefore, feed-related activities should also be the primary target in improving the environmental sustainability of broiler production. The significance of enhancing feed efficiency becomes evident, as this single component holds the potential to notably decrease GHG emissions attributed to poultry production; the processes related to feed production itself also offer further opportunities for reducing impacts.

As discussed above, the majority of the feed-related GHG emissions are related to the cultivation, processing and transportation of feed (Hietala et al. 2022a, Usva et al. 2023). Amongst these activities, the burning of fossil fuels is a major contributor to emissions, both in the production process of fertilizers, and also in the farming activities related to crop cultivation. Following the international agreements on the reduction of GHG emissions, it can be expected that the proportion of renewable energy used both in the fertilizer production and farming of feed crops will increase in the future, and this will lead to considerable reduction of the feed-related emissions. Therefore, it is likely that the relative role of the "non-fossil" components of feed production emissions, namely the nitrous oxide emissions from fertilizer application together the LUC emissions, will become even more significant and will be the major target in reduction of the feed production emissions in the future.

In addition to its influence on global warming potential, the quantity and composition of feed has a substantial influence on the eutrophication and acidification potentials related to poultry production, both directly and indirectly. The production of feed crops is tied to diverse direct emissions, including nitrate leaching into water and ammonia emissions into the atmosphere. Indirect effects related to feeding include emissions from poultry housing and manure management, similarly contributing to the eutrophication and acidification potential in poultry production. The scale of these emissions directly hinges on the nitrogen content in the manure which, in turn, is contingent upon the birds' nitrogen conversion efficiency, namely the proportion of nitrogen retained in the bird's body or eggs (in the form of protein) divided by the nitrogen acquired from the feed (also in the form of protein). A significant portion of this surplus nitrogen is released into the environment through various emissions, although some can be utilized as crop nutrients if the manure is employed as fertilizer on fields. Therefore, it is obvious that improved feed efficiency has multiple effects on the environmental performance of poultry production. Reduced feed consumption can simultaneously reduce the climate effects and other impacts of feed production and the emissions arising from housing and manure management.

It is important to keep in mind that some environmental impacts have a strong regional dimension. For example, in terms of biodiversity and water scarcity, the magnitude of the impact depends not only on the resource consumed (land use, water consumption) but also on the characteristics of the site. For example the water scarcity impact of poultry meat is strongly dependent on the geographical locations where feeds are produced, resulting in differences in impacts between poultry production systems. To improve environmental performance, it is important to trace the geographic origin of all feed ingredients and other inputs to quantify the actual impact of these materials.

Globally, soya bean meal serves as a primary protein source in poultry diets due to its favorable amino acid composition. However, this choice brings about challenges beyond the conventional environmental impacts associated with feed production. Notably, a considerable portion of the soya bean utilized for animal feed in the Europe is sourced from South America, where recent large-scale land use changes (LUC) have taken place. As mentioned earlier, these land use changes must be factored in when assessing the overall greenhouse gas emissions and biodiversity impact linked to crop production and its subsequent use as animal feed. Currently, soya production in Europe is unable to cover the demand from the European poultry sector. This situation is likely to be exacerbated by unstable conditions due to climate change, e.g. droughts and heat waves, which increase food insecurity. This suggests that e.g. novel plant breeding techniques solutions, including gene editing, should be further studied for their potential to produce more resilient crop varieties.

The significance of LUC emissions in broiler production was demonstrated e.g. by Usva et al. (2023), who reported 23% of the Finnish broiler chickens' global warming potential originating from feed crop production-related LUC. The impact was mainly generated in soya bean production in South America. Although South American soya bean meal accounted for only part of the soya bean meal in broiler chicken diets, its impact was notable. Besides South American soya bean meal, soya from other sources with much lower LUC can be utilized in broiler production. While the shift to different sources of soya bean can aid in reducing LUC emissions from the production, it was noted by Usva et al. (2023) that a trade-off was occurring with other environmental impacts, most notably with increased water scarcity impacts, e.g., due to irrigation and use of other inputs.

This example demonstrates a common strategy to mitigate the feed-related environmental impacts in poultry production through selecting feed ingredients with lower environmental footprints during their production phase. In addition to aiming for sustainable sources of soya bean meal, another approach towards more environmentally sustainable livestock diets involves substituting at least some of the soya bean meal with other, preferably local, protein sources (Leinonen et al. 2013). It is now becoming clear that alternative protein sources such as potato protein, corn gluten meal, rapeseed meal, sunflower meal and by-products of animal production, together with use of synthetic amino acid supplements (to improve nutrient use efficiency), have the potential to replace soya bean meal in commercial poultry diets (Leinonen et al. submitted).

It is also important to notice that even if the change of the origin of soya or the use of alternative protein sources can reduce direct LUC emissions of a particular broiler production system, the 'traditional' use of LCA may not capture all indirect LUC effects occurring globally. It can be argued that any use of agricultural crops will increase global land use pressure and thus contribute to LUC. It has therefore been suggested that these indirect LUC effects should also be included in LCAs for animal products (e.g. Audsley et al. 2009). In general, a completely different LCA approach, so called consequential LCA, may be needed to take into account all globally occurring changes in environmental

Published by Burleigh Dodds Science Publishing Limited, 2025.

impacts resulting from large scale system changes (e.g. Ekvall and Weidema 2004).

Considering alternative 'environmentally friendly' ingredients in poultry diets also requires evaluating their potential effects on poultry performance, including growth rate and feed efficiency. One requirement for high feed efficiency is am optimally balanced nutrient profile. Such a profile enables birds to efficiently utilize a substantial portion of their dietary nutrient intake, thereby reducing nitrogen and phosphorus excretion and also reducing overall feed consumption. Any changes in performance could significantly alter environmental outcomes. For instance, poorer feed conversion efficiency would increase environmental impacts per unit of product due to higher resource requirements, and such an effect has also been demonstrated for beef by Huuskonen et al. (2023) and dairy by Hietala et al. (2023). Similar effects would occur with increased mortality, as it would reduce the amount of output. As mentioned above, changes in bird performance would also impact manure management, with poorer efficiency leading to more nitrogen excretion, consequently affecting emissions to the environment.

In general, improving feed utilization through dietary optimization is an efficient way to gain positive environmental effects. In poultry production, optimal diets would decrease environmental impacts from both feed production and manure management. In addition to selecting optimal combination of feed ingredients, one way to achieve this is through feed processing techniques or additives like enzymes, to enhance digestion. For protein digestion in particular, improving efficiency is vital, as excessive protein use has various environmental consequences, as discussed above. One promising strategy to achieve this is to add the enzyme protease to feed to enhance protein utilization (Oxenboll et al. 2011, Leinonen and Williams 2015).

In addition to altering feeding strategies, genetic selection aimed at improving feed efficiency has been found to play a significant role in mitigating the environmental impacts of livestock production (Neeteson-van Nieuwenhoven et al. 2013, Hietala et al. 2022b, 2022c, Mehtiö et al. 2023). In broiler production, increased growth rate has been a prominent genetic trend over recent decades, resulting in reduced time to reach slaughter weight (Laughlin 2007). This trend leads to more metabolizable energy being retained as body protein and lipid, while less is released as heat. In practical terms, this is observed as reduced feed consumption by birds (de Beer et al. 2011; Laughlin 2007; Mussini 2012, Tallentire et al. 2016, 2018, Zuidhof et al. 2014).

Studies suggest that these genetic changes, especially the decrease in feed consumption, have substantially lowered greenhouse gas, ammonia, and nitrate emissions from poultry production (Defra 2008, Pelletier 2010). A recent study by Leinonen (submitted) compared a wide range of modern fast-growing and slow-growing broiler genotypes to quantify their global warming potential. It was found that the climate impact of the fastest-growing genotypes was significantly smaller than the impact of the slowest-growing genotypes. This was mainly a result of the fact that a faster growth rate reduces the time to reach slaughter weight, and therefore less energy is lost as heat during the whole growth cycle and less feed is also needed to reach slaughter weight.

In addition to improved feed energy efficiency as a result of faster growth, differences in the body composition (protein vs. lipid content) of modern broilers also have an effect on energy consumption of birds and on the corresponding global warming potential. However, this effect is much smaller than the effect of the high growth rate (Leinonen, submitted).

Although, in theory, reducing the protein concentration of the diet may reduce the GWP through reducing the inclusion of some high-emission ingredients, this effect is also dependent on the differences between the genotypes. In fast-growing genotypes, low feed protein concentration can cause adverse effects on the growth rate of the birds which can then increase environmental impacts. This effect is not as strong in slower-growing genotypes. It can be concluded that, in terms of environmental sustainability, it may be more beneficial to use low-protein diets for slow-growing birds than for fast-growing ones (Leinonen, submitted).

While conventional poultry systems have been able to improve their environmental performance due to their efficiency, where the main driver is the fast growth rate and short production cycles (Leinonen et al. 2012a, 2012b), modern poultry production aims not only for efficiency but also places significant emphasis on improving bird welfare. A commonly increasing trend, at least in some countries in Europe, is to favor slow-growing bird genotypes in broiler meat production (Chodova et al. 2021, Dal Bosco et al. 2014, Devatkal et al. 2019, Fanatico et al. 2007, 2008). As already discussed, the necessary consequence of this practice is lower feed efficiency, and the need to produce more feed to obtain a certain amount of meat which, in turn, increases various environmental impacts. As a result, when moving from conventional production to alternative options where slow-growing birds are used, it is even more important to find additional ways to reduce feed-related impacts. Options for this might be diets with lower protein content and more sustainable protein sources (Leinonen, submitted).

5.2 Addressing non-feed-related environmental impacts

Changes in animal housing can also promote improved welfare, but they can also have consequences on the environmental performance of production. Leinonen et al. (2014) quantified the environmental consequences of some new welfare-enhancing production systems introduced in the EU for broiler and egg production. Their conclusion was that high environmental performance can also be achieved in modern, less intensive production systems.

The impacts arising from direct electricity consumption in broiler farms have been significantly reduced in several countries, including Finland, as a result of selecting sustainable energy sources and the general reduction of the emissions intensity in national generation of electricity. However, in many countries, the energy consumption in farms still makes a major contribution to the overall environmental impacts of poultry production. Therefore, improving the energy efficiency of broiler housing, together with increased use of sustainable energy, will remain a major target in improving the environmental sustainability of poultry production.

As previously discussed, a significant portion of eutrophication and especially acidification in poultry systems stems from housing and manure management. While improvements in feed efficiency can reduce these emissions, as discussed above, it is also reasonable to expect that direct measures focused on enhancing manure management at both the farm and end-use stages could contribute to emission reduction. In several countries a common practice involves spreading manure directly onto fields as fertilizer. While this practice offers certain environmental benefits, such as decreasing the need for synthetic fertilizers and increasing soil carbon storage (resulting in CO_2 removal from the atmosphere), a major drawback is the substantial release of ammonia and nitrate into the environment. Hence, alternative strategies for manure management are being explored (Lynch et al. 2013; Huang et al. 2015).

As one option, the use of poultry litter as an energy source has been suggested. In an early study, Williams et al. (2016) evaluated the environmental consequences of utilization of poultry litter as a fuel source for electricity generation at power stations, using the LCA framework. Another option is to apply anaerobic digestion for poultry litter and utilize the bioenergy and digestate. According to an LCA study by Beausang et al. (2020), this option can potentially provide considerable environmental benefits.

As shown in above examples, the options for improving the environmental sustainability of broiler meat production may differ between systems. However, only a minority of global poultry meat is produced in extensive systems while intensive specialized broiler systems cover 81% of the global production (MacLeod 2013). The poultry industry is largely run by few multinational enterprises, and the best practices are spread widely around the world. The largest poultry meat-producing companies are often vertically integrated and located in North and South America and in Eastern and South-Eastern Asia (Windhorst 2017).

Differences in options for improving the environmental performance of broiler systems may not always be related to the production system itself but, rather, to the specific inputs used in production. Examples of such differences include the regional production patterns of protein crops used in feed, or the country-specific energy generation systems, affecting the emission arising from the energy used in the production.

6 Conclusion

In general, LCA is a strong method to assess the environmental performance of current poultry systems and potential improvement options. As a system-based method, in assessing future scenarios LCA can identify true improvements in the environmental performance, instead of just shifting emissions one stage of the production chain to another. However, some LCA methods still need to be further improved. For example, methods to assess impacts on water scarcity and eutrophication are not as developed as climate impact, and especially the biodiversity impact category is undeveloped. The inclusion of the indirect LUC emissions may also need further development, as discussed above. In addition, LCA is a data intensive method, and the lack of data may be a challenge in case general level LCA is conducted. The existing LCA results are not always comparable or usable in other assessments due to different scopes and modelling solutions. If the aim of an LCA study is to improve environmental performance of a specific production chains, the primary data may be more easily available, and the study can provide results that support the improvement in the chains.

Addressing the environmental sustainability of livestock systems is a multidimensional challenge that requires a comprehensive understanding of interconnected factors. Systems modeling-based LCA offers a valuable approach to systematically navigate these intricate interactions, helping to pinpoint areas where significant enhancements in the sustainability of livestock systems can be achieved.

Although poultry production is generally deemed more environmentally sustainable compared to many other livestock commodities, it still contributes to environmental impacts such as global warming, eutrophication, acidification, water use, and biodiversity. Therefore, there remain opportunities to mitigate these impacts through a combination of strategies, including altering feeding practices, optimizing feed composition, genetic selection, animal housing, and energy use.

Among the various stages within the poultry production chain, the production of feed stands out as the primary contributor to the global warming potential of poultry systems. Exploring alternative feed ingredients, such as locally sourced protein crops, holds promise for improving multiple environmental impact categories, provided that these new feeding strategies do not negatively impact bird performance.

Improving feed efficiency emerges as a pivotal approach to reducing environmental impacts. Especially, genetic selection has shown the potential to decrease resource demands in broiler production, particularly feed consumption. Fast and slow growing broiler genotypes provide somewhat different potential solutions for improving their environmental performance. Fast growth itself improves feed efficiency and reduces the impacts related to feed production. On the other hand, slow-growing birds may be able to better utilize "climate-friendly" feed ingredients for example in the form of low-protein diets.

7 Where to look for further information

- Costantini, M., Ferrante, V., Guarino, M. and Bacenetti, J. 2021. Environmental sustainability assessment of poultry productions through life cycle approaches: a critical review. *Trends Food Sci. Technol.* 110: 201–212.
- Hietala, S., Usva, K., Nousiainen, J., Vieraankivi, M.-L., Vorne, V. and Leinonen, I., 2022. Environmental impact assessment of Finnish feed crop production with methodological comparison of PEF and IPCC methods for climate change impact. *J. Clean. Prod.* 379 (1). https://doi.org/10.1016/j.jclepro.2022.134664.
- Tallentire, C.W., Leinonen I. and Kyriazakis I. 2018. Artificial selection for improved energy efficiency is reaching its limits in broiler chickens. *Scientific Reports* 8: 1168. doi:10.1038/s41598-018-19231-2.
- Tallentire, C.W, Leinonen, I. and Kyriazakis, I. 2016. Breeding for efficiency in the broiler chicken. A review. *Agronomy for Sustainable Development* 36: 66. doi:10.1007/s13593-016-0398-2.
- Usva K., Hietala, S., Nousiainen, J., Vorne V., Vieraankivi, M-L., Jallinoja, M., and Leinonen, I. 2023. Environmental life cycle assessment of Finnish broiler chicken production – focus on climate change and water scarcity impacts. *Journal of Cleaner Production* 410, 137097. https://doi.org/10.1016/j.jclepro.2023.13709

8 References

Audsley, E., Brander, M., Chatterton, J., Murphy-Bokern, D., Webster, C. and Williams, A. 2009. *How low can we go? An assessment of greenhouse gas emissions from the UK food system and the scope to reduce them by 2050.* WWF-UK.

Basset-Mens, C., Ledgard, S. and Boyes, B. 2009. Eco-efficiency of intensification scenarios for milk production in New Zealand. *Ecol. Econom.* 68: 1615–1625.

Beausang, C., McDonnell, K. and Murphy, F., 2020. Anaerobic digestion of poultry litter – A consequential life cycle assessment. *Sci. Total Environ.* 735: 139494.

Boggia, A., Paolotti, L. and Castellini, C. 2010. Environmental impact evaluation of conventional, organic and organic-plus poultry production systems using life cycle assessment. *World's Poult. Sci. J.* 66: 95–114.

Bos, J. F., de Haan, J., Sukkel, W. and Schils, R. L. 2014. Energy use and greenhouse gas emissions in organic and conventional farming systems in the Netherlands. *NJAS-Wagen. J. Life Sci.* 68: 61–70.

Boulay, A. M., Bare, J., Benini, L. *et al.* 2018. The WULCA consensus characterization model for water scarcity footprints: assessing impacts of water consumption based on available water remaining (AWARE). *Int. J. Life. Cycle. Assess.* 23: 368–378. https://doi.org/10.1007/s11367-017-1333-8

BSI. 2006. Environmental management: life cycle assessment – Principles and framework. http://www.iso.org/iso/iso_catalogue/catalogue_tc/catalogue_detail.htm?csnumber=37456

BSI. 2011. Specification for the assessment of the life cycle greenhouse gas emissions of goods and services. *PAS.* 2050: 2011. http://www.bsigroup.com/Standards-and-Publications/How-we-can-help-you/Professional-Standards-Service/PAS-2050/

Bulle, C., Margni, M., Patouillard, L. et al. 2019. IMPACT World+: a globally regionalized life cycle impact assessment method. *Int. J. Life Cycle Assess.* 24: 1653–1674. https://doi.org/10.1007/s11367-019-01583-0

Cesari, V., Zucali, M., Sandrucci, A., Tamburini, A., Bava, L. and Toschi, I. 2017. Environmental impact assessment of an Italian vertically integrated broiler system through a life cycle approach. *J. Clean. Prod.* 904–911. https://doi.org/10.1016/j.jclepro.2016.12.030

Chodová, D., Tůmová, E., Ketta, M. and Skřivanová, V. 2021. Breast meat quality in males and females of fast-, medium- and slow-growing chickens fed diets of 2 protein levels. *Poult. Sci.* 100(4): 100997.

Costantini, M., Ferrante, V., Guarino, M. and Bacenetti, J. 2021. Environmental sustainability assessment of poultry productions through life cycle approaches: a critical review. *Trends Food Sci. Technol.* 110: 201–212.

Dal Bosco, A., Mugnai, C., Amato, M.G., Piottoli, L., Cartoni, A. and Castellini, C. 2014. Effect of slaughtering age in different commercial chicken genotypes reared according to the organic system: 1. Welfare, carcass and meat traits. *Ital. J. Anim. Sci.* 13: 467–472.

de Beer, M., Elfick, D. and Emmerson, D. A. 2011. Is a feed conversion ratio 1:1 realistic and appropriate goal for broiler chicken in the next 10 years? *Recent Adv. Anim. Nutr. – Aust.* 18: 9–14.

Defra. 2008. *A study of the scope for the application of research in animal genomics and breeding to reduce nitrogen and methane emissions from livestock based food chains*. Defra project AC0204.

Dekker, S. E. M., de Boer, I. J. M., Vermeij, I., Aarnink, A. and Groot Koerkamp, P. W. G. 2011. Ecological and economic evaluation of Dutch egg production systems. *Livest. Sci.* 139: 109–121.

Devatkal, S. K., Naveena, B. M. and Kotaiah, T. 2019. Quality, composition, and consumer evaluation of meat from slow-growing broilers relative to commercial broilers. *Poult. Sci.* 98(11): 6177–6186.

de Vries, M. and de Boer, I. J. M. 2010. Comparing environmental impacts for livestock products: a review of life cycle assessments. *Livest. Sci.* 28: 1–11.

Ekvall, T. and Weidema, B. P. 2004. System boundaries and input data in consequential life cycle inventory analysis. *Int. J. Life Cycle Ass.* 9: 161–171.

Eriksson, I. S., Elmquist, H., Stern, S. and Nybrant, T. 2005. Environmental systems analysis of pig production – The impact of feed choice. *Int. J. Life Cycle Ass.* 10: 143–154.

European Commission 2021. C (2021) 9332 final: COMMISSION RECOMMENDATION on the use of the Environmental Footprint methods to measure and communicate the life cycle environmental performance of products and organisations. https://eur-lex.europa.eu/legal-content/EN/TXT/?uri=PI_COM:C(2021)9332

Fanatico, A. C., Pillai, P. B., Emmert, J. L. and Owens, C. M. 2007. Meat quality of slow- and fast-growing chicken genotypes fed low-nutrient or standard diets and raised indoors or with outdoor access. *Poult. Sci.* 86: 2245–2255.

Fanatico, A. C., Pillai, P. B., Hester, P. Y., Falcone, C., Mench, J. A., Owens, C. M. and Emmert, J. L. 2008. Performance, livability, and carcass yield of slow- and fast-growing chicken genotypes fed low-nutrient or standard diets and raised indoors or with outdoor access. *Poult. Sci.* 87: 1012–1021.

FAO. 2006. *Livestock's long shadow environmental issues and options*. FAO.

FAO. 2016. *Greenhouse gas emissions and fossil energy use from poultry supply chains: guidelines for assessment*. Livestock Environmental Assessment and Performance Partnership. FAO.

FAO. 2020. *Emissions due to agriculture. Global, regional and country trends 2000–2018*. FAOSTAT Analytical Brief Series No 18. Rome.

Gerber, P. J., Steinfeld, H., Henderson, B., Mottet, A., Opio, C., Dijkman, J., Falcucci, A. and Tempio, G. 2013. *Tackling climate change through livestock: a global assessment of emissions and mitigation opportunities.* Food and Agriculture Organization of the United Nations (FAO).

González-García, S., Gomez-Fernández, Z., Dias, A. C., Feijoo, G., Moreira, M. T. and Arroja, L. 2014. Life cycle assessment of broiler chicken production: a Portuguese case study. *J. Clean. Prod.* 74: 125–134.

Guinée, J. B., Gorrée, M., Heijungs, R., Huppes, G., Kleijn, R., Koning, A. de, Oers, L. van, Wegener Sleeswijk, A., Suh, S., Udo de Haes, H. A., Bruijn, H. de, Duin, R. van and Huijbregts, M. A. J. 2002. *Handbook on life cycle assessment. Operational guide to the ISO standards. I: LCA in perspective. IIa: Guide. IIb: Operational annex. III: Scientific background.* Kluwer Academic Publishers, ISBN 1-4020-0228-9, p. 692.

Haas, G., Wetterich, F. and Köpke, U. 2001. Comparing intensive, extensified and organic grassland farming in southern Germany by process life cycle assessment. *Agric. Ecosyst. Environ.* 83: 43–53.

Hietala, S., Heusala, H., Katajajuuri, J.-M., Järvenranta, K., Virkajärvi, P., Huuskonen, A. and Nousiainen, J. 2021. Environmental life cycle assessment of Finnish beef – cradle-to-farm gate analysis of dairy and beef breed beef production. *Agric. Syst.* 194: 103250.

Hietala, S., Usva, K., Nousiainen, J., Vieraankivi, M.-L., Vorne, V. and Leinonen, I. 2022a. Environmental impact assessment of Finnish feed crop production with methodological comparison of PEF and IPCC methods for climate change impact. *J. Clean. Prod.* 379(1). https://doi.org/10.1016/j.jclepro.2022.134664

Hietala, S., Leino, A. M., Nousiainen, J. and Huuskonen, A. 2022b. Improving efficiency of Finnish beef production through breeding with genomic selection effects climate impact of beef. In *Proceedings of 13th International Conference on Life Cycle Assessment of Food (LCA Foods 2022), 12-14 October 2022*, Lima, Peru. PELCAN-PUCP, Lima, Peru.

Hietala, S., Negussie, E., Astaptsev, A., Leino, A. M. and Lidauer, M. 2022c. Breeding towards efficiency in Finnish dairy and beef cattle improves environmental performance. In *Book of Abstracts of the 73rd Annual Meeting of the European Federation of Animal Science*. Wageningen Academic Publishers.

Hietala, S., Vanhatalo, A., Kuoppala, K., Kokkonen, T., Reinikainen, A., Timonen, K. and Välimaa, A-L. 2023. *Effects of increased forage to dairy cow diets on carbon footprint of raw milk*. Proceedings of 74th EAAP Annual Meeting, Joint International Congress on Animal Science co-organised by the European Federation of Animal Science (EAAP), the World Association for Animal production (WAAP) and Interbull On "Climate change, biodiversity and global sustainability of animal production" 26 August -1 September 2023, Lyon, France.

Huang, Y., Anderson, M., McIlveen-Wright, D., Lyon, G. A., McRoberts, W. C., Wang, Y. D., Roskilly, A. P. and Hewitt, N. J. 2015. Biochar and renewable energy generation from poultry litter waste: a technical and economic analysis based on computational simulations. *Appl. Energ.* 160: 656–663.

Huijbregts, M., Steinmann, Z., Elshout, P., Stam, G., Verones, F., Vieira, M., Zijp, M., Hollander, A. and Zelm, R. 2016. ReCiPe2016: a harmonised life cycle impact assessment method at midpoint and endpoint level. *Int. J. Life Cycle Assess.* 22. https://doi.org.10.1007/s11367-016-1246-y

Huuskonen, A., Hietala, S., Hyvönen, J., Leinonen, I. and Manni, K. 2023. Environmental impacts and animal performance of finishing bulls fed different silage-based total mixed rations. *Livest. Sci.* 268: 105166.

IPCC. 2006. IPCC Guidelines for National Greenhouse Gas Inventories. (2006 Guidelines). http://www.ipcc-nggip.iges.or.jp/

Kalhor, T., Rajabipour, A., Akram, A. and Sharifi, M. 2016. Environmental impact assessment of chicken meat production using life cycle assessment. *Inf. Process Agric.* 4: 262–271.

Kyttä, V., Hyvönen, T. and Saarinen, M. 2023. Land-use-driven biodiversity impacts of diets–a comparison of two assessment methods in a Finnish case study. *Int. J. Life Cycle Assess.* 28: 1104–1116. https://doi.org/10.1007/s11367-023-02201-w

Laughlin, K. 2007. *The evolution of genetics, breeding and production*. Temperton Fellowship Report No.15. Harper Adams University College.

Leinonen, I. (submitted). Comparing the global warming potential of modern fast and slow-growing broiler genotypes with alternative diets.

Leinonen, I. and Williams, A. G. 2015. Effects of dietary protease on nitrogen emissions from broiler production: a holistic comparison using Life Cycle Assessment. *J. Sci. Food Agric.* 95: 3041–3046.

Leinonen, I., Williams, A. G. and Kyriazakis, I. 2014. The effects of welfare-enhancing system changes on the environmental impacts of broiler and egg production. *Poult. Sci.* 93: 256–266.

Leinonen, I., Williams, A. G. and Kyriazakis, I. 2016. Comparing the environmental impacts of UK turkey production systems using analytical error propagation in uncertainty analysis. *J. Clean. Prod.* 112: 141–8.

Leinonen, I., Williams, A. G., Waller, A. H. and Kyriazakis, I. 2013. Comparing the environmental impacts of alternative protein crops in poultry diets: the consequences of uncertainty. *Agric. Syst.* 121: 33–42.

Leinonen, I., Williams, A. G., Wiseman, J., Guy, J. and Kyriazakis, I. 2012a. Predicting the environmental impacts of chicken systems in the UK through a life cycle assessment: broiler production systems. *Poult. Sci.* 91: 8–25.

Leinonen, I., Williams, A. G., Wiseman, J., Guy, J. and Kyriazakis, I. 2012b. Predicting the environmental impacts of chicken systems in the UK through a life cycle assessment: egg production systems. *Poult. Sci.* 91: 26–40.

Lynch, D., Henihan, A. M., Bowen, B., Lynch, D., McDonnell, K., Kwapinski, W. and Leahy, J. J. 2013. Utilisation of poultry litter as an energy feedstock. *Biomass Bioenergy*. 49: 197–204.

MacLeod, M., Gerber, P., Mottet, A., Tempio, G., Falcucci, A., Opio, C., Vellinga, T., Henderson, B. and Steinfeld, H. 2013. *Greenhouse gas emissions from pig and chicken supply chains: a global life cycle assessment*. Food and Agriculture Organization of the United Nations (FAO).

Martinelli, G., Vogel, E., Decian, M., Farinha, M. J. U. S., Bernardo, L. V. M., Borges, J. A. R., Gimenes, R. M. T., Garcia, R. G. and Ruviaro, C. F. 2020. Assessing the eco-efficiency of different poultry production systems: an approach using life cycle assessment and economic value added. *Sustain. Prod. Consum*. 181–193. https://doi.org/10.1016/j.spc.2020.07.007

McAuliffe, G. A., Takahashi, T., Beal, T., Huppertz, T., Leroy, F., Buttriss, J., Collins, A. L., Drewnowski, A., McLaren, S. J., Ortenzi, F., van der Pols, J. C., van Vliet, S. and Lee, M. R. F. 2023. Protein quality as a complementary functional unit in life cycle assessment (LCA). *Int. J. Life Cycle Assess*. 28(2): 146–155. https://doi.org/10.1007/s11367-022-02123-z

McLaren, S., Berardy, A., Henderson, A., Holden, N., Huppertz, T., Jolliet, O., De Camillis, C., Renouf, M., Rugani, B., Saarinen, M., van der Pols, J., Vázquez-Rowe, I., Antón Vallejo, A., Bianchi, M., Chaudhary, A., Chen, C., CooremanAlgoed, M., Dong, H., Grant, T., Green, A., Hallström, E., Hoang, H., Leip, A., Lynch, J., McAuliffe, G., Ridoutt, B., Saget, S., Scherer, L., Tuomisto, H., Tyedmers, P. and van Zanten, H. 2021. *Integration of environment and nutrition in life cycle assessment of food items: opportunities and challenges*. FAO.

Mehtiö, T., Mäntysaari, P., Negussie, E., Kempe, R., Sevón-Aimonen, M. L., Chegini, A., Hietala, S., Kostensalo, J. and Lidauer, M. 2023. *Breeding dairy cattle for resource efficiency and environmental sustainability: Final report of the A++ COW-project (2019-2023)*, Natural Resources Institute Finland.

Mussini, F. J. 2012. *Comparative response of different broiler genotypes to dietary nutrient levels*. PhD thesis, University of Arkansas, Fayetteville, AR.

Neeteson-van Nieuwenhoven, A. M., Knap, P. and Avendaño, S. 2013. The role of sustainable commercial pig and poultry breeding for food security. *Anim. Front*. 3: 52–57.

Opio, C., Gerber, P., Mottet, A., Falcucci, A., Tempio, G., MacLeod, M., Vellinga, T., Henderson, B. and Steinfeld, H. 2013. *Greenhouse gas emissions from ruminant supply chains: a global life cycle assessment*. Food and Agriculture Organization of the United Nations (FAO).

Oxenboll, K. M., Pontoppidan, K. and Fru-Nji, F. 2011. Use of a protease in poultry feed offers promising environmental benefits. *Int. J. Poult. Sci*. 10: 842–848.

Pelletier, N. 2008. Environmental performance in the US broiler poultry sector: life cycle energy use and greenhouse gas, ozone depleting, acidifying and eutrophying emissions. *Agric. Syst*. 98: 67–73.

Pelletier, N. 2010. *Breeding poultry for environmental performance: a life cycle-based supply chain perspective*. Proceedings of the 9th World Congress on Genetics Applied to Livestock Production Paper. 75. Leipzig, Germany.

Prudêncio da Silva, V., van der Werf, Hayo, M. G., Soares, S. R. and Corson, M. S. 2014. Environmental impacts of French and Brazilian broiler chicken production scenarios: an LCA approach. *J. Environ. Manag*. 222–231. https://doi.org/10.1016/j.jenvman.2013.12.011

Skunca, D., Tomasevic, I., Nastasijevic, I., Tomovic, V. and Djekic, I. 2018. Life cycle assessment of the chicken meat chain. *J. Clean. Prod.* 440–450. https://doi.org/10.1016/j.jclepro.2018.02.274

Steinfeld, H., Gerber, P., Wassenaar, T., Castel, V., Rosales, M. and de Haan, C. 2006. *Livestock's long shadow: environmental issues and options*. FAO.

Tallentire, C. W, Leinonen, I. and Kyriazakis, I. 2016. Breeding for efficiency in the broiler chicken: a review. *Agron. Sustain. Dev.* 36: 66. https://doi.org/10.1007/s13593-016-0398-2.

Tallentire, C.W., Leinonen I. and Kyriazakis I. 2018. Artificial selection for improved energy efficiency is reaching its limits in broiler chickens. *Sci. Rep.* 8: 1168. https://doi.org/10.1038/s41598-018-19231-2.

Usva K., Hietala, S., Nousiainen, J., Vorne V., Vieraankivi, M-L., Jallinoja, M. and Leinonen, I. 2023. Environmental life cycle assessment of Finnish broiler chicken production – focus on climate change and water scarcity impacts. *J. Clean. Prod.* 410: 137097. https://doi.org/10.1016/j.jclepro.2023.137097

van der Werf, H. M. G., Petit, J. and Sanders, J. 2005. The environmental impacts of the production of concentrated feed: the case of pig feed in Bretagne. *Agric. Syst.* 83: 153–177.

Williams, A. G., Audsley, E. and Sandars, D. L. 2006. *Determining the environmental burdens and resource use in the production of agricultural and horticultural commodities*. Main Report. Defra Research Project IS0205. www.agrilca.org

Williams, A. G., Audsley, E. and Sandars, D. L. 2010. Environmental burdens of producing bread wheat, oilseed rape and potatoes in England and Wales using simulation and system modeling. *Int. J. Life Cycle Assess.* 15: 855–868.

Williams, A. G., Leinonen, I. and Kyriazakis, I. 2016. Environmental benefits of using turkey litter as a fuel instead of a fertiliser. *J. Clean. Prod.* 113: 167–175.

Windhorst, H.-W. 2017. Chapter 1 - Dynamics and patterns of global poultry-meat production. In *Woodhead Publishing Series in Food Science, Technology and Nutrition, Poultry Quality Evaluation*. pp. 1–25. https://doi.org/10.1016/B978-0-08-100763-1.00001-5

Zuidhof, M. J., Schneider, B. L., Carney, V. L., Korver, D. R. and Robinson, F. E. 2014. Growth, efficiency, and yield of commercial broilers from 1957, 1978, and 2005. *Poult. Sci.* 12: 2970–2982.

Chapter 14

Use of alternative feed ingredients and additives to minimize the environmental impact of poultry production

Kelley G.S. Wamsley, Pratima Adhikari, and Timothy Boltz, Mississippi State University, USA

1 Introduction

The use of alternative feed ingredients is vital to the longevity and sustainability of animal protein production. Conventional ingredients used in North America are corn and soybean meal and, to a lesser extent, distillers' dried grains with solubles. Other cereals such as wheat and barley, as well as canola meal, are used for livestock feed in other parts of the world, like Canada, Europe, and South America (Kiarie and Nyachoti, 2009). With growing concern about the contribution of land use change (e.g. deforestation for crop cultivation) to climate change, the use of scarce land resources to grow crops for both food and feed is increasingly seen as environmentally unsustainable (Thronton, 2010). In addition, competition for these raw ingredients from other sectors such as biofuel processing and the food industry has increased, leading to an increased cost for these raw ingredients (Woyengo et al., 2014; Borowski et al., 2016). Since most diets are formulated on a least-cost basis, this makes the use of alternative protein sources important to both the environmental and economic sustainability of poultry production.

http://dx.doi.org/10.19103/AS.2024.0146.17

During processing operations, different industries, such as the biofuels and the food industries, produce co- and by-products that have little to no potential human use but are viable potential feed ingredients (Kiarie and Nyachoti, 2009). Other sources like insect meal have also attracted increasing interest due to ease of rearing, lower environmental impact, and good nutrient profile (Chadd, 2007; Churchward-Venne et al., 2017; Elahi et al., 2022). Another novel dietary source is algae. Two species of algae, *Chlorella* and *Spirulina*, are often used as these products have high protein, vitamin, mineral, and antioxidant content (Kovac et al., 2013; Evans et al., 2015; Grigorova et al., 2017; Madeira et al., 2017). With more diverse rotations becoming more common as part of the shift to more regenerative crop production, the use of nitrogen-fixing plants (such as various species of legumes) as cover crops has increased. These cover crops also provide a potential alternative to conventional soybean meal (Crépon et al., 2010; White et al., 2015).

The greater prevalence of smallholder production within short supply chains has allowed feeding alternative ingredients to remain a common practice in regions such as Asia and Africa. However, the increasing intensification, scale, and geographical specialization of poultry production in regions such as North America and Europe have made sourcing sufficient quantities of alternative feed sources of sufficiently high and consistent quality more challenging (Babatunde et al., 2021).

By definition, alternative feed sources come from a wide range of sources. Their nutritional composition may vary depending on various factors, including species/variety, seasonal factors, and growing/processing conditions. They may also contain toxins and anti-nutritional factors (ANFs), which compromise safety and reduce nutrient availability (Grosjean et al., 2000; Woyengo, et al., 2014). Phytate or phytic acids, for example, are non-digestible ANFs that can reduce nutrient digestibility in broilers by forming chelated forms of minerals, starch, amino acids (AAs), proteins, and carbohydrates.

However, there is now a growing body of research on these ingredients to identify their nutritional composition, nutrient utilization by various species of livestock and poultry, and optimal inclusion rates in diets (Crépon et al., 2010; Digiacomo and Leury, 2019; Babatunde et al., 2021). The major nutrient classes that alternative ingredients can provide include

- protein and
- energy and fiber.

These are discussed in more detail in the following sections.

2 Alternative protein-contributing ingredients

Protein is required by animals to supply AAs used in tissue synthesis to maintain optimal growth and health (Carbone and Pasiakos, 2019). The most common source of protein in diets currently comes from soybean meal. Increased demand and higher prices of soybeans have made alternative sources more commercially attractive. Protein sources come from one of two sources:

- plant-based sources and
- animal-based sources.

Plant-based sources are typically cheaper and more readily available than animal-based sources (Woyengo, et al., 2014; Babatunde et al., 2021). Many plant-based protein ingredients are sourced from leguminous crops as well as by-products from the food and oil industries (Crépon et al., 2010; Woyengo, et al., 2014; Babatunde et al., 2021). Common plant-origin protein sources include peas and beans, as well as by-products from food and oilseed processing such as oil palm kernel and cottonseed meal. All of these ingredients can be included in broiler diets at lower levels (6–9%) without negatively impacting bird performance (El Boushy et al., 1990; Laboissière et al., 2020).

Legumes are a good source of the essential AA lysine, but they are low in other essential AAs like methionine and tryptophan (Stein and de Lange, 2007). This may require supplementation with synthetic AAs to meet the nutritional requirements of the animal. The specific cultivar being grown can also have an impact on the overall nutritional quality of the product as different cultivars have varying nutrient compositions (Igbasan et al., 1997). These are important factors to consider when deciding whether an alternative source can and should be used in the diet.

Protein sources such as copra/coconut and oil palm kernel are byproducts of the oil extraction process and can be valuable feed ingredients. Copra and palm kernel meals can be found in tropical regions where these crops are grown (Stein et al., 2015). Both have relatively high levels of arginine and sulfur-containing AAs but have low levels of lysine (NRC, 1994; Sulabo et al., 2013). The use of palm kernel meal in poultry diets can be limited by its gritty and fibrous texture as well as its non-starch polysaccharide (NSP) content, hindering nutrient utilization by the animal (Knudsen, 1997; Alshelmani et al., 2017). Cotton is grown in multiple regions of the world (USDA-ERS, 2022). Cottonseed meal is known for a high crude protein (CP) level (41%) but can be difficult to use in formulation due to its high fiber level and the presence of the ANF gossypol (Leeson and Summers, 2005; Alshelmani, et al., 2021). Gossypol binds lysine, making this AA unavailable to the bird (Alshelmani, et al., 2021).

Gossypol can also produce undesirable colors in egg yolks (olive green) when fed to laying hens (Leeson and Summers, 2005; McDonald et al., 2010).

Anti-nutritional factors, such as tannins, protease inhibitors, lectins, and trypsin inhibitors, are also a major concern with plant-origin protein sources (Mustafa et al., 2000; Partanen et al., 2003; McNeill et al., 2004; Bampidis and Christodoulou, 2011). Thermal processing of these ingredients can decrease ANF content (James et al., 2005; Stein and Bohlke, 2007). However, care must be taken since, whilst thermal processing may aid in reducing ANF content, it may also reduce AA digestibility and vitamin concentration.

Livestock slaughter and processing plants are a source of protein-rich by-products, including feather and blood meals (Murray et al., 1997; Babatunde et al., 2021). Feather meal is a by-product of poultry processing. With around 7% of a bird's live weight being feathers, a large volume of feather meal can be generated (Zhang et al., 2014). Further processing is required to make the feathers digestible to animals, with steam hydrolyzation and enzymatic processing being the most common methods (Onifade et al., 1998; Riffel and Brandelli, 2002). Care must be taken to not overprocess feathers during steam hydrolyzation as this can damage the AAs present in the final product (Moritz and Latshaw, 2001; Queiroga et al., 2012).

Blood meal is another animal source of protein that can be used in diets, but it is not common in Europe and North America (M'Ncene et al., 1999). Blood meal is high in AAs, like lysine and methionine, but low in isoleucine (Khawaja et al., 2007; Babatunde et al., 2021). As with feather meal, how blood meal is processed can impact the composition and digestibility of AAs in the final product (Almeida et al., 2013).

As noted earlier, insect meal has been gaining attention recently as an alternative protein source in animal feed. While consuming insects is not new, large-scale, commercial rearing of insects as a feed source is recently new. Common species of insects include black soldier fly larvae, mealworms, crickets, and common houseflies (Churchward-Venne et al., 2017; Elahi et al., 2022). Insects can be fed to poultry live, dried/ground, or in paste form (Khan, 2018; Elahi et al., 2020). To improve the CP content and palatability of insect meals, a defatting process is required (Veldkamp and Bosch, 2015). Insect meals are high in CP (40–60%) and contain essential AAs, especially sulfur-containing AAs (Makkar et al., 2014; Kareem et al., 2018; DiGiacomo et al., 2019). The inclusion of black soldier fly meal at up to 15% of the overall feed has, e.g., been shown to improve broiler performance (Dabbou et al., 2018; Pieterse et al., 2019).

3 Alternative energy/fiber-contributing ingredients

When formulating diets, the bulk of the diet will contain ingredients that supply energy. Currently, the most common energy-contributing ingredient is corn, followed by other cereals such as wheat and barley, as well as Distillers' dried grains with solubles (Babatunde et al., 2021). The use of byproducts from other sectors and the cultivation and use of root crops have started to be investigated as alternative ingredients.

Cassava and sweet potatoes are some of the common root crops grown for human consumption that can also be included in feed for animals (Morgan and Choct, 2016; Babatunde et al., 2021; Harding et al., 2023). In regions that grow these crops, like Africa and South America for cassava and the United States for sweet potatoes, they can be viable energy components of diets. Cassava contains an ANF, cyanogenic glucosides, which is poisonous to animals, limiting its widespread use. This ANF can be reduced through a drying process, and varieties have been developed that contain lower cyanide content (Morgan and Choct, 2016). Pelleting cassava improves the digestibility and performance of broiler chickens (Adeyemi et al., 2008). Sweet potato byproducts (such as peel) or whole items rejected due to damage or other quality defects provide another feed source (Thibodeau et al, 2002). Sweet potatoes have been shown to be good sources of vitamins and β-carotene (Bromfield and Bovell-Benjamin, 2002; Byamukama et al., 2004).

Other grains such as sorghum and triticale can be used as a partial replacement for corn in diets. Sorghum has a similar nutrient value to corn but contains high levels of tannins that can negatively impact bird performance (NRC, 1994; Alshelmani et al., 2021). Tannins in sorghum can be minimized via processing, and use of lower-tannin varieties exist (Alshelmani et al., 2021). Triticale is a hybrid between wheat and rye that is grown all over the world (Sullivan et al., 2007). Its nutritional value is similar to rye, but there is significant variability due to the many cultivars grown, which limits widespread use in diets (Pourreza et al., 2007). Non-starch polysaccharides can also be found in triticale, leading to increased digesta viscosity and decreased bird performance.

Bakery meal is another energy source derived from baked goods manufacture. This product consists of waste as well as rejected pasta, cake, cookie, cereal, bread, and candy products that are ground, mixed, and dried. Depending on the sources used, bakery meal is typically low in protein and can vary in nutrient composition (Slominski et al., 2004; Zhang and Adeola, 2017). Variability between batches can be an issue. High concentrations of salt may also be present, which can lead to negative effects on broilers. Concern about the Maillard reaction during the pelleting of bakery meal is another issue (Dale, 1992; Bergeron et al., 2018). Inclusion of up to 25% bakery meal has been shown to have no negative effects on poultry performance and feed consumption (Saleh et al., 1996).

A considerable amount of food waste is generated throughout the food supply chain, providing a major potential feed source (Jaglo et al., 2021; Beckman et al., 2024). Food waste includes unused components such as peel, rejected products (e.g. those which fail to meet quality specifications), uneaten/ discarded food from catering outlets, and organic household waste. Using this waste is clearly beneficial on economic and environmental grounds in creating a more sustainable circular economy (Hall et al., 2009). Food waste sourced from a university campus has been shown to have similar nitrogen-corrected apparent metabolizable energy and apparent ileal digestibility of amino acids when compared to soybean meal (Beckman et al., 2024). As with other ingredients, a key problem is variability in nutrient content given the hugely varied nature of food waste. More research into optimal processing techniques and nutrient composition is needed before it can be accepted for commercial use.

Fiber is an important part of the diet as it stimulates gut movement and the health of the animal, as well as providing energy from volatile fatty acid production. Increased digesta viscosity can be an issue, but the hindgut microbes of the bird are well adapted to process fiber. Fiber ingredients are mostly byproducts such as hulls, pulp, and bran. Feeding sugar beet pulp to broilers, e.g., has been shown to improve the performance of young birds, though effects subsequently dissipate (Gonzalez-Alvarado et al., 2010). Sugar beet pulp contains NSPs, which increase digesta viscosity, affecting older birds. Inclusion of oat hulls has been shown to improve feed consumption and starch digestibility, without impacting weight gain or overall nutrient digestibility (Hetland and Svihus, 2001).

4 Exogenous feed enzyme additives: phytase, carbohydrase, and protease

Poultry produce a limited number of endogenous enzymes in their gastro-intestinal tract (GIT) to help improve feed conversion. Supplementation of endogenous enzymes can help to improve nutrient-use efficiency (and thus reduce feed costs) and reduce excreted waste with environmental benefits. The use of exogenous feed enzymes in poultry diets also helps to overcome the adverse effects of anti-nutritional factors (ANFs). Key enzyme classes are (Bedford and Cowieson, 2020):

- phytase,
- carbohydrase, and
- proteases.

These are discussed in more detail the following subsections.

4.1 Phytase

Phosphorus (P) is an essential component of poultry diets, critical for normal growth and bone development, egg formation, nutrient metabolism, and protein synthesis (Li et al., 2016). Poultry can only utilize around 25% of the total P in plant-based feedstuffs as it is primarily bound to phytate (Fernandes et al., 2019). Traditionally, inorganic phosphate (Pi) has been added to commercial broiler diets to ensure the P requirements of the birds are met. However, supplies of this mineral are finite, and higher levels of P excretion associated with its use are environmentally damaging (Lei et al., 2007; Slominski, 2011).

Use of exogenous phytase enzymes in poultry feed catalyzes hydrolytic splitting of phytate, liberating lower inositols of phosphate esters and Pi (Bedford et al., 2020; Selle et al., 2023). A range of 500–1000 FTU/kg has been used to mitigate the negative effect of phytate and increase the availability of P and other nutrients such as AA (Fernandes et al., 2019). Higher rates of inclusion from 1500 to 2000 FTU/kg have also been evaluated to increase nutrient availability, release inositol, increase apparent metabolizable energy, (ME) and AA digestibility, as well as for better bone mineralization (Gehring et al., 2013; Fernandes et al., 2019; Majeed et al., 2020; Ayres et al., 2021). It has been shown that Ca digestibility was higher at higher levels of up to 2000 FTU/kg (Bowen et al., 2022). Phytase addition has been associated with a reduction in the anti-nutritional effect of phytic acid on endogenous AA flow and digestibility (Cowesion and Ravindran, 2007). Phytase can result in as high as a 12% reduction in N excretion, associated with an increase in the ileal loss of AA and also an increase in the retention of N in skeletal muscles (Zhang et al., 2022). The effectiveness of phytase depends on a range of factors such as substrate composition, pH, and temperature, as well as animal breed and age (Lee et al., 2018).

4.2 Carbohydrase

Non-starch polysaccharides are complex chemical structures that poultry can struggle to digest. The amount of NSPs varies according to factors such as plant type, genetics, and environmental factors (Baker et al., 2021). Wheat, corn, and its by-product (corn DDGS) contain higher levels of NSPs that can have a detrimental effect on broiler performance. The NSP beta-glucan is found in both wheat and DDGS and contributes to viscosity in the poultry GIT, reducing absorption of nutrients and overall performance (Bedford and Classen, 1992). Antinutritional effects include altered chyme viscosity in the lumen, mechanical abrasion of the mucosa, and higher bulk density in the gut (Bedford et al., 2023).

Novel, plant-based options typically include NSPs with some levels of starch. Enzymes like carbohydrases reduce viscosity, optimizing the flow of digesta in the gut with improved nutrient absorption and utilization. Carbohydrase enzymes can be produced to target a specific substrate, including (Bedford and Classen, 1992):

- xylanase for xylan or arabinoxylan;
- B-mannanase for B-mannan, pectinase for pectin; and
- β-glucanase for β-glucan.

Carbohydrases can result in increases in metabolizable, digestible, and net energy values of between 50 kcal/kg and 150 kcal/kg and digestible AA values of 1–3% (Bedford et al., 2023).

4.3 Protease

Excessive AA and CP intake by broilers results in higher excretion of N and compounds such as ammonium with potentially negative environmental effects as well as lower nutrient use efficiency, which increases feed costs (Weir et al., 2017; Woyengo et al., 2023). The exogenous enzyme protease has been found to improve ileal AA and CP digestibility (Lee et al., 2018; Walk et al., 2018). More broadly, it promotes better bacterial composition in the GIT, improving overall gut function and health (Wilkie et al., 2005; Amer et al., 2021). The addition of protease has been found effective particularly in low-protein diets or protein of low quality (Leinonen et al., 2015). However, some studies have found more variable effects on ileal AA digestibility and overall broiler growth (Erdaw et al., 2017; Amer et al., 2021). Possible reasons include adverse effects on endogenous enzyme secretion, insufficient levels of proteases provided in feed, or present in the gut for protein digestion (Kaczmarek et al., 2014; Cardinal et al., 2019).

5 Additives to promote gut health for improved sustainability: prebiotics, probiotics, synbiotics, and precision biotics

The early establishment of beneficial bacteria in the chicken gut is critical for preventing pathogen colonization, enhancing immune system function, and healthy gut development. The growing public pressure to ban sub-therapeutic antibiotics from poultry diets has also increased the emphasis on improving gut health to prevent disease (MacDonald et al., 2011; Abdel-Moneim et al., 2022). A healthy intestine also allows for better utilization and absorption of nutrients and lower excretion of metabolites (Moita and Kim, 2022). Additives to promote gut health include:

- prebiotics;
- probiotics;
- synbiotics; and
- precision biotics.

These are discussed in the following subsections. These additives are either based on strains of beneficial bacteria (like in probiotics and synbiotics) or provide food for those beneficial bacteria (such as prebiotics).

5.1 Prebiotics

Prebiotics are defined as 'a non-digestive food ingredient that beneficially affects the host by stimulating selectively the growth and/or activity of one or a limited number of bacteria in the colon, and thus improves host health' (Gibson and Roberfroid, 1995). Prebiotics are non-digestible to the host but are available to host microorganisms in the GIT, such as *Bifidobacterium* spp. and *Lactobacillus* spp. The most common prebiotics are (Adhikari et al., 2020; Jahan et al., 2022):

- non-starch polysaccharides (NSP);
- oligosaccharides like xylo-oligosaccharides (XOS);
- fructo-oligosaccharides (FOS);
- mannan-oligosaccharides (MOS); and
- galacto-oligosaccharides (GOS).

These prebiotics can act to reduce pathogenic bacteria in the gut as well as increase digestibility and utilization of nutrients by modifying microbial metabolism or maintaining optimal pH in the poultry gut (Ricke, 2018; Abd El-hack et al., 2022). Some common genera of bacteria used as probiotics are *Bacillus*, *Lactobacillus*, *Enterococcus*, *Bifidobacterium*, and *Streptococcus* (Khan et al., 2020; Tabashum et al., 2023).

5.2 Probiotics

The mode of action of probiotics is in providing a balanced gut microbiota, both inhibiting the proliferation of pathogens and improving intestinal immunomodulation and nutrient processing (Anadon et al., 2019; Leone and Ferrante, 2023). Probiotics also provide different metabolites such as branched-chain fatty acids (BCFAs) and single-chain fatty acids (SCFAs), vitamins, and polyunsaturated fatty acids (PUFAs) that favor beneficial bacteria.

Probiotics, e.g., have been used for in-ovo supplementation at 12–18 days of incubation to optimize post-hatch populations of beneficial gut bacteria such as *Bifidobacterium* (Pilarski et al., 2005; Bednarczyk et al., 2011; Madej et al., 2016). Probiotics have been reported to decrease the colonization of *Campylobacter jejuni* and *Salmonella* Enteritidis, reducing risk of disease, boosting immune function, and improving feed conversion and weight gain (Neveling et al., 2020; Tabashum et al., 2023)

5.3 Synbiotics

Synbiotics are combinations of probiotics and prebiotics. Synbiotics have been shown to improve intestinal microbial composition (Bogucka et al., 2019), resulting in improved body weight gain and feed efficiency (Awad et al., 2009). Synbiotic preparations include *Lactobacillus* and yeast of *Saccharomyces cerevisiae*, and microencapsulated *Lactobacillus plantarum* and FOS (Awad et al., 2009; Markowiak et al., 2018). A synbiotic combination of inulin and *Enterococcus faecium*, e.g., was found to increase concentrations of cecum butyrate bacteria (Calik et al., 2017).

5.4 Precision biotics

In contrast to prebiotics, probiotics, and synbiotics, precision biotics are synthetically designed metabolites (also known as microbiome metabolic modulators (MMMs)) to promote assimilation pathways in the ceca, such as those for BCFA, SCFA, and AAs (Jacquier et al., 2022; Yan et al., 2023). When carbohydrates are consumed, they are broken down into sugar-based intermediates known as glycans, used in various biological functions including glycoprotein synthesis (Conte et al., 2021). By controlling metabolic pathways and consequentially their metabolites, these additives can improve energy efficiency and reduce intestinal ammonia production (Jacquier et al., 2022).

6 Conclusion

Life cycle assessments commonly identify feed production and transport as key source of emissions in poultry production, followed by manure. Manure produces ammonia, itself a pollutant, which can then be converted into nitrous oxide, one of the principal GHGs. In addition, ammonia can be a welfare issue for poultry, affecting eyes, causing skin damage, weight gain, and growth performance (Swelum et al., 2021). There is an urgent need to improve the environmental profile of feed production and to optimize feed efficiency to reduce manure production.

This chapter has discussed the potential use of more sustainable alternative feed ingredients in poultry feed. It has introduced a range of alternative protein-contributing ingredients, including crop sources such as legumes, peas and beans, as well as by-products from food and oilseed processing such as oil palm kernel and cottonseed meal, as well animal sources such as feather and blood meal. It has also examined potential energy/fiber-contributing ingredients such as bakery meal, food waste, and fiber co-products such as hulls, pulp, and bran.

Diet can be improved further by increasing feed digestibility. As this chapter suggests, this can be done by adding exogenous enzymes; e.g.

carbohydrase that can break down NSPs and reduce digesta viscosity, phytase that can aid in lowering Pi, and protease that can increase protein utilization. Gut health products, such as prebiotics, probiotics, synbiotics, and PBs, can lower the excretion of different toxic metabolites of nitrogen into the environment.

7 References

Abd El-Hack, M. E., et al. (2022). Alternatives to antibiotics for organic poultry production: types, modes of action and impacts on bird's health and production. *Poultry Science*, 101(4), 101696.

Abdel-Moneim, A. M. E., et al. (2022). Spirulina platensis and biosynthesized selenium nanoparticles improve performance, antioxidant status, humoral immunity and dietary and ileal microbial populations of heat-stressed broilers. *Journal of Thermal Biology*, 104, 103195.

Adeyemi, O. A., et al. (2008). Feeding broiler chicken with diets containing whole cassava root meal fermented with rumen filtrate. *Archiva Zootechnica*, 57, 247–258.

Adhikari, P. et al. (2020). An approach to alternative strategies to control avian coccidiosis and nectrotic enteritis. *J. Applied Poultry Res.* 29(2): 515–534.

Almeida, F. N., et al. (2013). Comparative amino acid digestibility in US blood products fed to weanling pigs. *Animal Feed Science and Technology*, 181, 80–86.

Alshelmani, M. I., et al. (2017). Effect of solid state fermentation on nutrient content and ileal amino acids digestibility of palm kernel cake in broiler chickens. *Indian Journal of Animal Sciences*, 87(9), 1135–1140.

Alshelmani, M. I., et al. (2021). Nontraditional feedstuffs as an alternative in poultry feed. In *Advances in poultry nutrition research*. IntechOpen.

Amer S. A., et al. (2021). Effects of different feeding regimens with protease supplementation on growth, amino acid digestibility, economic efficiency, blood biochemical parameters, and intestinal histology in broiler chickens. *BMC Veterinary Research*, 17(1), 283.

Anadón, A., et al. (2019). Prebiotics and probiotics in feed and animal health. In *Nutraceuticals in veterinary medicine* (pp. 261–285).Cham: Springer.

Awad, W. A., et al. (2009). Effects of dietary inclusion of probiotic and synbiotic on growth performance, organ weights, and intestinal histomorphology of broiler chickens. *Poultry Science*, 88(1), 49–56.

Ayres, V. E., et al. (2021). The effects of two corn-expressed phytases on broiler growth performance and tibia mineralization. *Journal of Applied Poultry Research*, 30(1), 100131.

Babatunde, O. O., et al. (2021). Nutritional potentials of atypical feed ingredients for broiler chickens and pigs. *Animals*, 11(5), 1196. https://doi.org/10.3390/ani11105119

Bampidis, V. & Christodoulou, V. (2011). Chickpeas (Cicer arietinum L.) in animal nutrition: a review. *Animal Feed Science and Technology*, 168, 1–20.

Beckman, N. M., et al. (2024). Nitrogen-corrected apparent metabolizable energy and apparent ileal digestibility of amino acids of thermally processed food waste for broilers. *Journal of Applied Poultry Research*, 33(1), 100379.

Bergeron, A. N., et al. (2018). The Effects of Diet Formulation and Thermal Processes Associated With Pelleting on 18-day Broiler Performance and Digestible Amino Acid Concentration. *Journal of Applied Poultry Research*, 27(4), 540–549.

Baker, J. T., et al. (2021). Friend or foe? Impacts of dietary xylans, xylooligosaccharides, and xylanases on intestinal health and growth performance of monogastric animals. *Animals*, 11(1), 28.

Bedford, M. R. & Classen, H. L. (1992). Reduction of intestinal viscosity in barley-fed broilers by β-glucanase: site of action and effect on bird performance. *Proceedings of the British Society of Animal Production*, 1972, 86.

Bedford, M. R. & Cowieson, A. J. (2020). Matrix values for exogenous enzymes and their application in the real world. *Journal of Applied Poultry Research*, 29, 15–22.

Bedford, M. R., et al. (2023). Dietary fibre effects and the interplay with exogenous carbohydrases in poultry nutrition. https://doi.org/10.1016/j.aninu.2023.09.007

Bednarczyk, M., et al. (2011). Field and in vitro study on prebiotic effect of raffinose family oligosaccharides in chickens. *Bulletin of the Veterinary Institute Pulawy*, 55, 465–469.

Bogucka, J., et al. (2019). Microstructure of the small intestine in broiler chickens fed a diet with probiotic or synbiotic supplementation. *Journal of Animal Physiology and Animal Nutrition*, 103(6), 1785–1791.

Borowski, S., et al. (2016). Co-digestion of poultry manure and residues from enzymatic saccharification and dewatering of sugar beet pulp. *Renewable Energy*, 99, 492–500.

Bowen, K. M. et al. (2022). Performance, carcass quality and mineral digestibility responses of Ross 708 broilers to two commercially-available phytase mixes. *J. Applied Poultry Res.* 31(3): article 100264.

Bromfield, E. M. & Bovell-Benjamin, A. (2002). Consumer acceptance and purchase intent of a sweetpotato beverage. SAE Technical Paper.

Byamukama, E., et al. (2004). Within-crop spread of sweet potato virus disease and the population dynamics of its whitefly and aphid vectors. *Crop Protection*, 23, 109–116.

Calik, A., et al. (2017). The effect of intra-amniotic and posthatch dietary synbiotic administration on the performance, intestinal histomorphology, cecal microbial population, and short-chain fatty acid composition of broiler chickens. *Poultry Science*, 96(1), 169–183.

Carbone, J. W. & Pasiakos, S. M. (2019). Dietary protein and muscle mass: translating science to application and health benefit. *Nutrients*, 11(5), 1136. https://doi.org/10.3390/nu11051136

Cardinal, K. M., et al. (2019). Growth performance and intestinal health of broilers fed a standard or low-protein diet with the addition of a protease. Revista Brasileira de Zootecnia, 48. http://dx.doi.org/10.1590/rbz4820180232

Chadd, C. (2007). *Future trends and developments in poultry nutrition*. Poultry in the 21st Century: Avian Influenza and Beyond, International Poultry Conference, 5–7 November, 2007, Bangkok, Thailand.

Churchward-Venne, A. T., et al. (2017). Consideration of insects as a source of dietary protein for human consumption. *Nutrition Reviews*, 75, 1035–1045.

Crépon, K., et al. (2010). Nutritional value of faba bean (Vicia faba L.) seeds for feed and food. *Field Crops Research*, 115, 329–339.

Conte, F., et al. (2021). Galactose in human metabolism, glycosylation and congenital metabolic diseases: time for a closer look. *Biochimica et Biophysica Acta*, 1865(8), 129898.

Cowieson, A. J. and Ravindran, V. (2007). Effect of Phytic acid and microbial phytase on the flow and amino acid composition of endogenous protein in the ileum of growing broilers. *Br. J. Nut.* 98(4): 745–752

Dabbou, S., et al. (2018). Black soldier fly defatted meal as a dietary protein source for broiler chickens: effects on growth performance, blood traits, gut morphology and histological features. *Journal of Animal Science and Biotechnology*, 9, 49. https://doi.org/10.1186/s40104-018-0266-9

Dale, N. (1992). Pelleting Effects on Lysine Bioavailability in Diets Containing Dried Bakery Product. *Journal of Applied Poultry Research*, 1(1), 84–87.

DiGiacomo, K., et al. (2019). Insects: a novel animal-feed protein source for the Australian market. *Animal Production Science*, 59, 2037–2045.

Digiacomo, K. & Leury, B. J. (2019). Review: insect meal: a future source of protein feed for pigs? *Animal*, 13, 3022–3030.

Erdaw, M. M., et al. (2017). Growth and physiological responses of broiler chickens to diets containing raw, full-fat soybean and supplemented with a high-impact microbial protease. *Asian-Australasian Journal of Animal Sciences*, 30(9), 1303.

El Boushy, A., et al. (1990). Feather meal–A biological waste: Its processing and utilization as a feedstuff for poultry. *Biological Wastes*, 32, 39–74.

Elahi, U., et al. (2020). Evaluation of yellow mealworm meal as a protein feedstuff in the diet of broiler chicks. *Animals*, 10. https://doi.org/10.3390/ani10020224

Elahi, U., et al. (2022). Insect meal as a feed ingredient for poultry. *Animal Bioscience*, 35(2), 332–346. https://doi.org/10.5713/ab.21.0435

Evans, A., et al. (2015). Effects of algae incorporation into broiler starter diet formulations on nutrient digestibility and 3 to 21 d bird performance. *Journal of Applied Poultry Research*, 24(2), 206–214.

Fernandes, J. I. M., et al. (2019). Effects of phytase superdosing on digestibility and bone integrity of broilers. *Journal of Applied Poultry Research*, 28(2), 390–398.

González-Alvarado, J., et al. (2010). Effect of inclusion of oat hulls and sugar beet pulp in the diet on productive performance and digestive traits of broilers from 1 to 42 days of age. *Animal Feed Science and Technology*, 162, 37–46.

Grigorova, S., et al. (2017). Influence of natural sources of biologically active substances on livestock and poultry reproduction. *Iranian Journal of Applied Animal Science*, 7(2), 189–195.

Grosjean, F., et al. (2000). Feeding value for poultry of isogenic fababeans (Vicia faba L.) involving zero-tannin and zero-vicine genes. *Science Technology Avic*, 32, 17–23.

Gehring, C. K., et al. (2013). Extra-phosphoric effects of phytase with and without xylanase in corn-soybean meal-based diets fed to broilers. *Poultry Science*, 92(4), 979–991.

Gibson, G. R. & Roberfroid, M. B. (1995). Dietary modulation of the human colonic microbiota: introducing the concept of prebiotics. *The Journal of Nutrition*, 125(6), 1401–1412.

Hall, K. D., et al. (2009). The progressive increase of food waste in America and its environmental impact. *PLoS One*, 4(11), e7940. https://doi.org/10.1371/journal.pone.0007940

Harding, K. L., et al. (2023). The effect of feeding a sweet potato and/or high-oleic peanut diet on layer performance and the quality and chemistry of eggs produced. *International Journal of Poultry Science*, 22, 73–83.

Hetland, H. & Svihus, B. (2001). Effect of oat hulls on performance, gut capacity and feed passage time in broiler chickens. *British Poultry Science*, 42, 354–361.

Jacquier, V., et al. (2022). Evaluation of a precision biotic on the growth performance, welfare indicators, ammonia output, and litter quality of broiler chickens. *Animals,* 12(3), 231.

Jahan, A. A., et al. (2022). Role of supplemental oligosaccharides in poultry diets. *World's Poultry Science Journal*, 78(3), 615–639.

Igbasan, F. A., et al. (1997). Field peas: chemical composition and energy and amino acid availabilities for poultry. *Canadian Journal of Animal Science*, 77, 293–300.

Jaglo, K., et al. (2021). *From farm to kitchen: the environmental impacts of U.S. Food Waste*. U.S. Environmental Protection Agency Office of Research and Development.

James, K. A. C., et al. (2005). The effects of cultivar and heat treatment on protein quality and trypsin inhibitor content of New Zealand field peas. *New Zealand Journal of Agricultural Research*, 48, 117–124.

Kaczmarek, S. A., et al. (2014). The effect of protease, amylase, and nonstarch polysaccharide-degrading enzyme supplementation on nutrient utilization and growth performance of broiler chickens fed corn-soybean meal-based diets. *Poultry Science*, 93(7), 1745–1753.

Khan, S., et al. (2020). The gut microbiota of laying hens and its manipulation with prebiotics and probiotics to enhance gut health and food safety. *Applied and Environmental Microbiology*, 86(13), e00600-20.

Kareem, K. Y., et al. (2018). Effect of feeding larvae meal in the diets on growth performance, nutrient digestibility and meat quality in broiler chicken. *Indian Journal of Animal Sciences*, 88, 1180–1185.

Khan, S. H. (2018). Recent advances in role of insects as alternative protein source in poultry nutrition. *Journal of Applied Animal Research*, 46, 1144–1157.

Khawaja, T., et al. (2007). Effect of different levels of blood meal on broiler performance during two phases of growth. *International Journal of Poultry Science*, 6(12), 860–865.

Kiarie, E. & Nyachoti, C.M. (2009). *Alternative feed ingredients in swine diets*. In Proceedings of the Saskatchewan Industry Symposium, Saskatoon, SK, Canada, 17–18 November 2009.

Knudsen, K. E. B. (1997). Carbohydrate and lignin contents of plant materials used in animal feeding. *Animal Feed Science and Technology*, 67, 319–338.

Kovač, D. J., et al. (2013). Algae in food and feed. *Food and Feed Research*, 40(1), 21–32.

Leone, F. & Ferrante, V. (2023). Effects of prebiotics and precision biotics on performance, animal welfare and environmental impact: a review. *Science of the Total Environment*, 901, 165951.

Leinonen, I. & Williams, A. G. (2015). Effects of dietary protease on nitrogen emissions from broiler production: a holistic comparison using life cycle assessment. *Journal of the Science of Food and Agriculture*, 95(15), 3041–3046.

Lee, S. A., et al. (2018). Superdosing phytase reduces real-time gastric pH in broilers and weaned piglets. *British Poultry Science*, 59(3), 330–339.

Lei, X. G., et al. (2007). Phytase: source, structure and application. In J. Polaina & A. P. MacCabe (Eds.), *Industrial enzymes: Structure, function and applications* (pp. 505–529). Dordrecht: Springer.

Li, X., et al. (2016). Phosphorus bioavailability: a key aspect for conserving this critical animal feed resource with reference to broiler nutrition. *Agriculture*, 6, 25.

Laboissière, M., et al. (2020). Feather and blood meal at different processing degrees in broiler pre-starter and starter diets. *Revista Brasileira de Zootecnia*, 49, 20190036.

Leeson, S. & Summers, J. (2005). *Commercial poultry nutrition*. Department of Animal and Poultry Science, University of Guelph. 3rd ed. University Book.

MacDonald, J. M. & Wang, S. L. (2011). Foregoing sub-therapeutic antibiotics: the impact on broiler grow-out operations. *Applied Economic Perspectives and Policy*, 33(1), 79–98.

Madej, J. P. & Bednarczyk, M. (2016). Effect of in ovo-delivered prebiotics and synbiotics on the morphology and specific immune cell composition in the gut-associated lymphoid tissue. *Poultry Science*, 95(1), 19–29.

Majeed, S., et al. (2020). Limestone particle size, calcium and phosphorus levels, and phytase effects on live performance and nutrients digestibility of broilers. *Poultry Science*, 99(3), 1502–1514.

Moita, V. H. C. & Kim, S. W. (2022). Nutritional and functional roles of phytase and xylanase enhancing the intestinal health and growth of nursery pigs and broiler chickens. *Animals*, 12(23), 3322.

Markowiak, P. & Śliżewska, K. (2018). The role of probiotics, prebiotics and synbiotics in animal nutrition. *Gut Pathogens*, 10(1), 1–20.

M'Ncene, W., et al. (1999). Nitrogen utilization and performance of pigs given diets containing a dried or undried fermented blood/molasses mixture. *Animal Feed Science and Technology*, 78, 239–247.

Madeira, M. S., et al. (2017). Microalgae as feed ingredients for livestock production and meat quality: a review. *Livestock Science*, 205, 111–121.

Makkar, H. P., et al. (2014). State-of-the-art on use of insects as animal feed. *Animal Feed Science and Technology*, 197, 1–33.

McDonald, P., et al. (2010). *Animal nutrition*. 7th ed. UK: Pearson.

McNeill, L., et al. (2004). Food intake, growth rate, food conversion and food choice in broilers fed on diets high in rapeseed meal and pea meal, with observations on sensory evaluation of the resulting poultry meat. *British Poultry Science*, 45, 519–523.

Morgan, N. K. & Choct, M. (2016). Cassava: nutrient composition and nutritive value in poultry diets. *Animal Nutrition*, 2, 253–261.

Moritz, J. S. & Latshaw, J. D. (2001). Indicators of nutritional value of hydrolyzed feather meal. *Poultry Science*, 80, 79–86.

Murray, S. M., et al. (1997). Raw and rendered animal by-products as ingredients in dog diets. *Journal of Animal Science*, 75, 2497–2505.

Mustafa, A. F., et al. (2000). Nutritional value of feed grade chickpeas for ruminants and pigs. *Journal of the Science of Food and Agriculture*, 80, 1581–1588.

Neveling, D. P., et al. (2020). Genetic and phenotypic characteristics of a multi-strain probiotic for broilers. *Current Microbiology*, 77, 369–387.

NRC. (1994). *Nutrient requirements of poultry*. 9th ed. Washington, DC: National Academy Press. ISBN-13: 9780309048927.

Onifade, A., et al. (1998). A review: potentials for biotechnological applications of keratin-degrading microorganisms and their enzymes for nutritional improvement of feathers and other keratins as livestock feed resources. *Bioresource Technology*, 66, 1–11.

Partanen, K., et al. (2003). Faba beans in diets for growing-finishing pigs. *Agricultural and Food Science*, 12, 35–47.

Pieterse, E., et al. (2019). Black soldier fly (Hermetia illucens) pre-pupae meal as a dietary protein source for broiler production ensures a tasty chicken with standard meat quality for every pot. *Journal of the Science of Food and Agriculture*, 99, 893–903. https://doi.org/10.1002/jsfa.9261

Pourreza, J., et al. (2007). Effect of supplemental enzyme on nutrient digestibility and performance of broiler chicks fed on diets containing triticale. *International Journal of Poultry Science*, 6, 115–117.

Pilarski, R., et al. (2005). Assessment of the effect of α-galactosides injected during embryogenesis on selected chicken traits. *Folia Biologica (Kraków)*, 53(1–2), 13–20.

Queiroga, A. C., et al. (2012). Potential use of wool-associated Bacillus species for biodegradation of keratinous materials. *International Biodeterioration & Biodegradation*, 70, 60–65.

Ricke, S. C. (2018). Focus: nutrition and food science: impact of prebiotics on poultry production and food safety. *The Yale Journal of Biology and Medicine*, 91(2), 151.

Riffel, A. & Brandelli, A. (2002). Isolation and characterization of a feather-degrading bacterium from the poultry processing industry. *Journal of Industrial Microbiology and Biotechnology*, 29, 255–258.

Saleh, E. A., et al. (1996). High-level usage of dried bakery product in broiler diets. *Journal of Applied Poultry Research*, 5, 33–38.

Slominski, B. A., et al. (2004). Wheat by-products in poultry nutrition. Part I. Chemical and nutritive composition of wheat screenings, bakery by-products and wheat mill run. *Canadian Journal of Animal Science*, 84, 421–428.

Stein, H. & de Lange, K. (2007). *Alternative feed ingredients for pigs*. In Proceedings of the 7th London Swine Conference, London, ON, Canada, 3–4 April, 2007.

Stein, H. H. & Bohlke, R. A. (2007). The effects of thermal treatment of field peas (Pisum sativum L.) on nutrient and energy digestibility by growing pigs. *Journal of Animal Science*, 85, 1424–1431.

Stein, H. H., et al. (2015). Nutritional value of high fiber co-products from the copra, palm kernel, and rice industries in diets fed to pigs. *Journal of Animal Science Biotechnology*, 6, 56. https://doi.org/10.1186/s40104-015-0056-6

Sulabo, R. C., et al. (2013). Amino acid digestibility and concentration of digestible and metabolizable energy in copra meal, palm kernel expellers, and palm kernel meal fed to growing pigs. *Journal of Animal Science*, 91, 1391–1399.

Sullivan, Z. M., et al. (2007). Effects of triticale-based diets on finishing pig performance and pork quality in deep-bedded hoop barns. *Meat Science*, 76, 428–437.

Selle P. H., et al. (2023). The contribution of phytate-degrading enzymes to chicken-meat production. *Animals (Basel)*, 13(4), 603.

Thornton, P. K. (2010). Livestock production: recent trends, future prospects. *Philosophical Transactions of the Royal Society B: Biological Sciences*, 365(1554), 2853–2867.

Slominski, B. A. (2011). Recent advances in research on enzymes for poultry diets. *Poultry Science*, 90(9), 2013–2023.

Swelum, A. A., et al. (2021). Ammonia emissions in poultry houses and microbial nitrification as a promising reduction strategy. *Science of the Total Environment*, 781, 146978.

Tabashsum, Z., et al. (2023). Alternative approaches to therapeutics and subtherapeutics for sustainable poultry production. *Poultry Science*, 102(7), 102750.

Thibodeau, M. S., et al. (2002). Health and production aspects of feeding sweetpotato to cattle. *Veterinary Clinics of North America: Food Animal Practice*, 18, 349–365.

USDA-ERS. (2022). Cotton sector at a glance. Accessed January 2024 https://www.ers.usda.gov/topics/crops/cotton-and-wool/cotton-sector-at-a-glance

Vieira, B. S., et al. (2018). Combination of phytase and organic acid for broilers: role in mineral digestibility and phytic acid degradation. *World Poultry Science*, 74, 711–726.

Veldkamp, T. & Bosch, G. (2015). Insects: A protein-rich feed ingredient in pig and poultry diets. *Animal Frontiers*, 5, 45–50.

Walk, C. L. et al. (2018). Evaluation of novel protease enzymes on growth performance and apparent ileal digestibility of amino acids in poultry. *Poultry Sci*. 97(6): 2123–2138.

Weir, J., et al. (2017). Characterizing ammonia emissions from horses fed different crude protein concentrations. *Journal of Animal Science*, 95(8), 3598–3608.

White, G., et al. (2015). Replacement of soya bean meal with peas and faba beans in growing/finishing pig diets: effect on performance, carcass composition and nutrient excretion. *Animal Feed Science and Technology*, 209, 202–210.

Wilkie, D. C., et al. (2005). Dietary amino acids affect intestinal Clostridium perfringens populations in broiler chickens. *Canadian Journal of Animal Science*, 85(2), 185–193.

Woyengo, T. A., et al. (2014). Nonruminant nutrition symposium: controlling feed cost by including alternative ingredients into pig diets: a review. *Journal of Animal Science*, 92, 1293–1305.

Woyengo, T. A., et al. (2023). Low-protein diets for broilers: Current knowledge and potential strategies to improve performance and health, and to reduce environmental impact. *Animal Feed Science and Technology*, 115574. https://doi.org/10.1016/j.anifeedsci.2023.115574

Yan L., et al. (2023). Microbiome modulation by a precision biotic in broilers chickens: a commercial study validation. *Poultry Science*, 102(5), 102596.

Zhang, F. & Adeola, O. (2017). Energy values of canola meal, cottonseed meal, bakery meal, and peanut flour meal for broiler chickens determined using the regression method. *Poultry Science*, 96, 397–404.

Zhang, Z., et al. (2014). Effects of partially replacing dietary soybean meal or cottonseed meal with completely hydrolyzed feather meal (defatted rice bran as the carrier) on production, cytokines, adhesive gut bacteria, and disease resistance in hybrid tilapia (Oreochromis niloticus ♀ × Oreochromis aureus ♂). *Fish & Shellfish Immunology*, 41, 517–525.

Zhang, Q. et al. (2022). Comparative effects of two phytases on growth performance, bone mineralization and nutrient digestibility of broilers. *J. Applied Poultry Res*. 31(2): article 100247.

Chapter 15

Managing poultry litter to improve safety and reduce environmental impact

Hongye Wang, California State University, East Bay, USA; Xiuping Jiang, Clemson University, USA; and Jingxue Wang, Ocean University of China, China

1 Introduction

On poultry farms, poultry litter is a mixture of feathers, dead birds, broken eggs, waste feed, and bedding materials. The U.S. poultry industry is characterized by its high production system. Over 65 billion pounds of live weight broilers and turkeys were produced in 2022 (USDA, 2023). Poultry litter accumulates continuously throughout the poultry life cycle, resulting in the ongoing production of substantial quantities of poultry litter. Taking Maryland's Eastern Shore as an example, nearly 550 000 tons of poultry litter were produced each year (Tidwell, 2012). The large-scale production of poultry litter highlights the importance of efficient management and the potential benefits of using poultry litter as a source of plant nutrients. Treated or untreated poultry litter is a low-cost, highly nutritious source of phosphorus and nitrogen that may be utilized as an organic soil fertilizer and organic amendment to boost crop yields and improve soil fertility (Gascho et al., 2006; Harmel et al., 2008). Its organic composition not only adds soil fertility but also provides beneficial microbial activity in the rhizosphere, thereby promoting sustainable agricultural practices

http://dx.doi.org/10.19103/AS.2024.0146.19

and crop yields. Application of poultry litter can contribute to significantly higher yields with cotton, corn, soybeans, peanuts, and other crops (Lin et al., 2018).

Composted pelletized poultry litter has less impact on the environment than chemical fertilizer (Kiss et al., 2021), indicating the suitability of this organic fertilizer as an alternative to chemical fertilizer. However, improper handling of poultry litter can cause environmental problems. For example, over-application of poultry litter to the field can overload agricultural fields or surrounding water with excessive nutrients, resulting in excessive plant growth and an adverse impact on water bodies. As the microbial community of poultry litter can be complex, harboring diverse microorganisms – including potential human pathogens – soil application of poultry litter can spread pathogens, create toxic substances that are harmful to plants, cause air pollution, and release greenhouse gases (Harmel et al., 2008; Stingone et al., 2011). The insufficient treatment or improper use of raw poultry litter or poultry litter-based compost may introduce foodborne pathogens to agricultural production, leading to public health risks. The application of poultry litter for crop production can also raise concerns related to antibiotic-resistant bacteria (ARBs) and antibiotic-resistant genes (ARGs) (Li et al., 2017). To mitigate the risks associated with using poultry litter as biofertilizer and ensure the safety of the final product, various treatments should be considered. These may include composting, physical heat treatment, addition of chemicals or competitive exclusion microorganisms, anaerobic digestion, and other appropriate methods.

Recognizing that poultry litter can be both a low-cost and nutrient-rich fertilizer for farming and a possible vector for environmental or microbiological hazards highlights the importance of proper treatment or handling the poultry litter. This chapter aims to explore the microbiological quality and safety aspects of poultry litter, along with various treatment methods to mitigate associated risks.

2 Microbiological quality and safety of poultry litter

Foodborne pathogens found in poultry litter include *Actinobacillus*, *Campylobacter*, *Clostridium*, Shiga toxin-producing *Escherichia coli* (STEC), *Listeria monocytogenes*, *Mycobacterium*, *Salmonella* spp., *Staphylococcus*, *Streptococcus*, and others (Larney et al., 2003; Franke-Whittle et al., 2005; Grewal et al., 2006; Slana et al., 2011; Khan et al., 2016; Kagambèga et al., 2018). Pathogen incidence and concentration in poultry litter are quite varied. It is essential to conduct regular monitoring and testing for pathogens to ensure the safety of agricultural products. An extensive review on the microbiological surveys on the prevalence of certain foodborne pathogens in poultry litter or poultry litter-based organic fertilizers has been published by Chen and Jiang

(2014). Clearly, there is a need for ongoing research into innovative treatment methods (physical, chemical, or biological) to ensure the microbiological safety of poultry litter-based fertilizers.

Assessing the amount and frequency of these pathogens in poultry litter or poultry litter-based fertilizers is crucial for establishing intervention methods to reduce food contamination on farms. Among pathogens, *Salmonella* is a predominant concern in poultry products. Between 2009 and 2018, over half of the foodborne disease outbreaks linked to *Salmonella* were traced back to meat/poultry products, with eggs also being a significant source (White et al., 2022). These data highlight why *Salmonella* is one of the most widely spreading pathogens in poultry litter. For instance, studies have shown that *Salmonella* levels in fresh animal manure can range between 3 $\log_{10}$ and 5 $\log_{10}$ colony-forming unit (CFU)/g across several types of livestock including cattle, pigs, poultry, and sheep (Hutchison et al., 2004a, 2004b). *Salmonella* has been found most prevalently in broiler litter when compared to swine, dairy, and beef manures in a comprehensive analysis of over 2500 manure samples in Canada (Flockhart et al., 2017).

As indicated earlier, *Salmonella* is more frequently isolated from poultry litter or fecal samples compared to other pathogens being investigated. The population of *Salmonella* in chicken litter can vary dramatically, ranging from 4 $\times 10^5$ to 1.1 $\times 10^5$ most probable number (MPN)/g litter, further indicating the challenge of controlling *Salmonella* in poultry litter (Chen and Jiang, 2014). A synthesis of studies showed that the prevalence of *Salmonella* in poultry litter could range anywhere from 1% to 70%, with geography, flock cohorts, and poultry houses playing crucial roles in these variations (Table 1) (Wang et al., 2023; Oladeinde et al., 2023). Besides *Salmonella*, other pathogens like *Listeria* spp., STEC, and *Campylobacter* have also been documented, underscoring the breadth of microbial hazards in poultry litter (Dahshan et al., 2016; Kassem et al., 2017; Moffo et al., 2022). *Listeria* spp. tested positive in 95 out of 200 litter samples (Dahshan et al., 2016), and the prevalence of *Campylobacter* was 2.43% (7/288) and 9.38% (27/288) in broiler litter collected from South Georgia, U.S., as reported by Kassem et al. (2017) and Oladeinde et al. (2023), respectively. Moffo et al. (2022) reported 68% (32/47) of broiler litter samples positive for *E. coli* in another survey.

However, conducting prevalence studies in agricultural fields is both time-consuming and expensive due to the variability in experimental design factors such as sampling protocols and methods, sample formats (feces, soil, litter), and sizes (English et al., 2015; Shang et al., 2018). This variability often leads to inconsistent results, highlighting the need for standardized methodologies in assessing the microbiological safety of poultry litter. It is important that these methodologies account for a comprehensive set of variables to ensure accurate and reliable results, thereby informing more effective interventions.

Table 1 Prevalence of *Salmonella* in poultry litter published from 2014 to 2024

Year/Location	Sample type	Sample size	Prevalence	References
2010–2011/U.S.	Clinical and environmental samples from farm	204	Manure storage pile composites: 45% Wide turkey feces: 22.5% Soil: 10.3%	Jay-Russell et al., 2014
N.A. Mid-Atlantic U.S.	Poultry litter	62	60% (41/62)	Sapkota et al., 2014
N.A./U.S.	Litter samples (swab)	9210	11.9% (1100/9210)	English et al., 2015
2006–2011/Canada	Fecal samples	2523	*Salmonella* prevalence: broiler manure (55%)	Flockhart et al., 2017
N.A./Nigerian	Litter (n = 67) and feces sample (n = 75)	142	12.8% for litter sample, and 14.3% for feces sample	Fagbamila et al., 2017
Nigeria	Poultry samples including litter	523	43.6% (228/523)	Fagbamila et al., 2017
2015–2016/South Korea	Poultry litter	396	11.1% (44/396)	Shang et al., 2018
N.A.	Poultry litter swab samples	1584	8.4% (133/1584)	Shang et al., 2018
2017/Burkina Faso	Poultry feces	103	52.42%	Kagambèga et al., 2018
2013–2014/Ethiopia	Poultry feces	549	4.7%	Eguale et al., 2018
2013–2015/Virginia, U.S.	Poultry litter	N.A.	26.9% in wet litter; 23.7% in dry litter	Gu et al., 2019
Brazil	Broiler litter	442	28.4% (120/442)	Voss-Rech et al., 2019
N.A.	Poultry feces	80	~50	Bailey et al., 2020
2017/Bangladesh	Broiler litter	35	~20%	Alam et al., 2020
2014–2017/Southeastern U.S.	Broiler feces	726	16.4% (119/726)	Hwang et al., 2020
2017–2018/Florida, U.S.	Poultry litter	54	61.1%	Gutierrez et al., 2020
Southeastern U.S.	Poultry litter	24	8% (2/24)	Rama et al., 2022
2017–2018/southern U.S.	Poultry litter	490	6.7% (33/490)	Dunn et al., 2022
2020/Colombia	Poultry litter	45	16 *Salmonella* isolates	Bonilla-Caballero et al., 2022
2018/South Georgia, U.S.	Broiler litter	288	10.41% (20/288)	Oladeinde et al., 2023
2020/São Paulo	Poultry litter	19	3/19	Franceschine et al., 2023
N.A.	Poultry samples (including litter)	200	47.5% (95/200)	Dahshan et al., 2016
2019/Cameroon	Poultry litter	47	68% (32/47)	Moffo et al., 2022
N.A.	Broiler litter	288	2.43% (7/288)	Kassem et al., 2017
2018/South Georgia, U.S.	Broiler litter	288	9.38% (27/288)	Oladeinde et al., 2023

Importantly, the application of untreated animal waste on crop fields introduces significant microbial hazards; therefore, raw animal manure should be incorporated into the soil more than 90 days prior to harvest for crops that have no direct contact with soil, and 120 days if the produce has direct contact with soil (Clements et al., 2019). Raw manure must not come into contact with produce during application, and the potential for raw manure contact with produce after application should be minimized (Clements et al., 2019). Clearly, this practice, coupled with the implementation of both biological and physical treatment options for animal waste, can significantly reduce the environmental release of pathogenic bacteria (Goberna et al., 2011). Utilizing treated manure not only mitigates microbial hazards but also recycles valuable nutrients back into the soil, demonstrating a sustainable approach to managing animal waste and ensuring food safety.

3 Composting of poultry litter

Composting is a controlled biological process in which aerobic microorganisms thermophilically decompose organic residues such as manure, animal carcasses, straw, and yard trimmings. The composting process usually has four stages: mesophilic, thermophilic, cooling, and maturation stages. These stages depend on the temperature and the activities from microorganisms. Due to temperature fluctuations, the dynamics of the microbial community change constantly during the composting process (Hassen et al., 2001; Ma et al., 2018). The microbial metabolism transforms the organic matter into a more uniform, stable, and nutrient-rich soil amendment, and the heat generated from the microbial activities can inactivate pathogens (Larney et al., 2003). There are other factors that should be well-controlled to ensure the microbial safety of the composting process. These key factors include a balanced nutrient level from raw ingredients (carbon to nitrogen ratio, C:N), proper moisture content, pH and O_2 levels of the compost mix, the heterogeneous nature due to heap turning, and the outdoor environment of the composting process (Erickson et al., 2009; Shepherd et al., 2010; Ceustermans et al., 2007; Sundberg et al., 2004; Green et al., 2004).

In addition to reducing manure volume by 50–65%, composting is effective at eliminating pathogens and is comparatively simple to manage (Flynn et al., 1996). In agricultural fields, the finished compost can be utilized as a humus-like soil amendment to enhance soil structure and stimulate enzymatic and microbial activities (Wu et al., 2024). The U.S. Environmental Protection Agency (USEPA, 2008) regulates the biosolids composting process in accordance with Title 40, Part 503, Biosolids Rule of the Code of Federal Regulations. While windrows must reach 55°C for 15 days with at least five turnings, static aerated composting or in-vessel composting must be kept at 55°C for 3 days in order to produce a Class A soil amendment appropriate for crop production. The composting operators should ensure that the compost is produced with the right combination of time/temperature for the type of composting system.

Poultry litter can contain pathogenic bacteria, posing a risk of zoonotic disease in public health. Some elements (heavy metals and essential nutrients) that are used as feed additives for poultry can be released into the poultry litter. The improper application of poultry litter to agricultural fields can impact crops and yields. Such elements can be a potential co-selection factor for ARGs by promoting the induction of those genes. Some alterations in the composting process can contribute to the reduction of pathogens in poultry litter compost and improve the quality of the final product. For example, the addition of an extra carbon source and mechanical aeration during poultry litter composting can reduce antimicrobial residues, enteroparasites, and antimicrobial-resistant *E. coli* (Okada et al., 2024).

Heat and microbial competition produced during the composting process can inactivate pathogenic bacteria (Gurtler et al., 2018). For instance, with an initial level of 6.5 $\log_{10}$ CFU/g, *Salmonella* spp. decreased by >5 $\log_{10}$ CFU/g after 3 days and 7 days at internal locations of windrow and static piles containing finished compost covering, respectively (Patel et al., 2015). Gulter et al. (2018) and Lepesteur (2022) have extensively reviewed pathogen reduction during composting. These review papers discussed a variety of factors affecting the pathogen killing mechanisms during composting, such as temperature, pH, oxygen, moisture content, C:N ratio, enzymes produced by microbiota, and compost ingredients. These parameters can influence the rate of pathogen inactivation during the thermophilic phase of aerobic composting. During the maturation composting phase, the microbial activity of indigenous microbiota is a major contributor to the elimination of pathogens (Gurtler et al., 2018).

The finished compost should be thoroughly decomposed and free of pathogenic bacteria. *Salmonella* in poultry litter compost can be eradicated through the implementation of efficient composting methods, such as covering the compost heaps with the finished compost (Shepherd et al., 2010). However, sporadic cases have been reported about the presence of foodborne pathogens in finished compost, indicating that finished compost made from animal waste is a potential source of pathogens (Miller et al., 2013; Dharmasena and Jiang, 2018). Due to improper composting processes or post-composting contamination, *Salmonella* continued to be detected following the composting process (Trimble et al., 2013). The following are some explanations as to why the pathogens survive the composting process: (1) bacteria that are heat-resistant or form spores, including *Salmonella*, *Bacillus* spp., and *Clostridium* spp.; (2) bacteria that survive thermophilic conditions with adaptation to the composting process; and (3) regrowth or re-contamination of bacteria during storage (Lepesteur, 2022). Following the Food Safety Modernization Act (FSMA) produce safety rule, biological soil amendments from animals should have less than 0.3 MPN of *E. coli* O157:H7 per gram or milliliter of the analytical portion, less than 3 MPN per 4 g or ml of total solids for *Salmonella* spp., and less than 1 CFU per 5 g or ml of the analytical portion for *L. monocytogenes* (FDA, 2018). To achieve these standards, alternative treatments for reducing or eliminating human pathogens should be incorporated into raw animal waste before land application (Clements et al., 2019).

In summary, as a controlled aerobic process, composting can efficiently decompose animal wastes, including poultry litter. The key to inactivating pathogens during the four composting phases is the heat generated by indigenous microbial metabolism. Recognized for its agronomic benefits, compost enhances soil structure when applied as a soil amendment. However, challenges such as the presence of pathogenic microorganisms either surviving the composting process or recontamination from the environment and heavy metals in poultry litter necessitate careful management to mitigate public health risks and ensure compliance with FDA standards for microbial levels in the finished compost. It is important for farmers and composting facilities to monitor and control these risks through proper handling and monitoring of composting processes.

4 Physical heat treatment of poultry litter

Physical heat treatment of poultry litter, such as pelletization, pasteurization, and dry heating, can stabilize organic matter by reducing the moisture content of poultry litter significantly after or without composting. In an industrial setting, physical heat treatment of poultry litter can be conducted through either continuous or batch processes. In a continuous operation system, poultry litter, with or without prior composting, is consistently fed into the drying tunnel, where the drying process is facilitated by the continuous rotation within the tunnel dryer. Conversely, in the batch operation system, poultry litter is dried in discrete batches (Wang et al., 2021). To produce a pathogen-free final product, some critical factors, including time-temperature, moisture content of poultry litter, heat sources, and monitoring and validation procedures, should be controlled. According to National Organic Program (NOP) guidance, the minimal requirement for time-temperature is >65°C for >60 min to decontaminate fecal coliforms, *Salmonella*, and *E. coli* O157:H7 (NOP, 2002). However, there is extremely limited research on the microbiological safety of these physically heat-treated animal waste products.

Table 2 summarizes laboratory-scale studies for evaluating the physical heat treatment required to kill pathogens in animal waste or animal waste-based compost. As a result of the scarcity of published research, various forms of animal waste, including poultry litter, have been investigated in the pursuit

Table 2 Lab-scale physical heat treatments of animal waste with or without composting for pathogenic bacteria inactivation

			Thermal study set up				
Animal waste	Tested microorganisms	Initial populations	Sample container	Heating source	Treatment duration	Significant results	Reference
Dairy compost	*Clostridium difficile* endospores	5.5 $\log_{10}$ CFU/g	Sealed Tyvek pouches	Environmental chamber	3–5 days	Endospore: reduced by <68% at 55°C and 65°C	Dharmasena et al., 2019
Dairy compost	*Clostridium difficile* endospores	5.5 $\log_{10}$ CFU/g	Sealed Tyvek pouches	Dry heat: environmental chamber.Moist heat: water bath	3– 5 days	Endospore: reduced by 80–84% and 75–99.9% at 55°C and 65°C	Dharmasena et al., 2019
Broiler litter	Desiccation-adapted *Salmonella*	6 $\log_{10}$ CFU/g	Aluminum pan (i.d., 10 cm)	Conventional oven	3 h	>5 $\log_{10}$ reduction at 75°C or 85°C	Wang et al., 2018
Aged chicken litter	Desiccation-adapted *Salmonella*	7 $\log_{10}$ CFU/g	Aluminum pan (i.d., 10 cm)	Conventional oven	3 h	At 70°C: 5 $\log_{10}$ reduction in aged chicken litter with 20% moisture content (MC). At 150°C, >5 $\log_{10}$ reduction for 50 min in chicken litter with 20% MC.	Chen et al., 2013
Broiler chicken litter with different storage periods	Desiccation-adapted *Salmonella*	7 $\log_{10}$ CFU/g	Aluminum pan (i.d., 10 cm)	Conventional oven	1.5 h	At 150°C, >5 $\log_{10}$ reduction after 45 min of heat treatment in 0-month litter. For 3-, 6-, and 9-month litter samples, <5 $\log_{10}$ reduction.	Chen et al., 2015a
Cow manure compost	*E. coli* and *Salmonella*	6 $\log_{10}$ CFU/g	Directly placed into machine	Drum rotation heat-treatment machine.	6–24 h	>5$\log_{10}$ reduction at 54°C to 68°C.	Gong et al., 2007

Dairy compost (fresh and finished)	Acid-adapted *E. coli* O157:H7	7 log_{10} CFU/g	Sealed Tyvek pouches	Environmental chamber with 70% humidity	>19 days	Survival duration: acid-adapted cells > control cells at 50°C.	Singh and Jiang, 2012
Woodchip poultry litter, corncob poultry litter with Co_{60} woodchip or Co_{60}	*E. coli*, *Salmonella pullorum*, and *Arizona* sp.	4–5 log_{10} CFU/g	Thermal death tube (13 × 17 mm)	Water bath	30 min	*E. coli* inactivated at 57.2°C or 68.3°C. *Salmonella* inactivated at 62.8°C after 30 min. *Arizona* sp. inactivated at 57.2°C after 30 min	Messer et al., 1971
Dairy compost	Heat-shocked *E. coli* O157:H7, *Salmonella*, and *L. monocytogenes*.	6.81–7.33 log_{10} CFU/g	Sealed Tyvek pouches	Environmental chamber	4 h	At 50°C, 99.9% reduction.	Singh et al., 2010
Broiler litter with pine shavings	*Salmonella* Typhimurium	5 log_{10} CFU/g	Perforated nursery flat (41.91 × 33.0 cm)	Soil steamer cart	5 min, 30 min, or 120 min	*Salmonella* was inactivated with steaming in 30 min or 120 min	Stringfellow et al., 2010
Broiler litter	*E. coli* and *Salmonella*	8–9 log_{10} CFU/g	50-ml plastic tube in a 100-mm diameter plastic container	Water bath	1 h	>99% inactivation at 55°C and 65°C	Wilkinson et al., 2011
Aged broiler litter and fresh layer litter	*Salmonella*	7 log_{10} CFU/g	Aluminum pan (i.d., 10 cm)	Conventional oven	6 h	>5 log_{10} reduction in aged broiler litter with 30% MC at 70°C, 75°C, or 80°C after 300 min, 165 min, and 75 min, respectively. >5 log_{10} reduction in fresh layer litter with 30% MC at 70°C, 75°C, or 80°C after 105 min, 90 min, and 60 min, respectively	Kim et al., 2012

(*Continued*)

Table 2 (*Contined*)

Animal waste	Tested microorganisms	Initial populations	Thermal study set up			Significant results	Reference
			Sample container	Heating source	Treatment duration		
Poultry sludge	Fecal coliforms, *Salmonella*	10.5 log MPN/g of total solids, 4.5 log MPN/g of total solids	Directly placed in reactor	Hermetically closed thermal reactor.	120 min	Fecal coliforms and *Salmonella* were both inactivated at 70°C in 120 min	Ruiz-Espinoza et al., 2012
Layer manure	Total aerobic bacteria, yeast and mold, indigenous *E. coli*, and *Salmonella*	9.7, 3.4, and 7.4 log_{10} CFU/g, respectively	Drying tray (1 cm, 2 cm, or 3 cm × 10 cm × 10 cm)	Oven	55–12 min	65.6–99.8% inactivation was achieved for total aerobic bacteria.74.1–99.6% inactivation was achieved for both yeast and mold.99.97% inactivation was achieved for *E. coli*.100% inactivation was achieved for *Salmonella*.	Ghaly and Alhattab, 2013
Anaerobically digested dairy manure	Indigenous *E. coli*, *Salmonella*, and fecal coliforms	3.9, 3.8, and 2.2 log MPN/g, respectively	Test tube	Digestion unit with an aluminum block	120 min	85–95% inactivation was achieved for fecal coliforms.87–96% inactivation was achieved for *E. coli*.100% inactivation was achieved for *Salmonella*.	Collins et al., 2013
Dairy manure	*Salmonella*, *Enterococcus faecalis*	6–7 log_{10} CFU/g, 6–7 log_{10} CFU/g	Tube	Block thermostat.	1.2 h	*D*-values: 0.37 h at 55°C; 22.52 h at 46.0°C; 0.45 h at 55.0°C to 14.50 h at 47.5°C for *Salmonella* and *Enterococcus faecalis*, respectively.	Elving et al., 2014
Ovine and bovine waste	*E. coli* O157:H7	6.5–7.0 log_{10} CFU/ml	Sterile microcentrifuge tube	Water bath	10 min	>5 log_{10} reduction at 60°C after 10 min	Avery et al., 2009

Source: Partially adapted from Chen and Jiang (2014).

of laboratory-scale heat treatment studies. The initial concentration of targeted microorganisms was between 2 $\log_{10}$ CFU/g and 10 $\log_{10}$ CFU/g, representing a wide variety of pathogens with varying degrees of heat resistance that were present in animal waste. Some key factors affecting the physical heat treatment of animal waste are discussed below.

4.1 Physiological status of pathogenic bacterial cells

Pre-adaptation of pathogenic cells to stresses, including heat shock, acid adaptation, and desiccation adaptation, resulted in a substantial increase in the thermal resistance of the pathogens present in raw animal waste or compost derived from animal waste. Heat shock can occur when the pathogenic cells are exposed to sub-lethal temperatures, i.e. during the mesophilic phase of composting (Singh and Jiang, 2012). The stress response of heat-adapted microorganisms can improve their survival following subsequent heat treatment. For example, at 55°C, heat-shocked cultures of *E. coli* O157:H7, *Salmonella*, and *L. monocytogenes* decreased by 1.2-, 1.9-, and 2.3-$\log_{10}$ CFU/g within 1 h, respectively, whereas the comparable control cultures had 4-, 5.6-, and 4.8-$\log_{10}$ CFU/g decreases, respectively (Singh et al., 2010).

As moisture levels can be lowered during the storage of raw animal waste, the significance and frequency of desiccation stress require special consideration for the safe production of heat-treated animal waste. Evaporation or heating resulting from a thermophilic process occurs at the surface of the compost pile (Chen and Jiang, 2017). Desiccation-adapted *Salmonella* cells were found to be more heat-resistant than non-adapted cells in poultry litter during physical heat treatment (Erickson et al., 2014; Chen et al., 2013), implying that desiccation-adapted surrogate cells should be used to validate the heat treatment conditions of animal waste, including poultry litter, to ensure a safety margin.

4.2 Properties of animal waste

The chemical, physical, and microbiological characteristics of poultry litter compost are altered with storage; hence, the efficiency of heat treatment differs between fresh and aged chicken litters (at least 3–4 months, depending on the farm; Wilkinson et al., 2011; Chen et al., 2013). Depending on the initial moisture content of fresh chicken litter, a 7-$\log_{10}$ decrease of desiccation-adapted *Salmonella* required heat exposure of fresh chicken litter for 80.5–100.8 min, 78.4–93.1 min, and 44.1–63 min at 70°C, 75°C, and 80°C, respectively (Chen et al., 2013). In comparison, in aged chicken litter with a 20% moisture content, a 5-$\log_{10}$ decrease of desiccation-adapted *Salmonella* spp. needed >360 min, >360 min, and 240–300 min of exposure at 70°C, 75°C, and 80°C, respectively (Chen et al., 2013). Chen et al. (2015) studied the effect of the chicken litter storage period and ammonia concentration on the thermal resistance of desiccation-adapted *Salmonella* spp. and found that thermal inactivation rates decreased with increased storage time (6 months). The study also found that the amount of ammonia in the chicken litter dropped significantly as the storage time increased. This suggests that ammonia may be another key factor affecting the thermal resistance of *Salmonella* cells in the chicken litter.

Physically heat-treated poultry litter, or poultry litter compost, is widely utilized as organic fertilizer on both organic and conventional farms. To reduce contamination from *Salmonella*, further validation studies on pathogen inactivation under physical heat treatment in industrial settings are urgently required. However, due to biosafety considerations, introducing pathogens into an industrial context is not suggested for plant validation experiments. As a result, indigenous and non-pathogenic bacteria should be exploited as surrogates and indicator microorganisms to better understand pathogen growth/survival behaviors in industrial environments (Harris et al., 2013). Wang et al. (2021) conducted the first in-plant validation study of the physical heat-treatment of poultry litter in two commercial processing plants with continuous or batched operations. The thermal processes in the two plants yielded 98.22% to >99.86% reductions in *Enterococcus faecium* as a surrogate microorganism for *Salmonella* spp. Even though the processing conditions varied among trials and plants, the processing conditions in both plants were sufficient to reduce the potential *Salmonella* contamination of processed poultry litter compost by at least 5 logs.

In summary, physical heat treatment is employed to stabilize poultry litter by reducing moisture, with or without the composting process. Factors influencing the effectiveness of heat treatment include the physiological status of pathogenic bacterial cells, which may exhibit increased thermal resistance when pre-adapted to stresses like heat shock or desiccation. Additionally, the properties of animal waste can affect the treatment's efficacy. Laboratory studies highlight the variability in pathogen reduction across different studies. Recent validation studies in commercial processing plants have shown promising results, achieving >5 $\log_{10}$ reductions in *E. faecium.* More validation studies need to be done in industrial settings, focusing on *Salmonella* or other pathogens that need to be killed. These studies should be done with several types of waste and heating sources. This will help ensure the effectiveness and safety of the processing methods. Additionally, these studies should also consider variations in processing time and temperature to determine the optimal conditions for pathogen inactivation.

5 Amendments to poultry litter and poultry litter-based compost

Poultry litter, high in nitrogen, is a leading source of ammonia emissions due to the microbial mineralization of organic nitrogen and uric acid (Cook et al., 2011). Volatilized ammonia inside poultry houses can be detrimental to the health of the poultry, causing eye and lung irritation, foot pad burns, and increasing susceptibility to respiratory disease, leading to poor bird performance, carcass degradation and condemnations, food safety concerns, and loss of profitability. In the meantime, the personal health and well-being of growers and those who work in poultry houses are adversely affected by exposure to elevated levels of ammonia. Since poultry litter is widely used in agriculture, the loss of nitrogen via ammonia volatilization reduces the value of litter as an organic fertilizer. In addition, the release of ammonia from poultry litter has the potential to negatively impact the air quality in nearby areas, posing a significant environmental issue.

As compared with other animal manures, poultry litter has an overall higher ammonia content. To control ammonia emission, an array of poultry litter treatments, including physical (such as ventilation control), chemical (acidifying products), and/or biological (competitive exclusion cultures or enzymes) methods, have been used. Cook et al. (2011) compared chemical, biological, or adsorbent amendments and concluded all these amendments reduced nitrogen loss (0–15%) in poultry litter as compared with no amendments (20%), but with acidified amendments outperforming other litter amendments. Acidifier amendments have shown the most promising and consistent results, with Al+Clear® (aluminum sulfate), Poultry Guard® (acidified clay), and Poultry Litter Treatment® (PLT) (sodium bisulfate) being the most widely used in the poultry industry (Cook et al., 2011). These amendments can reduce the pH, inhibit bacteria from converting litter nitrogen into ammonia, and reduce pathogen persistence (Payne et al., 2002; Pope and Cherry, 2000; Nahm et al., 2005).

Research has primarily focused on reducing ammonia emissions from poultry litter. Studies have shown that the use of chemicals can mitigate ammonia volatilization from poultry litter, thereby influencing the final properties of the litter. The addition of alum, for example, can decrease ammonia volatilization and improve the N/P ratio by increasing total soluble nitrogen in the litter. Moreover, a 58% reduction in volatilization was achieved with the addition of 200 g of ferrous sulfate per kilogram of litter, which also lowered pH and consequently reduced microbial activity (Moore et al., 1995). A comprehensive review on the use of chemical additives for poultry litter and livestock manure was published by Nahm et al. (2005).

Pre-treatment of poultry litter can enhance the nutritional value of the final composted product. For example, extracting ammonium nitrogen from poultry litter using water can increase the biogas production of poultry litter (Gaspareto et al., 2020). Studies have also highlighted the impact of additives on the microbial composition of poultry litter during composting. These additives improve microbial activity, reduce mobile gene elements, and decrease ARGs. Liu et al. (2021) showed that adding 5% chitosan to compost changes the bacteria composition by altering physicochemical properties like temperature, pH, and total nitrogen, as well as lowering the concentration of antibiotics and metals like Cu^{2+} and Zn^{2+}. Consequently, the risk of ARG transmission is reduced. Guo et al. (2020) discovered that adding microbial inoculum and ferric chloride to compost increased the activity of microbes, which helped reduce ARGs.

Due to their substantial surface area, porosity, and cation exchange capacity, biochar and palygorskite are efficacious additives for compost, capable of immobilizing heavy metals, reducing ammonia, and facilitating the composting process (Zhang et al., 2022; Awasthi et al., 2020; Pan et al., 2019). Zhang et al. (2022) reported that biochar addition promotes saprotrophic fungi while reducing pathogenic fungi during chicken manure composting. Additionally, the concentration of humic acids increased with the addition of biochar and palygorskite during the composting of corn stover and chicken manure. Similar findings were also reported by Zainudin et al. (2020), who observed that both nitrifying and denitrifying bacterial communities (*Pusillimonas*, *Halomonas*, and *Pseudofulvimonas*) increased after the addition of biochar during the composting of poultry manure and rice straw.

A few studies investigated the impact of litter amendments on pathogenic microbes in litter or birds. Pope and Cherry (2000) conducted two field trials to evaluate the presence of *Salmonella*, *Campylobacter*, and generic *E. coli* in litter and on broilers raised on the litter treated with PLT on two farms. Their results demonstrated the inhibitory effect of PLT on the survival of *Salmonella* and *E. coli* in litter and the reduction of both bacteria on bird rinse as compared with the control without PLT treatment. However, there was no difference in *Campylobacter* counts between treated and control groups. Another field study conducted by Payne et al. (2002) reported that an application rate of 100 lb/1000 ft^2 (0.49 kg/m^2) for the granulated sulfuric acid and sodium bisulfate litter amendments was required to significantly reduce the *Salmonella* population in litter. However, variables such as litter pH and age were critical when determining rates of litter amendments for pathogen control.

In addition to the abovementioned litter amendments, ozone has been explored as a treatment method. Chang et al. (2020) conducted a series of experiments investigating the effects of ozone concentration, exposure time, and moisture content of poultry litter on *S.* Typhimurium and *E. coli* O157:H7 reduction using ozonation. Their results showed that higher ozone concentrations and longer exposure times were required for substantial pathogen reduction in poultry litter as compared with other applications such as the decontamination of surfaces or food products. Furthermore, the moisture content of poultry litter is a critical factor for pathogen reduction in addition to ozone concentration and exposure time.

In summary, strategically adding certain chemicals and materials to poultry litter can significantly enhance its environmental and agronomic values. These additives not only reduce harmful ammonia emissions and pollutants but also contribute to the protection of poultry welfare, profitability, and the production of a more nutrient-rich organic fertilizer. Importantly, the use of these additives can also improve the microbial safety of poultry litter during composting, thereby reducing the risk of foodborne pathogen and ARG transmission and enhancing the overall quality of the compost.

6 Anaerobic digestion of poultry litter

Anaerobic digestion is a biological process that breaks down the organic matter in animal wastes and produces biogas such as methane and carbon dioxide through the actions of a consortium of microorganisms in the absence of oxygen. The biogas produced can be captured as an alternative clean energy source. The entire process progresses from hydrolysis of complex organic matter to acidogenesis, acetogenesis, methanogenesis, biogas production, and digestate formation (Salminen and Rintala, 2002). The anaerobic digestion can be classified into dry, semidry, and wet types based on the solid content. The technology works under either thermophilic (50–60°C) or mesophilic (32–49°C) conditions. There is extensive research done on fermentative microbes involved in anaerobic digestion and optimizing conditions for biogas production (Shapovalov et al., 2020; Wang et al., 2019). In this chapter, we will focus on the pathogen reduction of anaerobic digestion and the microbiological risk of the anaerobic digestate.

Anaerobic digestate, the semi-solid or solid fraction of effluent produced by anaerobic digestion of organic wastes, is commonly used as a biofertilizer and soil amendment for crops, animal bedding, horticultural products, turf, and landscape applications. Although anaerobic digestion is considered an effective method for waste management and energy production, ensuring microbiological safety in the digestate is crucial due to the potential risks associated with pathogens present in the waste. The high temperature, acidic environment, high volatile fatty acids or ammonia concentration, long retention time, and biological process involved in anaerobic digestion process can contribute to pathogen inactivation and reduction of microbial load in the digestate, but some pathogens of bacteria, especially spore-formers, ARB, viable but nonculturable bacteria or persisters, viruses, and parasites, may potentially survive or remain viable in the final digestate (Lin et al., 2022; Resende et al., 2014; Thomas et al., 2019). In general, under mesophilic conditions, anaerobic digestion results in not more than 1–2 $\log_{10}$ units of pathogen reduction within 24 hours retention time, whereas thermophilic digestion could achieve a 3–4 $\log_{10}$ reduction of pathogens within hours (Gibbs et al., 1995). Larsen et al.

(1994) monitored the digestates from five full-scale biogas plants (two used thermophilic, one used mesophilic, and two used a two-stage process involving thermophilic pretreatment followed by mesophilic digestion) for pathogen inactivation for 6–16 months. Their results demonstrated that thermophilic digestion and mesophilic digestion with thermophilic pretreatment resulted in sufficient reduction of both *E. coli* and enterococci pathogens and intestinal parasites, whereas mesophilic digestion used alone was not effective. According to a technical evaluation report compiled by OMRI for the USDA National Organic Program, foodborne pathogens in animal or agricultural wastes such as *Bacillus anthracis*, *Campylobacter jejuni*, *Clostridium* spp., *Escherichia* spp., *Klebsiella* spp., *L. monocytogenes*, *Salmonella* spp., *Shigella* spp., and *Yersinia enterocolitica* could survive the anaerobic digestion process (NOSB, 2017). Sometimes the pathogens may even grow under mesophilic digestion. For example, the population of *Clostridium perfringens* spores in the digestate was higher than those in the feedstocks following mesophilic anaerobic digestion (Lloret et al., 2013). In poultry litter, studies have documented the survival of foodborne pathogens such as *Clostridium* spp., *Escherichia* spp., *Klebsiella* spp., and *Shigella* spp. during the anaerobic digestion process, particularly those heat-resistant spore-forming pathogens (Alfa et al., 2014; NOSB 2017). However, limited research is available on the microbiological safety of the digestate when used as organic fertilizer for agricultural applications, particularly during the anaerobic digestion of poultry wastes (Lin et al., 2022).

A few studies have also reported the presence of antimicrobial-resistant bacteria (ARB) in the digestate of anaerobic digestion of animal wastes. Thomas et al. (2018) compared the reduction kinetics of two extended-spectrum-beta-lactamase (ESBL)-/AmpC-producing *E. coli* during lab-scale mesophilic (37°C, 42°C) and thermophilic (55°C) anaerobic digestion of chicken manure. Their results showed the decimal reduction times (D-value) of approximately 3–6 days at 37°C, 1.5 days at 42°C, and 48 min at 55°C for the two artificially inoculated ESBL-/AmpC-producing *E. coli* strains, suggesting temperature and retention time as the main inactivation factors. Resende et al. (2014) used a pilot-scale biogas reactor to digest cattle manure anaerobically under ambient temperature (between 25°C and 35°C) and reported a decay of all tested bacterial groups, including enterobacteria, nonfermenting Gram-negative rods, and Gram-positive cocci after 60 days of digestion. However, multidrug-resistant bacteria were detected both in the influent and effluent and exhibited the same pattern of resistance to different antibiotics. The authors emphasized the need to implement sanitary and microbiologically safe treatment of animal waste to avoid environmental contamination.

Anaerobic digesters of animal wastes should follow all federal, state, and local regulations. There are microbiological standards for digestate of anaerobic digestion, although they vary from country to country (Lin et al.

2022). According to the USEPA (40 CFR 503), anaerobic digestate from sewage sludge should have <10^3 MPN/g total solid for fecal coliform, <3 MPN/4 g total solid for *Salmonella*, <1 viable egg/4 g total solid for Helminth, and <1 PFU/4 g total solid for Enteric virus in Class A biosolid and <2×10^6 MPN/g total solid for fecal coliform in Class B biosolid (USEPA, 2008). Even though the digestate may meet the microbiological standard requirements, it may not represent the absence of potential pathogen risk. The recontamination and regrowth of a few surviving populations of enteric pathogens during storage and handling of the digestate could occur (Larsen et al., 1994). Therefore, there is a need for effective digestate treatment to prevent contamination of the agricultural environment.

The FDA FSMA Produce Safety Rule placed limitations on using raw manure and set microbial standards for composted manure on crops produced for direct human consumption (Clements et al., 2019). Although anaerobic digestate is not specifically described in the act, the microbial risk should be considered when used on crops intended for human consumption. According to the Organic Foods Production Act, animal manure-based anaerobic digestate can be used as raw manure if it does not meet the processing requirements of NOP §205.203(c)(2). Therefore, it is subject to days-to-harvest intervals after application to a crop destined for human consumption, i.e. ≥120 days between soil incorporation and the harvest of a product whose edible portion has direct contact with the soil surface, or ≥ 90 days for a product whose edible portion does not have direct contact with the soil surface.

In summary, anaerobic digestion is an effective approach for global waste management and renewable energy generation. While the resulting digestate retains valuable nutrients and can serve as an organic fertilizer and soil amendment, it may also harbor viable microorganisms, including human pathogens. Consequently, ensuring proper handling, treatment, and storage is critical to prevent environmental contamination when applying digestate to agricultural lands.

7 Surrogate and indicator microorganisms

It is important to confirm the findings obtained from laboratory experiments in a real-world context. Surrogate microorganisms, which are non-pathogenic strains capable of responding to a certain treatment in a way like a pathogenic strain, are frequently employed due to the need for biosafety (Hu et al., 2017). There are several ways to get surrogate microorganisms: changing the genetic makeup of the pathogens, isolating them from the original tested products, or adapting a surrogate that has been used before to a new study matrix (Table 3). Non-pathogenic *E. coli* or an avirulent strain of *S.* Typhimurium that contains no pathogenicity islands were used as surrogate microorganisms in

Table 3 Surrogate bacteria evaluated for *Salmonella* spp. in physically heat-treatment (food products) studies since 2000121°C, 149°C, or 177°C

Pathogens	Surrogates	Study matrix	Initial populations	Heat-treatment conditions: Heating source	Heat-treatment conditions: Treatment condition	Suggestion on surrogate	References
Salmonella	Commercial *Pediococcus acidilactici* starter culture	Beef jerky	8 $\log_{10}$ CFU/ strip for *Salmonella*, and 3.3–4.7 $\log_{10}$ CFU/ strip for surrogate	Wet-bulb spikes	54.4–76.7°C for up to 60 min	*P. acidilactici* starter culture can be a potential surrogate	Buege et al., 2006
Salmonella	15 nonpathogenic *E. coli*	Diluted bacterial suspension	Not describe	Techne submerged-coil heating apparatus	60°C	Five *E. coli* strains with less heat resistance were recommended as *Salmonella* surrogates	Eblen et al., 2005
Salmonella and *L. monocytogenes*	*E. faecium* NRRL B-2354, *P. parvulus* HP, and *P. acidilactici* LP	Ground beef	7.5 $\log_{10}$ CFU/g	Circulating water bath	58°C, 62°C, 65°C, or 68°C	*E. faecium* was a conservative surrogate	Ma et al., 2007
Salmonella	*E. coli* K12	Liquid egg	7.5 $\log_{10}$ CFU/ml	*Water bath*	52–60°C.	*E. coli* K12 may serve as a surrogate for pathogenic *Salmonella* in liquid egg	Jin et al., 2008
Salmonella	Commercial lactic acid bacteria	Ground-and-formed beef jerky	8 $\log_{10}$ CFU per beef strip	Dehydrator	68.3°C for 0 h, 3.5 h, and 7 h	LAB can be used as *Salmonella* surrogate for beef jerky process	Borowski et al., 2009
Salmonella	*E. faecium* NRRL B-2354	Almonds	Not describe	Computer-controlled moist-air convection oven.	121–204°C	*E. faecium* can be used as a conservative surrogate for SE PT30	Jeong et al., 2011

Salmonella	A surrogate cocktail combination of *E. faecium* ATCC 8459 and *E. faecium* ATCC 35667	Peanut	6.5–7.7 log_{10} CFU/g	Industrial setting	149°C, 163°C, 177°C, and 191°C	Roasted peanut process has been validated using surrogate microorganisms	Poirier et al., 2014
Salmonella	*P. acidilactici and Staphylococcus carnosus*	Frankfurter batter and ground beef	7 log_{10} CFU/g	Circulating water bath	55°C for up to 120 min	The use of *P. acidilactici* and *S. carnosus* as *Salmonella* surrogates to validate thermal processing interventions in ground beef and frankfurter batter is not recommended.	Vasan et al., 2014
Salmonella	*E. faecium* NRRL B-2354	*Carbohydrate-protein meal*	6 log_{10} CFU/g	Single-screw extruder	50–100°C	*E. faecium* can be a surrogate for eliminating *Salmonella*.	Bianchini et al., 2014
Salmonella	*E. faecium* NRRL B-2354	Talc powder	8–9 log_{10} CFU/g; using dry inoculum	Controlled water bath	85°C	*E. faecium* can be a conservative surrogate for *S. tennessee*.	Enache et al., 2015
Salmonella	*P. acidilactici* ATCC 8042 and *E. faecium* NRRL B-2354	Dry pet foods	Not describe	Circulating water bath	76.7°C and 87.8°C	*P. acidilactici* ATCC 8042 can be utilized as a surrogate for *Salmonella* in dry pet food	Ceylan et al., 2015
Salmonella	*E. faecium* NRRL B-2354 and *Saccharomyces cerevisiae*	Hamburger buns	8 log_{10} CFU/g	Conventional oven	218.3°C for 13 min with internal ~100°C in 8 min	*E. faecium* demonstrated greater thermal resistance compared with *Salmonella*, not *S. cerevisiae*	Channaiah et al., 2016

(*Continued*)

Table 3 (*Contined*)

Pathogens	Surrogates	Study matrix	Initial populations	Heat-treatment conditions		Suggestion on surrogate	References
				Heating source	Treatment condition		
Salmonella	*E. coli* 3A-I, *E. faecium* 2B-I, and ten different isolates from flour	Flour	5 log_{10} CFU/g	Agitation water bath	70°C, 75°C, and 80°C for 12–60 min	*Pantoea dispersa* strain JFS isolated from flour was suggested as a surrogate for *S.* Typhimurium PT 42	Fudge et al., 2016
Salmonella	Avirulent strain of *S.* Typhimurium	Diluted bacterial suspension	~7 log_{10} CFU/ml	Water batch	50°C for up to 10 min	The avirulent strain of *S.* Typhimurium contains no pathogenicity islands and shares similar thermal resistance with that of the wild type	de Moraes et al., 2016
Salmonella and *L. monocytogenes*	*E. faecium* NRRL B-2354	Low moisture foods	7–8 log_{10} CFU/g	Oil bath	70–140°C	*E. faecium* can be a suitable surrogate for *Salmonella* for low moisture foods, but not for the sugar containing confectionery formulation	Rachon et al., 2016
Salmonella	*E. coli ATCC 25922*	Almond powder	~3 log_{10} CFU/g	Controlled atmosphere/ heating block	65–80°C.	No directly comparison of thermal resistance between *Salmonella* and *E. coli* was provided	Cheng et al., 2017
Salmonella	*E. faecium* NRRL B-2354	Almond kernels	~8 log_{10} CFU/g	*Pilot-scale and moist-air impingement oven*		*E. faecium* can work as a surrogate for *Salmonella*	Jeong et al., 2017
Salmonella	*E. faecium* NRRL B-2354	Flaxseed, sunflower seeds, and peppercorn	~7–8 log_{10} CFU/g	Vacuum steam	75°C, 85°C, 95°C, and 105°C	*E. faecium* can be used as surrogate for *Salmonella* PT 30.	Shah et al., 2017

Salmonella	*E. faecium* NRRL B-2354	Wheat flour	~8 log_{10} CFU/g	Aluminum-test-cells	75°C, 80°C, and 85°C	*E. faecium* can be a conservative surrogate for *Salmonella* in thermal processing of wheat flour.	Liu et al., 2018a
Salmonella	*E. faecium* NRRL B-2354	SiO_2 particles	7.8 log_{10} CFU/g	*Ethylene glycol oil bath*	*80°C*	The thermal resistance of both *S.* Enteritidis and its surrogate increased with decreasing relative humidity.	Liu et al., 2018b
Salmonella	*E. faecium* NRRL B-2354	Oat flour	7–8 log_{10} CFU/g	Lab-scale single-screw extruder running at different screw speeds (75–225 rpm)	75°C, 85°C, and 95°C	Effect of different product and process parameters on inactivation of *E. faecium* was different from that of *Salmonella*.	Verma et al., 2018
Salmonella	Non-pathogenic *E. coli* surrogates from ATCC	Ground beef	6 log_{10} CFU/g	Water bath	54°C, 57°C, 60°C, and 63°C	*E. coli* surrogates can be used as *Salmonella* surrogates in ground beef	Redemann et al., 2018
Salmonella and *L. monocytogenes*	*E. faecium* NRRL B-2354	In-shell pecans	6.54–7.88 log_{10} CFU/g	Hot water	70°C, 80°C, or 90°C for 1 min, 2 min, 3 min, 4 min, or 5 min	*E. faecium* showed similar resistance as *S. enterica. L. monocytogenes*	Kharel et al., 2018
Salmonella	*E. faecium* NRRL-B2354	Almond meal	~8 log_{10} CFU/g	Water bath	80°C	*E. faecium* was more heat-resistant in dry inoculation methods, as compared to the wet inoculation	Ahmad et al., 2019

(*Continued*)

Table 3 (*Contined*)

Pathogens	Surrogates	Study matrix	Initial populations	Heat-treatment conditions		Suggestion on surrogate	References
				Heating source	Treatment condition		
Salmonella	*P. acidilactici* ATCC 8042and*E.faecium* NRRL-B2354	Toasted oats cereal and peanut butter	6 log_{10} CFU/g	Water bath	80°C, 85°C, 90°C, and 95°C for toasted oats cereal, and 63°C, 68°C, 73°C, and 77°C for peanut butter	*P. acidilactici* can only be used as *Salmonella* surrogate in toasted oats cereal when heat treated at >85°C	Deen et al., 2019
Salmonella	*E. faecium* NRRL-B2354	Cocoa powder	~8 log_{10} CFU/g	Ethylene glycol bath	70–80°C.	The suitability of *E. faecium* as a surrogate for *Salmonella* varies because of a_w of cocoa powder	Tsai et al., 2019
Salmonella and *L. monocytogenes*	4 non-pathogenic surrogates: *E. faecium* NRRL B-2354, *E. coli* P1, *E. coli* K12, and *Listeria innocua*	Black peppercorns	~8 log_{10} CFU/g	Steam treatment	70°C and 75°C for 5 min	*E. faecium* was the most suitable surrogate to validate mild steaming processes	Zhou et al., 2019
Salmonella and *L. monocytogenes*	*L. innocua* and*E. faecium* NRRL B-2354	Frozen blueberry	8.6 log_{10} CFU/g	Osmotically dehydrated and air-dried		After osmotic dehydration of blueberries, only *E. faecium* was detected	Bai et al., 2020

Salmonella	*E. coli* P1, *E. faecium* NRRL B-2354	Basil leaves	~6 $\log_{10}$ CFU/g	Hot air drying in the pilot scale dryer	60°C and 69°C for 90 min or at 100°C for 70 min	*E. coli P1* was a more suitable surrogate to validate hot-air drying processes compared to *E. faecium*.	Zhou et al., 2020
Salmonella	*E. faecium* NRRL B-2354	Pistachios	7.4–7.7 $\log_{10}$ CFU/g	Hot oil or water bath	Hot oil (121°C) or hot water (80°C) for 6 min	*E. faecium*is a good potential thermal-treatment surrogate for pistachios	Moussavi et al., 2020
Salmonella	*E. coli* AW1.7, *P. acidilactici*, *E. faecium* NRRL B-2354, and *S. carnosus*	Beef jerky	~8 $\log_{10}$ CFU/g	Water bath	60°C for 1 min, 4 min, or 15 min	*E. faecium* can be surrogate organisms for *Salmonella* after desiccation adaptation	Schultze et al., 2020

lab-scale studies (Eblen et al., 2005; Jin et al., 2008; de Moraes et al., 2016). The surrogate microorganism may replicate most characteristics exhibited by the specific pathogens, except for their ability to cause disease. However, it is not advisable to make such a selection in an industrial context due to the potential for pathogenicity reversion or the development of false positives in the end products (Hu et al., 2017). An alternative approach involves the isolation of surrogates from the preliminary study matrix. Fudge et al. (2016) examined ten isolates from flour samples in their study. The researchers proposed that *Pantoea dispersa* JFS, a Gram-negative bacterium often found in plants, soil, and water, might serve as an appropriate surrogate microorganism for *Salmonella* in wheat flour when subjected to heat treatment. In addition, several commercially available lactic acid bacterial cultures have been employed as a substitute for *Salmonella* spp. in the production of beef jerky or sausages (Buege et al., 2006; Borowski et al., 2009). Various bacterial species have been suggested as potential surrogate microorganisms for *Salmonella* in diverse matrices. One often-employed surrogate for *S. enterica* in the validation of thermal processing of nuts and carbohydrate-protein meal is *E. faecium* NRRL B-2354 (Jeong et al., 2011; Bianchini et al., 2014). In their comprehensive investigation, Kopit et al. (2014) examined the genetic safety of *E. faecium* NRRL B-2354 as a substitute and determined that this strain did not possess the genomic and phenotypic characteristics associated with nosocomial infections.

Before conducting a validation study in a commercial or open environment, the surrogate microorganisms should be evaluated in a controlled laboratory setting against the specific pathogens of interest. In various studies involving matrices such as ground beef and pecans, *Salmonella* has been identified as the most heat-resistant non-spore-forming bacterial pathogen (Ma et al., 2007; Den Besten et al., 2018; Kharel et al., 2018). This resistance to physical heat treatment has been consistently documented (Ma et al., 2007; Kharel et al., 2018). The heat resistance of bacterial cells is influenced by a multitude of factors, such as the physiological status of the culture (whether it is stress-adapted or not) and the method of inoculation (whether wet or dry) (O'Bryan et al., 2006). In validation studies that compare surrogate microorganisms to targeted pathogens, it is essential to consider these variables for both the surrogate and the pathogen (Ahmad et al., 2019; Schultze et al., 2020). Also, most validation studies use a high inoculation level (6–8 $\log_{10}$ CFU/g or ml) to ensure that heat treatment causes a significant $\log_{10}$ reduction. There have been few studies on the use of surrogates in animal waste. In poultry litter, both bacteriophages MS2 and FO were evaluated as surrogate microorganisms for avian influenza virus in chicken litter compost, and bacteriophage MS2 was found to have comparable thermal resistance to the reported resistance of avian influenza virus to bacteriophage FO (Wang et al., 2020). To make sure that the thermal processing of chicken litter was safe, *E. faecium* NRRL B-2354 was used as a surrogate microorganism for *Salmonella* spp. in another study done in the same lab (Wang et al., 2018). Future comprehensive studies are still needed to incorporate different animal waste, pathogenic bacteria, and diverse types of heating sources to further validate the thermal processing of poultry.

Monitoring indigenous bacterial species as indicator microorganisms can also help predict the survival of pathogens in animal waste. Studies suggested that indigenous enterococci or non-pathogenic *E. coli* could be a choice of indicator to validate the effectiveness of thermal processing for biosolid or animal waste (Côté et al., 2006; Graham et al., 2009; Chen et al., 2015). Indigenous enterococci were chosen over aerobic bacteria in poultry litter as a more effective way to evaluate the inactivation of desiccation-adapted *S.* Senftenberg 775/W in poultry litter (Chen and Jiang, 2017). Based on reports by Graham et al. (2009) and Chen et al. (2015), the populations of indigenous enterococci and mesophiles in chicken litter during composting or stockpiling varied between 3.0–7.5 $\log_{10}$ CFU/g and 6.6–8.9 $\log_{10}$ CFU/g, respectively. Raw poultry litter before physical heat treatment contained at least 3 $\log_{10}$ of naturally occurring enterococci, which can provide enough log reduction for predicting the survival behaviors of *Salmonella* in poultry litter compost during physical heat treatment.

Due to the presence of many types of human pathogens in the animal wastes destined for anaerobic digestion, indicator microorganisms are needed to monitor the reduction of pathogens in animal wastes treated in biogas plants. According to the findings of Côté et al. (2006), the anaerobic digestion of swine slurries led to a significant decrease in total coliforms by 97.94–100% and indigenous *E. coli* by 99.67–100%. Consequently, the levels of indigenous *Salmonella* in the slurries became undetectable. Watcharasukarn et al. (2009) evaluated *E. coli*, *Enterococcus faecalis,* and *C. perfringens* as indicator microorganisms for cow manure digestion at 37°C, 55°C, and 70°C anaerobically, and concluded the use of *E. faecalis* and *C. perfringens* at high temperatures of 55°C and 70°C and *E. coli* at 37°C represented a wide spectrum of potentially pathogenic bacteria due to the varied levels of heat resistance. Unlike aerobic composting, anaerobic conditions may allow some pathogens such as anaerobic or microaerophilic bacteria to survive. In Denmark, the indicator microorganisms used for quality assurance of digestate are *Salmonella*, *Listeria*, *Campylobacter*, and *Yersinia* (Sahlstrom, 2003).

In summary, both surrogate and indicator microorganisms can be used to validate the safety of food and animal waste treatments. For instance, *E. faecium* NRRL B-2354 can function as a surrogate for *Salmonella* in different matrices due to its genetic similarity and safety profile. Additionally, monitoring indigenous non-pathogenic species within the waste can help predict the survival of pathogens in the waste, serving as indicators of the thermal process's effectiveness. These validation studies are instrumental in contributing to microbial risk assessments for poultry litter products, ensuring their safety and suitability for agricultural use.

8 Conclusion and future trends

Overall, this chapter delves into the management of poultry litter to enhance microbial safety and mitigate environmental impacts. As outlined, poultry litter, a byproduct of the poultry industry, holds significant potential as an organic fertilizer and soil amendment. However, untreated or inadequately treated poultry litter can pose risks through harboring of human pathogens, contributing to antimicrobial resistance, and environmental degradation. The exploration of various treatment methods – composting, physical heat-treatment, addition of amendments, and anaerobic digestion – underscores the multifaceted approach required to ensure the microbiological safety and environmental sustainability of poultry litter utilization in agriculture. Recognizing existing research gaps in poultry litter waste management, the following suggestions for future studies are proposed:

1. The persistence and control of resistant microbial populations such as bacterial spore-formers, stress-adapted bacteria, multidrug-resistant bacteria, or parasites, especially under mesophilic conditions of aerobic composting and anaerobic digestion.
2. Identification of potential environmental factors contributing to pathogen regrowth in finished compost or anaerobic digestate.
3. Assessment of the impact of poultry litter amendments, such as biochar, clay, and chemical additives, on the fate of pathogens and antibiotic resistance genes during composting and after soil application.
4. The effectiveness and reliability of surrogate and indicator microorganisms for validating pathogen inactivation in different industrial settings, ensuring their suitability for monitoring microbial safety during poultry litter treatment processes.
5. Risk assessment of applying treated poultry litter to grow fresh produce in agricultural fields.

Innovative treatment technologies can be developed to optimize waste treatment methods, thereby enhancing the microbiological safety and fertilizer value of poultry litter. Moving forward, it is imperative to establish regulations and guidelines to ensure that poultry litter management positively contributes to public health, environmental protection, and sustainable agriculture.

9 Where to look for further information

9.1 Further reading

- An extensive review of the microbiological surveys on the prevalence of certain foodborne pathogens in poultry litter or poultry litter-based organic fertilizers by Chen and Jiang (2014).
- A good review on composting to inactivate foodborne pathogens for crop soil application by Gurtler et al. (2018).
- A study on on-farm agro-economic effects of fertilizing cropland with poultry litter by Harmel et al. (2008).
- A systematic review and meta-analysis of the sources of *Salmonella* in poultry production (pre-harvest) and their relative contributions to the microbial risk of poultry meat by Wang et al. (2023).
- Federal regulation: FSMA Produce Safety Rule: Documentation requirements for commercial soil amendment suppliers by Clements et al. (2019).
- A comprehensive technical report on anaerobic digestate by NOSB (2017).

9.2 Key journals/conferences

- Applied Environmental Microbiology: a high-impact journal publishing significant and novel studies on all aspects of applied and environmental microbiology.
- Bioresource Technology: high-impact journal publishing original research to advance and disseminate knowledge in all the related areas of biomass and biological waste.
- USCC Composting Council's Annual Conference is the world's largest composting conference and exhibition for the organics management industry.
- International Association for Food Protection (IAFP) is providing attendees with information on current and emerging food safety issues, including pre-harvest stage. https://www.foodprotection.org/annualmeeting/

10 References

Ahmad, N.H., et al. (2019), 'Effect of talc as a dry-inoculation carrier on thermal resistance of *Enterococcus faecium* NRRL B-2354 in almond meal', *J. Food Prot.*, 82(5), 1110–1115.

Alam, S.B., et al. (2020), 'Molecular detection of multidrug-resistant *Salmonella* species isolated from broiler farm in Bangladesh', *Pathog.*, 9(3), 201. https://doi.org/10.3390/pathogens9030201

Alfa, M.I., et al. (2014), 'Assessment of biofertilizer quality and health implications of anaerobic digestion effluent of cow dung and chicken droppings', *Renew. Energy.*, 63, 681–686.

Avery, L.M., et al. (2009), 'Heat and lime-treatment as effective control methods for *Escherichia coli* O157: H7 in organic waste', *Bioresour. Technol.*, 100(11), 2692–2698.

Awasthi, M.K., et al. (2020), 'Emerging applications of biochar: improving pig manure composting and attenuation of heavy metal mobility in mature compost', *J. Hazard. Mater.*, 389, 122116.

Bai, X., et al. (2020), 'Inactivation by osmotic dehydration and air drying of *Salmonella*, Shiga toxin-producing *Escherichia coli*, *Listeria monocytogenes*, hepatitis A virus and selected surrogates on blueberries', *Int. J. Food Microbiol.*, 320, 108522.

Bailey, M., et al. (2020), 'Prevalence and antimicrobial resistance of *Salmonella* from antibiotic-free broilers during organic and conventional processing', *J. Food Prot.*, 83(3), 491–496.

Bianchini, A., et al. (2014), 'Use of *Enterococcus faecium* as a surrogate for *Salmonella enterica* during extrusion of a balanced carbohydrate-protein meal', *J. Food Prot.*, 77(1), 75–82.

Bonilla-Caballero, M.A., et al. (2022), 'First report of multidrug-resistant *Salmonella Infantis* in broiler litter in Tolima, Colombia', *Vet. World*, 15(6), 1557.

Borowski, A.G., et al. (2009), 'Validation of ground-and-formed beef jerky processes using commercial lactic acid bacteria starter cultures as pathogen surrogates', *J. Food Prot.*, 72(6), 1234–1247.

Buege, D.R., et al. (2006), 'Lethality of commercial whole-muscle beef jerky manufacturing processes against *Salmonella* serovars and *Escherichia coli* O157: H7', *J. Food Prot.*, 69(9), 2091–2099.

Ceustermans, A., et al. (2007), 'Inactivation of *Salmonella* Senftenberg strain W 775 during composting of biowastes and garden wastes', *J. Appl. Microbiol.*, 103, 53–64.

Ceylan, E. and Bautista, D.A. (2015), 'Evaluating *Pediococcus acidilactici* and *Enterococcus faecium* NRRL B-2354 as thermal surrogate microorganisms for *Salmonella* for in-plant validation studies of low-moisture pet food products', *J. Food Prot.*, 78, 934–939.

Chang, R., et al. (2020), 'Assessment of gaseous ozone treatment on *Salmonella* Typhimurium and *Escherichia coli* O157:H7 reductions in poultry litter', *Waste Manag.*, 117, 42–47.

Channaiah, L.H., et al. (2016), 'Validation of baking to control *Salmonella* serovars in hamburger bun manufacturing, and evaluation of *Enterococcus faecium* ATCC 8459 and *Saccharomyces cerevisiae* as nonpathogenic surrogate indicators', *J. Food Prot.*, 79, 544–552.

Chen, Z., et al. (2013), 'Thermal inactivation of desiccation-adapted *Salmonella* spp. in aged chicken litter', *Appl. Environ. Microbiol.*, 79, 7013–7020. https://doi.org/10.1128/AEM.01969-13

Chen, Z. and Jiang, X. (2014), 'Microbiological safety of chicken litter or chicken litter-based organic fertilizers: A review', *Agriculture*, 4(1), 1–29.

Chen, Z. and Jiang, X. (2017), 'Selection of indigenous indicator microorganisms for validating desiccation-adapted *Salmonella* reduction in physically heat-treated poultry litter', *J. Appl. Microbiol.*, 122, 1558–1569.

Chen, Z. and Jiang, X. (2014), 'Microbiological safety of chicken litter or chicken litter-based organic fertilizers: A review', Agriculture, 4(1), 1–29.

Chen, Z., et al. (2015a), 'Effects of chicken litter storage time and ammonia content on thermal resistance of desiccation-adapted *Salmonella* spp.', *Appl. Environ. Microbiol.*, 81, 6883–6889. https://doi.org/10.1128/AEM.01876-15

Chen, Z., et al. (2015b), 'Developing a two-step heat treatment for inactivating desiccation-adapted *Salmonella* spp. in aged chicken litter', *Foodborne Pathogens and Disease*, 12(2), 104–109.

Cheng, T., et al. (2017), 'Influence of controlled atmosphere on thermal inactivation of *Escherichia coli* ATCC 25922 in almond powder', *Food Microbiol.*, 64, 186–194.

Clements, D., et al. (2019), 'FSMA produce safety rule: documentation requirements for commercial soil amendment suppliers'. Available at: https://ucanr.edu/sites/Small_Farms_/files/315707.pdf (Accessed on February 2024).

Collins, E., et al. (2013), 'Evaluating the effects of pasteurization temperature and treatment duration on pathogen inactivation in separated liquid dairy manure', In *Proceedings of the 2013 ASABE Annual International Meeting* (p. 1). American Society of Agricultural and Biological Engineers.

Cook, K.L., et al. (2011), 'Evaluation of nitrogen retention and microbial populations in poultry litter treated with chemical, biological or adsorbent amendments', *J. Environ. Manage.*, 92(7), 1760–1766.

Côté, C., et al. (2006), 'Reduction of indicator and pathogenic microorganisms by psychrophilic anaerobic digestion in swine slurries', *Bioresour. Technol.*, 97, 686–691.

Dahshan, H., et al. (2016), '*Listeria* species in broiler poultry farms: Potential public health hazards', *J. Microbiol. Biotechnol.*, 26(9), 1551–1556.

de Moraes, M.H., et al. (2016), 'Development of an avirulent *Salmonella* surrogate for modeling pathogen behavior in pre- and postharvest environments', *Appl. Environ. Microbiol.*, 82, 4100–4111. https://doi.org/10.1128/AEM.00898-16

Deen, B. and Diez-Gonzalez, F. (2019), 'Assessment of *Pediococcus acidilactici* ATCC 8042 as potential *Salmonella* surrogate for thermal treatments of toasted oats cereal and peanut butter', *Food Microbiol.*, 83, 187–192.

Den Besten, H.M., et al. (2018), 'Natural diversity in heat resistance of bacteria and bacterial spores: impact on food safety and quality'. *Annu. Rev. Food Sci. Technol.*, 9, 383–410.

Dharmasena, M., et al. (2019), 'Thermal resistance of *Clostridium difficile* endospores in dairy compost upon exposure to wet and dry heat treatments', *J. Appl. Microbiol.*, 127, 274–283.

Dharmasena, M. and Jiang, X. (2018), 'Isolation of toxigenic *Clostridium difficile* from animal manure and composts being used as biological soil amendments', *Appl. Environ. Microbiol.*, 84. https://doi.org/10.1128/AEM.00738-18

Dunn, L.L., et al. (2022), 'The prevalence and concentration of *Salmonella enterica* in poultry litter in the southern United States', *PLoS One*, 17(5), e0268231.

Eblen, D.R., et al. (2005), 'Studies to select appropriate nonpathogenic surrogate Escherichia coli strains for potential use in place of *Escherichia coli* O157: H7 and *Salmonel*la in pilot plant studies', *J. Food Prot.*, 68, 282–291.

Elving, J., et al. (2014), 'Thermal treatment for pathogen inactivation as a risk mitigation strategy for safe recycling of organic waste in agriculture', *J. Environ. Sci. Heal. B*, 49, 679–689.

Enache, E., et al. (2015), 'Development of a dry inoculation method for thermal challenge studies in low-moisture foods by using talc as a carrier for *Salmonella* and a surrogate (*Enterococcus faecium*)', *J. Food Prot.*, 78, 1106–1112.

English, K.G. (2015), 'Effect of biosecurity and management practices on the prevalence of *Salmonella*, *Campylobacter*, and *Clostridium perfringens* in a poultry production system' (Doctoral dissertation, Auburn University).

Eguale, T. (2018), 'Non-typhoidal *Salmonella* serovars in poultry farms in central Ethiopia: prevalence and antimicrobial resistance', *BMC Vet. Res.*, 14, 1–8.

Erickson, M.C., et al. (2009), 'Inactivation of *Salmonella* spp. in cow manure composts formulated to different initial C:N ratios', *Bioresour. Technol.*, 100, 5898–5903.

Erickson, M.C., et al. (2014), 'Inactivation of pathogens during aerobic composting of fresh and aged dairy manure and different carbon amendments', *J. Food Prot.*, 77(11), 1911–1918.

Fagbamila, I.O., et al. (2017), '*Salmonella* serovars and their distribution in Nigerian commercial chicken layer farms', *PLoS One*, 12(3), e0173097.

Flockhart, L., et al. (2017), 'Distribution of *Salmonella* in humans, production animal operations and a watershed in a FoodNet Canada sentinel site', *Zoonoses Public Health*, 64, 41–52.

Flynn, R.P. and Wood, C.W. (1996), 'Temperature and chemical changes during composting of broiler litter', *Compost Sci. & Util.*, 4(3), 62–70.

Franceschine, L., et al. (2023), 'Molecular and phenotypic detection of the resistance profile to β-Lactams and Colistin of *Salmonella* spp. isolated from broilers' litter', *Brazilian J. of Poultry Sci.*, 25, eRBCA-2022.

Franke-Whittle, I.H., et al. (2005), 'Design and application of an oligonucleotide microarray for the investigation of compost microbial communities', *J. Microbiol. Methods*, 62, 37–56.

Fudge, J., et al. (2016), 'The isolation and identification of *Pantoea dispersa* strain JFS as a non-pathogenic surrogate for *Salmonella* Typhimurium phage type 42 in flour', *Int. J. Food Microbiol.*, 219, 1–6.

Gascho, G.J. and Hubbard, R.K. (2006), 'Long-term impact of broiler litter on chemical properties of a Coastal Plain soil', *J. Soil Water Conserv.*, 61, 65–74.

Gaspareto, T.C., et al. (2020), 'Pre-treatment strategies for value addition in poultry litter', *Front. bioeng. biotechnol.*, 8, 477.

Ghaly, A. and Alhattab, M. (2013), 'Drying poultry manure for pollution potential reduction and production of organic fertilizer', *Am. J. Environ. Sci.*, 9, 88–102.

Gibbs, R.A., et al. (1995), 'Pathogen die-off in stored wastewater sludge', *Water. Sci. Tech.*, 31, 93–95.

Goberna, M., et al. (2011), 'Pathogenic bacteria and mineral N in soils following the land spreading of biogas digestates and fresh manure', *Appl. Soil Ecol.*, 49, 18–25.

Gong, C.M. (2007), 'Microbial safety control of compost material with cow dung by heat treatment', *J. Environ. Sci.*, 19, 1014–1019.

Graham, J.P., et al. (2009), 'Antibiotic resistant *Enterococci* and *Staphylococci* isolated from flies collected near confined poultry feeding operations', *Sci. Total Environ.*, 407, 2701–2710.

Green, S.J., et al. (2004), 'Similarity of bacterial communities in sawdust- and straw-amended cow manure composts', *FEMS Microbiol. Lett.*, 233, 115–123.

Grewal, S.K., et al. (2006), 'Persistence of *Mycobacterium avium* subsp. *paratuberculosis* and other zoonotic pathogens during simulated composting, manure packing, and liquid storage of dairy manure', *Appl. Environ. Microbiol.*, 72, 565–574.

Gu, G., et al. (2019), 'Diversity and dynamics of *Salmonella enterica* in water sources, poultry litters, and field soils amended with poultry litter in a major agricultural area of Virginia', *Front. Microbiol.*, 10, 2868.

Guo, H., et al. (2020), 'Elucidating the effect of microbial inoculum and ferric chloride as additives on the removal of antibiotic resistance genes from chicken manure during aerobic composting', *Bioresour. Technol.*, 309, 122802.

Gurtler, J.B., et al. (2018), 'Composting to inactivate foodborne pathogens for crop soil application: a review', *J. Food Prot.*, 81(11), 1821–1837.

Gutierrez, A., et al. (2020), 'Prevalence, concentration, and antimicrobial resistance profiles of *Salmonella* isolated from Florida poultry litter', *J. Food Prot.*, 83(12), 2179–2186.

Harmel, R.D., et al. (2008), 'On-farm agro-economic effects of fertilizing cropland with poultry litter', *J. Appl. Poultry Res.*, 17(4), 545–555.

Harris, L.J., et al. (2013), 'A framework for developing research protocols for evaluation of microbial hazards and controls during production that pertain to the application of untreated soil amendments of animal origin on land used to grow produce that may be consumed raw', *J. Food Prot.*, 76(6), 1062–1084.

Hassen, A., et al. (2001), 'Microbial characterization during composting of municipal solid waste', *Bioresour. Technol.*, 80, 217–225.

Hu, M. and Gurtler, J.B. (2017), 'Selection of surrogate bacteria for use in food safety challenge studies: a review', *J. Food Prot.*, 80, 1506–1536.

Hutchison, M.L., et al. (2004a), 'Levels of zoonotic agents in British livestock manure', *Lett. Appl. Microbiol.*, 39, 207–214.

Hutchison, M.L., et al. (2004b), 'Effect of length of time before incorporation on survival of pathogenic bacteria present in livestock waste applied to agricultural soil', *Appl. Environ. Microbiol.*, 70, 5111–5118. https://doi.org/10.1128/AEM.70.9.5111-5118.2004

Hwang, D., et al. (2020), 'Farm management practices that affect the prevalence of *Salmonella* in pastured poultry farms', *LWT*, 127, 109423.

Jay-Russell, M.T., et al. (2014), '*Salmonella* Oranienburg isolated from horses, wild turkeys and an edible home garden fertilized with raw horse manure', *Zoonoses Public Health*, 61(1), 64.

Jeong, S., et al. (2011), 'Quantifying the performance of *Pediococcus* sp. (NRRL B-2354: *Enterococcus faecium*) as a nonpathogenic surrogate for *Salmonella* Enteritidis PT30 during moist-air convection heating of almonds', *J. Food Prot.*, 74, 603–609.

Jeong, S., et al. (2017), 'Comparing thermal process validation methods for *Salmonella* inactivation on almond kernels" *J. Food Prot.*, 80, 169–176.

Jin, T., et al. (2008), 'Thermal resistance of *Salmonella enteritidis* and *Escherichia coli* K12 in liquid egg determined by thermal-death-time disks', *J. Food Eng.*, 84, 608–614.

Kagambèga, A., et al. (2018), '*Salmonella* spp. and *Campylobacter* spp. in poultry feces and carcasses in Ouagadougou, Burkina Faso', *Food Sci. Nutr.*, 6(6), 1601–1606.

Kassem, I.I., et al. (2017), 'Nonculturability might underestimate the occurrence of *Campylobacter* in broiler litter', *Foodborne Pathog. Dis.*, 14(8), 472–477.

Khan, N., et al. (2016), 'Physical and chemical properties of biochars co-composted with biowaste and incubated with a chicken litter compost', *Chemosphere*, 142, 14–23.

Kharel, K., et al. (2018), 'Hot water treatment as a kill-step to inactivate *Escherichia coli* O157: H7, *Salmonella enterica*, *Listeria monocytogenes*, and *Enterococcus faecium* on in-shell pecans', *LWT*, 97, 555–560.

Kim, J., et al. (2012), 'Validating thermal inactivation of *Salmonella* spp. in fresh and aged chicken litter', *Appl. Environ. Microbiol.*, 78, 1302–1307.

Kiss, N.É., et al. (2021), 'Assessment of composted pelletized poultry litter as an alternative to chemical fertilizers based on the environmental impact of their production', *Agriculture*, 11(11), 1130.

Kopit, L.M., et al. (2014), 'Safety of the surrogate microorganism *Enterococcus faecium* NRRL B-2354 for use in thermal process validation', *Appl. Environ. Microbiol.*, 80, 1899–1909.

Larney, F.J., et al. (2003), 'Fate of coliform bacteria in composted beef cattle feedlot manure', *J. Environ. Qual.*, 32, 1508–1515.

Larsen, H.E., et al. (1994), 'Use of indicators for monitoring the reduction of pathogens in animal waste treated in biogas plants', *Zentralbl Hyg Umweltmed*, 195(5–6), 544–555.

Lepesteur, M. (2022), 'Human and livestock pathogens and their control during composting', *Crit. Rev. Environ. Sci. Technol.*, 52(10), 1639–1683.

Li, K., et al. (2017), 'Impact of built-up-litter and commercial antimicrobials on *Salmonella* and *Campylobacter* contamination of broiler carcasses processed at a pilot mobile poultry-processing unit', *Front. vet. sci.*, 4, 88.

Lin, M., et al. (2022), 'Challenges of pathogen inactivation in animal manure through anaerobic digestion: a short review', *Bioengineered*, 3(1), 1149–1161.

Lin, Y., et al. (2018), 'Influence of poultry litter on crop productivity under different field conditions: A meta-analysis', *Agron. J.*, 110(3), 807–818.

Liu, H., et al. (2021), 'Chitosan as additive affects the bacterial community, accelerates the removals of antibiotics and related resistance genes during chicken manure composting', *Sci. Total Environ.*, 792, 148381.

Liu, S., et al. (2018a), '*Enterococcus faecium* as a *Salmonella* surrogate in the thermal processing of wheat flour: Influence of water activity at high temperatures', *Food. Microbiol.*, 74, 92–99.

Liu, S., et al. (2018b), 'Exponentially Increased Thermal resistance of *Salmonella* spp. and *Enterococcus faecium* at reduced water activity', *Appl. Environ. Microbiol.*, 84, e02742-17.

Lloret, E., et al. (2013), 'Semi full-scale thermophilic anaerobic digestion (TAnD) for advanced treatment of sewage sludge: stabilization process and pathogen reduction', *Chem. Eng. J.*, 232, 42–50.

Ma, L., et al. (2007), 'Development of thermal surrogate microorganisms in ground beef for in-plant critical control point validation studies', *J. Food. Prot.*, 70, 952–957.

Ma, S., et al. (2018), 'Bacterial community succession during pig manure and wheat straw aerobic composting covered with a semi-permeable membrane under slight positive pressure', *Bioresour Technol.*, 259, 221–227.

Messer, J.W., et al. (1971), 'An assessment of some public health problems resulting from feeding poultry litter to animals: Microbiological and Chemical Parameters', *Poult. Sci.*, 50, 874–881.

Miller, C., et al. (2013), 'Analyzing indicator microorganisms, antibiotic resistant *Escherichia coli*, and regrowth potential of foodborne pathogens in various organic fertilizers', *Foodborne Pathog. Dis.*, 10, 520–527.

Moffo, F., et al. (2022), 'Associations between antimicrobial use and antimicrobial resistance of *Escherichia coli* isolated from poultry litter under field conditions in Cameroon', *Prev. Vet. Med.*, 204, 105668.

Moore, P.A. Jr. (1995), 'Reducing ammonia volatilization from poultry litter with aluminum sulfate', *Proc. Meeting Arkansas Nutrition Conference*, Clarin Inn, Fayetteville, AR.

Moussavi, M., et al. (2020), 'Thermal resistance of foodborne pathogens and *Enterococcus faecium* NRRL B-2354 on inoculated pistachios', *J. Food Prot.*, February 21.

Nahm, K.H. (2005), 'Environmental effects of chemical additives used in poultry litter and swine manure', *Crit. Rev. Environ. Sci. Technol.*, 35(5), 487–513.

National Organic Standards Board (2002), '2002 Compost Task Force Recommendation, as amended by NOP', Available at: http://www.ams.usda.gov/AMSv1.0/getfile?dDocName=STELPRDC5058471.

National Organic Standards Board (2017), 'Technical Report: Anaerobic Digestate', *Agricultural Marketing Service (USDA)*, Available at: https://www.ams.usda.gov/sites/default/files/media/AnaerobicDigestateTR.pdf.

O'Bryan, C.A., et al. (2006), 'Heat resistance of *Salmonella* spp., *Listeria monocytogenes*, *Escherichia coli* O157:H7, and *Listeria innocua* M1, a potential surrogate for *Listeria monocytogenes*, in meat and poultry: A review', *J. Food Sci.*, 71, R23–R30.

Okada, E., et al. (2024), 'Effect of on-farm poultry litter composting processes on physicochemical, biological, and toxicological parameters and reduction of antibiotics and antibiotic-resistant *Escherichia coli*', *Waste Manag.*, 174, 310–319.

Oladeinde, A., et al. (2023), 'Management and environmental factors influence the prevalence and abundance of food-borne pathogens and commensal bacteria in peanut hull-based broiler litter', *Poult. Sci.*, 102(2), 102313.

Pan, J., et al. (2019), 'Influence of palygorskite addition on biosolids composting process enhancement', *J. Clean. Prod.*, 217, 371–379.

Patel, J.R., et al. (2015), 'Physical covering for control of *Escherichia coli* O157:H7 and *Salmonella* spp. in static and windrow composting processes', *Appl. Environ. Microbiol.*, 81(6), 2063–2074.

Payne, J.B., et al. (2002), 'Evaluation of litter treatments on *Salmonella* recovery from poultry litter', *J. Appl. Poultry Res.*, 11, 239–243.

Poirier, D., et al. (2014), '*Salmonella* surrogate reduction using industrial peanut dry roasting parameters', *Peanut. Sci.*, 41, 72–84.

Pope, M.J. and Cherry, T.E. (2000), 'An evaluation of the presence of pathogens on broilers raised on poultry litter treatment (PLT) treated litter', *Poult. Sci.*, 79, 1351–1355.

Rachon, G., et al. (2016), 'Inactivation of *Salmonella*, *Listeria monocytogenes*, and *Enterococcus faecium* NRRL B-2354 in a selection of low moisture foods', *Int. J. Food Microbiol.*, 231, 16–25.

Rama, E.N., et al. (2022), 'Prevalence and antimicrobial resistance of *Salmonella* in conventional and no antibiotics ever broiler farms in the United States', *Food Control*, 135, 108738.

Redemann, M.A., et al. (2018), 'Evaluation of thermal process lethality for non-pathogenic *Escherichia coli* as a surrogate for *Salmonella* in ground beef', *LWT*, 90, 290–296.

Resende, J.A., et al. (2014), 'Prevalence and persistence of potentially pathogenic and antibiotic resistant bacteria during anaerobic digestion treatment of cattle manure', *Bioresour. Technol.*, 153, 284–291.

Ruiz-Espinoza, J.E., et al. (2012), 'Effect of low temperature thermal pre-treatment on the solubilization of organic matter, pathogen inactivation and mesophilic anaerobic digestion of poultry sludge', *J. Environ. Sci. Health A*, 47, 1795–1802.

Sahlstrom, L. (2003), 'A review of survival of pathogenic bacteria in organic waste used in biogas plants', *Bioresour. Technol.*, 87, 161–166.

Salminen, E. and Rintala, J. (2002), 'Anaerobic digestion of organic solid poultry slaughterhouse waste – a review', *Bioresour. Technol.*, 83(1), 13–26.

Sapkota, A.R., et al. (2014), 'Lower prevalence of antibiotic-resistant *Salmonella* on large-scale US conventional poultry farms that transitioned to organic practices', *Sci. Total Environ.*, 476, 387–392.

Schultze, D.M., et al. (2020), 'Lethality of high-pressure carbon dioxide on Shiga toxin-producing *Escherichia coli*, *Salmonella* and surrogate organisms on beef jerky', *Int J Food Microbiol*, 321, 108550.

Shah, M.K., et al. (2017), 'Efficacy of vacuum steam pasteurization for inactivation of *Salmonella* PT 30, *Escherichia coli* O157: H7 and *Enterococcus faecium* on low moisture foods', *Int. J. Food Microbiol.*, 244, 111–118.

Shang, K., et al. (2018), 'Distribution and dissemination of antimicrobial-resistant *Salmonella* in broiler farms with or without enrofloxacin use', *BMC Vet. Res.*, 14, 1–4.

Shapovalov, Y., et al. (2020), 'Dry anaerobic digestion of chicken manure: a review', *Appl. Sci.*, 10, 7825. https://doi.org/10.3390/app10217825

Shepherd, M.W., et al. (2010), 'Effect of heat-shock treatment on the survival of *Escherichia coli* O157:H7 and *Salmonella enterica* Typhimurium in dairy manure co-composted with vegetable wastes under field conditions', *Bioresour. Technol.*, 101, 5407–5413.

Singh, R., et al. (2010), 'Thermal inactivation of heat-shocked *Escherichia coli* O157: H7, *Salmonella*, and *Listeria monocytogenes* in dairy compost', *J. Food Prot.*, 73, 1633–1640.

Singh, R. and Jiang, X. (2012), 'Thermal inactivation of acid-adapted *Escherichia coli* O157: H7 in dairy compost', *Foodborne. Pathog. Dis.*, 9(8), 741–748.

Slana, I., et al. (2011), 'Persistence of *Mycobacterium avium* subsp. *paratuberculosis* at a farm-scale biogas plant supplied with manure from paratuberculosis-affected dairy cattle', *Appl. Environ. Microbiol.*, 77, 3115–3119. https://doi.org/10.1128/AEM.02407-10.

Soni, A., et al. (2022). 'A comprehensive review of variability in the thermal resistance (D-values) of food-borne pathogens–a challenge for thermal validation trials', *Foods*, 11(24), 4117.

Stingone, J.A. and Wing, S. (2011), 'Poultry litter incineration as a source of energy: reviewing the potential for impacts on environmental health and justice', *New Solut.: A Journal of Environmental and Occupational Health Policy*, 21(1), 27–42.

Stringfellow, K., et al. (2010), 'Pasteurization of chicken litter with steam and quicklime to reduce *Salmonella* Typhimurium', *J. Appl. Poult. Res.*, 19, 380–386.

Sundberg, C., et al. (2004), 'Low pH as an inhibiting factor in the transition from mesophilic to thermophilic phase in composting', *Bioresour. Technol.*, 95, 145–150.

Thomas, C., et al. (2019), 'Inactivation of ESBL-/AmpC-producing *Escherichia coli* during mesophilic and thermophilic anaerobic digestion of chicken manure', *Waste Manag.*, 84(1), 74–82.

Tidwell, M. (2012), 'The climate impact of poultry waste disposal practices and manure-to-energy power systems', *Chesapeake Climate Action Network*. Available at: https://chesapeakeclimate.org/the-climate-impact-of-poultry-waste-disposal-practices-and-manure-to-energy-power-systems/

Trimble, L.M., et al. (2013), 'Prevalence and concentration of *Salmonella* and *Campylobacter* in the processing environment of small-scale pastured broiler farms', *Poult. Sci.*, 92, 3060–3066.

Tsai, H., et al. (2019), 'Evaluation of *Enterococcus faecium* NRRL B-2354 as a surrogate for *Salmonella* during cocoa powder thermal processing', *Food. Microbiol.*, 82, 135–141.

U.S. Environmental Protection Agency (2008), 'Aerated static pile composting', Available at: http://www.epa.gov/osw/conserve/rrr/composting/static.htm.

U.S. FDA (2018), 'Produce Safety Rule Biological Soil Amendments of Animal Origin', Available at: https://www.fda.gov/files/food/published/Fact-Sheet--Biological-Soil-Amendments-of-Animal-Origin_Download.pdf Accessed Oct 2024.

USDA (2023), 'Poultry - Production and Value 2022 Summary', (April 2023). https://downloads.usda.library.cornell.edu/usda-esmis/files/m039k491c/wm119387d/5138kw352/plva0423.pdf.

Vasan, A., et al. (2014), 'Thermal tolerance of O157 and non-O157 Shiga toxigenic strains of *Escherichia coli*, *Salmonella*, and potential pathogen surrogates, in frankfurter batter and ground beef of varying fat levels', *J. Food. Prot.*, 77, 1501–1511.

Verma, T., et al. (2018), 'Evaluation of *Enterococcus faecium* NRRL B-2354 as a surrogate for *Salmonella* during extrusion of low-moisture food', *J. Food. Sci.*, 83, 1063–1072.

Voss-Rech, D., et al. (2019), 'Longitudinal study reveals persistent environmental *Salmonella* Heidelberg in Brazilian broiler farms', *Vet. Microbiol.*, 233, 118–123.

Wang, D., et al. (2018), 'Comparative persistence of *Salmonella* and *Escherichia coli* O157: H7 in loam or sandy loam soil amended with bovine or swine manure', *Can. J. Microbiol.*, 64, 979–991.

Wang, H., et al. (2018), 'Testing a nonpathogenic surrogate microorganism for validating desiccation-adapted *Salmonella* inactivation in physically heat-treated broiler litter', *J. Food. Prot.*, 81(9), 1418–1424.

Wang, H., et al. (2020), 'Comparing and modeling the thermal inactivation of bacteriophages as pathogenic viruses surrogates in chicken litter compost', *Compost. Sci. & Util.*, 28(2), 87–99.

Wang, H., et al. (2021), 'Plant-scale validation of physical heat treatment of poultry litter composts using surrogate and indicator microorganisms for *Salmonella*', *Appl. Environ. Microbiol.* 87(5), e02234-20.

Wang, J., et al. (2023), 'A systematic review and meta-analysis of the sources of *Salmonella* in poultry production (pre-harvest) and their relative contributions to the microbial risk of poultry meat', *Poult. Sci.*, Feb 9, 102566.

Wang, N.X., et al. (2019), 'A comprehensive review of anaerobic digestion of organic solid wastes in relation to microbial community and enhancement process', *J. Sci. Food. Agric.*, 99(2), 507–516.

Watcharasukarn, M., et al. (2009), 'Screening *Escherichia coli*, *Enterococcus faecalis*, and *Clostridium perfringens* as indicator organisms in evaluating pathogen-reducing capacity in biogas plants', *Microb. Ecol.*, 58(2), 221–230.

White, A.E., et al. (2022), 'Foodborne illness outbreaks reported to national surveillance, United States, 2009–2018', *Emerg. Infect. Dis.*, 28(6), 1117.

Wilkinson, K., et al. (2011), 'Effect of heating and aging of poultry litter on the persistence of enteric bacteria', *Poult. Sci.*, 90, 10–18.

Wu, X., et al. (2024), 'Biotic and abiotic effects of manganese salt and apple branch biochar co-application on humification in the co-composting of hog manure and sawdust', *J. Chem. Eng.*, 482, 149077.

Zainudin, M.H., et al. (2020), 'Biochar enhanced the nitrifying and denitrifying bacterial communities during the composting of poultry manure and rice straw', *Waste. Manag.*, 106, 240–249.

Zhang, W., et al. (2022), 'Additives improved saprotrophic fungi for formation of humic acids in chicken manure and corn stover mix composting', *Bioresour. Technol.*, 346, 126626.

Zhou, Z., et al. (2019), 'Effect of mild steaming treatment on the inactivation of *Salmonella*, *Listeria monocytogenes*, *Escherichia coli* O157: H7 and their surrogates on black peppercorns', *Food Control*, 106, 106726.

Zhou, Z., et al. (2020), 'Decontamination effect of hot-air drying against bacterial pathogen and surrogate strains on basil leaves, from laboratory to pilot scale settings', *LWT*, 122, 109035.

Chapter 16

Valorizing byproducts from broiler processing: biologics and sustainability

S.C. Ricke, V.A. Leone and A.A. Tarcin, University of Wisconsin-Madison, USA

1 Introduction

Balancing human nutrient needs and ensuring environmental sustainability continue to be a challenge (Ederer and Leroy, 2023). This also involves the agricultural sector. The concept of improving the sustainability of agricultural production has become a critical issue in the twenty-first century as environmental impacts have received more scrutiny. Food waste or loss can occur in many forms throughout the entire supply chain, including retail. Meat and animal products are one of the main contributors, accounting for 11 to 12% of food losses globally (Jones et al., 2021), and include those derived from the meat and poultry industry during processing. While various meat products for human consumption result from processing, a wide range of products not usable for human consumption are also generated (i.e. byproducts or waste) that require disposal. Depending on the nature of the byproduct, the type of disposal system can vary and in turn, have different environmental impacts, which can range from further processing to generate usable by-products

http://dx.doi.org/10.19103/AS.2024.0146.20

for nonhuman foods to disposal of non-usable products directly into the environment.

In general, byproducts consist of animals in their entirety, parts of animals, and animal-derived products that are not meant for human consumption (Lasekan et al., 2013). Classification of meat products generated during processing can be further categorized into essentially four types (Fallows and Verner Wheelock, 1982; Morawicki, 2010; Bah et al., 2013) as follows:

- high-value edible meats;
- high-value inedible byproducts;
- low-value edible byproducts; and
- processing byproducts.

Obviously, the primary products of interest are those considered high-value edible meats (either raw products or further processed) for human consumption. Edible meats are followed by a series of inedible byproducts which, depending on the type of compound, have different paths for their further utilization. Some inedible byproducts such as hides, bones, hooves, and blood can be repurposed for industrial applications, whereas low-value byproducts, including offal and meat meal may have unique paths. Finally, processing byproducts, such as digestive contents, currently have no apparent value and must simply be disposed of directly, which is a potential environmental pollutant in terms of foodborne pathogens in addition to being an expensive waste for disposal. While meat processing products can be broadly classified into edible and inedible components, the level of value is subject to change as technological advances are introduced. Currently, there are limited commercial uses for some of the more recalcitrant byproducts, resulting in major contributions to the waste stream and challenges to agricultural sustainability.

The term 'byproduct' is general terminology for inedible parts of meat processing, while more recent terms include co-products (Morawicki, 2010). In this review to avoid confusion the term 'byproduct' will be retained to represent both historically published literature and discuss more recent developments. However, the specialized concept of 'biologics' as a research focus offers the opportunity to deepen the meaning and shift the overall concept of byproducts. In general, biologics have been described as large heterogenous protein molecules containing hundreds of amino acids, but this definition can be broadened to include any biologically derived molecule (Morrow and Felcone, 2004). Given this general definition, the origin of these types of biologic compounds are relatively diverse. Thus, animal biologics can be defined as molecules or tissues derived from animal processing which have the potential to provide high-value products for enhancing animal and/or human health. Essentially, the biologics-based research field involves discovery of unique properties and functions associated with a particular molecule such as an antimicrobial peptide or tissues which can improve or benefit animals and/or humans depending on the properties of the biologic candidate.

In general, the concept of biologics can be applied to the entire meat production sector with the potential for overlap in some of the research advances. However, the focus of this review will be on poultry production and processing specifically because there are opportunities to expand the range of edible products originating from the poultry industry. The high level of vertical integration in the poultry industry allows for more centralized decisions on production and processing practices and potentially more rapid adaptation to new models of production. Consequently, the opportunity to further refine technology to isolate and/or generate biologic compounds and tissues may be easier to implement than in other meat animal commodities. In this review, current poultry processing practices will be described along with the current outcomes for inedible byproducts followed by potential future opportunities to enhance the value of these byproducts as sources of high-value biologic compounds and tissues.

2 Poultry processing: general concepts

The U.S. commercial poultry industry has become a major agricultural commodity producer that is vertically integrated from hatcheries to final retail products across several large corporations and represents a multibillion-dollar enterprise (Dittoe et al., 2020). In 2017, broiler production for meat products in the United States alone reached over 30 billion dollars in sales (USDA, 2018; Dittoe et al., 2020). The capacity for the poultry industry to rapidly expand can largely be attributed to a combination of advances in poultry genetics and improved nutritional formulation along with other factors (Wythe et al., 2023). Improvements in housing and veterinary care have also contributed to the economic expansion of the commercial broiler industry (Dittoe et al., 2020). More recently, the introduction of a wide range of biotic-based and chemical feed additives such as prebiotics, probiotics, postbiotics, organic and botanical additives, among others has offered further opportunities to improve gastrointestinal tract (GIT) health, mitigate foodborne pathogens, and enhance broiler performance (Calo et al., 2015; Dittoe et al., 2018; Ricke, 2018a; Perumalla et al., 2023; Ricke et al., 2023).

As the production and number of broilers have increased, the processing technologies have also advanced to accommodate the nearly 9 billion birds produced in 2017 alone (USDA, 2018; Dittoe et al., 2020). Levels of production will only expand further as poultry consumption increases globally, with the United States currently exporting one-third of the world's poultry supply (Chai et al., 2017). Technological advances, such as introducing more automation in processing lines, will serve to further expedite processing capabilities, especially as computer systems become more sophisticated in creating additional interfaces with data gathering and high-throughput computational interpretation (Chai et al., 2014; Thompson et al., 2017). Furthermore, efforts to increase processing line speeds along with improved uniformity have been introduced (Ricke et al., 2022). However, these additions also add to the complexity of poultry processing operations.

Poultry processing plants have the capability of handling over 100,000 birds daily depending on the type of plant, taking them through multiple stages from slaughter to final meat products (Luján-Rhenals et al., 2017a). Converting live birds to meat products during the initial stage of poultry processing in the United States involves a multitude of individual steps starting with the entry of live birds to the processing plant followed by stunning and bleeding, de-feathering, GIT viscera and crop removal, washing, and immersion chilling (Owens et al., 2010; Handley et al., 2015; Blevins et al., 2017; Singh and Rama, 2017; Ricke et al., 2022). Following chilling, whole carcasses are deboned and skinned to generate various parts including breasts, wings, thighs, and depending on the type of product; further seasoned or marinated (Owens et al., 2010; Handley et al., 2015; Blevins et al., 2017; Ricke et al., 2022). After further cutting, coating, and cooking, a final stage of products is generated that can include anything from ready-to-eat nuggets, patties, and breaded tenders, to coated meat products (Owens et al., 2010; Handley et al., 2015; Ricke et al., 2022).

During poultry processing, waste products consisting primarily of processing water and organic byproducts are also generated (Luján-Rhenals et al., 2017b; Micciche et al., 2018), where some of the organic byproducts, such as protein and non-protein components, can be used as ingredients for pet foods (Morawicki, 2010; Luján-Rhenals et al., 2017a). Many of these downstream byproducts must be further processed to be recycled or utilized in other streams.

3 Treatment of poultry byproducts and waste: live bird production

As poultry production continues to expand globally, the sheer volume of byproducts will also greatly increase (Meeker and Meisinger, 2015). These byproducts and their ultimate destinations are dependent on several factors and this journey starts at the farm. During live poultry production, the primary byproducts are whole birds due to mortality caused by a variety of reasons, such as disease. Mortality losses on the farm can range anywhere from 3% to 5% of the broiler flock depending on management, environmental conditions, and exposure to disease-carrying vectors. The resulting carcasses require some form of disposal (Edwards and Daniels, 1992).

Traditionally, pit disposal, burial, and incineration have been used, but other approaches such as composting with poultry straw-litter combinations, acid preservation, and *Lactobacilli*-mediated fermentation have also been suggested (Edwards and Daniels, 1992; Cai and Pancorbo, 1994; Cai et al. 1995). Shafer et al. (2000, 2001) explored the application of alkaline hydroxide-preserved whole-poultry carcasses to generate dry byproduct meal. Silage-type fermentations involving chopped broiler carcasses combined with different sugars or industrial carbohydrate sources along with GIT *Lactobacilli* have been shown to achieve pathogen reduction and retain sufficient nutrients for reprocessing and refeeding (Murphy and Silbert, 1992).

However, in cases such as outbreaks of avian influenza that require depopulation of entire flocks, bird disposal can present considerable epidemiological and environmental challenges (Simancas-Racines et al., 2023). In these more dramatic cases where massive carcass numbers are being generated in a short amount of time, either burial or incineration is recommended with specific requirements to minimize exposure. While incineration is considered one of the more effective thermal destruction methods for decimation of infectious agents, it is not recommended for large populations of birds due to expense, environmental contamination, and volatility of the remains (Ritter and Chirnside, 1995; Simancas-Racines et al., 2023).

4 Treatment of poultry byproducts and waste: rendering

Once poultry carcasses have been processed into various meat products for human consumption, the remaining inedible components are directed toward rendering plants where they are subjected to thermal processing (Morawicki, 2010). Rendering is a thermal process involving a dedicated cooking apparatus that is operated to release fat from the protein of the starting materials (Morawicki, 2010). Residual bird offal destined for rendering can include not only condemned whole birds both from processing and occasionally from farms but a diverse set of materials including protein, fat tissue, viscera, heads, bones, blood, and other portions that may not be cost-effectively further processed for human consumption (Morawicki, 2010; Meeker and Meisinger, 2015). Dry rendering is considered the gold standard in the United States for heat application and involves delivery of heat through a steam-jacketed cylindrical cooker or agitator shaft resulting in dehydrated fats and proteins upon completion of thermal processing (Morawicki, 2010). Once fats and proteins are separated, several commercial products can be generated including a variety of poultry meals as well as poultry fats (Morawicki, 2010; Meeker and Meisinger, 2015). Fats can in turn be converted into sources for cooking oils, soap, detergents, and cosmetics or to generate methane-based biogas in anaerobic fermenters (Lasekin et al., 2013). Rendered dried protein products have commercial value for livestock, poultry, aquaculture, and companion animal feeds (Meeker and Meisinger, 2015).

In general, rendering facilities are operated to minimize bacterial and viral pathogens that could contaminate feeds (Meeker and Meisinger, 2015). This is an important concern since any foodborne pathogens remaining in livestock feed proteins can be a source of contamination in animal as well as pet food diets and can be spread directly and indirectly to humans (Ge et al., 2013; Nemser et al., 2014; Meeker and Meisinger, 2015; Molina et al., 2016; Ricke 2018b; Olson et al., 2023). Of particular concern with animal-based feed protein supplementations is the presence of foodborne *Salmonella*, which if poultry flocks are exposed to and colonized by, the contaminated feed can result in widespread dissemination in susceptible birds (Maciorowski et al., 2004, 2006; Jones, 2011). Thermal processes both in rendering and pelleting at the feed mill can generally eliminate *Salmonella* spp. if lethal temperatures are sufficiently implemented, but recontamination can occur post pelleting (Jones, 2011). However, the utilization of animal byproducts for animal feeds has come under stricter control in more recent times due to specific transmission risks. This is, in part, related to potential risk exposure due to the potential for transmitting certain pathogens and toxins as well as environmental concerns over odorous volatiles generated during rendering (Lasekan et al., 2013). Environmental concerns over odorous volatile compounds involve the passage of noncondensable gases under atmospheric temperatures and pressures through scrubbers to remove these compounds and discharge the remaining air (Morawicki, 2010). However, the primary concern originated from the emergence of bovine spongiform encephalopathy and the accompanying concerns with feeding animal-based byproducts to food animals ultimately resulting in the European ban on feeding these byproducts to ruminants (Lasekan et al., 2013).

5 Biologics from broiler processing: general concepts

With the increased generation of waste products (da Silva Ararújo et al., 2021), solutions to limit their impact on the environment represent an ongoing challenge. While rendering offers a viable economic outlet for poultry processing solid wastes, bottlenecks and limitations have encouraged the development of alternative applications and economical repurposing of these byproducts. However, given the wide variation in the types of poultry byproducts generated, identifying repurposing strategies represents a challenge. For instance, while offal from poultry comprised of organs such as the heart, liver, and gizzard are traditionally considered edible and can be sold commercially as individual products in specialty markets (Morawicki 2010; da Silva Ararújo et al., 2021), chicken feet, feathers, head, comb, wattle, earlobe, beak, and skin are compositionally diverse and therefore vary in the types of processes that can be used for extraction and retrieval of valuable compounds (da Silva Ararújo et al., 2021). A similar case can also be made for blood, keel tissue, bones, viscera, and offal that are classified as byproducts and require further processing (da Silva Ararújo et al., 2021).

Some poultry processing components can be fractionated and utilized for bulk applications. For example, fats can be incorporated into animal feed, serve as a source for chemical industry products, or simply burned as fuel (Salminen and Rintala, 2002). Viscera can potentially be used to produce polyunsaturated fatty acids (Patil and Nag, 2011). More specific molecules can be derived from certain chicken parts. In the United States, keel tissue has been examined as a source of chondroitin sulfate that can serve as a nutraceutical supplement to relieve osteoarthritis in humans (Morawicki, 2010). Likewise, rooster combs and wattles contain several useful biomolecules, including hyaluronic acid for applications to treat degenerative joints and for ophthalmic applications (Morawicki, 2010). Sialic acids containing glycopeptide fractions isolated via papain hydrolysis and gel chromatography have also been detected in comb and wattle tissues and are considered important in human physiology (Nakano et al., 1994; Wang and Brand-Miller, 2003). Sialic acids are α-keto acids comprised of a nine-chain carbon backbone that serve as terminal monosaccharides of glycolipids or glycoproteins attached to the surfaces of cells and soluble proteins (Wang and Brand-Miller, 2003; Ling et al., 2022). High levels of sialic acid referred to as N-acetylneuraminic acid are present in the human brain and are associated with ganglioside structures involved in neural transmission synaptogenesis (Brand-Miller, 2003). More recently, the characterization of sialic acid structural complexities has revealed that multiple forms exist and suggest a much more diverse range of biological functions in human nutrition including brain and cognition recognition, enhancement of immune function, skin whitening, and anti-cancer and -hypertension activities (Ling et al., 2022). Natural sources of sialic acid found in poultry wattles and combs represent an example of the potential for retrieving high-value specific biological molecules from a poultry processing byproduct.

These highly specific compounds are examples of the potential for harvesting clinically or nutritionally important individual compounds from relatively minor parts of the broiler carcass such as the keel tissue which constitutes only 0.3% of the live bird weight (Romero-Garay et al., 2022). However, there are much greater quantities of waste generated from poultry processing that are a challenge from a sheer bulk standpoint. For example, skin, blood, bones, and feathers collectively represent almost a quarter of the live bird weight with feathers alone accounting for nearly 8% of the byproducts generated during processing (da Silva Ararújo et al., 2021). The protein fraction alone of each of these major byproduct proteins varies in composition and presents different challenges for collecting and harvesting due to their high molecular weights, extensive cross-linking, and high hydrophobicity (da Silva Ararújo et al., 2021). Potential biologically active compounds derived from byproducts are also quite variable depending on the original functionality of tissue in the bird and the type of compound of interest. This is also impacted by the extraction and hydrolysis process including the types and combinations of proteases used for hydrolysis (Romero-Garay et al., 2022). As a result, some byproducts have been examined more extensively than others as sources of biologically useful compounds. For example, bones have been used as a source of collagen but are still being explored for identifying compounds with more specific biomaterial properties, such as a source of hydroxyapatite (Lasekan et al., 2013; da Silva Ararújo et al., 2021; Foroutan et al., 2021; Ghosh and Pramanik, 2023). In some cases, the biologic compound can be generated by bacterial fermentation of poultry tissue (Romero-Garay et al., 2022). The resulting hydrolysates and peptides can exhibit a wide range of biological activity such as antihypertensive, antioxidant, antibacterial, or anti-inflammatory properties among others (Romero-Garay et al., 2022). In the following sections, the discussion will be focused on the main solid wastes generated in a poultry slaughter plant where there has been more substantial work conducted on the identification and potential application of biologic-based substances from blood, feathers, and skin.

6 Biologics from blood

Nearly 3% of the bird's live weight consists of blood and it is a major environmental disposal concern (da Silva Ararújo et al., 2021). Blood, comprised of cells and plasma, is generated during broiler processing; after lyophilization, this component consists of nearly 88% proteins, 9% ash, 2% moisture, 1% carbohydrates, and 0.15% lipids (da Silva Ararújo et al., 2021). Because of the optimally balanced amino acids present in blood, along with the presence of other elements such as iron, this tissue serves as an ideal protein supplement in animal feeds and pet foods (Bah et al., 2013; da Silva Ararújo et al., 2021). Blood meal is produced by coagulation to separate the red cells from the plasma by steam injection with sufficient holding times to allow proteins to coagulate and aggregate, followed by centrifugation and drying in a ring dryer (Morawicki, 2010).

While blood collected from processing may provide a high-quality protein source, utility beyond animal feed supplements has historically been limited. Human food use has traditionally been impaired due to a variety of reasons including generally negative public perceptions, restrictive religious practices, and biological safety concerns (Johnson et al., 2010; Bah et al., 2013). However, whole blood has been used for human food products such as blood sausages, black puddings, and blood tofu while plasma proteins have been incorporated as binders in restructured meat products and fat replacers in fermented Bologna sausages (Cofrades et al., 2000; Tseng et al., 2006; Bah et al., 2013). More specifically, chicken blood has shown potential in foods to support stable emulsions along with optimal foaming responses, serving as replacements for soy protein in sausage making and as potential protease inhibitors in surimi gels (Rawdkuen et al., 2004; Oro et al., 2018; Bah et al., 2013; Alves et al., 2020; da Silva Ararújo et al., 2021).

Several components of animal blood are sold as commercial products. For example, hemoglobin is commercially marketed for pet foods and aquatic diets while bovine fibrinogen is used as a reagent in microbial media and as a medical reagent (Bah et al., 2013). More recently, specific molecules from blood fractions have received interest for potential applications. For example, some of the bioactive peptides that consist of anywhere from 2 to 30 amino acids and molecular weights of 400 to 2000 Da have been explored for their utility as biologic molecules (Bah et al., 2013; Romero-Garay et al., 2022). In an extensive review of the literature, Romero-Garay et al. (2022) summarized the biological activities of various peptides isolated from poultry byproducts. Among the activities noted were positive human health impacts that included antihypertensive, antioxidant, antidiabetic, antimicrobial, and anti-inflammatory characteristics (Romero-Garay et al., 2022). While a range of poultry byproducts can serve as sources of biologically active peptides, certain protein hydrolysates and peptides from chicken blood have shown potential as antioxidant ingredients for functional food formulas and can also exhibit cryoprotective activities (Zheng et al., 2018; Hamzeh et al., 2019; da Silva Ararújo et al., 2021). Future research will likely lead to the discovery of more biologically active peptides from blood, but this will require further refinement of purification procedures and a better understanding of their functionality in the live bird. In addition, the development of high-throughput functional screening tools would expedite the discovery process.

7 Biologics from skin

Skin primarily serves as a protective outer layer while also controlling internal body temperature and water levels (da Silva Ararújo et al., 2021). Chicken skin, consisting mostly of lipids and protein, accounts for nearly 7% of the live bird weight and possesses properties uniquely distinct from mammals such as reduced thickness and feathers among other characteristics (da Silva Ararújo et al., 2021). In poultry, skin is commercially recovered as three fractions: breast skin without the neck skin, thigh and leg skin (which are collected together), and complete body skin minus the neck region (Morawicki, 2010). Poultry skin possesses a high level of unsaturated fatty acids that have nutritional value including oleic, linoleic, eicosapentaenoic, and docosahexaenoic acids (Dalziel et al., 2015; da Silva Ararújo et al., 2021). Chicken skin protein consists primarily of hydrophobic and acidic amino acids and contains collagen, which exhibits several chemical and textural attributes that make it attractive as a byproduct for meat products, i.e. sausages, as well as a film coating for food (Loo and Sarbon, 2020; da Silva Ararújo et al., 2021).

Collagen and related proteins from chicken skin have potential beyond food applications such as pharmaceuticals and biomedicines (Mecham, 2008; Munasinghe et al., 2014; Nadalian et al., 2019; da Silva Ararújo et al., 2021). For example, chicken skin gelatin peptides have been associated with antioxidant activity and have been demonstrated to potentially exhibit antihypertension capabilities (Onuh et al., 2015; Sarbon et al., 2018; da Silva Ararújo et al., 2021). Other useful protein polymers have been identified in chicken skin. For example, Nadalian et al. (2019) isolated the insoluble elastic fibrous protein elastin from broiler hen skin using the enzyme elastase to achieve hydrolysis. Antioxidant properties of the elastin-based hydrolysate were detected using radical scavenging assays based on 2,2-diphenyl-1-picryl-hydrazyl-hydrate and 2,2′-azino-bis (3-ethylbenzothiazoline-6-sulphonic acid) activities along with metal chelation. To purify antioxidant peptides from elastin, reversed-phase high-performance liquid chromatography followed by amino acid sequence analysis with MALDI-TOF/TOF-MS was required. The authors concluded that all elastin hydrolysates did exhibit antioxidant activity, suggesting that the higher level of activity was associated with peptide fractions of greater hydrophobicity.

These initial studies would suggest that a series of isolated and purified peptides from chicken skin could be further screened to identify antioxidant activities with more precise functions that could be assigned to specific amino acid compositional profiles and structural characteristics. Other biological activities are likely attributable to chicken skin tissues in general, and there may be some specificity depending on the location the skin samples originate from, such as the breast versus the thigh and leg. Other factors including breed of bird, housing environment, nutrition, and age may also be factors that dictate the relative antioxidant properties of the skin tissues.

8 Feathers as a poultry byproduct

One of the key characteristics of all avian species, including domesticated chickens, is the ubiquitous presence of an outer feather covering. In general, feathers provide external environmental protection to the bird as well as body temperature regulation and protection from physical injury among other functions (Leeson and Walsh, 2004; da Silva Ararújo et al., 2021). Feather development in chickens happens early with initial growth occurring at approximately day 5 of incubation followed by keratinization at days 2 to 3 post hatch (Leeson and Walsh, 2004). Adult birds are covered with 7000 to 9000 feathers and in market-age broilers, this results in approximately 50 g of feathers per bird (Leeson and Walsh, 2004). Feathers grow in distinct tracts that vary in size and structure but generally consist of a central shaft with the shaft base attached to a skin follicle and vanes on either side of the shaft (Leeson and Walsh, 2004). Feather pigmentation for domestic chickens is generally white, brown, or a mixture, with the color originating from melanin pigment (Leeson and Walsh, 2004). This consistent feather pigmentation results from natural, sexual, or as artificial selection, and multiple genes have been identified that are associated with pigmentation diversity (Wang et al., 2022). However, as Wang et al. (2022) discuss, some of the genetics of feather pigmentation are still not entirely understood.

Over 90% of the feather consists of protein and 85 to 90% of the feather protein is keratin, a high sulfur-containing fibrous protein with cysteine making up 7 to 13% of the amino acids (Leeson and Walsh, 2004; Saucedo-Rivalcoba et al., 2011; da Silva Ararújo et al., 2021). Elemental analyses of feathers reveal a compositional profile of 2 to 5% sulfur, 15 to 18% nitrogen, 45 to 50% carbon, 6 to 7% hydrogen, 25 to 30% oxygen, along with various trace minerals (Chukwunonso Ossai et al., 2022). Given its high protein content, feather keratin has been of considerable interest as a source of amino acids. However, keratin in its native form is relatively resistant to degradation. This is because keratin possesses cross-linking and covalent bonds that make it considerably recalcitrant to proteolytic degradation by enzymes such as pepsin, papain, and trypsin (Leeson and Walsh, 2004; Li et al., 2019; Chukwunonso Ossai et al., 2022; Saba et al., 2022). Consequently, keratin is essentially a structural fibrous protein formed via disulfide bonds, hydrogen bonds, and hydrophobic interactions (Li, 2019). Because of the strong cross-linked nature of keratin, it is also highly chemically resistant to chemical or enzymatic hydrolysis (Mokrejs et al., 2011). Keratin polymers, in addition to being resistant to proteolytic enzymes and chemical degradation, are insoluble in water and organic solvent matrices (Li, 2019).

Given the recalcitrant nature of poultry feathers, their use as a byproduct poses an environmental challenge on a global scale with 8 to 10 Mt produced annually worldwide (Chukwunonso Ossai et al., 2022). Compared to other keratin-based wastes, poultry feathers are the most abundant at 40% with the United States producing nearly 0.9 to 1.1 billion kg of feather wastes annually (Khardenavis et al., 2009; McGovern, 2000; Mokrejs et al., 2011; Chukwunonso Ossai et al., 2022). Poultry slaughter plants produce nearly 30 000 kg of feathers per day (McGovern, 2000). Most keratinous wastes are simply disposed of by incineration or loaded into landfills, which involve not only operational costs, pathogen transmission, and environmental issues, but loss of a potentially viable recyclable biomaterial (Li, 2019; Chukwunonso Ossai et al., 2022). While incineration does destroy potential pathogens, the generation of various gases such as hydrogen sulfide and ammonia can contribute to overall pollution (Chukwunonso Ossai et al., 2022). Landfill disposal can result in leachates that in turn contribute to increased nitrogen and phosphorus levels in soil, groundwater, and freshwater ecosystems (Chukwunonso Ossai et al., 2022). In freshwater ecosystems, the increased levels of nitrogen and phosphorus can lead to algae blooms, toxicity to aquatic animals, and acidification of surface water (Bhari et al., 2021). Aerobic composting and anaerobic digestion have also been employed as potential methods for feather disposal, but the risk of pathogen survival and environmental emissions concerns limit their practical usefulness and effectiveness (Bhari et al., 2021).

8.1 Recovering value from feather byproducts

Despite the challenges of feather keratin being highly recalcitrant, recovering value from poultry feathers has been an active area of interest for several decades. Classical uses of feathers include bedding, decoration, manufacture of fibers, sporting equipment, and others (Fonkwe et al., 2001). Some interest has also developed in using feathers to produce biodiesel (Kondamudi et al., 2009). Furthermore, raw feathers and various derivatives have been examined for their potential as support fibers and biosorbent materials. Ahn et al. (2011) characterized the biodegradability of injection molded poultry feather fiber-based bioplastic pots used in composting and concluded that the poultry feather fibers were not degraded during the composting time frame and would likely require the inclusion of keratin degrading organisms to achieve sufficient composting. More recently, several studies have explored various modifications of feathers to develop active biosorption surfaces. Pu et al. (2022) created a phosphonate-functionalized bio-adsorbent from raw feather fibers to extract uranium from seawater. The authors demonstrated that feather fibers with phosphate grafting resulted in a biosorbent material that not only provided optimal adsorption performance for enriching uranium from seawater but was also selective and recyclable.

Feathers have also been chemically modified to further develop their biosorption capabilities. Rahmani-Sani et al. (2020) used chicken feathers as a precursor to generate an eggshell-based chicken feather composite for the synthesis of an activated carbon material followed by iron oxide magnetization. By being magnetized, the resulting activated carbon biosorbent material was not only capable of removing several heavy metals from water but could easily be separated from the water sample. Using a different strategy, Zubair et al.

(2022) created a nano-modified feather keratin biosorbent by cross-linking graphene oxide to increase the surface area of the keratin proteins and was able to demonstrate improved sorption capacity for simultaneous removal of multiple heavy metals from synthetically contaminated wastewater.

Given the high protein content, reducing the feather keratin recalcitrance by some form of hydrolysis to release the amino acids for either use in animal feeds or other applications such as biofertilizers has been an area of extensive research efforts (Onifade et al., 1998; Dalev, 1994; El Boushy and van der Poel, 1990; Bhari et al., 2021; Li, 2019, 2022). However, while either physical, chemical, or enzymatic hydrolysis is required to improve digestibility, there are inherent risks associated with these processes that end up yielding a product with reduced amino acid availability and decreased nutritive quality (Papadopoulos, 1985; Bhari et al., 2021). Steam cooking of intact feathers can result in overexposure to extreme conditions and the generation of non-nutritive amino acids such as lysinoalanine and lanthionine (Bhari et al., 2021). Acid, alkali, or solvents may reduce the amino acid content and digestibility of feather meal (Bhari et al., 2021). The effectiveness of chemical hydrolysis in retaining the biological activity of the feather composition depends on the chemicals used and the extraction approach. For example, Sharma et al. (2017) used sodium sulfide for keratin alkaline hydrolysis extraction and iso-electric precipitation to create biofunctional keratin microparticles that exhibited antioxidant and anti-cancerous activities.

As opposed to chemical hydrolysis, keratin hydrolysis by proteolytic enzymes offers the advantage of generating a hydrolysate that possesses high concentrations of peptides, proteins, and amino acids (Bhari et al., 2021). Although some of the common proteases are unable to hydrolyze the keratin in feathers, biological degradation of feathers does occur in nature. This degradation is likely due to the presence of microorganisms that possess proteases with keratinolytic capabilities that could access feather keratin (Papadopoulos, 1985). Environments such as poultry processing plants, feather disposal sites, and slaughterhouse/tannery ecosystems represent potential habitats for keratinolytic organisms (Bhari et al., 2021). Over 40 years ago, a feather-degrading bacterium, *Bacillus licheniformis,* was isolated from a poultry waste digester that contained feathers and identified as an organism with the capability to hydrolyze keratin and produce free amino acids (Williams and Shih, 1989; Williams et al., 1990, 1991; Lin et al., 1992; Shih, 1993). In summarizing the literature, Jagadeesan et al. (2024) concluded that feather hydrolysate protein profiles generated via *Bacillus* spp. enzymes were maximized compared to physical and chemical extraction methods. Not surprisingly, feather meal subjected to this type of keratinase treatment was also demonstrated to serve as an alternative protein source in nutritional feeding studies (Onifade et al., 1998; Kim and Patterson, 2000; Kim et al., 2002; Odetallah et al., 2003, 2005).

Other keratinolytic *Bacillus* species and non-*Bacillus* bacteria with the capacity to degrade feathers have also been isolated from a variety of environments, such as soils and poultry processing plants (Zaghloul et al., 1998; Riffel and Brandelli, 2002; Khardenavis et al., 2009; Bhari et al., 2021; Li, 2021). An even wider range of environments have served as additional sources of keratinase-producing bacteria and fungi (Srivastava et al., 2020). While these organisms all possess keratinases, they vary in their hydrolytic effectiveness and some optimization may still be required. For example, combinations of alkaline-enzymatic hydrolysis of feathers as a two-stage approach have been examined to achieve a more efficient hydrolysis process (Mokrejs et al., 2011). The type of high-value product generated from feather meal also impacts the keratin treatment approach and whether some form of pretreatment is required. For example, Forgács et al. (2011) used autoclaving to pretreat chicken feather waste prior to incubation with *Bacillus megaterium* to generate a fermentation broth hydrolysate for anaerobic batch production of methane. In a different two-stage process, pretreatment with a keratinase-secreting *Pseudomonas* followed by co-digestion with carbon-rich substrates has been employed to improve biogas production and avoid ammonia inhibition (Saba et al., 2022). There is some evidence that enhanced keratin degradation can occur in multiple species containing microbial communities, suggesting that synergistic metabolic mechanisms may be involved (Nasipuri et al., 2020).

8.2 Advances in feather meal enzymatic hydrolysis

As additional keratinolytic organisms have been identified and isolated over the years, opportunities to understand the mechanisms of keratin enzymatic hydrolysis and the potential to improve the process have become more apparent. It is now known that keratinase activity is essentially a two-step process involving the combination of a disulfide reductase (sulfitolysis) to cleave the disulfide bonds in keratin followed by proteolysis of the peptides (Bhari et al., 2021; Li, 2019, 2022). Sequence analyses have allowed for the identification and classification of keratinases based on amino acid composition and optimization of enzyme activity (Li, 2021). Further examination with whole genomic sequencing and proteomics of the keratinolytic strain of *Bacillus paralicheninformis* has revealed that there are multiple proteases that serve as part of a protease degradomic consortium required for feather meal degradation (SanthaKalaikumari et al., 2021). As the genetics of keratinase mechanisms have become better understood, cloning and recombinant production of keratinases in nonkeratinolytic organisms has allowed for genetic modification to generate mutants with properties such as improved enzymatic activities and stability that are more amendable for further characterization and industrial applications (Li, 2019, 2021; Srivastava et al., 2020).

Optimizing keratinase activity to maximize complete feather hydrolysis remains a key requirement to develop additional practical applications. Current limitations of known keratinases include low efficiency on complex feather matrices and low stability (Sharma et al., 2022). Expanding the utility of keratinases for robust feather degradation requires several strategies to improve enzyme activities. Part of this strategy is dependent on gaining a better understanding of the feather keratin matrix as an enzyme substrate. Given the complexity of feather matrices, Valencia et al. (2019) proposed a proteolytic susceptibility mathematical model assessment to provide a quantitative index as a kinetic-based ratio of enzyme concentration and temperature, divided by substrate concentration. When the authors compared subtilisin hydrolysis of turkey feather keratin versus turkey feather denatured with a Na_2SO_3 solution followed by autoclaving, the susceptibility of the denatured feather keratin was increased eight-fold versus the raw feather keratin. It would appear that thermal treatment of feather keratins would enhance their susceptibility to enzymatic hydrolysis, but most wild-type keratinases lack thermostability (Su et al., 2022).

To instill thermal stability and retain hydrolytic activity, modifications in amino acid composition and keratinase enzyme protein structure have been identified. Sharma et al. (2022) isolated keratinase-producing organisms from a poultry feather dumping site in India, specifically identifying *Bacillus pacificus*. When the purified keratinase activity was assessed over a wide temperature and pH range, full enzymatic stability was detected at pH 9 and the optimal temperature was 60°C with slight loss in activities in temperature increases up to 80°C. Using *in silico* functional-structural analyses of the keratinase, the authors detected distinct amino acid residues along with a high-affinity calcium-binding site. When the keratinolytic isolate was introduced into a 5-l fermenter, over 90% feather hydrolysis occurred within 24 h. Su et al. (2022) used a combination of *in silico* computational analyses for initial genetic target predictions followed by genetic engineering to improve the thermostability of a *Bacillus subtilis* keratinase enzyme. Modifications of the high-flexible-loop and calcium-binding loop regions and replacement of non-consensus amino acids with consensus amino acids to improve structural stability resulted in a mutant expressing a computationally redesigned keratinase with an optimal temperature increase from 40 to 60°C, a half-life increase of 17 to 66 min at 60°C, with the mutant capable of doubling the feather degradation rate (Su at al., 2022). Su et al. (2022) also suggested that this bioinformatic computational approach offered a more targeted and predictable strategy for optimizing thermostability engineering.

As more keratinase-producing organisms are isolated and sequenced, further bioinformatic refinement will be possible for more precise protein engineering of keratinases. This, coupled with quantitative assessment and indexing of the susceptibilities of various keratin protein sources as described by Valencia et al. (2019) may lead to a much more specific series of uniquely engineered keratinases that can be used for generating hydrolysates that yield peptides with highly specific functions. Since microbial community synergism appears to enhance feather degradation, this may indicate that combinations of keratinase-producing organisms and perhaps non-keratolytic organisms are needed to optimize complete feather degradation. In-depth microbiome sequencing and metabolomics of microbial communities inhabiting feather-containing environments may possibly reveal some of these potential synergistic relationships.

While most of the feather hydrolysate products have been developed for nutritional and biofertilizer applications, other functions such as antioxidant activities have been identified with *Bacillus* spp. keratinolytic hydrolysates (Lemes et al., 2016; Callegaro et al., 2018). Additional biological functions have also been associated indirectly with feathers. For example, Peralta-Sánchez et al. (2013) suggested feathers lining barn swallow nests may impact the microbial load on the shells of laid eggs by selectively supporting bacteria on feathers with antimicrobial activities. They noted that antimicrobial activity was associated with feather pigments with white feathers exhibiting greater antimicrobial activity compared to black pigmented feathers, hinting that this could reduce the incidence of egg microbial contamination. Whether this occurs in domesticated chickens remains to be determined; however, the introduction of aviary and free-range nesting systems for laying hens could be environments that would support this type of activity. While not directly an indicator of such activities, preliminary results based on 16S rDNA sequencing suggest that transitioning layer hens from cages to aviaries does impact their feather microbial diversity over time (Tarcin et al., 2024). Factors such as the breed of laying hen and feather pigmentation could also influence the level of antibacterial activity associated with the feather microbial communities.

9 Conclusion and future trends

Commercial poultry broiler production continues to expand in response to market demands for poultry meat. With this expansion also comes an increase in poultry byproducts and potential negative impacts on the environmental footprint associated with poultry production. While achieving a reduced environmental impact and enhancing sustainability would appear to be a challenge, the vertical integration of the poultry industry offers an opportunity to institute major changes on a large scale. However, efforts to improve the environmental sustainability of broiler production and processing must be balanced with other factors such as economics, intensive manual labor demands, and food safety regulations. In addition, the introduction of new technologies, such as automation in poultry processing, may add further factors for consideration when assessing opportunities to improve environmental sustainability. While automation could reduce manual labor costs, it may introduce additional environmental inputs that have not yet been anticipated. Finally, developing economic motivation for the poultry industry to better control their energy consumption, water use, and waste disposal to further reduce the environmental footprint is needed.

Certainly, reduction of energy usage and managing water utilization are important management tools, but opportunities to repurpose waste materials to avoid outright disposal should also be considered to further reduce the environmental impact of poultry production. The concept of harvesting and/or converting poultry waste byproduct materials, particularly the large contributors such as bones, skin, blood, and feathers into valued biologics is an attractive option. Even much less prominent poultry byproducts such as keel tissue and combs may serve as sources for specific biologic molecules with potential benefits. The contributions of these poultry byproduct biologics can be quite varied and can be intrinsically valuable. For example, certain molecules and compounds such as collagen, biologically active peptides, and sialic acid can elicit human health benefits in the form of antioxidant, antidiabetic, antimicrobial, and anti-inflammatory activities among others. The amino acids recovered from blood and feathers represent quality animal and pet feed protein sources as well as potential supplements for companion animals. Other byproducts such as viscera remain relatively under investigated at this time. However, metabolically active tissues, such as the GIT lining, likely possess numerous functional tissues and molecules that have biological potential. Likewise, while often overlooked, the rich diversity of GIT microorganisms has great utility for multiple applications. Some of these organisms have already been used as sources of probiotics and antimicrobial compounds to enhance broiler performance and prevent pathogen GIT establishment. However, other GIT microbial activities may have benefits that have not yet been identified. High-throughput screening of these organisms for specific functions or metabolites would be of interest, especially with the multiple omics methodologies now available.

Despite the presence of high-value compounds such as amino acids and peptides in various poultry byproducts, extraction, and recovery without losing biological activity can be a challenge. This is particularly evident with feathers. Feathers are a high-protein poultry byproduct with several amino acids that are nutritionally important. However, feathers are composed primarily of fibrous keratin protein which is relatively resistant to hydrolysis. Several chemical methods have been devised to release the amino acids from the feather keratin but, depending on the approach can result in loss of amino acids and general decreases in nutritional availability. Initial isolation of organisms that possessed keratinases capable of hydrolyzing recalcitrant feather keratin led the way to a more effective means to retain most of the amino acids during hydrolysis. Since the initial discovery of the keratinase enzyme from a few organisms, a much wider range of keratinolytic organisms have been identified over the years. As more organisms are characterized and sequenced, it has become possible to develop a better understanding of the mechanisms and variety of keratinases present in different organisms. This has led, in turn, to the ability to genetically modify keratinase genes and introduce properties such as thermal stability to enhance their efficiency for keratin hydrolysis resulting in a more complete digestion of feathers and feather meal.

While considerable progress has been made in the development of biologicals derived from poultry waste byproducts, much remains to be done. Part of this involves the need to further introduce and develop the concept of biologics as an economically viable option for poultry waste byproducts. While avenues such as rendering for certain poultry byproducts exist for conversion into protein sources for food for animals and pets, these processes fail to take advantage of the rich biological diversity from the discarded tissues generated during processing. The advances in our understanding of avian physiology and genetics provide a unique opportunity to identify specific functions and/or molecules in tissues with the possibility to harness this knowledge by recovering these components as active compounds during processing. The challenges to make this economically viable are three-fold. First, extraction methods must be sufficiently nondestructive to ensure retention of as much of the biological activity as possible. Second, workflows that can be scaled up and incorporated into current poultry processing lines will be needed to make the recovery of these products economically viable in the poultry industry. Finally, commercial markets will need to be identified to ensure a positive cost/benefit for the biologic compound of interest. While mature markets may already have processes in place for some biologics, a certain level of entrepreneurship and support of startup companies may be necessary for other biologic compounds. Taken together, deeper integration of biologics capabilities into the poultry processing space offers an attractive economical option to not only reduce the environmental footprint of poultry production and enhance agricultural sustainability but also provide opportunities to diversify the commercial markets for poultry processing byproducts.

10 Where to look for further information

Key articles exploring different aspects of the topic are:

- da Silva Ararújo, I.B., et al. 2021. Chapter 25. Recovery and application of bioactive proteins from poultry by-products. In R. Bhat (Ed.) *Valorization of agri-wastes and by-products*. https://doi.org/10.1016/B978-0-12-824044-1.00032-5.
- Lasekan, A., et al. 2013. Potential of chicken by-products as sources of useful biological resources. *Waste Manag.* 33: 552–565. https://doi.org/10.1016/j.wasman.2012.08.001.
- Li, Q. 2021. Structure, application, and biochemistry of microbial keratinases. *Front. Microbiol.* 12: 674345. https://doi.org/10.3389/fmicb.2021.674345.
- Li, Q. 2019. Progress in microbial degradation of feather waste. *Front. Microbiol.* 10: 2717. https://doi.org/10.3389/fmicb.2019.02717.
- Morawicki, R. 2010. Chapter 19. Coproducts and by-products from poultry processing. In C.M. Owens, C.M., et al. (Eds.). *Poultry meat processing*. 2nd edition. CRC Press, Boca Raton, FL. pp. 381–396.
- Romero-Garay, M.G., et al. 2022. Bioactivity of peptides obtained from poultry by-products: a review. *Food Chem. X* 13: 100181. https://doi.org/10.1016/j.fochx.2021.100181.
- Srivastava, B., et al. 2020. Microbial keratinases: an overview of biochemical characterization and its ecofriendly approach for industrial applications. *J. Cleaner Prod.* 252: 119847. https://doi.org/10.1016/j.jclepro.2019.119847.

11 References

Ahn, H.K., et al. 2011. Biodegradability of injection molded bioplastic pots containing polylactic acid and poultry feather fiber. *Bioresource Technol.* 102: 4930–4933. https://doi.org/10.1016/j.biortech.2011.01.042.

Alves, F.E.S.B., et al. 2020. Functional and antioxidant properties of a chicken blood meal hydrolysate. *Brazilian J. Dev. and Appl. Sci.* 6: 21149–21162. https://doi.org/10.34117/bjdv6n4-330.

Bah, C.S.F., et al. 2013. Slaughterhouse blood: am emerging source of bioactive compounds. *Compr. Revs. Food Sci. and Food Safety.* 12: 314–331. https://doi.org/10.1111/1541-4337.12013.

Bhari, R., et al. 2021. Chicken feather waste hydrolysate as a superior biofertilizer in agroindustry. *Curr. Microbiol.* 78: 2212–2230. https://doi.org/org/10.1007/s00284-021-02491-z.

Blevins, R.E., et al. 2017. Chap. 18. Historical, current and future prospects for food safety in poultry product processing systems. In: S.C. Ricke, et al. (Eds.) *Food and Feed*

Safety Systems and Analysis. Elsevier Inc., San Diego, CA. pp. 323–345. https://doi.org/10.1016/B978-0-12-811835-1.00018-X.

Cai, T. and Pancorbo, O.C. 1994. Stabilization of poultry-processing by-products and waste and poultry carcasses through lactic acid fermentation. *J. Appl. Poultry Res.* 3:17–25.

Cai, T., et al. 1995. Stabilization of poultry-processing by-products and poultry carcasses through direct chemical acidification. *Bioresource Technol*. 52(1): 69–77.

Callegaro, K., et al. 2018. Feathers as bioresource: microbial conversion into bioactive protein. *Process Biochem*. 75: 1–9. https://doi.org/10.1016/j.procbio.2018.09.002.

Calo, J.R., et al. 2015. Essential oils as antimicrobials in food systems: a review. *Food Control*. 54: 111–119. https://doi.org/10.1016/j.foodcont.2014.12.040.

Chai, S.J., et al. 2017. Poultry: the most common food in outbreaks with known pathogens, United States, 1998–2012. *Epidemiol. Infect*. 145: 316–325.

Chao, K., et al. 2014. Control interface and tracking control system for automated poultry inspection. Computer Standards & Interfaces 36: 271–277. https://doi.org/10.1016/j.csi.2011.03.006.

Chukwunonso Ossai, I., et al. 2022. Valorisation of keratinous wastes: a sustainable approach towards a circular economy. *Waste Manag*. 151: 81–104. https://doi.org/10.106/j.wasman.2022.07.021.

Cofrades, S., et al. 2000. Plasma protein and soy fiber content effect on Bologna sausage properties as influenced by fat level. *J. Food Sci*. 63; 281–287. https://doi.org/10.1111/j.1365-2621.2000.tb15994.x.

Dalev, P. 1994. Utilization of waste feathers from poultry slaughter for production of a protein-concentrate. *Bioresource Technol*. 48(3): 265–267.

Dalziel, C.J., et al. 2015. Fat and fatty acid composition of cooked meat from UK retail chickens labelled as from organic and non-organic production systems. *Food Chem*. 179: 103–108.

da Silva Ararújo, I.B., et al. 2021. Chapter 25. Recovery and application of bioactive proteins from poultry by-products. In R. Bhat (Ed.) *Valorization of Agri-Wastes and By-Products*. https://doi.org/10.1016/B978-0-12-824044-1.00032-5.

Dittoe, D.K., et al. 2020. Chapter 1. Commercial poultry production and gut function – Historical perspective. In S.C. Ricke (Ed.) *Improving Gut Function in Poultry*. Burleigh Dodd Publishing, Cambridge, UK. pp. 3–30. https://doi.org/10.19103/AS.2019.0059.01.

Dittoe, D.K., et al. 2018. Organic acids and potential for modifying the avian gastrointestinal tract and reducing pathogens and disease. *Front. Vet. Sci*. 5: 216. https://doi.org/10.3389/fvets.2018.00216.

Ederer, P. and Leroy, F. 2023. The societal role of meat – What the science says. *Anim. Front*. 13: 3–8. https://doi.org/org. /10.1093/af/vfac098.

Edwards, D.R. and Daniel, T.C. 1992. Environmental impacts of on-farm waste disposal. *Bioresource Technol*. 41: 9–33.

El Boushy, A.R. and van der Poel, A.F.B. 1990. Feather meal-a biological waste: its processing and utilization as a feedstuff for poultry. *Biol. Wastes* 32: 39–74.

Fallows, S.J. and Verner Wheelock, J. 1982. By-products from the U.K. food system 2. The meat industry. *Conserv. Recycl*. 5: 172–182.

Fonkwe, L. G., et al. 2001. Utilization of poultry processing wastes. *J. Food Sci. Nutr*. 6(4): 257–262.

Forgács, G., et al. 2011. Biological treatment of chicken feather waste for improved biogas production. *J. Environ. Sci.* 23: 1747–1753. https://doi.org/10.1016/S1001-0742(10)60648-1.

Foroutan, R., et al. 2021. Hydroxyapatite biomaterial production from chicken (femur and beak) and fishbone waste through a chemical less method for Cd+ removal from ship building wastewater. *J. Hazard. Mater.* 413: 1–13. https://doi.org/10.106/j.jhazinat.2021.125428.

Ge, B., et al. 2013. Retrospective analysis of *Salmonella*, *Campylobacter*, *Escherichia coli*, and *Enterococcus* in animal feed ingredients. *Foodborne Pathogens Dis.* 10: 684–691.

Ghosh, S. and Pramanik, K. 2023. Unveiling the secrets of food waste derived biomaterials in remediation of environmental pollutants – A review. *Bioresource Technol.* Reports 22: 101460. https://doi.org/10.1016/j.biteb.2023.101469.

Hamzeh, A., et al. 2019. Cellular and chemical antioxidant activities of chicken blood hydrolysates as affected by in vitro gastrointestinal digestion. *Poultry Sci.* 98: 6138–6148. https://doi.org/10.3382/ps/pez283.

Handley, J.A., et al. 2015. Chapter 6. *Salmonella* and the potential role for microbial process indicators on chicken carcasses. In: S.C. Ricke, et al. (Eds.) *Food Safety: Emerging Issues, Technologies and Systems*. Elsevier, Oxford, UK, pp. 81–104. https://doi.org/10.1016/B978-0-12-800245-2.00006-X.

Jagadeesan, Y., et al. 2024. A scientific version of understanding "why did the chicken cross the road"? – A guided journey through *Bacillus* spp. towards sustainable agriculture, circular economy and biofortification. *Environ. Res.* 244: 117907. https://doi.org/10.1016/j.envres.2023.117907.

Johnson E. S., et al. 2010. Cancer mortality in poultry slaughtering/processing plant workers belonging to a union pension fund. *Environ. Res.* 110: 588–594.

Jones S.L., et al. 2021. Critical factors and emerging opportunities in food waste utilization and treatment technologies. *Front. Sustain. Food Syst.* 5: 781537. https://doi.org/10.3389/fsufs.2021.781537.

Jones, F.T. 2011. A review of practical *Salmonella* control measures in animal feed. *J. Appl. Poultry Res.* 20: 102–113.

Khardenavis, A. A., et al. 2009. Processing of poultry feathers by alkaline keratin hydrolyzing enzyme from *Serratia* sp. HPC 1383. *Waste Manage.* 29(4): 1409–1415.

Kim, W.K., et al. 2002. Effect of enzymatic and chemical treatments on feather solubility and digestibility. *Poultry Sci.* 81: 95–98.

Kim, W.K. and Patterson, P.H. 2000. Nutritional value of enzyme- or sodium hydroxide-treated feathers from dead hens. *Poultry Sci.* 79: 528–534.

Kondamudi, N., et al. 2009. A green process for producing biodiesel from feather meal. *J. Agric. Food Chem.* 57(14): 6163–6166.

Lasekan, A., et al. 2013. Potential of chicken by-products as sources of useful biological resources. *Waste Manage.* 33: 552–565. https://doi.org/10.1016/j.wasman.2012.08.001.

Leeson, S. and Walsh, T. 2004. Feathering in commercial poultry I. Feather growth and composition. World's *Poultry Sci. J.* 60: 42–51.

Lemes, A.C., et al. 2016. Simultaneous production of proteases and antioxidant compounds from agro-industrial by-products. *Bioresource Technol.* 222: 210–216. https://doi.org/10.1016/j.biortech.2016.10.001.

Li, Q. 2019. Progress in microbial degradation of feather waste. *Front. Microbiol.* 10:2717. https://doi.org/10.3389/fmicb.2019.02717.

Li, Q. 2021. Structure, application, and biochemistry of microbial keratinases. *Front. Microbiol.* 12: 674345. https://doi.org/10.3389/fmicb.2021.674345

Li, Q. 2022. Perspectives on converting keratin-containing wastes into biofertilizers for sustainable agriculture. *Front. Microbiol.* 13: 918262. https://doi.org/10.3389/fmicb.2022.918262.

Lin, X., et al. 1992. Purification and characterization of a keratinase from a feather-degrading *Bacillus licheniformis* strain. *Appl. Environ. Microbiol.* 58: 3271–3275.

Ling, A.J.W., et al. 2022. Review of sialic acid's chemistry, sources, extraction and functions with special reference to edible bird's nest. *Food Chem.* 367: 130755. https://doi.org/10.1016/j.foodchem.2021.130755.

Loo, C.P.Y. and Sarbon, N.M. 2020. Chicken skin gelatin films with tapioca starch. *Food Bioscience*. 35: 100589. https://doi.org/org/10.106/j.fbio.2020.100589.

Luján-Rhenals, D., et al. 2017a. Chapter 19. Energy and water use in poultry processing. In: S.C. Ricke (Ed.) *Achieving Sustainable Production of Poultry Meat- Volume 1 Safety, Quality and Sustainability*. Burleigh Dodd Publishing, Cambridge, UK, pp. 389–409. https://doi.org/10.19103/AS.2016.0010.22.

Luján-Rhenals, D., et al. 2017b. Chapter 20. Waste management and emissions in poultry processing. In: S.C. Ricke (Ed.) *Achieving Sustainable Production of Poultry Meat- Volume 1 Safety, Quality and Sustainability*. Burleigh Dodd Publishing, Cambridge, UK, pp. 411–431. https://doi.org/10.19103/AS.2016.0010.23

Maciorowski, K.G., et al. 2006. Animal production and contamination by foodborne *Salmonella*. *J. Consumer Prot. Food Safety* 1: 197–209.

Maciorowski, K.G., et al. 2004. Incidence and control of food borne *Salmonella* spp. in poultry feeds: a review. *World's Poultry Sci. J.* 60: 446–457.

McGovern, V. 2000. Recycling poultry feathers: more bang for the cluck. *Environ. Health Perspect.* 108(8): A366–A369.

Mecham, R.P. 2008. Methods in elastic tissue biology: ēlastin isolation and purification. *Methods* 45: 32–41.

Meeker, D.L. and Meisinger, J.L. 2015. Rendered ingredients significantly influence sustainability, quality, and safety of pet food. *J. Anim. Sci.* 93: 835–847. https://doi.org/10.2527/jas2014-8524.

Micciche, A.C., et al. 2018. The implementation and food safety issues associated with poultry processing reuse water for conventional poultry production systems in the United States. *Frontiers Sustain. Food Syst.* 2:70. https://doi.org/10.3389/fsufs.2018.00070.

Mokrejs, P., et al. 2011. Processing poultry feathers into keratin hydrolysate through alkaline-enzymatic hydrolysis. *Waste Manag. Res.* 29(3): 260–267.

Molina, A., et al. 2016. Vigilance for *Salmonella* in feedstuffs available in Costa Rica: prevalence, serotyping and tetracycline resistance of isolates obtained from 2009 to 2014. *Foodborne Pathogens Dis.* 13: 119–127.

Morawicki, R. 2010. Chapter 19. Coproducts and by-products from poultry processing. In C.M. Owens, et al. (Eds.). *Poultry Meat Processing*. 2nd edition. CRC Press, Boca Raton, FL, pp. 381–396.

Morrow, T. and Felcone, L.H. 2004. Defining the difference: what makes biologics unique. *Biotechnol. Healthcare*. 1(4): 24–29.

Murphy, D.W. and Silbert, S.A. 1992. Preservation of and nutrient recovery from poultry carcasses subjected to lactic acid bacteria fermentation. *J. Appl. Poultry Sci.* 1: 66–74.

Munasinghe, K.A., et al. 2014. Chicken collagen from low market value by-products as an alternative source. *J. Food Process.* 2114: Article 298295. https://doi.org/10.1155/2014/298295.

Nadalian, M., et al. 2019. Isolation, purification and characterization of antioxidative bioactive elastin peptides from poultry skin. *Food Sci. Anim. Resources.* 39: 966–979. https://doi.org/org./10.5851/kosfa.2019.e90.

Nakano, K., et al. 1994. Sialic acid contents in chicken eggs and tissues. *Can. J. Anim. Sci.* 74: 601–606.

Nasipuri, P., et al. 2020. Community-intrinsic properties enhance keratin degradation from bacterial consortia. *PLoS One.* 15: e0228108. https://doi.org/10.1371/journal.pone.0228108.

Nemser, S.M., et al. 2014. Investigation of *Listeria*, *Salmonella*, and toxigenic *Escherichia coli* in various pet foods. *Foodborne Pathogens Dis.* 11: 706–709.

Odetallah, N.H., et al. 2005. Versazyme supplementation of broiler diets improves market growth performance. *Poultry Sci.* 84: 858–864.

Odetallah, N.H., et al. 2003. Keratinase in starter diets improves growth of broiler chicks. *Poultry Sci.* 82: 664–670.

Olson, E.G., et al. 2023. Chapter 27. Microbial pathogen contamination of animal feed. In: L. Anelich, et al. (Eds.). *Present Knowledge in Food Safety: A Risk-Based Approach Through the Food Chain.* International Life Sciences Institute, Academic Press, Elsevier, London, pp. 378–393. https://doi.org/10/1016/B978-0-0-12-819470-6.00023-8.

Onifade, A.A., et al. 1998. A review: potentials for biotechnological applications of keratin-degrading microorganisms and their enzymes for nutritional improvement of feathers and other keratins as livestock feed resource. *Bioresour. Technol.* 66: 1–11.

Onuh, J.O., et al. 2015. Kinetics of in vitro renin and angiotensin converting enzyme inhibition by chicken skin protein hydrolysates and their blood pressure lowering effects in spontaneously hypertensive rats. *J. Functional Foods.* 14: 133–143. https://doi.org/1016/j.jff.2015.01.031.

Oro, C.E.D., et al. 2018. Formulation of chicken sausages with broiler blood proteins and dye. *J. Food Sci. Technol.* 55; 4694–4699. https://doi.org/10.1007/s13197-018-3403-8.

Owens, C.M., et al. eds. 2010. *Poultry Meat Processing*, 2nd Edition. CRC Press, Boca Raton, FL.

Papadopoulos, M.C. 1985. Processed chicken feathers as feedstuff for poultry and swine: a review. *Agric Wastes* 14: 275–290.

Patil, D. and Nag, A. 2011. Production of PUFA concentrates from poultry and fish processing waste. *J. Amer. Oil Chem. Soc.* 88 (4): 589–593.

Peralta-Sánchez, J.M., et al. 2013. Eggshell bacterial load Is related to antimicrobial properties of feathers lining barn swallow nests. *Microbial Ecol.* 67: 480–487. https://doi.org/10.1007/s00248-013-0338-5.

Perumalla, A.V.S., et al. 2023. Chapter 7. Probiotics in poultry preharvest food safety: Historical developments and current prospects. In T.R. Callaway and S.C. Ricke. (Eds.) *Direct Fed Microbials/Prebiotics for Animals: Science and Mechanisms of Action*, 2nd Edition, Springer Science, New York, NY, pp. 127–166. https://doi.org/10.1007/978-3-031-40512-9_6.

Pu, Y., et al. 2022. Waste feather fiber based extraction capacity bio-adsorbent for sustainable uranium extraction from seawater. *Int. J. Biol. Macromolecules* 206: 699–707. https://doi.org/10.106/j.ijbiomac.2022.03.019.

Rahmani-Sani, A., et al. 2020. Use of chicken feather and eggshell to synthesize a novel magnetized activated carbon for sorption of heavy metals. *Bioresource Technol.* 297: 122452. https://doi.org/10.1016./jbiortech.2019.122452.

Rawdkuen, S., et al. 2004. Chicken plasma protein: Proteinase inhibitory activity and its effect on surimi gel properties. *Food Res. Int.* 37: 156–165.

Ricke, S.C. 2018a. Impact of prebiotics on poultry production and food safety. *Yale J. Biol. Med.* 91: 151–159.

Ricke, S.C. 2018b. Chap. 8. Feed Hygiene. In: J. Dewulf and F. Van Immerseel (Eds.) *Biosecurity in Animal Production and Veterinary Medicine – From Principles to Practice*. ACCO (Academische Coöperative Vennootschap cvba), Leuven, Belgium, pp. 177–209.

Ricke, S.C., et al. 2023. Chapter 6. Prospects for prebiotics and postbiotics in poultry. In T.R. Callaway and S.C. Ricke (Eds.) *Direct Fed Microbials/Prebiotics for Animals: Science and Mechanisms of Action*, 2nd Edition, Springer Science, New York, NY, pp. 103–124. https://doi.org/10.1007/978-3-031-40512-9_6.

Ricke, S.C., et al. 2022. Practical opportunities for microbiome analyses and bioinformatics in poultry processing. *Poultry Sci*. 101:101787. https://doi.org/10.1016/j.psj.2022.

Riffel, A. and Brandelli, A. 2002. Isolation and characterization of a feather-degrading bacterium from the poultry processing industry. *J. Industrial Microbiol. & Biotechnol.* 29: 255–258.

Ritter, W. F. and Chirnside, A.E.M. 1995. Impact of dead bird disposal pits on groundwater quality on the Delmarva peninsula. *Bioresource Technol*. 53(2): 105–111.

Romero-Garay, M.G., et al. 2022. Bioactivity of peptides obtained from poultry by-products: a review. *Food Chem. X* 13: 100181. https://doi.org/10.1016/j.fochx.2021.100181.

Saba, M., et al. 2022. Microbial pretreatment of chicken feather and its co-digestion with rice husk and green grocery waste for enhanced biogas production. *Front. Microbiol.* 13:792426. https://doi.org/10.3389/fmicb.2022.792426

Salminen, E. and Rintala, J. 2002. Anaerobic digestion of organic solid poultry slaughterhouse waste: a review. *Bioresour. Technol*. 83: 13–26.

SanthaKalaikumari, S., et al. 2021. Whole-genome sequencing and mining of protease coding genes in *Bacillus paralicheninformis* MKU3, and its degradomics in feather meal medium. *Curr. Microbiol.* 78: 206–217. https://doi.org/10.1007/s00284-020-002271-1.

Sarbon, N.M., et al. 2018. Purification and characterization of antioxidative peptides derived from chicken skin gelatin hydrolysate. *Food Hydrocolloids*. 85: 311–320. https://doi.org/org/10.1016/j.foodhyd.2018.06.048.

Saucedo-Rivalcoba, V., et al. 2011. (Chicken feathers keratin)/polyurethane membranes. *Appl. Phys. A-Mater. Sci. Process*. 104(1): 219–228.

Shafer, D.J., et al. 2000. Chemical preservation of whole broiler carcasses utilizing alkaline hydroxides. *Poultry Sci.* 79(11): 1517–1523.

Shafer, D.J., et al. 2001. Characterization of alkaline hydroxide-preserved whole poultry as a dry byproduct meal. *Poultry Sci.* 80(11): 1543–1548.

Sharma, C., et al. 2022. Chicken feather waste valorization into nutritive protein hydrolysate: Role of novel thermostable keratinase from *Bacillus pacificus* RSA27. *Front. Microbiol.* 13: 882902. https://doi.org/10.3389/fmicb.2022.882902.

Sharma, S., et al. 2017. Characterization of keratin microparticles from feather biomass with potent antioxidant and anticancer activities. *Int. J. Biol. Macromol.* 104: 189–196. https://doi.org/10.1016/j.ijbiomac.2017.06.015.

Shih, J.C.H. 1993. Recent development in poultry waste digestion and feather utilization: a review. *Poult. Sci.* 72: 1617–1620.

Simancas-Racines, A., et al. 2023. Influenza: strategies to manage an outbreak. *Pathogens* 12: 610. https://doi.org/10.3390/ pathogens12040610.

Singh, M. and Rama, E.N. 2017. Chapter 6. Safety management and pathogen monitoring in poultry slaughterhouse operations: the case of the United States. In: S.C. Ricke (Ed.) *Achieving Sustainable Production of Poultry Meat- Volume 1 Safety, Quality and Sustainability*. Burleigh Dodd Publishing, Cambridge, UK, pp. 199–139. https://doi.org/10.19103/AS.2016.0010.08.

Srivastava, B., et al. 2020. Microbial keratinases: an overview of biochemical characterization and its ecofriendly approach for industrial applications. *J. Cleaner Prod.* 252: 119847. https://doi.org/10.1016/j.jclepro.2019.119847.

Su, C., et al. 2022. A combination of bioinformatics analysis and rational design strategies to enhance keratinase thermostability for efficient biodegradation of feathers. Science Total Environ. 818: 151824. https://doi.org/org/10.1016/j.scitenv.2021.151824.

Tarcin, A.A., et al. 2024. *Elucidating the feather microbiota composition of conventional and aviary housed laying hens over time.* Poultry Science Association Annual Meetings. July 15th–18th, Louisville, KY.

Thompson, D.R., et al. 2017. Chap. 20. Student cross-training opportunities for combining food, transportation, and critical infrastructure cybersecurity into an academic food systems education program. In: S.C. Ricke, e tal. (Eds.) *Food and Feed Safety Systems and Analysis*. Elsevier Inc., San Diego, CA, pp. 375–391. https://doi.org/10.1016/B978-0-12-811835-1.00020-8.

Tseng, T.-F., et al. 2006. Porcine blood plasma transglutaminase combined with thrombin and fibrinogen as a binder in reconstructed meat. *Asian-Austr. J. Anim. Sci.* 19: 1054–1058. https://doi.org/10.5713/ajas.2006.1054.

U.S. Dept. of Agriculture, National Agricultural Statistics Service. 2018. Poultry Production and Value 2017 Summary. United States Department of Agriculture.

Valencia, P.L., et al. 2019. Proteolytic susceptibility of food by-product proteins: an evaluation by means of a quantitative index. *Process Biochem.* 77: 63–69. https://doi.org/10.1016/j.procbio.2018.11.009.

Wang, B. and Brand-Miller, J. 2003. The role and potential of sialic acid in human nutrition. *European J. Clin. Nutr.* 57: 1351–1369.

Wang, H., et al. 2022. Candidate pigmentation genes related to feather color variation in an indigenous chicken breed revealed by whole genome data. *Front. Genet.* 13: 985228. https://doi.org/10.3389/fgene.2022.985228.

Williams, C.M. and Shih, J.C.H. 1989. Enumeration of some microbial groups in thermophilic poultry waste digesters and enrichment of a feather-degrading culture. *J. Appl. Microbiol.* 67: 25–35.

Williams, C.M., et al. 1990. Isolation, identification, and characterization of a feather-degrading bacterium. *Appl. Environ. Microbiol.* 56: 1509–1515.

Williams, C.M., et al. 1991. Evaluation of a bacterial feather fermentation product, feather-lysate, as a feed protein. *Poultry Sci.* 70: 85–94.

Wythe, L.A., et al. 2023. Chapter 4. Advancements in poultry nutrition and genetics, The role of antibiotic growth promoters, and the introduction of feed additive alternatives. In T.R. Callaway and S.C. Ricke (Eds.) *Direct Fed Microbials/Prebiotics for Animals: Science and Mechanisms of Action*, 2nd Edition, Springer Science, New York, NY, pp. 59–79. https://doi.org/10.1007/978-3-031-40512-9_4.

Zaghloul, T.I., et al. 1998. Isolation, identification, and keratinolytic activity of several feather-degrading bacterial isolates. *Applied Biochem. Biotechnol.* 70–72: 207–213.

Zheng, Z., et al. 2018. A novel antioxidative peptide derived from chicken blood corpuscle hydrolysate. *Food Res. Int.* 106: 410–419. https://doi.org/10.1016,j.foodres.2017.12.078.

Zubair, M., et al. 2022. Nano-modified feather keratin derived green and sustainable biosorbents for the remediation of heavy metals from synthetic wastewater. *Chemosphere* 308: 136339. https://doi.org/10.1016/j.chemosphere.2022.136339.

Index

www.ingramcontent.com/pod-product-compliance
Ingram Content Group UK Ltd.
Pitfield, Milton Keynes, MK11 3LW, UK
UKHW021825100926
8556IPUK00003B/154

9 781801 467896